Activation of Small Molecules

Edited by
William B. Tolman

Related Titles

Jaouen, G. (ed.)

Bioorganometallics
Biomolecules, Labeling, Medicine

2006
ISBN 3-527-30990-X

Dyker, G. (ed.)

Handbook of C-H Transformations
Applications in Organic Synthesis

2005
ISBN 3-527-31074-6

Bäckvall, J.-E. (ed.)

Modern Oxidation Methods

2004
ISBN 3-527-30642-0

Messerschmidt, A. (ed.)

Handbook of Metalloproteins
3 volume set

2004
ISBN 0-470-86981-X

Activation of Small Molecules

Organometallic and Bioinorganic Perspectives

Edited by
William B. Tolman

WILEY-VCH Verlag GmbH & Co. KGaA

The Editor

Prof. William B. Tolman
Department of Chemistry
University of Minnesota
207 Pleasant St. SE
Minneapolis, MN 54555
USA

Library of Congress Card No.: applied for

British Library Cataloguing-in-Publication Data
A catalogue record for this book is available from the British Library.

Bibliographic information published by the Deutsche Nationalbibliothek
The Deutsche Nationalbibliothek lists this publication in the Deutsche Nationalbibliografie; detailed bibliographic data are available in the Internet at http://dnb.d-nb.de.

Typsetting K+V Fotosatz GmbH, Beerfelden
Printing Strauss GmbH, Mörlenbach
Bookbinding Litges & Dopf Buchbinderei GmbH, Heppenheim
Cover Design Grafik-Design Schulz, Fußgönheim

Printed in the Federal Republic of Germany
Printed on acid-free paper

ISBN-13: 978-3-527-31312-9
ISBN-10: 3-527-31312-5

Contents

Activation of Small Molecules. Edited by William B. Tolman

ISBN: 3-527-31312-5

Preface

Familiar to anyone who has taken a basic chemistry course, simple small molecules like H_2, N_2 and O_2 are ubiquitous reservoirs of chemical energy. These energy sources may be used in many ways, such as for fueling biological systems and as synthons for the construction of more complex molecules. Moreover, they may serve as signaling agents in biological systems to trigger complex protein expression and regulation processes. Such small molecules are generally quite thermodynamically stable and their successful utilization depends on surmounting often quite significant kinetic barriers. It has long been recognized that metal ions play an important role in providing low-barrier reaction pathways through binding and activation events. Much fundamental chemistry research has therefore been aimed at addressing questions such as:

- How do metal ions coordinate to and modulate the reactivity of small, often rather inert molecules?
- What is the basis for the selectivity of natural and synthetic metal-containing systems for specific small-molecule substrates?
- Can one use knowledge of metal/small-molecule chemistry for the development of new catalytic processes useful in the laboratory and/or in industry?

These and other related questions have been addressed in the past 10–15 years through the application of creative synthetic strategies and advanced experimental and theoretical techniques. The aim of this book is to provide highlights of recent research, with emphasis on newly discovered fundamental chemistry involved in metal-mediated binding and activation of CO_2, CO, NO, N_2O, N_2, H_2, CH_4, H_2O and O_2. This work has led to a deep understanding that has significantly impacted the fields of bioinorganic, organometallic and catalytic chemistry. By bringing together concise, yet detailed reviews by experts in these wide-ranging fields into a single volume, cross-disciplinary insights are provided in a valuable resource for students and researchers. Importantly, by organizing each chapter by small molecule, organometallic and bioinorganic perspectives are discussed together, in comparative fashion, thus bridging the two disciplines and generating new, unifying concepts.

Industrial processes for the utilization of the greenhouse gas CO_2 are emphasized in Chapter 1 by M. Aresta, with comparisons also drawn to biological pro-

Activation of Small Molecules. Edited by William B. Tolman

ISBN: 3-527-31312-5

cesses. Chapter 2 by D. H. Lee, B. Mondal and K. D. Karlin focuses on NO and N_2O reduction using examples drawn broadly from coordination and organometallic chemistry, as well as biological systems. Mechanisms of reduction of N_2 by nitrogenase and synthetic model systems are discussed by J. Peters and M. Mehn in Chapter 3, while in Chapter 4, J. W. Tye and M. Hall provide a tutorial on H_2 binding and reduction by metal complexes and metalloproteins. The broad topic of O_2 coordination and activation is presented in Chapters 5 and 6 by C. N. Cornall and M. Sigman and A. Borovik, P. J. Zinn and M. K. Zart, respectively. Complementary views of metal–O_2 chemistry are provided in these chapters through discussion of mechanisms, oxidation catalysis and reactive intermediate characterization. The enormously important problem of methane functionalization is tackled in Chapter 7 by R. Periana. The role of metal–aquo complexes in bioinorganic catalysis is emphasized in Chapter 8 by L. Berreau, before turning back to industrial applications in the discussion of CO chemistry by P. W. N. M van Leeuwen and Z. Freixa in Chapter 9. Taken together, these chapters illustrate the diversity of, and provide detailed mechanistic insights into, metal systems that are used in the laboratory and by nature to capture and use small molecules. Challenges for the future are laced throughout, and it is hoped that the reader will be stimulated to address them in future research.

W. B. Tolman
Department of Chemistry and Center for Metals in Biocatalysis
University of Minnesota
Minneapolis, MN 55455
USA
June 2006

List of Contributors

Michele Aresta
Department of Chemistry
University of Bari
70126 Bari
Italy

Andrew S. Borovik
Department of Chemistry
1102 Natural Sciences II
University of California
Irvine, CA 92697-2025
USA

Lisa M. Berreau
Department of Chemistry
and Biochemistry
Utah State University
Logan, UT 84322-0300
USA

Brian Conley
Donald P. and Katherine B. Loker
Hydrocarbon Research Institute
Department of Chemistry
University of Southern California
Los Angeles, CA 90089
USA

Candace N. Cornell
University of Utah
Department of Chemistry
Salt Lake City, Utah 84112-0850
USA

Zoraida Freixa
Institut Catalá d'Investigació Química
43007 Tarragona
Spain

Kenneth D. Karlin
Department of Chemistry
Johns Hopkins University
Baltimore, MD 21218-2685
USA

Somesh Ganesh
Donald P. and Katherine B. Loker
Hydrocarbon Research Institute
Department of Chemistry
University of Southern California
Los Angeles, CA 90089
USA

William A. Goddard, III
Materials and Process Simulation
Center
Beckman Institute
Division of Chemistry and Chemical
Engineering
California Institute of Technology
Pasadena, CA 91125
USA

Activation of Small Molecules. Edited by William B. Tolman

ISBN: 3-527-31312-5

Jason Gonzales
Materials and Process Simulation Center
Beckman Institute
Division of Chemistry and Chemical Engineering
California Institute of Technology
Pasadena, CA 91125
USA

Michael B. Hall
Department of Chemistry
Texas A & M University
College Station, TX 77842-3255
USA

Dong-Heon Lee
Department of Chemistry
Chonbuk National University
Jeonju 561-756
Korea

Mark P. Mehn
Arnold and Mabel Beckman Laboratories of Chemical Synthesis
Department of Chemistry and Chemical Engineering
California Institute of Technology
Pasadena, CA 91125
USA

Steve Meier
Donald P. and Katherine B. Loker Hydrocarbon Research Institute
Department of Chemistry
University of Southern California
Los Angeles, CA 90089
USA

Biplab Mondal
Department of Chemistry
IIT – Guwahati
North Guwahati
Assam 781039
India

Jonas Oxgaard
Materials and Process Simulation Center
Beckman Institute
Division of Chemistry and Chemical Engineering
California Institute of Technology
Pasadena, CA 91125
USA

Roy A. Periana
Donald P. and Katherine B. Loker Hydrocarbon Research Institute
Department of Chemistry
University of Southern California
Los Angeles, CA 90089
USA

Jonas C. Peters
Arnold and Mabel Beckman Laboratories of Chemical Synthesis
Department of Chemistry and Chemical Engineering
California Institute of Technology
Pasadena, CA 91125
USA

Matthew Sigman
University of Utah
Department of Chemistry
Salt Lake City, Utah 84112-0850
USA

William J. Tenn, III
Donald P. and Katherine B. Loker Hydrocarbon Research Institute
Department of Chemistry
University of Southern California
Los Angeles, CA 90089
USA

Jesse W. Tye
Department of Chemistry
Texas A & M University
College Station, TX 77842-3255
USA

Piet W. N. M. van Leeuwen
Van't Hoff Institute for Molecular Sciences
Faculty of Sciences
Universiteit van Amsterdam
1018 Amsterdam
The Netherlands

Kenneth J. H. Young
Donald P. and Katherine B. Loker Hydrocarbon Research Institute
Department of Chemistry
University of Southern California
Los Angeles, CA 90089
USA

Matthew K. Zart
Department of Chemistry
University of Kansas
Lawrence, KD 66047
USA

Paul J. Zinn
Department of Chemistry
University of Kansas
Lawrence, KD 66047
USA

1
Carbon Dioxide Reduction and Uses as a Chemical Feedstock

Michele Aresta

1.1
Introduction

The utilization of carbon dioxide (CO_2) as a source of carbon in synthetic chemistry has been a practice exploited at the industrial level since the second half of the 19th century for the synthesis of urea [1 a] and salicylic acid [1 b, c]. CO_2 has also long been used for making inorganic carbonates and pigments. A renewed interest in the industrial utilization of CO_2 as a source of carbon arose after the 1973 oil crisis. The topic has been comprehensively reviewed by several authors [2]. A critical assessment of CO_2 utilization is also available [3].

CO_2 is ubiquitous – it can be either extracted pure from natural wells or recovered from various industrial sources. For instance, quite pure CO_2 is recovered from urea synthesis. Several other industries, such as those using fermentation or similar methods, also provide a convenient source of pure CO_2 (above 99%) at low recovery cost, but their seasonality prevents full exploitation so that several million tons per year (Mt $year^{-1}$) of pure CO_2 are vented. Today, there is a growing interest in recovering CO_2 from power station flue gases that contain around 14% of CO_2, but the separation techniques are quite expensive [4] and are seldom applied on a large scale [5].

In recent years, CO_2 has also found growing application as a technological fluid in several industrial sectors, such as a cleaning fluid, in refrigeration, air conditioning and fire extinguishing equipment, as a solvent for reactions, as a solvent for nano-particle production, and in separation techniques and water treatment, as well as in the food and agro-chemical industries (packaging, additive to beverages, fumigant) [6, 7]. In all such technological applications, CO_2 is not converted and can be recovered at the end of the application or vented to the atmosphere. Most of the CO_2 used for such applications is currently extracted from natural wells, yet it is highly desirable to substitute the extracted CO_2 with that which is recovered from power stations or industrial processes. This would be in line with the need to reduce its emission into the atmosphere – a worrying accumulation since the beginning of the industrial era [8]. How

Activation of Small Molecules. Edited by William B. Tolman

ISBN: 3-527-31312-5

much CO_2 is used in the chemical industry or other applications? Close to 110 Mt_{CO_2}/y are either converted into chemicals [9] such as urea (70 Mt_{CO_2} $year^{-1}$), inorganic carbonates and pigments (around 30 Mt_{CO_2} $year^{-1}$) or used as additives to CO in the synthesis of methanol (6 Mt_{CO_2} $year^{-1}$). Other chemicals such as salicylic acid (20 kt_{CO_2} $year^{-1}$) and propylene carbonate (a few kilotons per year) comprise a minor share of the market. In addition, 18 Mt_{CO_2} $year^{-1}$ are used [7] as technological fluids, and in the food and agro-chemical industries (see above). Among industrial uses, the synthesis of urea and salicylic acid are purely thermal processes, the latter being influenced by the nature of the Group 1 cation of the original phenolate reacted with CO_2. Conversely, both the carboxylation of epoxides and, more importantly, the synthesis of methanol are driven by catalysts, mainly metal systems. The development of new catalytic conversions of CO_2 requires the knowledge of the properties of metal systems.

Whether or not the utilization of CO_2 can effectively contribute to reducing its emission/accumulation into the atmosphere and, thus, if it should be considered as a technology for the control of global warming is under assessment [2e]. As effective reduction of CO_2 emission into the atmosphere requires the elimination of a few gigatons per year ($1\ Gt=10^9\ t$), technologies for disposal in natural fields appear better suited [10]. Nevertheless, such technologies are now seldom applied and are limited, in the best of cases, to megaton-scale geological disposal; others (e.g. ocean disposal) have yet to be demonstrated. At the moment, the best approach to reducing CO_2 emission into the atmosphere would be to make a selection of a number of technologies, each able to reduce the emission by a fraction of gigaton per year. According to this perspective, the utilization of over 130 Mt $year^{-1}$ of CO_2 could well represent a technology that could significantly contribute to the reduction of atmospheric loading by recycling carbon and reducing the emission at its source. Such sustainable production technologies would reduce waste and make a better use of energy and carbon.

The evaluation of how much CO_2 is avoided when it is used in chemical or technological processes is not a simple task. The avoided fraction is not represented only by the amount of fixed CO_2 – one must consider the whole reaction cycle based on the emission [11]. Life cycle assessment (LCA) is the only methodology that can give an answer to such a question [12].

In the following sections, the CO_2 molecule will be considered. Its interaction with metal centers and the conversion paths that already find utilization or may find industrial exploitation in the near future are discussed, with emphasis on the most useful processes.

1.2
Properties of the CO_2 Molecule

1.2.1
Molecular Geometry

In its ground state, CO_2, a $16e^-$ molecule, is linear and belongs to the $D_{\infty h}$ point group. This makes the molecule nonpolar, although it contains two polar C–O bonds; the vectors associated with charge separation in the C–O bonds are equal in intensity and opposite in direction (Scheme 1.1).

$$O{=}C{=}O \leftrightarrow {}^{\delta-}O\text{-}C^{+\delta}{=}O \leftrightarrow O{=}C^{\delta+}\text{-}O^{\delta-} \leftrightarrow {}^{+\delta}O{\equiv}C\text{-}O^{\delta-} \leftrightarrow {}^{\delta-}O\text{-}C{\equiv}O^{\delta+}$$

Scheme 1.1 Polarity of the CO_2 molecule in its ground state.

Nevertheless, CO_2 maintains all the characteristics of a species containing polar bonds, with two sites that behave quite differently. The carbon atom is electrophilic, while the oxygen atoms are nucleophilic. Consequently, CO_2 often requires bifunctional catalysis [13] for its activation or conversion. It is noteworthy that the electrophilicity of carbon is higher than the nucleophilicity of each of the oxygen atoms, so CO_2 prevalently behaves as an electrophile.

The Walsh diagram [14] (Fig. 1.1) shows the energy of the molecular orbitals in the ground and excited states. Any distortion of the molecule from linearity causes the variation of the molecular energy and C–O bond length, due to the repulsive interactions generated among electrons. Figure 1.1 also shows that the energy of the molecular orbitals varies according to the plane in which the bending of the molecule occurs. In a similar way, any excitation of the molecule or interaction with electron donors that causes the population of the lowest unoccupied molecular orbital (LUMO) will also cause a distortion of CO_2 from linearity. Consequently, electronically excited CO_2, the radical anion $CO_2^{\bullet-}$ or the adduct of CO_2 with an electron-rich species, such as B-CO_2, will have a bent geometry with the O–C–O angle close to 133°, a value that minimizes the electron repulsion and the molecular energy. This is clearly shown by solid-state structural determinations – in all forms in which the carbon atom of CO_2 is bonded to a third atom, the O–C–O angle is close to 133°.

1.2.2
Spectroscopic Properties

1.2.2.1 Vibrational

Table 1.1 shows infrared (IR) and Raman data [15] of gaseous and solid CO_2. Due to its nonpolar character, in the ground state the symmetric stretching of the C=O bond is not IR active. In the Raman spectrum this vibration is found at 1285–1388 cm^{-1}. The IR properties of the molecule are used for many purposes, including the quantification of the amount of CO_2 in the atmosphere (by

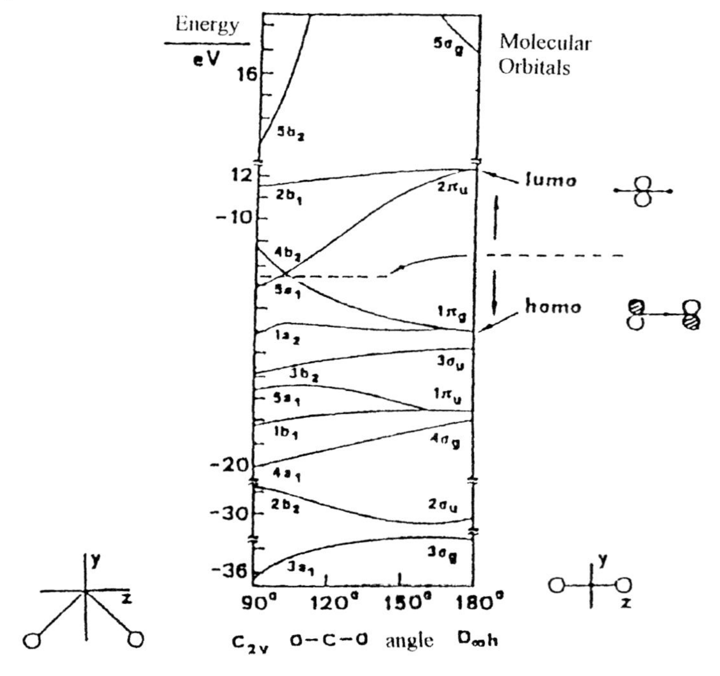

Fig. 1.1 Walsh diagram for CO_2: the energy of the MOs changes with the molecular geometry.

Table 1.1 IR and Raman data for CO_2

	$Sym_{C=O}$	Bending	$Asym_{C=C}$
Gaseous	1285–1388 (Raman)	667	2349
Aqueous solution			2342
Solid		660, 653	2344

nondispersive IR), and as a diagnostic tool for identifying CO_2 and its mode of bonding in metal systems.

1.2.2.2 UV-Vis

The UV-Vis spectrum [16] of gaseous CO_2 presents absorption bands of various intensities in the range 1700–3000 Å. The UV-Vis spectrum has not been used as extensively as the IR.

1.2.2.3 ^{13}C-Nuclear Magnetic Resonance (NMR)

CO_2 dissolved in a nonpolar solvent such as benzene or toluene shows a resonance at 126 ppm. In aqueous solutions the resonance is close to 125 ppm [17] and can be used for the quantification of free CO_2. The ^{13}C-NMR resonance is often used as a diagnostic tool for identifying the CO_2 moiety in a compound.

1.2.3 Energy Data and Reaction Kinetics Relevant to CO_2 Conversion

CO_2 is, with water, the thermodynamic end-product of the combustion of materials containing carbon and hydrogen. In fact, CO_2 is the most thermodynamically stable of all carbon-containing binary neutral species (Table 1.2). Carbonates, both organic and inorganic, that contain the "CO_3" moiety are even more stable than CO_2. The stability of CO_2 has generated the common belief that it is "nonreactive" and that any transformation of it will require an energy input that will render the use of CO_2 for the synthesis of chemicals inconvenient. The "inertness" of CO_2 is important with respect to oxidants such as O_2; indeed, CO_2 behaves as a great combustion regulator or suppressor. Conversely, there are a number of reactions in which there is no need for an external energy supply, because the coreagent brings enough energy for the reaction with CO_2 to occur at room temperature or lower (e.g. the reaction of CO_2 with hydroxide, amines or olefins). It is important to distinguish the thermodynamic from the kinetic aspects of reactions involving CO_2. In fact, quite exergonic reactions, such as

Table 1.2 Energy of formation of some chemicals relevant to CO_2 chemistry [18]

Compound	ΔH_f^0 (kJ mol^{-1})	ΔG_f^0 (kJ mol^{-1})	S_f^0 (cal K^{-1})
CO (g)	–110.53	–137.2	197.7
CO_2 (g)	–393.51	–394.4	213.8
CO_2 (l)		–386	
CO_2 (aq)	–413.26		
CO_3^{2-} (aq)	–675.23		
CaO (s)	–634.92		
HCO_3^- (aq)	–689.93	–603.3	38.1
H_2O (l)	–285.83		
H_2O (g)	–241.83		
$CaCO_3$ (s) (calcite)	–1207.6	–1129.1	91.7
$CaCO_3$ (s) (aragonite)	–1207.8	–1128.2	88
$COCl_2$ (g)	–219.1	–204.9	283
CS_2 (l)	89	64.6	151.3
CS_2 (g)	116.6	67.1	237.8
HCN (l)	108.9	125.0	112.8
HCN (g)	135.1	124.7	201.8
CH_2O (g)	–108.6	–102.5	218.8
HCOOH (l)	–424.7	–361.4	129
HCOOH (g)	–378.6		129
CH_4 (g)	–74.4	–50.3	186.3
CH_3Cl (g)	–81.9		
H_2NCONH_2 (s)	–333.6		
CH_3OH (l)	–239.1	–166.6	126.8
CH_3OH (g)	–201.5	–162.6	239.8

the formation of inorganic carbonates from CO_2 and oxides (see Table 1.2), are characterized by a high kinetic barrier that makes them proceed slowly. For example, the natural weathering of silicates [19] that converts silicates into carbonates (Eq. 1) and free silica is a very slow process that requires activation to occur in solution:

$$M_2SiO_4 + 2\,CO_2 \rightarrow 2\,MCO_3 + SiO_2 \qquad (1)$$

In general, the reactions of CO_2 can be classified into two categories, according to their energetics:

- Reactions that do not require an external energy input, such as those that incorporate into a chemical the whole CO_2 moiety, or, more generally, those in which the carbon atom maintains the formal +4 oxidation state. Such reactions produce carboxylates and lactones (RCOOR′), carbamates (RR′NCOOR″), ureas (RR′NCONRR′), isocyanates (RNCO), and carbonates [ROC(O)OR′].
- Reactions that generate reduced forms of CO_2, such as: $HCOO^-$ (formates), $[C(O)O]_2^{2-}$ (oxalates), H_2CO (formaldehyde), CO, CH_3OH and CH_4.

The latter require energy that is provided as heat (thermal processes), electrons (electrochemical processes) or irradiation (photochemical processes). The convenience of developing a process based on CO_2 does not depend on its endo- or exergonicity. As a matter of fact, several processes are today on-stream that are strongly endothermic, consume significant amounts of energy and produce large quantities of waste. Therefore, the convenience of developing a process based on CO_2 for substituting an existing one must be evaluated by comparing the two processes by applying the LCA methodology – the use of CO_2 will be convenient if it minimizes the material and energy consumption and the CO_2 emission.

Any one of the above reactions will require a catalyst, which often is a metal system. After the discovery of the first transition metal complex of CO_2 [20], emphasis has been on the study of the coordination chemistry of CO_2 with the aim of discovering new catalysts for CO_2 chemical utilization.

1.3 CO_2 Coordination to Metal Centers and Reactivity of Coordinated CO_2

1.3.1 Modes of Coordination

The modes of coordination of CO_2 to a metal center(s) are classified in Table 1.3. While the η^2-C,O [20] and η^1-C [22a, b] coordination modes have been known for a long time, the η^1-O mode was only recently demonstrated [23]. Moving from the η^1 to the μ_4-η^5 mode [40], the C–O bond order decreases and the length of the bond increases. Nevertheless, it is not straightforward to say that coordination increases the "reactivity" of the CO_2 molecule. Whether the co-

Table 1.3 Modes of bonding of CO_2 to metal centers

Mode of bonding	Structural features of the adduct	M [reference]	ν_{asym}	ν_{sym}	C–O bond length (Å)
η^1-C	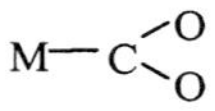 M—C(O)O	Ir [22a], Rh [22b]	1610	1210	1.20(2), 1.25(2)
η^1-O	M-O=C=O	U [23]	2188		1.122(4), 1.277(4)
η^2-C,O	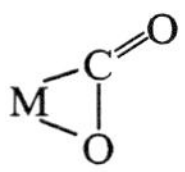 M(C=O)(O)	Ni [24], Rh [25], Fe [26], Pd [27]	1740	1140–1094	1.17, 1.22
μ_2-η^2	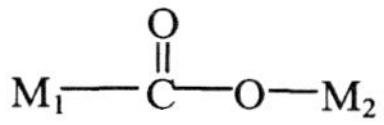 M_1—C(=O)—O—M_2	Pt [28], Ir/Zr [29], Ir/Os [30], Rh [31], Ru [32]	1495	1290–1190	1.229(12), 1.306(12)
μ_2-η^3, class I	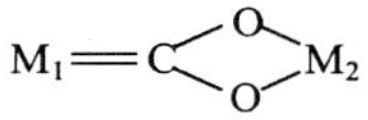 M_1=C(O)(O)M_2	Re/Zr [33], Ru/Zr [34], Ru/Ti, Fe/Zr, Fe/Ti [34]	1348 1348	1288 1290	1.285(5), 1.281(5)
μ_2-η^3, class II	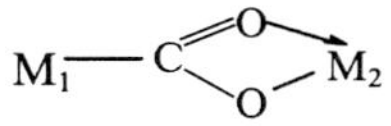 M_1—C(=O→)(O)M_2	Re/Sn [35], Fe/Sn [36]	1395 1450	1188 1152	1.269(11), 2.257(7), 1.252(3), 2.394(2)
μ_3-η^3	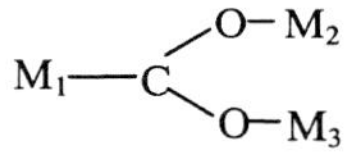 M_1—C(O—M_2)(O—M_3)	Os [37], Re [38]			1.276(5), 1.322(5), 1.28, 1.25
μ_3-η^4	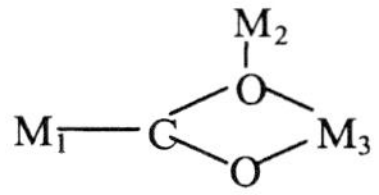 M_1—C(O(M_2))(O)M_3	Co [39]			1.20(2), 1.24(2)
μ_4-η^4	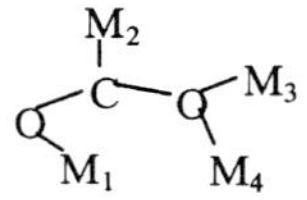 M_2; C; O–M_1; O–M_3, M_4	Ru [21]			1.283(15), 1.245(16)
μ_4-η^5	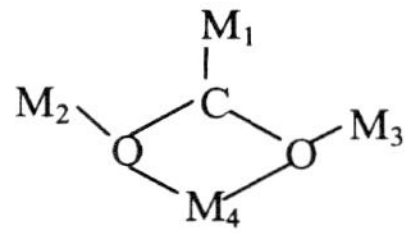 M_1; C; M_2–O, O–M_3; M_4	Rh/Zn [40]			1.29(14), 1.322(14)

ordinated CO_2 is an activated form of CO_2 is a question that requires a detailed investigation [25b, 41]. The reduction of the C–O bond order and energy upon coordination to metal centers represents *per se* an activation of CO_2, but such an activated form may not be ready for further conversion, due to the high energy of the bonds formed with metal centers: the metal complexes may sometimes behave as "stable forms of activated CO_2".

A consequent question is whether or not the coordination of CO_2 to a metal center is a prerequisite for its conversion into other chemicals. In fact, it depends on the kind of reaction CO_2 has to undergo. The data reported in Section 1.3.3 seem to suggest that the coordination to a metal center is necessary only if the reduction of CO_2 to CO is considered. In coupling reactions (C–C or C–E bond formation), CO_2 may react directly with nucleophiles produced in the reaction medium by the metal catalyst (e.g. activated olefins); pre-coordination to metal centers may not be necessary.

1.3.2
Interaction of CO_2 with Metal Atoms at Low Temperature: Stability of the Adducts

The interaction of the CO_2 molecule with main group and transition metal atoms has been studied using matrix isolation Fourier transform IR spectroscopy [42], which recently has been coupled with density functional theory (DFT) [43]. The latter has played a key role in determining the mode of coordination of the cumulene in the complex, predicting the equilibrium properties of the identified species and describing the bonding. Different behavior has been demonstrated for late transition metal atoms [Fe, Co, Ni, Ag and Cu form 1:1 $M(CO_2)$ complexes] compared to early transition elements (Ti, V and Cr insert spontaneously into one of the C=O bonds yielding oxo-carbonyl species) [43]. Isotopic experiments with $^{13}CO_2$ and $C^{18}O_2$ have permitted the spectroscopic identification of the bonding modes in organometallic species. Interestingly, the coordination of CO_2 to a metal atom is influenced by the gas matrix. For example, it has been shown that the cumulene binds in side-on fashion to nickel to afford a 1:1 complex with a binding energy equal to 75 kJ mol^{-1} in a pure CO_2 matrix, while in argon diluted matrices, no reaction occurs. Conversely, if N_2 is added to the rare gas matrix, the coordination of CO_2 occurs to preformed "NiN_2" with a binding energy equal to 133.8 kJ mol^{-1} [43]. The combination of experimental studies with DFT calculations has been fruitful for explaining the behavior of such systems and the role of N_2. Additionally, using DFT it has been shown that Ti inserts with no "energy barrier" into one of the CO bonds of CO_2, to afford a OTiCO species, which is more stable than any of the possible $Ti(CO_2)$ complexes [43].

1.3.3
Reactivity of CO_2 Coordinated to Transition Metal Systems

While several examples demonstrate that coordinated CO_2 undergoes electrophilic attack by protons or other similar reagents at the ^{2}-bonded oxygen, there is little evidence [15, 25 b, 41] that coordination promotes the formation of a C–C bond, e.g. between CO_2 and an olefin. In the latter case, it is more likely that CO_2 interacts with a M(olefin)-adduct. Alternatively, a three-molecular mechanism involving the metal center, the olefin and CO_2 may operate (see Section 1.4.1.1). Scheme 1.2 gives an overview of the documented reactions of coordinated CO_2.

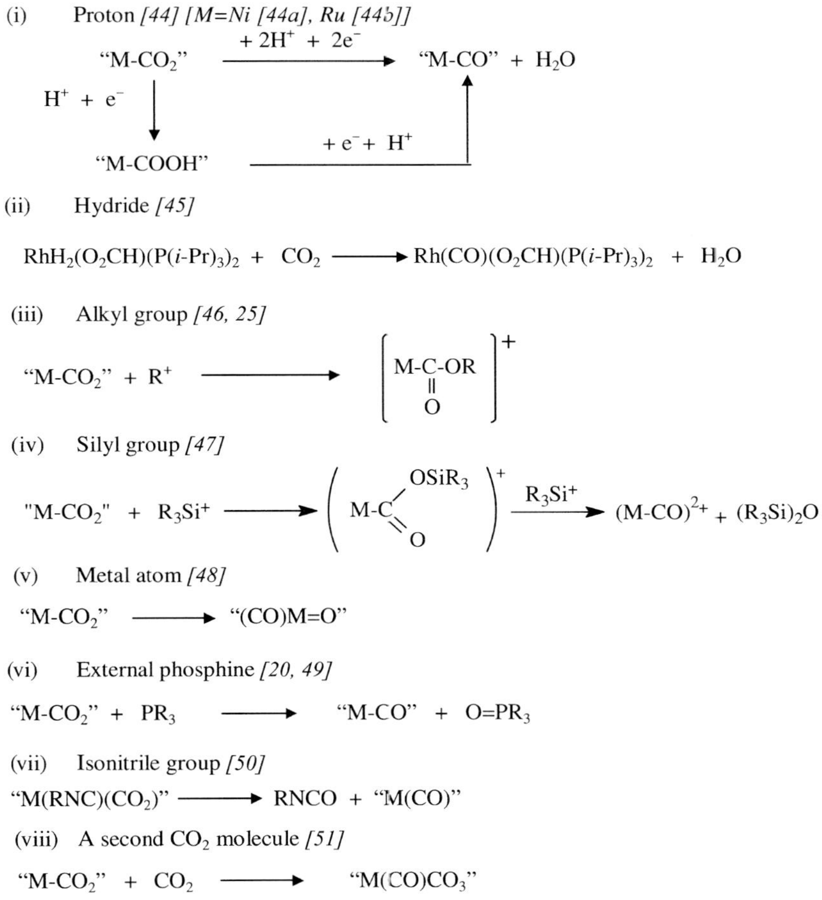

(i) Proton [44] [M=Ni [44a], Ru [44b]]

$$\text{“M-CO}_2\text{”} \xrightarrow{+2H^+ + 2e^-} \text{“M-CO”} + H_2O$$

$$\text{“M-CO}_2\text{”} \xrightarrow{H^+ + e^-} \text{“M-COOH”} \xrightarrow{+e^- + H^+} \text{“M-CO”}$$

(ii) Hydride [45]

$$RhH_2(O_2CH)(P(i\text{-Pr})_3)_2 + CO_2 \longrightarrow Rh(CO)(O_2CH)(P(i\text{-Pr})_3)_2 + H_2O$$

(iii) Alkyl group [46, 25]

$$\text{“M-CO}_2\text{”} + R^+ \longrightarrow [\text{M-C(=O)-OR}]^+$$

(iv) Silyl group [47]

$$\text{"M-CO}_2\text{"} + R_3Si^+ \longrightarrow (\text{M-C(=O)OSiR}_3)^+ \xrightarrow{R_3Si^+} (\text{M-CO})^{2+} + (R_3Si)_2O$$

(v) Metal atom [48]

$$\text{“M-CO}_2\text{”} \longrightarrow \text{“(CO)M=O”}$$

(vi) External phosphine [20, 49]

$$\text{“M-CO}_2\text{”} + PR_3 \longrightarrow \text{“M-CO”} + O{=}PR_3$$

(vii) Isonitrile group [50]

$$\text{“M(RNC)(CO}_2\text{)”} \longrightarrow RNCO + \text{“M(CO)”}$$

(viii) A second CO_2 molecule [51]

$$\text{“M-CO}_2\text{”} + CO_2 \longrightarrow \text{“M(CO)CO}_3\text{”}$$

Scheme 1.2 Reactions of coordinated CO_2

1.4 CO_2 Conversion

The utilization of CO_2 for the synthesis of compounds (e.g. carboxylates) containing the entire CO_2 moiety is an example of a process that follows the “sustainable chemistry” principles [2d, e]. In fact, with respect to processes on-stream, it reduces the production of waste at source, uses less starting materials, recycles carbon, diversifies the raw materials, and may make less use of solvents if CO_2 is used as solvent ($scCO_2$) and reagent. Because of the more direct synthetic procedure, there may also be an associated reduction of energy consumption. Such benefits are rigorously assessed by making use of the LCA methodology, applied to the CO_2-based process and to the process that is being substituted [12]. Conversely, if CO_2 is reduced to other C1 molecules, an energy input may be necessary. The real benefit in this case can be evaluated by comparing the CO_2 reduction to synthesis gas (syngas) production, which represents the current route to any reduced form of carbon-based products.

1.4.1 Carboxylation Reactions

The incorporation of CO_2 into an organic substrate to afford C-COOH, C-COOC, E-COO-C (E = N, O) or C-OC(O)O-C moieties is of great importance from the industrial point of view as it would allow the implementation of direct methodologies in place of those on-stream that do not respond to the energy- or atom-economy principles. The formation of a terminal "carboxylic moiety" C-CO_2 is today achieved through quite lengthy and waste-producing procedures. Thus, the oxidation of an organic moiety (i.e. CH_3 or benzene skeleton) or the hydration of CN groups are typically used, via multistep procedures, with production of waste and loss of carbon [52]. Alternatively, Grignard reagents can be reacted with CO_2 with loss of 1 mol Mg per mole of carboxylate produced. Even more complex routes are used for the synthesis of cyclic compounds containing a "CO_2" moiety. The catalytic carboxylation of olefins, or other organic substrates, would be of great value in this case and would represent a step forward towards sustainable processes. Among the carboxylation reactions, the synthesis of carboxylic acids or lactones, the carbamation of amines and the synthesis of carbonates are particularly important due to the large market for the products (of the order of several megatons per year). Special attention will be dedicated to such compounds in the following sections.

1.4.1.1 C–C Bond Formation

The carboxylation of an organic substrate such as a saturated hydrocarbon (Eq. 2) or benzene (Eq. 3) can be considered as a formal insertion of CO_2 into the C–H bond. The enthalpy of such a reaction is in general favorable, although dependent on the reagents. In fact, the carboxylation of methane or benzene is characterized by a negative change of enthalpy. Nevertheless, it must be emphasized that the dependence on entropy is quite different in the two cases, so the free energy change may be quite different:

$$CH_4(g) + CO_2(g) \rightarrow CH_3COOH(l) \quad (2)$$
$$\Delta H = -16.6\,\text{kJ mol}^{-1}$$
$$\Delta G_{298} = +71.17\,\text{kJ mol}^{-1}$$

$$C_6H_6(l) + CO_2(g) \rightarrow C_6H_5COOH(s) \quad (3)$$
$$\Delta H = -40.7\,\text{kJ mol}^{-1}$$

The key step in such reactions is the heterolytic C–H bond splitting that produces a carbanion that easily reacts with CO_2 to afford a carboxylate (Eq. 4):

$$R\text{-}H \longrightarrow R^- + H^+ \xrightarrow{CO_2} RCOO^- + H^+ \longrightarrow RCOOH \quad (4)$$

The differences and similarities of natural and artificial processes will be summarized and analyzed in the next sections.

1.4.1.1.1 **Natural Processes**

C-carboxylation reactions in nature use either CO_2 or its hydrated form, HCO_3^- (Scheme 1.3), depending on whether the enzyme active site is hydrophobic or not. Often a metal cation is required as cofactor. The most used metal ions are Mg^{2+}, Mn^{2+}, Co^{2+} and Fe^{2+}, with some evidence for the involvement of +3 cations such as Co^{3+}, Al^{3+} and Fe^{3+} (Table 1.4). The size of the cations, and their coordination number and charge density may play a key role in the stabilization of enzymes and in driving their catalytic activity. Cations with ionic radii in the range 85–110 pm [53] with a coordination number of 6 (octahedral geometry) are frequently encountered as cofactors in phosphoenolpyruvate carboxylases and other enzymes. The most abundant carboxylation enzyme in nature is ribulose 1,5-bis(phosphate)-carboxylase-oxidase (RuBisCO) [54], which is found in all eukaryotes and the majority of prokaryotes.

Table 1.4 Enzymes, substrates and metal cations implied in natural carboxylation reactions

Enzymes	Substrates	Metal cations	Products	Occurrence
RuBisCO	ribulose	Mg^{2+}	glucose	C3-plants (also higher)
Phosphoenolpyruvate carboxylase (PEPC)	pyruvic acid	Mg^{2+}, Mn^{2+}	oxaloacetate	C4-plants (mais, sugar cane, sorghum, etc.)
Phosphoenolpyruvate (PEP) carboxykinase	pyruvic acid	Mg^{2+}, Mn^{2+}	oxaloacetate	
Acetyl-CoA carboxylase	acetyl-CoA	Mg^{2+}, Mn^{2+}	malonyl-CoA	
Proprionyl-CoA carboxylase	proprionyl-CoA	Mg^{2+}, Mn^{2+}	methyl-malonyl-CoA	
Pyruvate carboxylase	pyruvate	Mg^{2+}, Mn^{2+}	oxaloacetate	
Vitamine-K-dependent carboxylases	first 10 glutamic acid residues in the N-terminal region of the precursor of prothrombin	Mn^{2+}	γ-carboxy-glutamic acid	

$$\text{R-C-H} + \text{HCO}_3^- = \text{R-C-COO}^- + \text{H}_2\text{O} \quad \text{(a)}$$

$$\text{R-C-H} + \text{CO}_2 \longrightarrow \text{R-C-COO}^- + \text{H}^+ \quad \text{(b)}$$

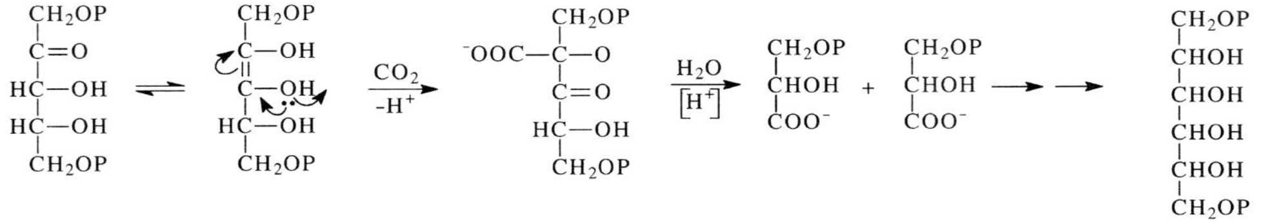

$$\begin{array}{c}\text{CH}_2\text{OP}\\ |\\ \text{C=O}\\ |\\ \text{HC—OH}\\ |\\ \text{HC—OH}\\ |\\ \text{CH}_2\text{OP}\end{array} \rightleftharpoons \begin{array}{c}\text{CH}_2\text{OP}\\ |\\ \text{C—OH}\\ \|\\ \text{C—OH}\\ |\\ \text{HC—OH}\\ |\\ \text{CH}_2\text{OP}\end{array} \xrightarrow[-\text{H}^+]{\text{CO}_2} \begin{array}{c}\text{CH}_2\text{OP}\\ |\\ ^-\text{OOC—C—O}\\ |\\ \text{C=O}\\ |\\ \text{HC—OH}\\ |\\ \text{CH}_2\text{OP}\end{array} \xrightarrow[\text{[H}^+\text{]}]{\text{H}_2\text{O}} \begin{array}{c}\text{CH}_2\text{OP}\\ |\\ \text{CHOH}\\ |\\ \text{COO}^-\end{array} + \begin{array}{c}\text{CH}_2\text{OP}\\ |\\ \text{CHOH}\\ |\\ \text{COO}^-\end{array} \longrightarrow \longrightarrow \begin{array}{c}\text{CH}_2\text{OP}\\ |\\ \text{CHOH}\\ |\\ \text{CHOH}\\ |\\ \text{CHOH}\\ |\\ \text{CHOH}\\ |\\ \text{CH}_2\text{OP}\end{array} \quad \text{(c)}$$

Scheme 1.3 C-carboxylation reactions (P = phosphate).

RuBisCO is constituted of a large and a small subunit, and performs both the carboxylation at C2 of ribulose and its oxidation at the same site with 50% selectivity [55]. The mechanism of action is quite complex – it has been shown that for both carboxylase and oxidase functions RuBisCO has an active site constituted by a lysine residue and a catalytic site, both placed in the large subunit. Once the two C3 moieties are formed (Scheme 1.3 c) the carboxylic functionalities are reduced and the two C3 moieties coupled to afford glucose. Formally the process consists of a CO_2 reduction to a "HCOH" moiety, inserted into a C–C bond of a C5 sugar to afford a C6 compound. This process uses some tens of gigatons per year of carbon of CO_2 in the natural carbon cycle.

The exploitation of biotechnologies for the utilization of CO_2 as source of carbon is an interesting approach to developing new, environmentally friendly synthetic technologies based on CO_2. In principle, both carboxylation and reduction reactions can be carried out, under mild conditions and using water as reaction medium, that would greatly improve the environmental quality of new processes with respect to those actually on stream [56, 58].

1.4.1.1.2 **Artificial Processes**

Despite the great industrial value of the formation of a C–C bond using CO_2, the only industrial application is represented by the synthesis of 2(or 4)-hydroxybenzoic acid, known for more than a century (Kolbe Schmitt reaction [1 b]). This reaction has been reconsidered [57] using other substrates such as 1- and 2-naphthol or hydroxypyridines.

A biotechnological synthesis has also been demonstrated to be possible [58]. *Thauera aromatica* bacteria can use phenol as the only source of carbon under anaerobic conditions; phenol is eventually converted into CO_2 and water. The first step of the degradation path is the carboxylation of phenolphosphate to 4-hydroxybenzoic acid which is then dehydroxylated to benzoic acid [59] (Scheme 1.4). The carboxylation of phenol is carried out by a phenolcarboxylase enzyme, a new type of lyase [60]. The isolation of the enzyme from the cytoplasmic portion of the cell allows its use *in vitro*. In order to extend the lifetime of the enzyme, its supported form on low melting agar can be used [61]. Cut-off membranes (that allow the passage of macromolecules of a given size) [58] can be

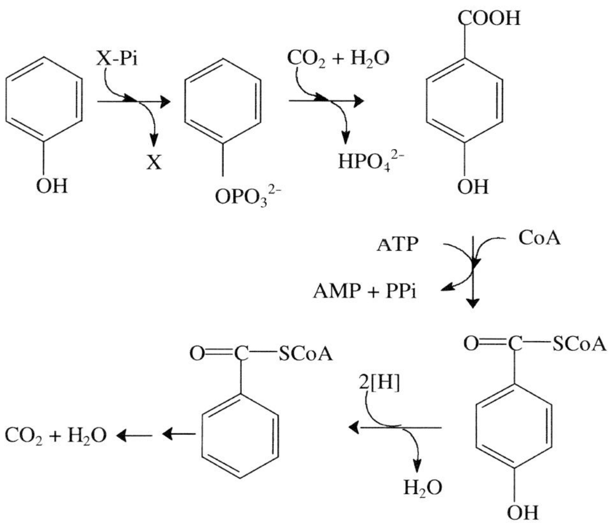

Scheme 1.4 Carboxylation of phenolphosphate to 4-hydroxybenzoic acid.

also used, with an easy separation of products from the mother liquid phase containing the enzyme and nutrients. Interestingly, the enzyme can also work in $scCO_2$ [62].

The direct carboxylation of hydrocarbons has been achieved only in the case of molecules containing active hydrogens using phenolate anion (PhO^-) as CO_2-transfer agent [63]. More recently, the 2-carboxylated form of imidazolium salts (Eq. 7) have been used for CO_2 transfer to molecules containing active hydrogens [64]. The resulting R^1R^2ImX can be recycled.

$$R^1R^2Im\text{-}2\text{-}CO_2 + \text{substrate-H} + MX \rightarrow R^1R^2ImX + \text{substrate-COOM} \quad (7)$$

$R^1R^2Im\text{-}2\text{-}CO_2$ = 1,3-dialkylimidazolium-2-carboxylate

Substrate = $PhC(O)CH_3$, CH_3OH

MX = $NaBF_4$, $NaBPh_4$, KPF_6

All the reactions presented above have as a common drawback, which is the use of 1 mol Group 1 metal cation per mole of carboxylated product, resulting, thus, in a process formally similar to the carboxylation of a Grignard reagent. For practical applications, metal cations would be better substituted with protons, but that is not an easy process.

The direct carboxylation of methane to acetic acid, a process of great industrial interest, has been achieved in low yield using a two step process with Rh and Pd catalysts [65].

The carboxylation of alkenes has been attempted using several transition-metal systems, such as Ni(0) [66], Ti [67], Fe(0) [68], Mo(0) [69] and Rh [70] as cata-

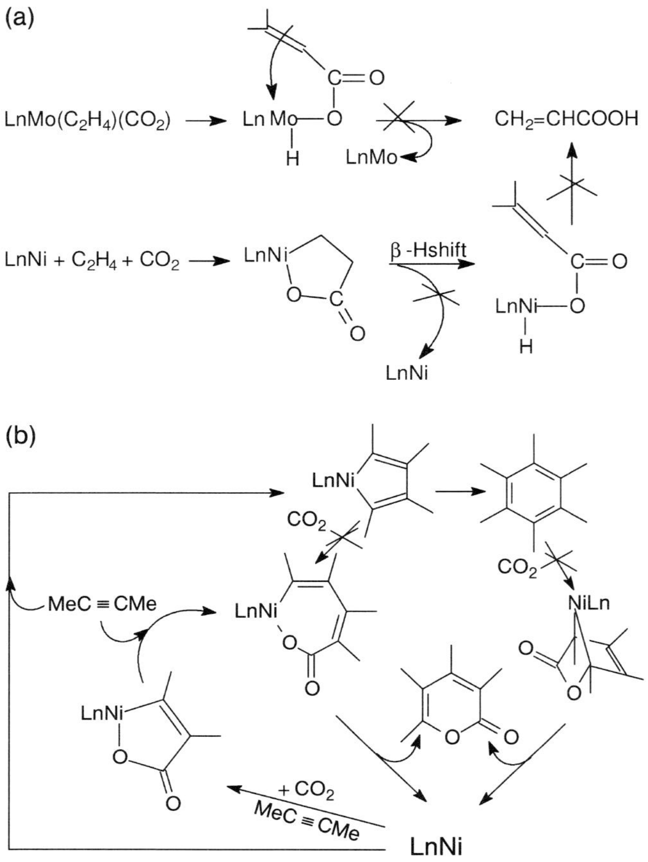

Scheme 1.5 (a) Modes of carboxylation of an olefin: the carboxylate is released upon protonation. (b) Conversion of terminal and internal alkynes into pyrones.

lysts (Scheme 1.5 a). In all cases, a stoichiometric amount of metal atoms was used to afford stable carboxylates such as metallacycle or hydrido-acrylate, the product of formal insertion of CO_2 into the C–H bond of ethylene. These reactions have been very recently revisited using DFT calculations that have shown the existing barriers in a catalytic cycle with high turnover numbers (TONs) [71, 72]. In particular, tailored coligands may assist the elimination of acrylic acid [71]. Such information can be very useful for designing active catalysts for the synthesis of carboxylates and, in particular, acrylic acids derivatives that have a large use in the polymer industry.

The carboxylation of alkynes (Scheme 1.5 b) and dienes has been successful, with both cumulated and conjugated systems being used (Schemes 1.6 and 1.7). In all cases high TONs have been obtained. The carboxylation of butadiene

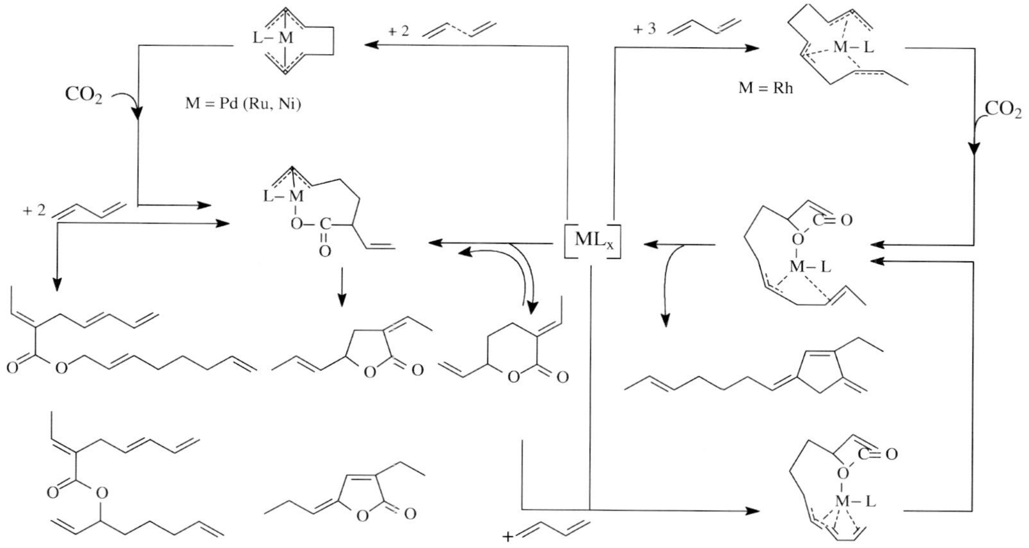

Scheme 1.6 Conversion of butadiene into lactones and linear esters using various metal catalysts.

to six-membered lactones has been performed with high selectivity and yield by using Pd systems with (i-C_3H_7)P$(CH_2)_n$CN (n=2–5) phosphane ligands in various solvents, including pyridine [73]. The reaction also proceeds under electrochemical catalysis [73b], and it has a good selectivity in $scCO_2$ using Pd(dba)$_3$ [73c] as catalyst.

Allene has been converted into pyrones or linear esters [74] by using Ni or Rh catalysts (Scheme 1.7a and b). Interestingly, allene and CO_2 undergo a formal "2+2" addition [75] to afford a four-membered lactone (Scheme 1.7c). Both the four- and six-membered lactones have industrial application, such as for antibiotics or fragrances, respectively.

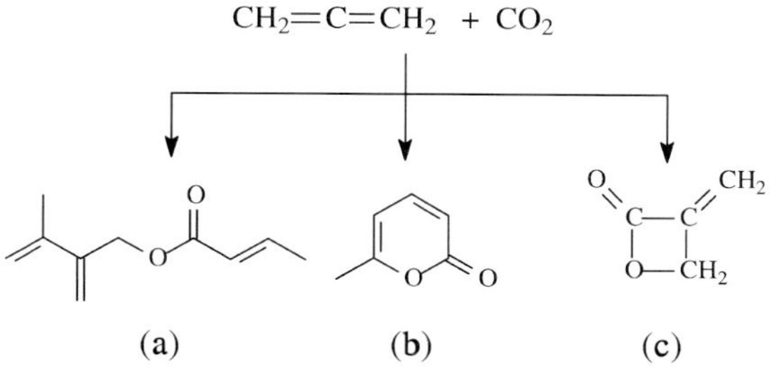

$$CH_2{=}C{=}CH_2 + CO_2$$

(a) (b) (c)

Scheme 1.7 Conversion of allene into linear esters or pyrones.

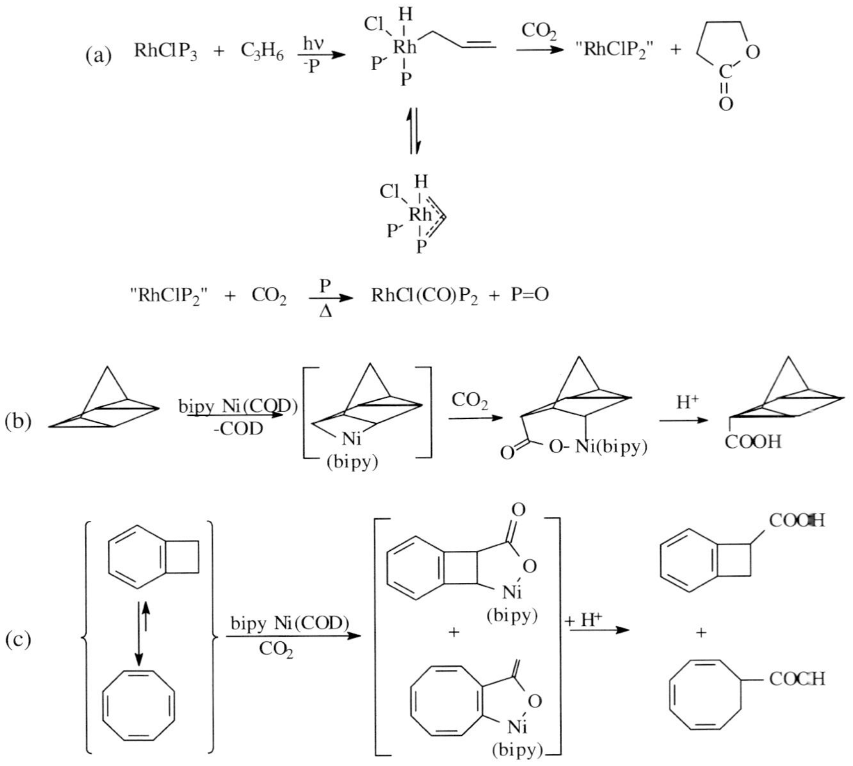

(a) $RhClP_3$ + C_3H_6 $\xrightarrow[-P]{h\nu}$ Cl, H, Rh, P, P $\xrightarrow{CO_2}$ "$RhClP_2$" + (lactone: C=O, O)

Cl, H, Rh, P, P

"$RhClP_2$" + CO_2 $\xrightarrow[\Delta]{P}$ $RhCl(CO)P_2$ + P=O

(b) bipy Ni(COD) / -COD; Ni (bipy); CO_2; O; O- Ni(bipy); H^+; COOH

(c) bipy Ni(COD) / CO_2; O; O; Ni (bipy); +; O; Ni (bipy); + H^+; COOH; +; COCH

Scheme 1.8 Carboxylation of strained rings.

Strained rings can be carboxylated by using transition metal catalysts [76]. The reactivity depends on the size of the ring, the metal used and the reaction conditions (Scheme 1.8). In the cases in Scheme 1.8 (b and c), 1 mol metal is consumed per mole of carboxylated product formed, which is a serious drawback to exploitation of the process.

1.4.1.2 **N–C Bond Formation**

The formation of the N–CO_2 bond is relevant to industrial processes as it would allow the synthesis of carbamic acid derivatives avoiding the use of phosgene (Scheme 1.9).

The synthesis of a labile carbamic acid has been achieved only very recently [77] using either benzylamine or $PhP(OCH_2CH_2)_2NH$. Both carbamic acids have been isolated as solid dimers and characterized in the solid state by X-ray diffraction. A common feature is the existence of the dimeric moiety represented below with an O–HO···O distance of 122 pm, very similar to that found in di-

(a) $RR'NH + CO_2 = \frac{1}{2}[RR'N\text{-}COOH]_2$
$[R,R' = PhCH_2\text{- or } PhP(OCH_2CH_2\text{-})_2]$

(b) $RR'NH + CO_2 + R''X + B = RR'N\text{-}COOR'' + BHX$

Scheme 1.9 Synthesis of carbamic acid (a) and carbamates (b) from amines and CO_2.

meric carboxylic acids. The monomer RR′N-COOH does not exist free, and decomposes back to the free amine and CO_2. Evidence for the formation of carbamic acid in solution has also been provided for amines like ω-(1-naphthyl)alkylamines [78].

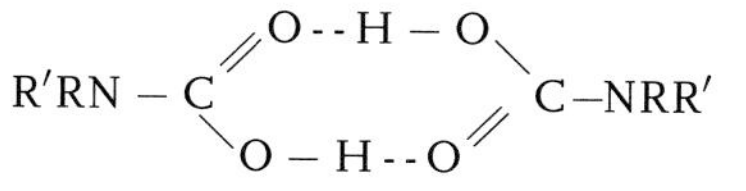

O--H – O
R′RN – C   C–NRR′
O – H--O

The reaction of amines with CO_2 in the presence of an alkylating agent and a base (Scheme 1.9 b) is important industrially as it produces organic carbamates that find large application in the chemical [79], pharmaceutical [80] and agrochemical industries [81]. The many attempts to use transition metals as catalysts in the 1970s [82] led to the discovery that the metal-carbamato complexes LnM-OC(O)-NRR′ reacted with the alkylating agent RX to undergo an electrophilic attack by the alkyl group at the nitrogen rather than at the oxygen atom, with a net alkylation of the amine (Scheme 1.10 a).

More recently, carbamic esters have been synthesized successfully under very mild conditions by using either Group 1 metal or ammonium carbamates in the presence of a crown-ether [83] (Scheme 1.10 b). The latter interacts with the

(a) $LnMx + 2RR'NH = LnM\text{-}NRR' + RR'NH_2X$
$LnM\text{-}NRR' + CO_2 = LnM\text{-}OC(O)\text{-}NRR'$
$LnM\text{-}OC(O)\text{-}NRR' + R''X = LnMX + RR'NR'' + CO_2$

(b) .

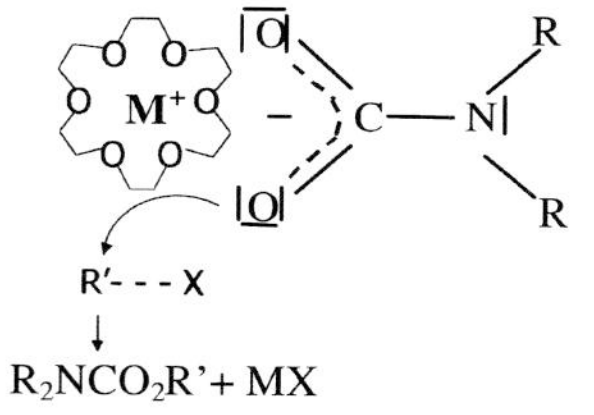

M⁺ (crown ether) O – C – N(R)R
R′- - - X
↓
$R_2NCO_2R' + MX$

(c) $RR'NH + (R''O)_2CO = RR'N\text{-}COOR'' + R''OH$
(d) $RR'NH + (R''O)_2CO = RR'R''N + R''OH + CO_2$

Scheme 1.10 Carboxy-alkylation of amines to carbamic esters.

metal or ammonium cation, increases the nucleophilicity of the oxygen and promotes the O-attack of the alkyl cation. Strong bases such as diazabicycloundecene (DBU) [84] have also been used. As the crown-ether can be easily recovered and recycled, the process is an easy, selective and high-yield route to carbamates at room temperature by directly using amines, CO_2 and alkylating agents. An alternative method for the carbamation of amines is the direct carboxy-alkylation by using carbonates (Scheme 1.10c). This reaction is of great industrial interest. It can be promoted by metal systems or other catalysts [85] with the major drawback being the alkylation of the amine, a process that occurs at higher temperature than the carboxy-alkylation (Scheme 1.10d). Either homogeneous [86] or heterogeneous [87] catalysts have been developed that work at low temperature and are very selective towards the carboxy-alkylation of amines. This route to organic carbamic esters is quite interesting as it may represent a phosgene-alkyl halides-free route if carbonates can be prepared from CO_2 and alcohols or by any other phosgene-free route.

A biomimetic catalyst has been developed for the carbamation of aromatic diamines [86h] under mild conditions.

1.4.1.3 O–C Bond Formation

The O–C bond formation is relevant to the synthesis of organic carbonates characterized by the O-C(O)O moiety. Both linear and cyclic carbonates are of industrial interest (Table 1.5). The semicarbonate species RO-C(O)OH is labile, as is the analogous RR′N-C(O)OH. CH_3O-C(O)OH has only recently been generated in solution and characterized by IR [88a] and NMR spectroscopy [88b].

1.4.1.3.1 Cyclic Carbonates

The most common route to cyclic carbonates is the reaction of epoxides with CO_2, which is promoted by a variety of homogeneous, heterogeneous and supported catalysts; either cyclic carbonates or polymers are obtained [89]. Main group metal halides [90a] and metal complexes [90b], ammonium salts [91] and supported bases [92], phosphines [93], transition metal systems [88, 94], metal oxides [95], and ionic liquids [96] have been shown to afford monomeric carbonates. Al porphyrin complexes [97] and Zn salts [89, 94, 98] copolymerize olefins and CO_2.

The carboxylation of epoxides is strongly dependent on the reaction conditions such as temperature and solvent. The use of ionic liquids as reaction medium seems to accelerate the reaction with respect to any other organic solvent, most probably because ionic liquids promote the formation of and/or stabilize polar or ionic intermediates. Heterogeneous catalysts such as oxides [95] or supported ammonium salts [92] or metal complexes [94] work well under these conditions. The solvent can play a key role in such reactions. Amides such as dimethylformamides or dialkylacetamides can themselves promote the carboxylation of epoxides, albeit with a low TON [100]. Most likely this is due to the abil-

Table 1.5 Linear and cyclic carbonates and their market and use (total market 18 Mt $year^{-1}$)

Carbonates					Uses
Linear	$(CH_3O)_2CO$ DMC dimethyl carbonate	$(CH_2CH{=}CH_2O)CO$ DAC diallyl carbonate	$(EtO)_2CO$ DEC diethyl carbonate	$(PhO)_2CO$ DPC diphenyl carbonate	solvents, reagents (for alkylation or acylation reactions), additive for gasoline
Cyclic	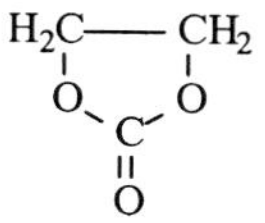 EC EC ethylene carbonate	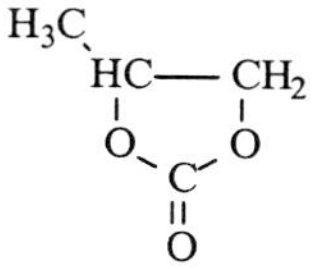 PC PC propylene carbonate	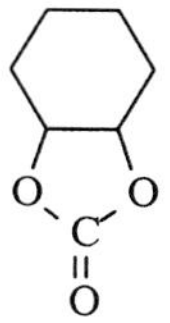 CC CC cyclohexene carbonate	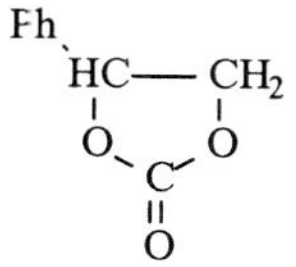 SC SC styrene carbonate	monomers for polymers, synthesis of hydroxyesters and hydroxyamines, component of special materials

ity of such species to activate either CO_2 or the epoxide (Scheme 1.11). $scCO_2$ also favors [99] the formation of the cyclic carbonate and copolymers.

Metal oxides have also been used for the synthesis of optically active carbonates [101] from pure enantiomers of the parent epoxide with total retention of configuration. The synthesis of optically active carbonates from a racemic mixture has been achieved with 22% *ee* in the best of cases, using Nb(IV) complexes with optically active N, O or P donor atom ligands. An NMR study [101] has demonstrated that such low percentage *ee* is caused by the de-anchoring of the ligand from the metal center, with loss of induction of asymmetry.

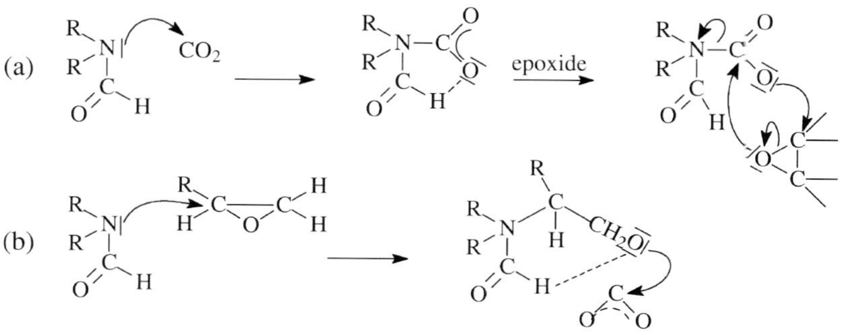

Scheme 1.11 Activation of CO_2 or the epoxide by an amide.

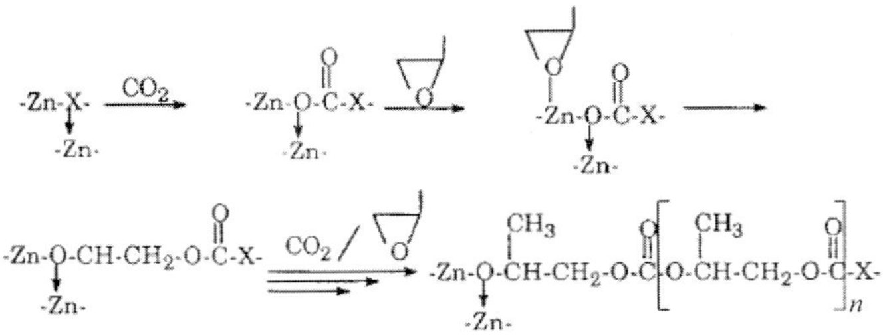

Scheme 1.12 Mechanism of copolymerization of propene and CO_2 with Zn complexes.

As noted above, when Al-porphyrin complexes [97] or Zn compounds [98] are used as catalysts for the carboxylation of epoxides, the formation of polymers is observed. Al catalysts are now used in a plant in China. The mechanism of the polymerization reaction has been studied and the most credited mechanism when Zn compounds are used is shown in Scheme 1.12. The molecular mass of the polymers varies with the catalyst. Primarily propene oxide and styrene oxide have been used so far, with some interesting applications of cyclohexene oxide. It is wished to enlarge the use of substrates in order to discover new properties of the polymers.

The limiting factor in the commercial development of the carboxylation of epoxides for the synthesis of monomeric or polymeric carbonates is the unavailability of large amounts of the parent epoxide. Such compounds are today prepared by using several techniques, some of which generate pollution [102]. The best route to epoxides is based on the use of H_2O_2 [103] that has as the main drawback the limited amount and the cost of H_2O_2. Finding a route to cyclic carbonates that is decoupled from H_2O_2 is of fundamental importance for developing the large volume industrial production of cyclic carbonates.

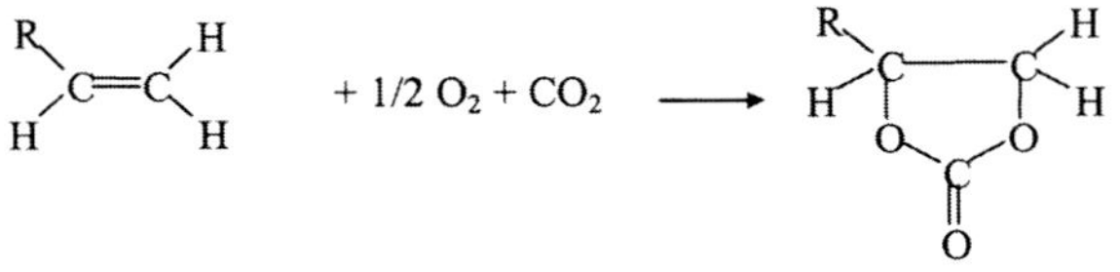

$$\mathrm{RHC{=}CH_2} + 1/2\ \mathrm{O_2} + \mathrm{CO_2} \longrightarrow \text{cyclic carbonate} \qquad (8)$$

A reaction of great interest in this direction is the "oxidative carboxylation" of olefins that converts cheap products, such as olefins, and a waste, such as CO_2, into valuable compounds (Eq. 8). Such a reaction has been performed using different catalysts on substrates such as propene and styrene. Little information is available on an early process that uses a complex catalytic mixture [104], although in later studies the reaction mechanism has been elucidated for both homogeneous [105] and heterogeneous [106] catalysts. Two reactions are observed (Scheme 1.13): "two-oxygen" addition across the double bond that causes the splitting of the olefin to afford two aldehydes (compounds 5 and 6 in Scheme 1.13) and "one-oxygen" transfer to the olefin that produces the epoxide,

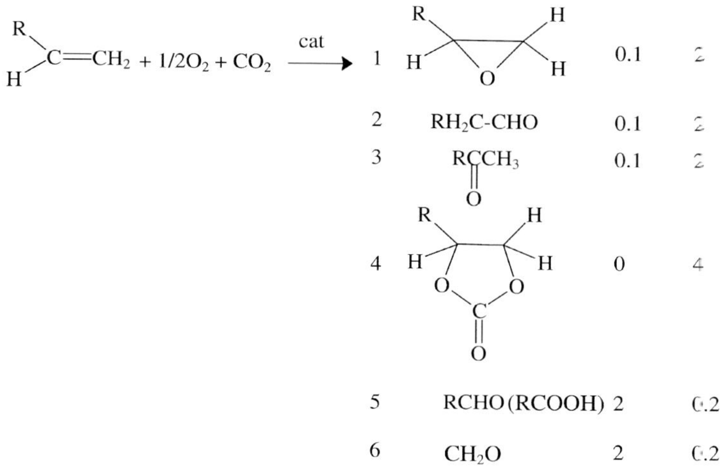

$RHC{=}CH_2 + 1/2O_2 + CO_2 \xrightarrow{cat}$

	Product		
1	epoxide	0.1	2
2	$RH_2C\text{-}CHO$	0.1	2
3	$RC(O)CH_3$	0.1	2
4	cyclic carbonate	0	4
5	RCHO (RCOOH)	2	0.2
6	CH_2O	2	0.2

Scheme 1.13 Oxidative carboxylation of olefins with homogeneous and heterogeneous catalysts.

that is either converted into the carbonate or isomerized to a terminal aldehyde (compound 2 in Scheme 1.13) or a ketone (compound 3 in Scheme 1.13).

Using $RhCl(PEt_2Ph)_3$ as catalyst a maximum TON of 2–4 towards the carbonate was observed [105 a–d]. A more detailed study [105 e] showed that the reaction proceeds via formation of a peroxocarbonate that is the real oxidant and converts into the relevant Rh carbonate (Scheme 1.14). Such a pathway does not explain the observed TON, as the carbonate should not be an active catalyst. Using $^{18}O_2$, $^{13}C^{18}O_2$ or $^{13}C^{16}O_2$ it has been possible to demonstrate by accurate IR study and calculation of vibrational frequencies for the peroxocarbonate com-

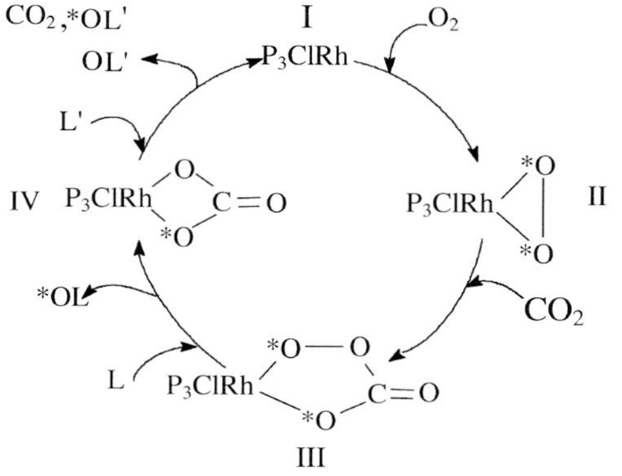

Scheme 1.14 Catalytic cycle for the formation of carbonate via oxidative carboxylation of olefins.

plex [107] that the formation of the peroxocarbonate occurs via insertion of CO_2 into the O–O bond of the dioxygen–Rh complex. The formation of an asymmetrically labeled *O–O peroxo bond allowed the subsequent oxygen transfer reaction [107] to an oxophile (like an olefin) to be followed. The resulting carbonate can be converted into a Rh(I) complex via deoxygenation of the carbonate [105 e] by a free phosphine (either added or released by the complex). The resulting Rh(I) complex can restart the process. The progressive release of phosphine causes the conversion of the original Rh(I) catalyst into a species bearing a phosphine oxide ligand that is not able to promote the epoxidation of the olefin anymore. If external phosphine is added, the yield of carbonate is not improved as the free phosphine is preferentially oxidized with respect to the olefin. Using unsaturated Rh(I) complexes of the formula "RhCl(L-L)" (L-L = bidentate phosphines or N-ligands) does not improve the yield as the catalyst is deactivated via an intermolecular ligand exchange as depicted in Eq. (9). The resulting "$RhCl(L\text{-}L)_2$" or "$RhCl(olefin)_n$" species are not able to promote the epoxidation of the olefin.

$$2''\mathrm{RhCl(L\text{-}L)}'' + n(\mathrm{olefin}) \rightarrow \mathrm{RhCl(L\text{-}L)_2} + \mathrm{RhCl(olefin)}_n \quad (9)$$

Therefore, such Rh catalysts are not useful for practical applications. Using Co, Cr or Mn analogs, the yield in carbonate is always low with no real improvement of the TON.

The use of Group 1 or 2 metal oxides [106] or of transition metal oxides gives catalysts with a longer life. It must be emphasized that in the oxidative carboxylation, the catalyst must perform two roles: the oxidation of the olefin using O_2 and the carboxylation of the epoxide. This makes the selection of the catalyst more difficult. For instance, Table 1.6 shows that metal oxides that behave as oxidants are not good carboxylation catalysts (see, e.g. Ag_2O). Table 1.6 also shows that, in a nonoptimized system, the main reaction is olefinic double-bond cleavage, suggestive of a radical reaction promoted by the metal oxide. A detailed study has identified the role of P_{CO_2} and P_{O_2}, temperature, solvent, and cocatalyst in the double-bond cleavage reaction, enabling the reaction to be performed so that it is not the main process anymore and the carbonate can be synthesized with more than 50% selectivity [106 c].

1.4.1.3.2 **Linear Carbonates**

The most interesting route to linear carbonates is the direct carboxylation of alcohols (Eq. 10):

$$2\,\mathrm{ROH} + \mathrm{CO_2} \rightarrow (\mathrm{RO})_2\mathrm{CO} + \mathrm{H_2O} \quad (10)$$

This reaction has an atom efficiency higher than the phosgene route, and is much safer and cleaner than the ENIChem and UBE processes that feature a comparable use of atoms (Scheme 1.15). The existing limitation to the exploitation of the reaction is the low yield at equilibrium that ranges between 1 and

Table 1.6 Products of the "oxidative carboxylation" of styrene using several oxides [a)]

Catalyst	Styrene conversion	Selectivity towards			
		Styrene carbonate (%)	Styrene oxide (%)	Benzal-dehyde (%)	Benzoic acid (%)
Molecular sieves 5 Å [b)]	16	3.1	11.8	67.5	3.1
SiO_2 anhydrous [b)]	23	9.1	15.2	45.6	19.1
SiO_2 hydrated [b)]	22	1.1	17.3	50.9	24.1
Ag_2O [b)]	28	–	16.4	50.8	24.5
MgO [b)]	14	6.8	13.6	58.8	8.9
Fe_2O_3 [b)]	28	10.3	1.8	46.4	33.9
MoO_3 [b)]	27	6.3	5.9	54.8	25.2
Ta_2O_5 [b)]	27	2.9	**17**	48.5	24.4
La_2O_3 [b)]	26	2.7	12.7	43	32.3
Nb_2O_5 [b)]	27	**16.6**	4.4	46.3	24.1
V_2O_5 [b)]	34	7.3	5	55.3	27
ZnO	12.5	1.3	14.2	41.1	36

a) Each entry is the average of three tests. The average deviation is ±5%. The operating conditions were the same in all tests. Catalyst: 7×10^{-4} mol; styrene: 1.75×10^{-2} mol; *N,N*-dimethylformamide as solvent: 10 mL; temperature: 393 K; reaction time: 5 h.

b) $P_{O_2} = 5$ atm; $P_{CO_2} = 45$ atm.

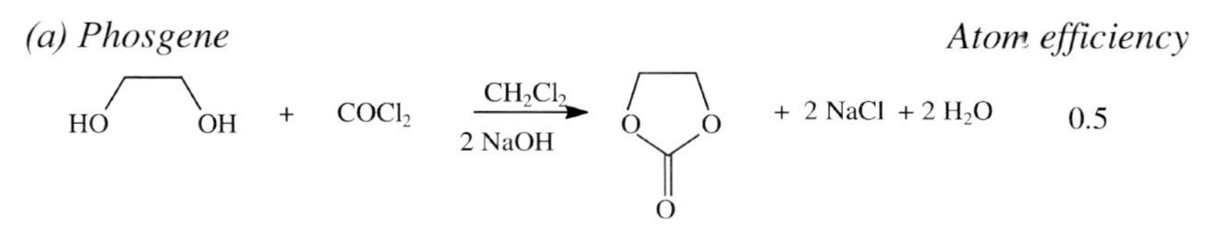

(b) Oxidative carbonylation

ENIChem

$2\,CH_3OH + \frac{1}{2}\,O_2 + 2\,CuCl \rightarrow 2\,Cu(OCH_3)Cl + H_2O$ — 0.8

$2\,Cu(OCH_3)Cl + CO \rightarrow (CH_3O)_2CO + 2\,CuCl$

Ube

$2\,CH_3OH + 2\,NO + \frac{1}{2}\,O_2 \rightarrow 2\,CH_3ONO + H_2O$ — 0.8

$2\,CH_3ONO + CO \rightarrow (CH_3O)_2CO + 2\,NO$

(c) Carboxylation of alcohols

$2\,ROH + CO_2 \rightarrow (RO)_2CO + H_2O$ — 0.8

Scheme 1.15 Routes to linear carbonates and their atom efficiency.

Table 1.7 Thermodynamic properties for the direct carboxylation of alcohols

Carbonate	ΔH (kJ mol^{-1})
$(MeO)_2CO$	–4.00
$(EtO)_2CO$	–3.80
(allyl-O)$_2$CO	–3.91
$[CH_3(CH_2)_nO]_2CO$ ($n > 2$)	–4.17
$(PhO)_2CO$	+12.06

Calculated according to Ref. [108].

2% of converted alcohol, using both homogeneous and heterogeneous catalysts. Table 1.7 gives the thermodynamic properties of some carbonates (aliphatic and aromatic). The values of ΔH show that the process is not very favored for aliphatic alcohols and is disfavored with phenol and other aromatic alcohols. Nonetheless, the low equilibrium concentration may not be a drawback for process development as the reagents can be recycled. Attempts have been made to use chemical water traps in order to displace the equilibrium to the right. Molecular sieves cannot be used at the reaction temperature as the formed surface OH groups are acidic enough to protonate the carbonate and reverse the reaction. Organic water traps are better suited: aldols (Eq. 11) [109, 111], ketals (Eq. 12) [110] and dicyclohexylcarbodiimide (DCC) [111] have been used as such.

$$(CH_3O)_2C(H)Ph + scCO_2 \xrightarrow{\text{cat.}} CH_3O\text{–}C(=O)\text{–}OCH_3 + O{=}C(H)Ph \tag{11}$$

$$(CH_3O)_2C(Me(Et))Me(Et) + scCO_2 \xrightarrow{\text{cat.}} CH_3O\text{–}C(=O)\text{–}OCH_3 + O{=}C(Me(Et))Me(Et) \tag{12}$$

In particular, the dimethyl cyclohexanone ketal also has been reacted with ethyleneglycol to afford a cyclic carbonate and cyclohexanone [112], plus methanol (Scheme 1.16). The use of DCC as water trap deserves comment. A detailed study has shown that it is a promoter of the carboxylation in addition to being a simple water removal agent. Combining experimental studies and DFT calculations, the reaction mechanism has been completely elucidated, as shown in Scheme 1.17 [113]. Several carbonates have been produced with very high yields (90–96%) and selectivity (close to 100%). The latter is highly influenced by the temperature as above 335 K the favored reaction is the formation of carbamate (Scheme 1.17 A). With DCC, using methanol and phenol it has been possible to produce the mixed methyl-phenyl-carbonate, (MeO)(PhO)CO [113].

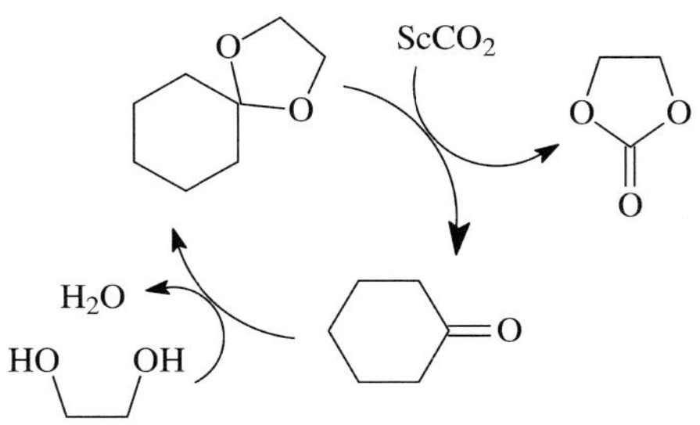

Scheme 1.16 Use of cyclohexanone-ketal in a two-step process for carbonate formation, avoiding the formation of water in the reaction medium containing the carbonate.

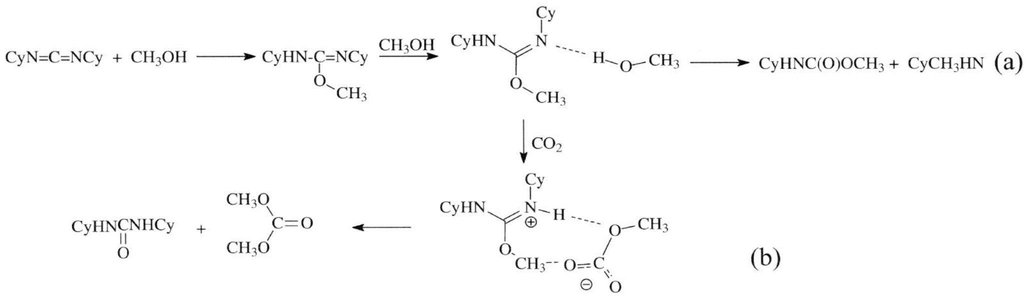

Scheme 1.17 Reaction mechanism of carboxylation of alcohols promoted by DCC.

The carboxylation of alcohols is an interesting reaction for the synthesis of carbonates that requires a better understanding in order to avoid catalyst deactivation by water. The reaction mechanism has been investigated for the Sn, Nb and DCC systems. Scheme 1.18 shows two different possible intra- and intermolecular mechanisms. The intramolecular mechanism that operates with Sn and DCC is based on a "double base-activation" of CH_3OH and produces an E=O double bond (E=C) that reduces the activity of the catalyst or generates an inert polymer (E=Sn). The intermolecular mechanism, that seems to be operative with Nb systems, can follow two routes that differ with respect to the intermediacy of one or two alcohol molecules. In the latter case, the reaction follows a "base plus acid activation" of methanol, and the catalysts perform much better and do not lose activity over several cycles [114]. The water formed in the reaction must be eliminated in order to push the equilibrium to right and avoid the destruction of the catalysts.

Heterogeneous catalysts also have been used [114c] that do not show better performances than homogeneous ones.

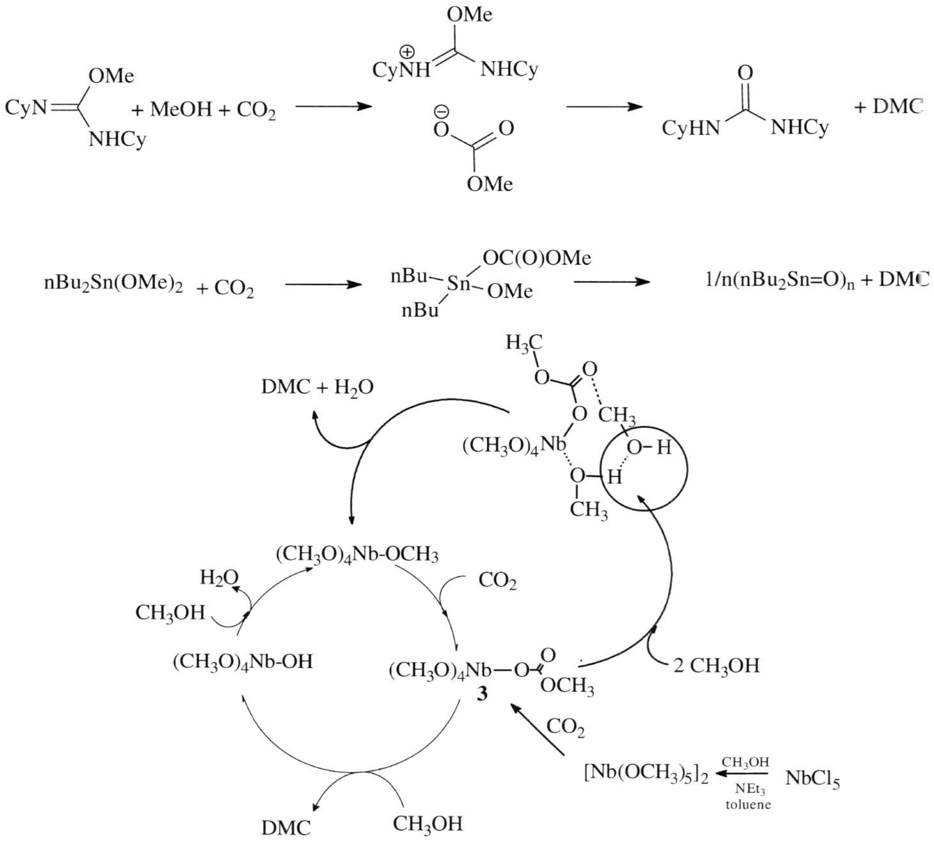

CyN=C(OMe)NHCy + MeOH + CO_2 → $[CyNH{=}C(OMe)NHCy]^+$ $[^-OC(O)OMe]$ → CyHNC(O)NHCy + DMC

$nBu_2Sn(OMe)_2$ + CO_2 → $nBu_2Sn(OMe)(OC(O)OMe)$ → $1/n(nBu_2Sn{=}O)_n$ + DMC

DMC + H_2O

H_3C O O O CH_3 $(CH_3O)_4Nb$ O–H O–H CH_3

$(CH_3O)_4Nb\text{-}OCH_3$

H_2O

CO_2

CH_3OH

$(CH_3O)_4Nb\text{-}OH$

$(CH_3O)_4Nb{-}O{-}C(O)OCH_3$ **3**

2 CH_3OH

CO_2

$[Nb(OCH_3)_5]_2$ ← (CH_3OH, NEt_3, toluene) $NbCl_5$

DMC CH_3OH

Scheme 1.18 Intra- (a and b) and inter-molecular (c) mechanism of formation of linear carbonates.

1.4.1.4 Use of Urea as an Active-CO_2 Form

Urea (H_2NCONH_2) is produced on a large scale (95 Mt $year^{-1}$) [9] from ammonia and CO_2 (Eq. 13):

$$2\,NH_3 + CO_2 \rightarrow H_2NCONH_2 + H_2O \qquad (13)$$

It can be used as an active form of CO_2 in reactions with alcohols to afford first the relevant urethane (Eq. 14) and then the carbonate (Eq. 15) [115]:

$$H_2NCONH_2 + ROH \rightarrow H_2NCOOR + NH_3 \qquad (14)$$

$$H_2NCOOR + ROH \rightarrow ROC(O)OR + NH_3 \qquad (15)$$

Such reactions are usually carried out in two steps and the intermediate urethane can be quantitatively isolated if necessary [116]. Urea has also been used for the synthesis of cyclic carbonates by reaction with diols (R=CH_3 [117]):

$$H_2NCONH_2 + HOCHRCH_2OH \longrightarrow \text{(cyclic carbonate: RHC—CH}_2\text{, O–C(=O)–O ring)} + 2\,NH_3 \quad (16)$$

In these reactions, the released ammonia can be easily recovered and re-converted into urea by reaction with CO_2. Such a procedure can be useful for the synthesis of a number of carbonates. It must be emphasized that the separation cost of the carbonate depends on the length of the carbon chain of the alcohol and decreases with increasing the number of carbon atoms. Such technology is not yet applicable for the synthesis of dimethylcarbonate from methanol, but is more suitable for the preparation of carbonates of higher alcohols.

1.4.1.5 **Transesterification Reactions**

Any carbonate can be transformed into another one by transesterification:

$$(RO)_2CO + 2\,R'OH \rightarrow (R'O)_2CO + 2\,ROH \quad (17)$$

$$\text{(cyclic carbonate: H}_2\text{C—CH}_2\text{, O–C(=O)–O ring)} + 2\,R'OH \longrightarrow (R'O)_2CO + HOCH_2CH_2OH \quad (18)$$

This process is already used for the industrial synthesis of diphenylcarbonate from DMC or ethylene carbonate, using $Ti(OPh)_4$ as catalyst [118]. Other catalysts have been developed [119] that produce a similar or better yield. In general, the reaction is very selective. This means that if a single process of those mentioned above is developed for industrial exploitation, then by transesterification other carbonates can be produced. The drawback to such a process is the coproduced alcohol that must be recycled or find utilization with a market comparable to that of carbonates (tens of megatons per year). The production system must be accurately designed so as to avoid the generation of undue amounts of byproducts. For example, one could imagine coupling two processes – the direct conversion of an olefin into the relevant cyclic carbonate and the conversion of the latter into a linear carbonate with formation of ethylene glycol. The latter could be reacted with urea to afford back ethylene carbonate and ammonia, recycled to produce urea.

1.4.2
Reduction Reactions

1.4.2.1 Energetics of the Reactions

As reported in Table 1.2, the conversion of CO_2 into other C_1 (or C_n) molecules, characterized by a lower O/C ratio or a higher H/C ratio, requires an energy input. Such energy can be provided in various forms: heat, electrons, radiation or chemicals. The convenience of the use of CO_2 as carbon source in the synthesis of products such as fuels (high H/C ratio) or intermediates (oxygenates) requires a detailed LCA study. While CO_2 lies in a deep potential well and its conversions require an external energy input, most of the routes that produce CO or syngas (H_2/CO=2 mixture) from more reduced carbon [coal, liquid natural gas (LNG), HC (liquid hydrocarbons) today are quite low in energy efficiency and the reactions used are strongly endothermic. In the following sections natural and artificial processes for CO_2 reduction will be considered and compared, when possible.

1.4.2.1.1 Natural Processes

Apart from the formal reduction of CO_2 to "H_2CO" that occurs upon RuBisCO catalysis, the reduction of CO_2 to other C1 molecules with the carbon in a lower oxidation state takes place in a number of microorganisms such as algae and bacteria. The processes of CO_2 reduction most often encountered are listed in Table 1.8, with the name of the enzyme that catalyzes the reduction and the metal present as prosthetic group.

Much interest has been placed on understanding the enzymatic reaction pathways so that they can be reproduced *in vitro*. In particular, CO dehydrogenase has been studied in detail and the cycle of production of acetic acid from CO_2 [126] is quite well understood (Scheme 1.19). The active site contains a S_4Fe_4-Ni cluster. The role of the different metal centers in the process has been a matter of discussion. Through spectroscopic studies on the enzyme, using Fe-labeled species, the Fe atom (Scheme 1.19a [127]) was suggested to be involved in the

Table 1.8 Processes of reduction of CO_2 in microorganisms, the enzyme implied and the metal present as prosthetic group

Reduction process	Name of enzyme	Metal	Reference
$CO_2+H^++2e^- \rightarrow HCOO^-$	formate dehydrogenase	W, Mo	120
$CO_2+2H^++2e^- \rightarrow CO+H_2O$	carbon monoxide dehydrogenase	Ni, Fe	121
$HCO_2H+2H^++2e^- \rightarrow H_2CO+H_2O$	formaldehyde dehydrogenase		122
$H_2CO+2H^++2e^- \rightarrow CH_3OH$	methanol dehydrogenase		123
$CO_2+CH_4 \rightarrow CH_3COOH$	methanogen	Ni, Fe, Co	124
$CO_2+8H^++8e^- \rightarrow CH_4+2H_2O$	tetrahydrofolate	none	125

CODH-mediated synthesis of the acetyl moiety

(a)

(CH_3)Ni-Fe(CO) → Ni-Fe-C(O)CH_3

Ni-Fe-C(O)CH_3 → H_3C(O)C-Ni-Fe

(b)

(CO)Ni-Fe) → (CO)(CH_3)Ni-Fe

(CH_3)(CO)Ni-Fe → H_3C(O)C-Ni-Fe

Corrin-CH_3 + (CO)Ni-Fe → Corrin + *CH_3(CO)* Ni-Fe
CH_3C(O)-Ni-Fe + $CoAS^-$ → Ni-Fe + CoAS-*C(O)CH_3*
CoA-S-C(O)CH_3 → CoASH + HOC(O)CH_3

Scheme 1.19 Building up the acetyl moiety.

CO_2 reduction to CO and CH_3 uptake from Ni for building the acetyl moiety. Alternatively, using biomimetic synthetic systems it was shown that Ni (Scheme 1.19b) [128b] could behave as the unique active center where both the CO_2 reduction and the methyl uptake, with consequent formation of the acetyl moiety, occurred. The involvement of Fe in this part of the process has now been discarded [129] and it has been agreed that Ni is the site in the S_4Fe_4-Ni enzyme where both the CO_2 to CO reduction and the coupling of CO with CH_3 to afford the acetyl moiety take place. Considering that the methyl moiety is made from CO_2 converted under tetrahydrofolate catalysis, this route to acetic acid would be of great industrial interest, if the correct energy source can be found.

1.4.2.1.2 **Artificial Processes**

Formic Acid and its Derivatives The hydrogenation of CO_2 to formic acid and its derivatives such as formamides has been investigated by several research groups. Active catalysts have been discovered that may hydrogenate CO_2 in organic solvents or water or else directly in scCO_2 and ionic liquids [130f]. Among the formic acid derivatives, formamides can be produced in high yield using homogeneous catalysts [130f]. Conversely, methods for the production of methyl formate have not been developed yet. Interestingly, the hydrogenation can occur in water and bicarbonate or carbonate can be hydrogenated to formate salts at very high rates, with TOF per hour values as high as 9600 at 80 °C and 1364 at room temperature. The hydrogenation of aqueous bicarbonate is sensitive to CO_2 pressures that most likely influence the HCO_3^- concentration. Whether CO_2 itself is hydrogenated instead of HCO_3^- is not clear and requires further mechanistic studies. These reductions show limited selectivity, as products other than formate, i.e. C1 molecules such as CO or CH_4, and C2 species (glycols and others) are also formed. Among the metal complexes used as catalysts, Ru and Rh systems in aqueous and organic solvents have received great attention. In general, the production of formic acid is promoted by a base added to the reaction medium. Its role is to convert formic acid into an ammonium formate that (i) has favorable thermodynamics and (ii) converts formic acid into a weaker acid (the ammonium cation has a K_a 4–5 orders of magnitude lower than the acid) reducing the probability of a back addition to the metal center. Ligands bearing amine functional groups as pendant arms were found to improve the catalytic activity of metal complexes in the absence of an external base, but the catalytic activity was found to be much lower than with added amines.

Ru(II)-trimethylphosphine complexes, being very active [130a,b], have been extensively investigated also from the kinetic point of view. The rate of formation of formic acid is first order in both H_2 and CO_2, which explains the influence of pressure and why scCO_2 as a solvent has a high positive influence [130b] on the reaction rate. Catalysts active in water have been extensively studied [130c]. It is highly probable that a cationic mechanism is operating and the active cationic species is generated with the assistance of added base or alcohol that promotes the loss of halide. This point has not been demonstrated yet, but only inferred on the basis of the fact that the effectiveness of various alcohols in promoting the hydrogenation matches their ability to promote the formation of cationic Ru complexes [130b]. Among Rh(I) catalysts, Rh(H)(diphosphine) compounds are the most active [130c]. Several other metal complexes have shown at least some activity, but the best, $NiCl_2$(dcpe), still has yields and rates well below those of Ru and Rh. Because the catalysts have been tested under very different conditions, it is difficult to make a ranking of them on the basis of TOF per hour values. If one assumes that the rates are first order in both H_2 and CO_2 pressures, and that they roughly double by increasing the temperature by 10 °C, then the most active catalyst precursors for formic acid production are RhCl$(TPPTS)_3$ [130d], Rh(hfacac)(dcpb) and RuCl(OAc)$(PMe_3)_4$ [130e]. That water may have a role is demonstrated by the fact that the aqueous-phase catalysts are the most active, which, based only on mass transfer rates, one would have not predicted. Methods have been discovered to prepare dialkylformamides [130f,g] and even phenylformamide [130h] from CO_2 and aniline with the assistance of DBU.

Production of CO and Syngas The production of syngas is of great importance in the energy and chemical industry. In fact, syngas can be used for the synthesis of methanol or long-chain hydrocarbons, such as gasoline, and other oxygenates. Therefore a great effort has been made to find catalysts and technologies that make use of CO_2 for producing CO or syngas, by using either methane (thermal routes) or water (photochemical conversion, Section 1.4.2.3) as the hydrogen source.

The interaction of CO_2 with a catalyst (metal or metal oxide) surface is relevant to its activation and conversion. Such chemistry has been investigated by several authors [131], who have studied the adsorption and reaction of CO_2 on clean Rh, Pd, Pt, Ni, Fe, Cu, Re, Al, Mg and Ag metal surfaces using spectroscopic methods. The formation of CO_2^- is a key step: depending on the metal, it can either dissociate to CO and O or convert into adsorbed CO_3^{2-} and CO. Adsorbed CO_2 dissociates on Fe, Ni, Re, Al and Mg, and remains unchanged over Pt and Cu surfaces. If oxygen is pre-adsorbed on the metals, the formation of stable carbonate structures is favored.

The production of syngas, used for the synthesis of methanol from methane, occurs via its partial oxidation (Eq. 19), steam reforming (Eq. 20) and CO_2 reforming (Eq. 21), also called dry reforming:

$$CH_4 + \frac{1}{2}O_2 \rightarrow CO + 2\,H_2 \qquad \Delta H^0_{298} = -36\ \text{kJ mol}^{-1} \tag{19}$$

$$CH_4 + H_2O \rightarrow CO + 3\,H_2 \qquad \Delta H^0_{298} = 206\ \text{kJ mol}^{-1} \tag{20}$$

$$CO_2 + CH_4 \rightarrow 2\,CO + 2\,H_2 \qquad \Delta H^0_{298} = 247\ \text{kJ mol}^{-1} \tag{21}$$

$$C + H_2O \rightarrow CO + H_2 \qquad \Delta H^0_{298} = 131\ \text{kJ mol}^{-1} \tag{22}$$

$$CO + H_2O \rightarrow CO + H_2 \qquad \Delta H^0_{298} = -41\ \text{kJ mol}^{-1} \tag{23}$$

Only the partial oxidation of methane (Eq. 19) is exothermic, and can, also if only partially, compensate the endothermic reactions (20) and (21).

The combination of the three reactions (19)–(21) is known as "tri-reforming" [132] that produces the proper CO/H_2 ratio (1.7) for methanol or higher hydrocarbons synthesis [133]. CO_2 reforming of methane produces syngas with lower residual CH_4 with respect to steam reforming, which leaves up to 2% of unreacted methane. The great existing interest in dry reforming is justified by the fact that it can be used in remote natural gas fields to convert on site CO_2-rich LNG streams into liquid fuels [gas-to-liquid (GTL) fuels], which are more easily transportable than gas and require less energy. Tri-reforming can advantageously be integrated into the direct re-use of flue gas from carbon-based power plants. The recent discovery of catalysts (Ni-La_2O_3) that prevent coke formation may push forward the exploitation of the process. An innovative approach is the utilization of catalysts under cold-plasma conditions for methane dry reforming [134] that can also be used for the direct synthesis of oxy-fuels [135]. In 2002, a pilot plant for the conversion of natural GTL fuel under plasma conditions began to be operated in Alberta, Canada [136]. The so called "tri-reforming" seems to be a good option for CO_2 and methane co-utilization [137]. Such technology converts two "greenhouse gases" into useful chemicals. CO_2 can be also used for the synthesis of ethanol (C_2) or higher (C_n) alcohols, for which there is a need for clean selective technologies. Efficient catalysts for ethanol formation from H_2/CO_2 have been developed [138] and higher alcohols can be produced in the same way [139]. Nevertheless, the yield is lower than using H_2/CO and the technology requires further development for practical application.

The catalytic hydrogenation of CO_2 has been applied to the selective synthesis of C_{5+} olefins using iron carbide (Fe_5C_2) as catalyst (Exxon Corporation process) and C_2–C_4 olefin synthesis using a Fe-K/alumina catalyst. A selectivity of about 44% at a CO_2 conversion of 68% under 2 MPa at 400 °C has been reported [140]. Alternatively, in a two-stage reactor, C_2–C_4 olefins are produced with over 90% overall selectivity by first making methanol using a Cu-ZnO catalyst, which is then converted using a solid acid catalyst in a second stage [141].

Another interesting use of CO_2 is as selective oxidant towards hydrocarbons. CO_2 is conveniently used as dehydrogenating agent in the conversion of $C_6H_5CH_2CH_3$ into styrene, $C_6H_5CH{=}CH_2$ [142]. The reaction is represented as:

$$C_6H_5CH_2CH_3 + CO_2 \rightarrow C_6H_5CH{=}CH_2 + CO + H_2O \quad (24)$$

An aspect of this reaction that deserves attention is that the coproducts CO and H_2O can be converted into CO_2 and H_2 by using a water-gas-shift catalyst:

$$H_2O + CO \rightarrow H_2 + CO_2 \quad (25)$$

In this way, the net result would be the use of CO_2 as mild oxidant with production of the unsaturated hydrocarbon and hydrogen, a process of great interest for a selective dehydrogenation of hydrocarbons.

Production of Methanol Remarkable progress has been made in the last 10 years in terms of catalyst development for the hydrogenation of CO_2 to methanol, such that 100% selectivity and high TOF have been observed. The excellent performance, most likely due to a different reaction mechanism [139] than with CO, compensates for the extra amount of dihydrogen needed for CO_2 reduction (Eq. 27) with respect to CO (Eq. 26), that is currently used:

$$CO + 2\,H_2 \rightarrow CH_3OH \quad (26)$$

$$CO_2 + 3\,H_2 \rightarrow CH_3OH + H_2O \quad (27)$$

CO_2 is currently added (up to 30% of total C) to syngas that is converted using Cu-ZnO-based oxide at 250–300 °C and 5–10 MPa (ICI process). Interestingly, the addition of CO_2 to the H_2/CO feed significantly improves the methanol yield and the energy balance. In such a system, one could imagine that CO_2 may be converted into CO by the water-gas-shift reaction and the latter is then converted into methanol that would still be a production of methanol based on syngas. That this is not the case has been demonstrated by tracer analysis [143]. The direct methanol synthesis from a CO_2/H_2 feed has been reviewed by several authors [144]. Experimental results show a higher yield of methanol from H_2/CO_2 at 260 °C, with respect to H_2/CO, with a further improvement when Pd modified Cu-ZnO [145] is used. The first pilot plant (50 kg day^{-1}) has been built in Japan [146], using a SiO_2^- modified Cu-ZnO catalyst. Recycling the feed produces a space–time yield of methanol of around 600 g/L h, with 99.9% selectivity during 8000 h operation at 250 °C and 5 MPa. Such a performance is several times that of conventional catalysts for syngas conversion. As a matter of fact, the methanol production from H_2/CO_2 is technically competitive with the industrial production from syngas, albeit economically still less convenient.

The synthesis of methanol from CO_2 and water can be carried out by using a cascade of enzymes that allow the conversion of CO_2 into $HCOO^-$ (formate dehydrogenase), CH_2O (formaldehyde dehydrogenase) and CH_3OH (methanol dehydrogenases) [122, 123]. All enzymes are easily available and can be used either separately or co-encapsulated into a unique system that makes the one-pot-three-step conversion of CO_2 into methanol [123b]. The limiting factor is the electron source: the $NADPH^+$/NADP couple has been used so far. Cheap reducing agents or solar energy should be used to generate the electrons necessary for the reduction of CO_2 to methanol. This process would be of great importance as the production of methanol under such conditions would represent the solution to recycling CO_2 and using it as source of carbon in the energy and chemical industry.

1.4.2.1.3 **Photoelectrochemical Reduction**

Transition metal complexes have been used as catalysts in photochemical conversions of CO_2 since they can absorb a significant part of the solar spectrum, have long-lived excited states, and can promote the activation of small molecules. Co complexes with N-macrocycles and $Ru(bpy)_2(CO)X_n$ complexes (X=CO, Cl, H) have shown quite interesting activity [147].

It is noteworthy that the potential for the reduction of CO_2 to the radical anion $CO_2^{\bullet -}$ is −2.2 V versus normal hydrogen electrode (NHE) in strictly anhydrous, aprotic solvents, making the one-electron reduction highly unfavorable. There is a large kinetic "overvoltage" because of the structural differences between linear CO_2 and bent $CO_2^{\bullet -}$. This causes the yield of photochemical conversion under transition metal catalysis to be low. In contrast, proton-assisted multielectron reductions are much more favorable as shown in Table 1.9. Therefore, multielectron reduction to any reduced form of CO_2 requires a considerably lower potential than the one-electron reduction to the radical anion. Consequently, electrolysis in the presence of catalysts can be carried out at reasonable voltages, as shown in Table 1.9.

Table 1.9 One-electron versus multielectron reduction of CO_2 assisted by protons (pH 7 in aqueous solution versus NHE) [147]

$CO_2 + e^- = CO_2^{\bullet -}$	E^0/V = −2.1
$CO_2 + 2H^+ + 2e^- = HCO_2H$	−0.61
$CO_2 + 2H^+ + 2e^- = CO + H_2O$	−0.53
$CO_2 + 4H^+ + 4e^- = C + 2H_2O$	−0.20
$CO_2 + 4H^+ + 4e^- = HCHO + H_2O$	−0.48
$CO_2 + 6H^+ + 6e^- = CH_3OH + H_2O$	−0.38
$CO_2 + 8H^+ + 8e^- = CH_4 + 2H_2O$	−0.24

Since the early work of Lehn and Ziessel [148] several metal complexes have been used with the aim of developing an effective photocatalyst for CO_2 reduction. In general, only low light efficiencies have been reached that are often coupled with low selectivity, which leaves open the need of research for developing effective systems that may find viable applications. Among the tested transition metals, encouraging results have been obtained with Co and Ru complexes. $CoHMD_2^+$ [147 b] has been used successfully as a catalyst for photochemical reduction of CO_2 because of the small $Co^{II}HMD^{2+}/Co^{I}HMD^{+}$ reorganization energy, the fast CO_2 binding to $Co^{I}HMD^{+}$ (1.7×10^8 M^{-1} s^{-1}) and the large K_{CO_2} (HMD = 5,7,7,12,14,14-hexamethyl-1,4,8,11-tetraazacyclotetradeca-4,11-diene). X-ray absorption near edge structure studies suggest that the Co(I) species can promote the two-electron transfer to the bound CO_2 with formation of CO_2^{2-} and a consequent easy reduction of CO_2. The slow step in the photoreduction of CO_2 has been suggested to be the C–O bond rupture of the bound carboxylic acid, produced by protonation of $[S\text{-}Co^{III}HMD(CO_2^{2-})]^-$. This hypothesis could not be tested because the UV-Vis spectrum of $[S\text{-}Co^{III}HMD(CO_2^{2-})]^-$ lacks features amenable to the study of the proton dependence of the reduction process [147 b]. Interestingly, it has been found that the doubly reduced species $[Ru^{I}(bpy)(bpy^{\bullet -})(CO)]$ is able to react with CO_2 in CH_3CN to produce $[Ru(bpy)_2(CO)(COO)]$ together with $[Ru(bpy)_2(CO)(COOH)]^-$ stabilized by a Group 1 metal cation [147 b]. Re systems have also been tested [147 c, d] that are much slower than the Co and Ni systems, most probably because of Re–Re bond formation.

Alternatively to homogeneous catalysts, semiconductors have been used in an attempt to reduce CO_2 in water under sunlight irradiation. HCOOH, HCHO and CH_3OH are produced by reduction of CO_2 with H_2O under solar irradiation of an aqueous suspension of a variety of semiconductors such as TiO_2 and $SrTiO_3$ [149]. The barrier to the exploitation of this reaction is the low quantum yield [144 c], that may be improved by using sacrificial hole traps or electron donors, such as *n*-propanol, tertiary amines or ethylenediaminetetraacetic acid. This solution is not economically convenient as the organic materials may be more valuable than the CO_2 reduction products. An interesting photocatalytic system for the reduction of CO_2 with H_2O has been reported [150] that has a selectivity of 30% for ethanol production using a Ti-modified mesoporous silica catalyst, compared to 1.4% over bulk TiO_2. This area is of great interest for carbon recycling and warrants further investigation.

1.5 Conclusions

CO_2 is a suitable source of carbon in many synthetic applications or can be used as a technological fluid with great advantage over other possible solutions that have a high greenhouse gas potential. In the short term, the use in carboxylation processes (synthesis of carbonates, carbamates and carboxylates, including cyclic compounds) appears to be the most promising synthetic application. Addition to CO for making methanol and use as a mild oxidant are other interesting applications. All such uses, in addition to recycling CO_2, reduce CO_2 emissions, in part because they are more efficient than existing technologies.

Currently, some 130 Mt_{CO_2} $year^{-1}$ are used in several applications: for a correct estimate of the amount of CO_2 not emitted into the atmosphere, LCA methodology must be applied. A fair estimate of the amount of CO_2 that may be avoided in the short-medium term is 300 Mt_{CO_2} $year^{-1}$, should all options of CO_2 utilization be implemented, including the use in air conditioners [151]. The photochemical reduction of CO_2 in water under solar light irradiation is an option that would greatly enlarge carbon recycling. A key point is the cost of CO_2 that does not have a natural origin, but should be recovered from industrial or power plants. The existing separation techniques are quite expensive, while large amounts of very pure CO_2 are vented. However, should the capture of CO_2 be implemented on a large scale, large amounts of CO_2 would be available for utilization. Concerted policies are necessary that make less random the research, and focus science and technology upon fields that may better contribute to reducing the CO_2 emission. As a matter of fact, the utilization of CO_2 is the only technology that may produce profit out of recovered CO_2, while contributing to reducing its global emission.

References

1 (a) The first report was by Bassarov in 1870, see D. Fromm, D. Lutzov, *Chem Uns Zeit* **1979**, *13*, 78; (b) H. Kolbe, E. Lautemann, *Ann* **1869**, *113*, 125; (c) R. Schmitt, E. Burkard, *Ber* **1877**, *20*, 2699.

2 (a) M. Aresta, G. Forti (Eds.), *Carbon Dioxide as a Source of Carbon: Chemical and Biochemical Uses*, NATO-ASI Series, Reidel, Dordrecht, **1987**, C206; (b) M. Aresta, J. Schloss (Eds.), *Enzymatic and Model Reduction and Carboxylation Reactions for Carbon Dioxide Utilization*, NATO-ASI Series, Kluwer, Dordrecht, **1990**, C314; (c) A. Behr (Ed.), *Carbon Dioxide Activation by Metal Complexes*, VCH, Weinheim, **1988**; (d) C. Song, A. M. Gaffney, K. Fujimoto (Eds.), *CO_2 Conversion and Utilization, ACS Symp Ser 809*, ACS, Washington, DC, **2002**; (e) M. Aresta (Ed.), *Carbon Dioxide Recovery and Utilization*, Kluwer, Dordrecht, **2003**; (f) C. J. Liu, R. Mallinson, M. Aresta (Eds.), *Utilization of Greenhouse Gases, ACS Symp Ser 852*, ACS, Washington, DC, **2003**.

3 M. Aresta, J. N. Armor, M. A. Barteau, E. J. Beckman, A. T. Bell, J. E. Bercaw, C. Creutz, E. Dinjus, D. A. Dixon, K. Domen, D. L. Dubois, J. Eckert, E. Fujita, D. H. Gibson, W. A. Goddard, D. W. Goodman, J. Keller, G. J. Kubas, H. H. Kung, J. E. Lyons, L. E. Manzer, T. J. Marks, K. Morokuma, K. M. Nicholas, R. Periana, L. Que, J. Rostrup-Nielsen, W. M. H. Sachtler, L. D. Schmidt, A. Sen, G. A. Somorjai, P. C. Stair, B. R. Stults, W. Tumas, *Chem Rev* **2001**, *101*, 953–996.

4 R. J. Allam, R. Bredesen, E. Drioli, in *Carbon Dioxide Recovery and Utilization*, M. Aresta (Ed.), Kluwer, Dordrecht, **2003**, pp. 53–118.

5 (a) A. Mathisen, in *Proc ICCDU VIII*, June 20–23, **2005**, p. 2; (b) H. Kongsjordan, O. Karstad, T. A. Torp, *Waste Manag* **1997**, *17*, 303–308.

6 J. De Simone, in *Proc ICCDU VI*, September 9–14, **2001**, p. 3[illegible].

7 J. Vansant, in *Recovery and Utilization of Carbon Dioxide*, M. Aresta (Ed.), Kluwer, Dordrecht, **2003**, pp. 3–50.

8 B. Mertz, O. Davidson, R. Swart, J. Pan (Eds.), *The Third Assessment Report of Intergovernmental Panel on Climate Change*, Cambridge University Press, Cambridge, **2001**.

9 M. Ricci, in *Recovery and Utilization of Carbon Dioxide*, M. Aresta (Ed.), Kluwer, Dordrecht, **2003**, pp. 395–402.

10 J. Gale, Y. Kaya (Eds.), *Greenhouse Gases Technologies*, Pergamon Press, Oxford, **2002**.

11 M. Aresta, M. Galatola, *J Cleaner Prod* **1999**, *7*, 181–193.

12 M. Aresta, A. Caroppo, A. Dibenedetto, M. Narracci, in *Environmental Challenges and Greenhouse Gas Control for Fossil Fuel Utilization in the 21st Century*, M. Maroto Valer, C. Song, Y. Soong (Eds.), Kluwer, Dordrecht, **2002**, pp. 331–347.

13 M. Aresta, A. Dibenedetto, in *CO_2 Conversion and Utilization, ACS Symp Ser 809*, C. Song, A. M. Gaffney, K. Fujimoto (Eds.), ACS, Washington, DC, **2002**, pp. 54–70.

14 J. W. Rabalais, J. M. McDonald, V. Scherr, S. P. McGlynn, *Chem Rev* **1971**, *71*, 73–108 and references therein.

15 M. Aresta, E. Quaranta, I. Tommasi, P. Giannoccaro, A. Ciccarese, *Gazz Chim Ital* **1995**, *125*, 509–539 and references therein.

16 D. E. Shemansky, *J Chem Phys* **1972**, *56*, 1582–1587.

17 G. Liger-Belair, E. Prost, M. Parmentier, P. Jeandet, J.-M. Nuzillard, *Agric Food Chem* **2003**, *51*, 7560–7563.

18 D. R. Lide, *Handbook of Chemistry and Physics*, 74th edn, CRC Press, Boca Raton, FL, **1993/1994**.

19 K. S. Lackner, C. H. Wendt, D. P. Butt, E. L. Joyce, D. H. Sharp, *Energy* **1995**, *20*, 1153–1170.

20 M. Aresta, C. F. Nobile, V. G. Albano, E. Forni, M. Manassero, *Chem Commun* **1975**, *15*, 636–637.

21 D. H. Gibson, *Chem Rev* **1996**, *96*, 2063–2095.

22 (a) T. Herskovitz, *J Am Chem Soc* **1977**, *99*, 2391–2392; (b) J. C. Calabrese, T. Herskovitz, J. B. Kinney, *J Am Chem Soc* **1983**, *105*, 5914–5915.

23 I. Castro-Rodriguez, H. Nakai, L. Zakharov, A. L. Rheingold, K. Meyer, *Science* **2004**, *305*, 1757–1759.

24 (a) M. Aresta, C. F. Nobile, *Dalton Trans* **1977**, *7*, 708–711; (b) M. G. Mason, J. A. Ibers, *J Am Chem Soc* **1982**, *104*, 5153–5157; (c) A. Doehring, P. W. Jolly, C. Krueger, M. Romao, *Z Naturforsch* **1985**, *40B*, 484–488.

25 (a) M. Aresta, C. F. Nobile, *Inorg Chim Acta* **1977**, *24*, L49–L50; (b) M. Aresta, E. Quaranta, I. Tommasi, *New J Chem* **1994**, *18*, 133–142.

26 (a) H. H. Karsch, *Chem Ber* **1977**, *110*, 2213; (b) S. Komiya, M. Akita, N. Kasuga, M. Hirano, A. Fukuoka, *Chem Commun* **1994**, *9*, 1115–1116.

27 M. Sakamoto, L. Shimizu, A. Yamamoto, *Organometallics* **1994**, *13*, 407–409.

28 (a) S. M. Tetrick, F. S. Thom, A. R. Cutler, *J Am Chem Soc* **1998**, *119*, 6193–6194; (b) S. M. Tetrick, C. Xu, J. R. Pinkes, A. R. Cutler, *Organometallics* **1998**, *17* 1861–1867.

29 T. A. Hanna, A. M. Baranger, R. G. Bergman, *J Am Chem Soc* **1995**, *117*, 3292–3293.

30 J. D. Audett, T. J. Collins, B. D. Santarsiero, G. H. Spies, *J Am Chem Soc* **1982**, *104*, 7352–7353.

31 C. P. Kubiak, C. Woodcock, R. Eisenberg, *Inorg Chem* **1982**, *21*, 2119–2126.

32 (a) J. S. Field, R. J. Haines, J. Sundermeyer, S. F. Woollam, *Chem Commun* **1990**, *14*, 985–988; (b) J. S. Field, R. J Haines, J. Sundermeyer, S. F. Woollam, *Dalton Trans* **1993**, *18*, 2735–2748.

33 C. T. Tso, A. R. Cutler, *J Am Chem Soc* **1986**, *108*, 6069.

34 (a) J. C. Vites, B. D. Steffey, M. E. Giuseppetti-Dery, A. R. Cutler, *Organometallics* **1991**, *10*, 2827–2834; (b) J. R. Pinkes B. D. Steffey, J. C. Vites, A. R. Cutler, *Organometallics* **1994**, *13*, 21–23.

35 D. R. Senn, J. A. Gladysz, K. Emerson, R. D. Larsen, *Inorg Chem.* **1987**, *26*, 2737–2739.

36 D. H. Gibson, M. Ye, B. A. Sleadd, J. M. Mehta, O. P. Mbadike, J. F. Richardson, M. S. Mashuta, *Organometallics* **1995**, *14*, 1242–1255.

37 (a) C. R. Eady, J. J. Guy, B. F. G. Johnson, J. Lewis, M. C. Malatesta, G. M. Sheldrick, *Chem Commun* **1976**, 602; (b) G. R. John,

B. F. G. Johnson, J. Lewis, K. C. Wong, *J Organomet Chem* **1979**, *169*, C23.

38 B. K. Balbach, F. Helus, F. Oberdorfer, M. L. Ziegler, *Angew Chem* **1981**, *93*, 479–480.

39 (a) C. Floriani, G. Fachinetti, *Chem Commun* **1974**, *15*, 615–616; (b) G. Fachinetti, C. Floriani, P. F. Zanazzi, *J Am Chem Soc* **1978**, *100*, 7405–7407; (c) S. Gambarotta, F. Arena, C. Floriani, P. F. Zanazzi, *J Am Chem Soc* **1982**, *104*, 5082–5092.

40 (a) D. H. Gibson, *Coord Chem Rev* **1999**, *185–186*, 335–355; (b) D. H. Gibson, *Comp Coord Chem II* **2004**, *1*, 595–602.

41 E. G. Lundquist, J. C. Huffman, K. Folting, B. E. Mann, K. G. Caulton, *Inorg Chem* **1990**, *29*, 128–134.

42 J. Mascetti, M. Tranquille, *J Phys Chem* **1988**, *92*, 2177–2184.

43 J. Mascetti, F. Galan, I. Papai, *Coord Chem Rev* **1999**, *190–192*, 557–576.

44 (a) M. Aresta, E. Quaranta, I. Tommasi, *Chem Commun* **1988**, *7*, 450–452; (b) K. Tanaka, D. Ooyama, *Coord Chem Rev* **2002**, *226*, 211–218.

45 T. Yoshida, D. Thorn, T. Okano, J. A. Ibers, A. Yamamoto, *J Am Chem Soc* **1979**, *101*, 4212–4221.

46 J. C. Calabrese, T. Herskovitz, J. B. Kinney, *J Am Chem Soc* **1983**, *105*, 5914–5915.

47 J.-C. Tsai, M. Khan, K. M. Nicholas, *Organometallics* **1989**, *8*, 2967–2970.

48 (a) B. Demerseman, G. Bouquet, M. Bigorgne, *J Organomet Chem* **1978**, *145*, 41–48; (b) H. Felkin, P. J. Knowles, B. Meunier, *J Organomet Chem* **1978**, *146*, 151–153.

49 (a) K. M. Nicholas, *J Organomet Chem* **1980**, *188*, C10–C12; (b) C. Bianchini, A. Meli, *J Am Chem Soc* **1984**, *106*, 2698–2699.

50 (a) T. Tsuda, S. I. Sanada, T. Saegusa, *J Organomet Chem* **1976**, *116*, C10–C12; (b) D. L. Delact, R. Del Rosario, P. E. Fanwick, C. P. Kubiak, *J Am Chem Soc* **1987**, *109*, 754–755; (c) J. Wu, P. E. Fanwick, C. P. Kubiak, *Organometallics* **1987**, *6*, 1805–1807.

51 (a) J. Chatt, M. Kubota, G.-J. Leigh, F. C. March, R. Mason, D. J. Yarrow, *Chem. Commun.* **1974**, *24*, 1033–1034; (b) H. H. Karsch, *Chem Ber* **1977**, *110*, 2213; (c) G. Fachinetti, C. Floriani, A. Chiesi-Villa, C. Guastini, *J Am Chem Soc* **1979**, *101*, 1767–1775; (d) C. Burkhart, H. Hoberg, *Angew Chem Int Ed* **1982**, *21*, 76; (e) C. Bianchini, C. Mealli, A. Meli, M. Sabat, *Inorg Chem* **1984**, *23*, 2731–2732; (f) G. Fachinetti, G. Fochi, T. Funaioli, P. F. Zanassi, *Chem Commun.* **1987**, *2*, 89–90; (g) G. R. Lee, J. M. Maher, N. J. Cooper, *J Am Chem Soc* **1987**, *109*, 2956–2962; (h) K. A. Belmore, R. A. Vanderpool, J.-C. Tsai, M. A. Kahn, K. M. Nicholas, *J Am Chem Soc* **1988**, *110*, 2004–2005; (i) J. Ruiz, V. Guerchis, D. Astrc, *Chem Commun* **1989**, 812.

52 M. Aresta, A. Dibenedetto, in *CO_2 Conversion and Utilization, ACS Symp Ser 809*, C. Song, A. M. Gaffney, K. Fujimoto (Eds.), ACS, Washington, DC, **2002**, pp. 54–70.

53 J. V. Schloss, G. H. Lorimer *J Biol Chem* **1982**, *257*, 4691–4694.

54 T. Akazawa, A. Incharoensakdi, T. Takabe, in *Carbon Dioxide as a Source of Carbon: Chemical and Biochemical Uses*, M. Aresta, G. Forti (Eds.), NATO-ASI Series, Reidel, Dordrecht, **1987**, pp. 83–91.

55 (a) T. J. Andrews, G. H. Lorimer, *FEBS Lett* **1978**, *90*, 1–9; (b) M. K. Morell, K. Paul, H. J. Kane, T. J. Andrews, *Aust J Bot* **1992**, *40*, 431–441; (c) Z. Swab, P. Hjdukiewicz, P. Maliga, *Proc Natl Acad Sci USA* **1990**, *87*, 8526–8530; (d) M. P. Reynolds, M. van Ginkel, J. M. Ribaut, *J Exp Bot* **2000**, *51*, 459–473; (e) T. J. Andrews, S. M. Whitney, *Arch Biochem Biophys* **2003**, *414*, 159–169; (f) M. A. J. Parry, P. J. Andralojc, R. A. C. Mitchell, P. J. Magdwick, A. J. Keys, *J Exp Bot* **2003**, *54*, 1321–1333; (g) M. W. Finn, F. R. Tabita, *J Bacteriol* **2004**, *186*, 6360–6366.

56 H. Wu, S. Huang, Z. Jiang *Catal Today* **2004**, *98*, 545–552.

57 (a) T. Yatsuka, A. Ito, O. Manabe, M. Dehara, H. Hiyama, *Yuki Gosei kagaku kyokai Shi* **1972**, *30*, 1030–1034; (b) W. Bachmann, C. Gnabs, K. Janecka, E. Mudlos, T. Papenfuhs, G. Waese, *Ger Offen 2,426,850*, **1976**; (c) F. Mutterer, C. D. Weis, *J Hetero-cycl Chem* **1976**, *13*, 1103–1104; (d) Z. Weglinski, T. Talik, *Rocz Chem* **1977**, *51*, 2401–2409; (e) R. Ueno, M. Kitayama, R. Otsuka, T. Shirai, *Pat Appl WO-JP2554 20040302*, **2004**; (f) S. Muradov, B. N. Khamidov,

Ch. Sh. Kadyrov, *O'zbekiston Kimyo Jurnali* **2004**, 25–29.
58 M. Aresta, A. Dibenedetto, *Rev Mol Biotechnol* **2002**, *90*, 113–128.
59 M. Aresta, A. Lack, I. Tommasi, G. Fuchs, *Eur J Biochem* **1991**, *197*, 473–479.
60 K. Schuehle, G. Fuchs, *J Bacter* **2004**, *186*, 4556–4567.
61 M. Aresta, E. Quaranta, R. Liberio, C. Dileo, I. Tommasi, *Tetrahedron* **1998**, *54*, 8841–8846.
62 M. Aresta, A. Dibenedetto, C. Pastore, R. Lonoce, *Environ Chem Lett* **2006**, *3*, 145–148.
63 (a) G. Bottaccio, G.P. Chiusoli, *Chem Commun* **1966**, *17*, 618; (b) G. Bottaccio, G.P. Chiusoli, M. Marchi, *Ger Offen 2514571*, **1975**.
64 (a) M. Aresta, I. Tkatchenko, I. Tommasi, in *Ionic Liquids as Green Solvents: Progress and Prospects, ACS Symp Ser 856*, R.D. Rogers, K.R. Seddon (Eds.), ACS, Washington, DC, **2003**, 93–99; (b) I. Tommasi, F. Sorrentino, *Tetrahedron Lett* **2005**, *46*, 2141–2145.
65 L.-H. Yin, Z.-H. Gao, W. Huang, K. Xie, *Taiyuan Ligong Daxue Xuebao* **2004**, *35*, 318–320.
66 (a) H. Hoberg, D. Schaefer, *J Organomet Chem* **1982**, *236*, C28–C30; (b) H. Hoberg, D. Schaefer, *J Organomet Chem* **1983**, *251*, C51–C53; (c) E. Dinjus, D. Walther, H. Schueltz, *Z Chem* **1983**, *23*, 408–409; (d) D. Walther, E. Dinjus, J. Sieler, L. Andersen, O. Lindqvist, *J Organomet Chem* **1984**, *276*, 99–107; (e) H. Hoberg, Y. Peres, A. Michelreit, *J Organomet Chem* **1986**, *307*, C41–C43; (f) H. Hoberg, Y. Peres, A. Michelreit, *J Organomet Chem* **1986**, *307*, C38–C40; (g) H. Hoberg, Y. Peres, C. Krueger, Y.-H. Tsai, *Angew Chem Int Ed* **1987**, *99*, 799–800.
67 S.A. Cohen, J.E. Bercaw, *Organometallics* **1985**, *4*, 1006–1014.
68 H. Hoberg, K. Jenni, K. Angermund, C. Krueger, *Angew Chem Int Ed* **1987**, *26*, 153.
69 R. Alvarez, E. Carmona, D.J. Cole-Hamilton, A. Galindo, E. Gutierrez-Puebla, A. Monge, M.I. Poveda, C. Ruiz, *J Am Chem Soc* **1985**, *107*, 5529–5531.
70 M. Aresta, E. Quaranta, *J Organomet Chem* **1993**, *463*, 215–221.
71 G. Schubert, I. Papai, *J Am Chem Soc* **2003**, *125*, 14847–14858.
72 I. Papai, G. Schubert, I. Mayer, G. Besenyei, M. Aresta, *Organometallics* **2004**, *23*, 5252–5259.
73 (a) S. Pitter, E. Dinjus, *J Mol Catal A* **1997**, *125*, 39–45; (b) F. Koster, E. Dinjus, E. Dunach, *Eur J Org Chem* **2001** *18*, 3575; (c) F. Gassner, V. Haack, A. Janssen, A. Elsagir, E. Dinjus, *EP Appl 103297 20000218*, **2000**.
74 (a) A. Doehring, P.W. Jolly, *Tetrahedron Lett* **1980**, *21*, 3021–3024; (b) M. Aresta, E. Quaranta, A. Ciccarese, *C1 Mol Chem* **1985**, *1*, 283–295; (c) P. Albano, M. Aresta, *J Organomet Chem* **1980**, *190*, 243–246.
75 M. Aresta, A. Dibenedetto, I. Papai, G. Schubert, *Inorg Chim Acta* **2002**, *334*, 294–300.
76 A. Behr, G. Thelen, *C1 Mol Chem* **1984**, *1*, 137–153.
77 (a) M. Aresta, D. Ballivet-Tkatchenko, M.C. Bonnet, R. Faure, H. Loiseleur, *J Am Chem Soc* **1985**, *107*, 2994–2995; (b) M. Aresta, D. Ballivet-Tkatchenko D. Belli Dell'Amico, M.C. Bonnet, D. Boschi, F. Calderazzo, R. Faure, L. Labella, F. Marchetti, *Chem Commun* **2000** 1099–1100.
78 K. Masuda, Y. Ito, M. Horiguchi, H. Fujita, *Tetrahedron* **2005**, *61*, 213–229.
79 M. Aresta, E. Quaranta, *ChemTech* **1997**, *27*, 32–40.
80 J. Barthelemy, *Lyon Phar* **1986**, *37*, 249–263.
81 (a) T.-T. Wu, J. Huang, N.D. Arrington, G.M. Dill, *J Agric Food Chem* **1987**, *35*, 817–823; (b) K. Suzuki, T. Kato, J. Takahashi, K. Kamoshita, *Nippon Noyaku Gakkaishi* **1984**, *9*, 497–501; (c) U. Romano, F. Rivetti, G. Sasselli, *EP 125726*, **1985**.
82 M.H. Chisholm, M.W. Extine, *J Am Chem Soc* **1977**, *99*, 782–792.
83 (a) M. Aresta, E. Quaranta, *Tetrahedron* **1992**, *48*, 1515–1530; (b) M. Aresta, E. Quaranta, *Ital Pat 1237208*, **1993**.
84 W.D. McGhee, D.P. Riley, *Organometallics* **1992**, *11*, 900–907.
85 (a) F. Porta, S. Cenini, M. Pizzotti, C Crotti, *Gazz Chim Ital* **1985**, *115*, 275; (b) R. Garcia Deleon, A. Kobayashi, T. Yamauchi, J. Ooishi, T. Baba, S. Masaki, F. Hiarata, *Appl Catal A* **2002**, *225*, 43–49; (c) M. Curini, F. Epifano, F. Maltese, O. Rosati, *Tetrahedron Lett* **2002**, *43*, 4895–4897; (d) C. Calderoni, F. Mizia, F. Rivetti,

U. Romano, *EP 391473*, **1990**; (e) S. Carloni, D.E. De Vos, P.A. Jacobs, R. Maggi, G. Sartori, *J Catal* **2002**, *205*, 199–204; (f) T. Baba, A. Kobayashi, T. Yamauchi, H. Tanaka, S. Aso, M. Inomata, Y. Kawanami, *Catal Lett* **2002**, *82*, 193–197.

86 (a) M. Aresta, E. Quaranta, *Ital Pat 1198206*, **1988**; (b) M. Aresta, E. Quaranta, *Tetrahedron* **1991**, *47*, 9489–9502; (c) M. Aresta, E. Quaranta, *Ital Pat 1237207*, **1993**; (d) M. Aresta, C. Berloco, E. Quaranta, *Tetrahedron* **1995**, *51*, 8073–8078; (e) M. Aresta, A. Bosetti, E. Quaranta, *Ital Pat 002202*, **1996**; (f) M. Aresta, A. Dibenedetto, E. Quaranta, *Tetrahedron* **1998**, *54*, 14145–14156; (g) M. Aresta, A. Dibenedetto, E. Quaranta, *Green Chem* **1999**, *1*, 237–242; (h) M. Aresta, A. Dibenedetto, *Chem A Eur J* **2002**, *8*, 685–690; (i) M. Distaso, E. Quaranta, *Tetrahedron* **2004**, *60*, 1531–1539; (j) M. Distaso, E. Quaranta, *J Catal* **2004**, *228*, 36–42.

87 (a) S. Carloni, D.E. De Vos, P.A. Jacobs, R. Maggi, G. Sartori, R. Sartorio, *J Catal* **2002**, *205*, 199–204; (b) S.P. Gupte, A.B. Shivarkar, R.W. Chaudari, *Chem Commun* **2001**, 2620–2621; (c) I. Vauthey, F. Valot, C. Gozzi, F. Fache, M. Lemaire, *Tetrahedron Lett* **2000**, *41*, 6347–6350; (d) Z.-H. Fu, Y. Ono, *J Mol Catal* **1994**, *91*, 399–405; (e) T. Baba, M. Fujiwara, A. Oosku, A. Kobayashi, R.G. Deleon, Y. Ono, *Appl Catal A* **2002**, *227*, 1–6.

88 (a) G. Gattow, W. von Behrendt, *Angew Chem Int Ed* **1972**, *11*, 534–535; (b) A. Dibenedetto, M. Aresta, P. Giannoccaro, C. Pastore, I. Pàpai, G. Schubert, *Eur J Inorg Chem* **2006**, *5*, 908–913.

89 D.J. Darensbourg, M.W. Holtcamp, *Coord Chem Rev* **1996**, *153*, 155–174.

90 (a) T. Sakai, N. Kihara, T. Endo, *Macromolecules* **1995**, *28*, 4701–4706; (b) M. Sone, T. Sako, C. Kamisawa, *Jpn Kokai Tokkyo Koho, JP 11335372*, **1999**.

91 (a) H. Yasuda, L.N. He, T. Sakakura, C.W. Hu, *J Catal* **2005**, *233*, 119–122; (b) Y. Du, F. Cai, D.L. Kang, L.N. He, *Green Chem* **2005**, *7*, 518–523.

92 (a) Y. Li, X.Q. Zhao, Y.J. Wang, *Appl Catal A* **2005**, *279*, 205–208; (b) X. Zhang, W. Wei, Y. Sun, in *Proc ICCDU VIII*, June 20–23, **2005**, p. 68.93

93 R.J. De Pasquale, *Chem Commun* **1973**, 157–158.

94 D.J. Darensbourg, M.S. Zimmer, *Macromolecules* **1999**, *32*, 2137–2140.

95 (a) T. Yano, H. Matsui, T. Koike, H. Ishiguro, H. Fujihara, M. Yoshihara, T. Maeshima, *Chem Commun* **1997**, 1129–1130; (b) K. Yamaguchi, K. Ebitani, T. Yoshida, H. Yoshida, K. Kaneda, *J Am Chem Soc* **1999**, *121*, 4526–4527; (c) M. Aresta, A. Dibenedetto, L. Gianfrate, C. Pastore, *J Mol Catal A* **2003**, *204/205*, 245–252.

96 J. Sun, S.-I. Fujita, M. Arai, *J Organomet Chem* **2005**, *690*, 3490–3497.

97 (a) H. Sugimoto, S. Inoue, *J Pol Sci A Pol Chem* **2004**, *42*, 5561–5573; (b) H. Sugimoto, H. Ohtsuka, S. Inoue, *Stud Surf Sci Catal* **2004**, *153*, 243–246.

98 M. Super, E. Berluche, C. Costello, E. Beckman, *Macromolecules* **1997**, *30*, 368–372.

99 (a) D.J. Darensbourg, N.W. Stafford, T. Katsurao, *J Mol Catal* **1995**, *104*, L1–L4; (b) S. Mang, A.I. Cooper, M.E. Colclough, N. Chauhan, A.B. Holmes, *Macrocromolecules* **2000**, *33*, 303–308; (c) M. Van Schilt, M. Kemmere, J. Keurentjes, in *Proc ICCDU VIII*, June 20–23, **2005**, p. 54; (d) T. Endo, in *Proc ICCDU VIII*, June 20–23 **2005**, p. 60.

100 M. Aresta, A. Dibenedetto, L. Gianfrate, C. Pastore, *J Mol Catal A* **2003**, *204/205*, 245–252.

101 M. Aresta, A. Dibenedetto, L. Gianfrate, C. Pastore, *Appl Catal A* **2003**, *255*, 5–11.

102 (a) S.J. Ainsworth, *Chem Eng News* **1992**, *2*, 9; (b) R.O. Kirk, T.J. Dempsey, in *Kirk–Othmer Encyclopedia of Chemical Technology*, M. Grayson, D. Eckroth, H.F. Mark, D.F. Othmer, C.G. Overberger, G.T. Seaborg (Eds.), Wiley, New York, **1982**, *19*, p. 46; (c) H.P. Wulff, F. Wattimenu, *US Patent 4,021,454*, **1977**, to Shell Oil Company; (d) N.W. Cant, W.K. Hall, *J Catal* **1978**, *52*, 81–87; (e) M.G. Clerici, P. Ingallina, *Catal Today* **1988**, *41*, 351–363 and references therein; (f) V. Duma, D. Honicke, *J Catal* **2002**, *191*, 93–104.

103 (a) G.F. Thiele, E. Roland, *J Mol Catal A* **1997**, *117*, 351–356; (b) M.G. Clerici,

G. Belussi, U. Romano, *J Catal* **1991**, *129*, 159–167.

104 (a) S.E. Jacobson, *EP 118248*; *EP 117147*, **1984**.

105 (a) M. Aresta, A. Ciccarese, E. Quaranta, *C1 Mol Chem* **1985**, *1*, 267–281; (b) M. Aresta, A. Ciccarese, E. Quaranta, *J Mol Catal* **1987**, *41*, 355–359; (c) M. Aresta, C. Fragale, E. Quaranta, I. Tommasi, *Chem Commun* **1992**, *4*, 315–317; (d) M. Aresta, A. Dibenedetto, I. Tommasi, *Eur J Inorg Chem* **2001**, 1801–1806.

106 (a) M. Aresta, A. Dibenedetto, I. Tommasi, *Appl Organomet Chem* **2000**, *14*, 799–802; (b) M. Aresta, A. Dibenedetto, *J Mol Catal* **2002**, *182/183*, 399–409; (c) M. Aresta, A. Dibenedetto, in *Proc ISHHC XII*, July 18–22, **2005**, p. 40.

107 M. Aresta, I. Tommasi, E. Quaranta, C. Fragale, J. Mascetti, M. Tranquille, F. Galan, M. Fouassier, *Inorg Chem* **1996**, *35*, 4254–4260.

108 S.W. Benson, N. Cohen, *Chem Rev* **1993**, *93*, 2419–2438.

109 T. Sakakura, Y. Saito, M. Okano, J.-C. Choi, T. Sako, *J Org Chem* **1998**, *63*, 7095–7096.

110 (a) T. Sakakura, Y. Saito, J.-C. Choi, T. Masuda, T. Sako, T. Oriyama, *J Org Chem* **1999**, *64*, 4506–4508; (b) T. Sakakura, Y. Saito, J.-C. Choi, T. Sako, *Polyhedron* **2000**, *19*, 573–576.

111 N.S. Isaacs, B. O'Sullivan, C. Verhaelen, *Tetrahedron* **1999**, *55*, 11949–11956.

112 M. Aresta, A. Dibenedetto, E. Amodio, C. Dileo, I. Tommasi, *J Sup Fluid* **2003**, *25*, 177–182.

113 M. Aresta, A. Dibenedetto, E. Fracchiolla, P. Giannoccaro, C. Pastore, I. Pápai, G. Schubert, *J Org Chem* **2005**, *70*, 6177–6186.

114 (a) M. Aresta, A. Dibenedetto, C. Pastore, in *Proc ICCDU VIII*, June 20–23, **2005**, p. 62; (b) M. Aresta, A. Dibenedetto, C. Pastore, *Topics Catal* **2006**, in press; (c) K. Tomishige, K. Kunimori, *Appl Catal A* **2002**, *237*, 103–106.

115 P. Ball, H. Fuellmann, W. Heitz, *Angew Chem* **1980**, *92*, 742–743.

116 M. Aresta, A. Dibenedetto, C. Devita, O.A. Bourova, O.N. Chupakhin, *Stud Surf Sci Catal* **2004**, *153*, 213–220.

117 (a) Q. Li, N. Zhao, W. Wei, Y. Sun, *Stud Surf Sci Catal* **2004**, *153*, 573–576; (b) Q. Li, W. Zhang, N. Zhao, W. Wei, Y. Sun, in *Proc ICCDU VIII*, June 20–23, **2005**, p. 164.

118 (a) G. Illuminati, U. Romano, R. Tesei (Snam Progetti SPA), *DE Patent 75-2528412*, **1985**; (b) G. Illuminati, U. Romano, R. Tesei (Snam Progetti SPA), *RO Patent 75-82648*, **1980**.

119 M. Aresta, A. Dibenedetto, C. Pastore, *Stud Surface Sci Catal* **2004**, *153*, 221–226.

120 (a) R.A. Schmitz, S.P.J. Albracht, R.K. Thauer, *FEBS Lett* **1992**, *309*, 78–81 (b) *Eur J Biochem* **1992**, *209*, 1013–1018; (c) R.A. Schmitz, M. Richter, D. Linder, R.K. Thauer, *Eur J Biochem* **1992**, *207*, 559–565; (d) J.J.G. Moura, C.D. Brondino, J. Trincão, M.J. Romão, *J Biol Inorg Chem* **2004**, *9*, 791–799.

121 (a) J.L. Kraft, *Biochemistry* **2002**, *41*, 1681–1688; (b) C.L. Drennan, T.I. Doukov, S.W. Ragsdale, *J Biol Inorg Chem* **2004**, *9*, 511–515.

122 R. Obert, B.C. Dace, *J Am Chem Soc* **1999**, *121*, 12192–12193.

123 (a) Z.Y. Jiang, H. Wu, S.W. Xu, S.F. Huang, *Chin J Catal* **2002**, *23*, 162; (b) Z. Jiang, S. Wu, H. Wu, *Stud Surf Sci Catal* **2004**, *153*, 475–480.

124 R.S. Wolfe, *Trends Biol Sci* **1985**, *10* 396–399.

125 G. Fuchs, in *Carbon Dioxide as a Source of Carbon: Chemical and Biochemical Uses*, M. Aresta, G. Forti (Eds.), NATO-ASI Series, Reidel, Dordrecht, **1987** pp. 263–274.

126 (a) S.W. Ragsdale, J.E. Clark, L.G. Ljungdahl, L. Lundie, H.L. Drake, *J Biol Chem* **1983**, *258*, 2364–2369; (b) M. Kumar, W.-P. Lu, S.W. Ragsdale, *Biochemistry* **1994**, *33*, 9769–9777.

127 D. Qiu, M. Kumar, S.W. Ragsdale, T.G. Spiro, *Science* **1994**, *264*, 817–819.

128 (a) M. Aresta, A. Dibenedetto, *Inorg Chim Acta* **1998**, *272*, 38–42; (b) M. Aresta, A. Dibenedetto, in *Proc ICCDU VIII*, June 20–23, **2005**, p. [illegible].

129 (a) S.W. Ragsdale, I. Tzanko, C.L. Drennan, *J Biol Inorg Chem* **2004**, *9*, 511–515; (b) S.W. Ragsdale, T. Craft,

Y.-C. Horng, *J Am Chem Soc* **2004**, *126*, 4068–4069.

130 (a) P.G. Jessop, T. Ikaryia, R. Noyori, *Nature* **1994**, *368*, 231–233; (b) P.G. Jessop, T. Ikaryia, R. Noyori, *Chem Rev* **1995**, *95*, 259–272; (c) P.G. Jessop, F. Joò, C.-C. Tai, *Coord Chem Rev* **2004**, *248*, 2425–2442; (d) F. Gassun, W. Leitner, *Chem Commun* **1993**, 1465; (e) P. Munshi, A.D. Main, J. Linehan, C.-C. Tai, P.G. Jessop, *J Am Chem Soc* **2002**, *124*, 7963–7971; (f) O. Krocker, R.A. Koppel, M. Froba, A. Baiker, *J Catal* **1998**, *178*, 284–298; (g) P.G. Jessop, H. Hsiao, T. Ikariya, R. Noyori, *J Am Chem Soc* **1994**, *116*, 8851–8852; (h) P. Munshi, D. Heldebrandt, E. McKoon, P.A. Kelly, C.-C. Tai, P.G. Jessop, *Tetrahedron Lett* **2003**, *44*, 2725–2726.

131 (a) F. Solymosi, G. Kliveny, *Surf Sci* **1994**, *315*, 255–268; (b) F. Solymosi, L. Bugyi, *Faraday Trans 1* **1987**, *83*, 2015–2033; (c) D.W. Goodman, D.E. Peably, J.M. White, *Surf Sci* **1984**, *140*, L239–L243.

132 C. Song, in *CO_2 Conversion and Utilization, ACS Symp Ser 809*, C. Song, A.M. Gaffney, K. Fujimoto (Eds.), ACS, Washington, DC, USA, **2002**, pp. 2–30.

133 (a) J.M. III Fox, *Catal Rev Sci Eng* **1993**, *35*, 169–212; (b) J.R. Rostrup-Nielsen, in *Natural Gas Conversion II*, H.E. Curry-Hyde, R.F. Howe (Eds.), Elsevier, Amsterdam, **1994**, p. 25; (c) J.R.H. Ross, A.N.J. van Keule, M.E.S. Hegarty, K. Seshan, *Catal Today* **1996**, *30*, 193–199.

134 B. Eliasson, U. Kogelschatz, *IEEE Trans Plasma Sci* **1991**, *19*, 1063–1077.

135 Y. Zhang, Y. Li, Yu Wang, C. Liu, B. Eliasson, *Fuel Process Technol* **2003**, *83*, 101–109.

136 A. Czernichowski, M. Czernichowski, P. Czernichowski, T.E. Cooley, *Fuel Chem Div Prepr* **2002**, *47*, 280–281.

137 M.M. Halmann, A. Steinfeld, in *Proc ICCDU VIII*, June 20–23, **2005**, p. 24.

138 M. Takagawa, A. Okamoto, H. Fujimura, Y. Izawa, H. Arakawa, in *Advances in Chemical Conversions for Mitigating Carbon Dioxide*, Elsevier, Amsterdam, **1998**, pp. 525–528.

139 R. Kieffer, M. Fujiwara, L Udron, Y. Souma, *Catal Today* **1997**, *36*, 15–24.

140 (a) R.A. Fiato, S.L. Soled, G.B. Rice, S. Miseo, *US Patent 5140049*, **1992**; (b) M.J. Choi, K. Kikim, H. Lee, S.B. Kim, J.S. Nam, K.W. Lee in *Proc 5th Int Conf on Greenhous Gas Control Technologies*, **2000**, pp. 607–612.

141 T. Inui, *Catal Today* **1996**, *29*, 329–337.

142 S.-E. Park, J.S. Yoo, *Stud Surf Sci Catal* **2004**, *153*, 303–314.

143 A. Rozovskii, *Russ Chem Rev* **1989**, *58*, 41–56.

144 (a) M. Saito, T. Fujitani, M. Takeuchi, T. Watanabe, *Appl Catal A* **1996**, *138*, 311–318; (b) H. Arakawa, in *Advances in Chemical Conversions for Mitigating Carbon Dioxide*, Elsevier, Amsterdam, **1998**, pp. 19–30; (c) M.M. Halmann, M. Steinberg (Eds.), *Greenhouse Gas Carbon Dioxide Mitigation Science and Technology*, Lewis, Boca Raton, FL, **1999**.

145 T. Inui, T. Takeguchi, K. Yanagisawa, M. Inoue, *Appl Catal A* **2000**, *192*, 201–209.

146 K. Ushikoshi, K. Mori, T. Watanabe, M. Takeuchi, M. Saito, in *Advances in Chemical Conversions for Mitigating Carbon Dioxide*, Elsevier, Amsterdam, **1998**, pp. 357–362.

147 (a) R. Ziessel, in *Carbon Dioxide as a Source of Carbon: Chemical and Biochemical Uses*, M. Aresta, G. Forti (Eds.), NATO-ASI Series, Reidel, Dordrecht, **1987**, pp. 113–138 and reference therein; (b) E. Fujita *Coord Chem Rev* **1994**, *185/186*, 373–384; (c) Y. Heyashi, S. Kita, B.S. Brunschwig, E. Fujita, *J Am Chem Soc* **2003**, *125*, 11976–11987; (d) E. Fujita, J.T. Muckerman, *Inorg Chem* **2004**, *43*, 7636–7647.

148 T. Inoue, A. Fujisima, S. Konishi, K. Honda, *Nature* **1979**, *277*, 637–638.

149 A. Mackor, A.H.A. Tinnemans, T.P.M. Koster, in *Carbon Dioxide as a Source of Carbon: Chemical and Biochemical Uses*, NATO-ASI Series, M. Aresta, G. Forti (Eds.), Reidel, Dordrecht, **1987**, pp. 393.

150 M. Anpo, S.G. Zhag, Y. Fujii, H. Yamashita, K. Koyano, K.T. Tatsumi, *Chem Lett* **1997**, 659–660.

151 M. Aresta, A. Dibenedetto, *Catal Today* **2004**, *98*, 455–462.

2
Nitrogen Monoxide and Nitrous Oxide Binding and Reduction

Dong-Heon Lee, Biplab Mondal, and Kenneth D. Karlin

2.1
Introduction

The chemistry of nitrogen monoxide (NO; also most commonly referred to as nitric oxide) and nitrous oxide (N_2O) has historically been a topic of great interest to inorganic chemists, and these areas continue to attract considerable attention since they impact on many diverse fields. NO is well known to have adverse effects on the environment, as it is implicated in ozone (O_3) layer depletion, formation of photochemical smog and acid rain [1]. More recently, NO has attracted much attention because of its versatile roles in biological systems [2–4], with its actions ranging from blood pressure control, neuronal communication, cytotoxic effects, peroxynitrite (a powerful nitrating and/or oxidizing agent) generation, etc. [5, 6]. N_2O [7, 8] plays an important role in atmospheric chemistry, as in the stratosphere it reacts with singlet oxygen to produce NO, which then participates in ozone decomposition. These reactions are of concern because of the possibility that increasing N_2O concentrations resulting from fossil fuel use, nylon production and also from denitrification of excess fertilizer may contribute to a decrease in stratospheric O_3 with the consequent potential for adverse impacts on ecosystems and human health. Also of concern is the fact that N_2O absorbs long-wavelength radiation and therefore serves as a potent greenhouse gas that may contribute to global warming [9].

As metal ions can or do mediate some of the above-mentioned processes, their interactions with NO and N_2O have been of great interest, even from the earliest days of the development of coordination and organometallic chemistries. NO became an important metal ligand due to its ability to stabilize low-valent metal ions (i.e. as part of classical organometallic chemistry research) [10] and a great deal of effort has been spent in the understanding of M–NO bonding because of the diverse structures and redox nature of the interaction (see below). Metalloporphyrin binding of NO has a rich history, in part because NO has been used as a "surrogate" in the study of O_2 binding to hemes (i.e. iron porphyrinates) [11–15]. As mentioned, it is now known that NO is an important

Activation of Small Molecules. Edited by William B. Tolman

ISBN: 3-527-31312-5

biological agent (also referred as an "endothelial-derived relaxing factor") whose binding to heme (and nonheme [16]) metal centers provides a basis for both "sensing" and signaling which leads to vasodilation/muscle relaxation [2–4]. Thus, there is considerable biomedical importance in detecting and/or sensing NO concentrations in biological fluids [3, 4, 17], or in developing reagents which release NO as a drug [18–22]. The use of metal ion chemistry for such purposes is a very active area of research, i.e. to sense biological NO [23–27], deliver it as drug [28–38] or modulate NO concentrations (as in septic shock therapy [39, 40]). Another form of cellular signaling is thought to also involve NO, but with biological thiols [41], consisting of nitrosylation–denitrosylation chemistry, i.e. $RSH + NO \leftrightarrow RSNO + e^-$. In fact, this process can be mediated by copper ions (via Cu^I/Cu^{II} redox shuttling) [42–44] and thus represents another example of chemical or biological manipulation of NO by metal ions.

Metal N_2O interactions are far less well studied [7, 8], but the area is quite important, as N_2O may potentially serve as a clean "green" oxidant [e.g. as an "O"-atom source, releasing N_2 (g)] and its manipulation or removal as a pollutant greenhouse gas is important, as mentioned above. Moreover, there is a metalloprotein that reduces it at a copper cluster ion active site and this will be described.

Thus, the purpose of this chapter is to review metal ion chemistry with NO and N_2O by highlighting basic aspects of bonding, known structures and reactivity, emphasizing reduction chemistry. Notably, we present this information in a somewhat different fashion than is usual, crossing broadly though the discipline of inorganic chemistry by including examples from coordination and organometallic chemistries as well as cases from biochemistry.

2.2 NO

2.2.1 Bonding and Structures of Metal Nitrosyls

NO in its gaseous state is a stable free radical and it exists primarily as a monomeric species. In air, NO reacts spontaneously with oxygen to form nitrogen dioxide (NO_2), although the reaction is not exceptionally fast because the mechanism is a third-order kinetic process (second order in NO) [45]. It combines rapidly with both organic and oxygen-centered radicals to yield a variety of highly reactive intermediates. As mentioned, NO can also interact with transition metals to produce an enormous range of complexes that have both theoretical significance and practical importance [12, 13, 46–55].

NO has 11 valence electrons and two possible Lewis structures can be drawn as follows:

$$\dot{\underset{\cdot\cdot}{N}}=\ddot{\underset{\cdot\cdot}{O}} \longleftrightarrow \ddot{\underset{\cdot\cdot}{N}}=\dot{\underset{\cdot\cdot}{O}}$$

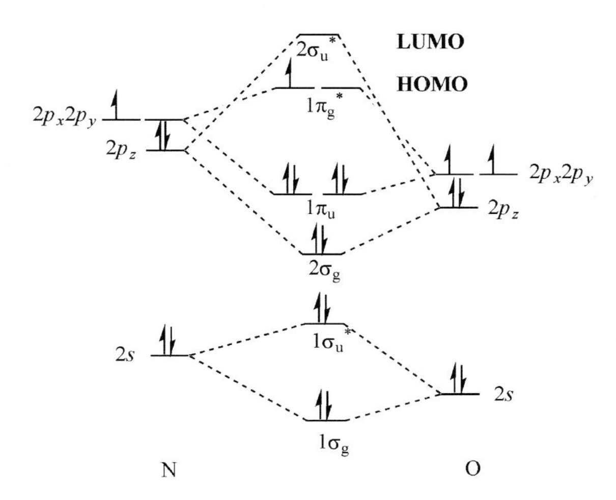

Fig. 2.1 MO ciagram for NO.

Both of these indicate the presence of a double bond, but the actual N–O bond length (about 1.14 Å) suggests an even higher bond order. This can be explained by the molecular orbital (MO) diagram of NO (Fig. 2.1), which reveals the unpaired electron to be located in an antibonding π^* orbital, resulting in an overall bond order of 5/2.

The highest occupied MO (HOMO) of NO possesses mostly N nitrogen character; the greater electronegativity of the O-atom lowers the energy of the O orbitals (Fig. 2.1). For this reason, NO as a ligand prefers to bind to metal ion centers via the N-atom. The electronic structure of NO should be compared to those species related to it by the loss or gain of one electron, i.e. NO^+ (nitrosonium cation) or NO^- (nitroxyl anion), respectively:

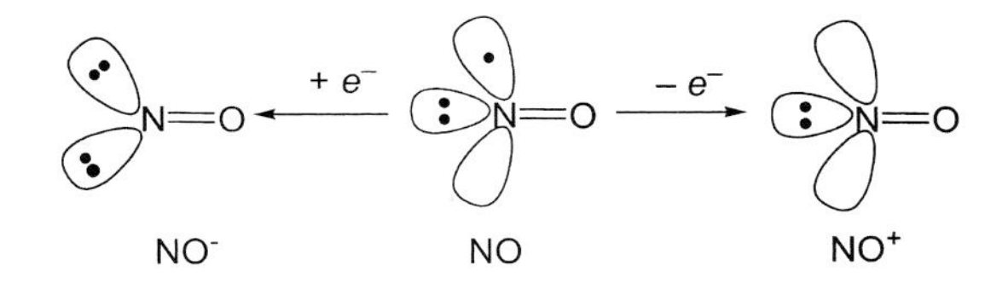

The nitrosonium cation is isoelectronic to CO, having the lone pair electrons in sp hybrid orbitals; thus, a linear M–N–O moiety is anticipated with an entity which is formally NO^+. On the other hand, ^-NO is isoelectronic to dioxygen (O_2) and can be written as N=O with the N-atom having the sp^2 hydrid orbitals, for which a M–N–O bond angle of around 120° is expected.

In the binding of NO to metal ions to afford metal-nitrosyl compounds (as they are referred to in a general manner), electrons are formally transferred from the NO ligand to empty d orbitals, while electron density from the metal ion d electrons can back-donate to NO π^* orbitals, providing for multiple metal–N-atom bonding character, just as in classical metal-carbonyl compounds [56]. Enemark and Feltham [46] developed an incredibly useful and widely used

Table 2.1 The typical coordination geometry of NO complexes predicted by the Enemark-Feltham electron counting formalism.

Coordination number (*CN*)	Idealized geometry			
CN=4	Linear 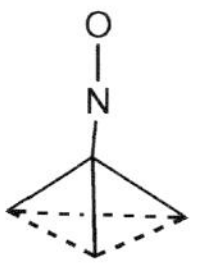 tetrahedral $n = 4, 8, 10$	Bent 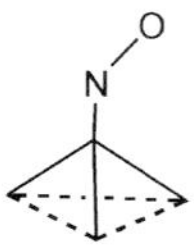 tetrahedral $n = 11$	Bent 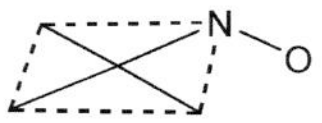 square planar $n = 8$	
CN=5	Linear 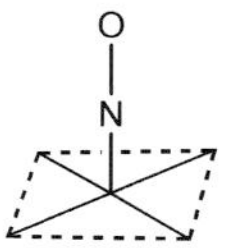 square pyramidal $n = 5, 6$	Bent 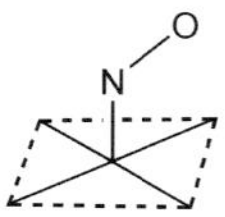 square pyramidal $n = 7, 8$	Linear 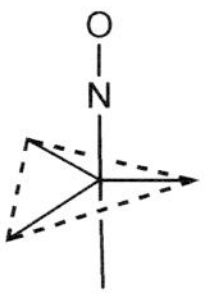 trigonal bipyramidal $n = 4, 8$	Bent 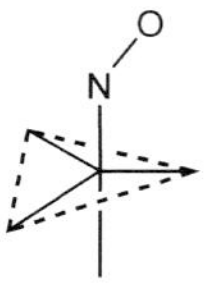 trigonal bipyramidal $n = 8$
CN=6	Linear 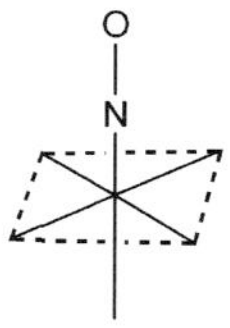 octahedral $n \leq 6$	Bent 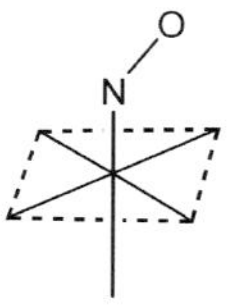 octahedral $n = 7, 8$		

description and notation for metal-nitrosyl complexes, which embodies aspects of the bonding and structures. Here, $\{MNO\}^n$ represents the sum of the metal (M) d electrons and the nitrosyl π^* electrons; for a simple addition reaction, this is the d electron count + 1, the latter being the NO π^* odd electron (see above). Alternatively, since metals (especially in positive oxidation states) are good electron acceptors compared to ligands, one readily considers formal electron transfer from $^{\bullet}NO$ to the metal ion. Thus, the n in $\{MNO\}^n$ represents the number

Table 2.2 Fe–NO bond distances and Fe–N–O angles for selected nitrosyl iron porphyrinate complexes [59].

Complex	Fe–NO (Å)	Fe–N–O (deg)	ν(N–O) (cm^{-1})
$Fe^{II}(OEP)(NO)$	1.726	143.6	1666
$Fe^{II}(TPP)(NO)$	1.717	149.2	1670
$Fe^{II}(TPP)(NO)(4MePip)$	1.740	143.7	1653
$Fe^{II}(T_{Piv}PP)(NO)$	1.716	143	1665
$Fe^{II}(T_{Piv}PP)(ONO)(NO)^-$	1.802	138.3	1616
Hb(NO)	1.74	145	
$Fe^{II}(TPP)(NO_2)(NO)$	1.743	142.1	1625
$Fe^{III}(TPP)(H_2O)(NO)^+$	1.652	174	1848
$Fe^{III}(OEP)(NO)^+$	1.644	146.9	1862

TPP = 5,10,15,20-tetraphenylporphyrinate; $T_{piv}PP$ = *α,α,α,α*-tetrakis-(*o*-pivalanidophenyl)-porphyrinate; OEP = octaethylporphyrinate; Hb = hemoglobin.

of d electrons on the metal when the NO is formally considered to be NO^+. The geometry of MNO units can then be predicted in terms of coordination number, *n*, and a Walsh diagram [46, 57]. This approach and summary is very successful in predicting the coordination geometries of common metal-nitrosyl complexes (Table 2.1).

For example, in mononitrosyl complexes, most M–N–O interactions are linear for hexacoordinate complexes containing the configuration $\{M(NO)\}^{1-6}$. Thus, $\{Fe(NO)\}^6$ possesses a linear coordination [53, 58] and such a compound would form with an Fe(III) porphyrinate (where iron is $3d^5$) which binds to NO. By contrast, the primary NO adduct of Fe(II) porphyrinates ($3d^6$), such as is found in the reduced heme which binds O_2 in blood or muscle (deoxyhemoglobin or deoxymyoglobin, respectively) possesses the $\{Fe(NO)\}^7$ configuration and such species are bent; see examples with structural and spectroscopic data (Table 2.2).

2.2.1.1 Heme Proteins: Guanylate Cyclase – NO Binding and *Trans*-bond Labilization

As mentioned, NO binding to hemes (iron porphyrinates) is of considerable interest because of its biological importance and physiological functions performed such as in blood vessel dilation, host response to infection and signaling [13, 58, 60]. Heme–NO interactions occur in blood-sucking insects which deliver NO to their victim's tissue [61] and heme Fe–NO chemistry is widespread in denitrifying anaerobic bacteria where metabolically produced reducing equivalents are used with metalloenzymes to effect the following series of reactions: NO_3^- (aq) → NO_2^- (aq) → NO (g) → N_2O (g) → N_2 (g) [54, 62]. The nitrite to NO conversion can be effected by either a heme or copper enzyme (*vide infra*) and the NO to N_2O transformation also involves hemes (see below). Further, relevant syn-

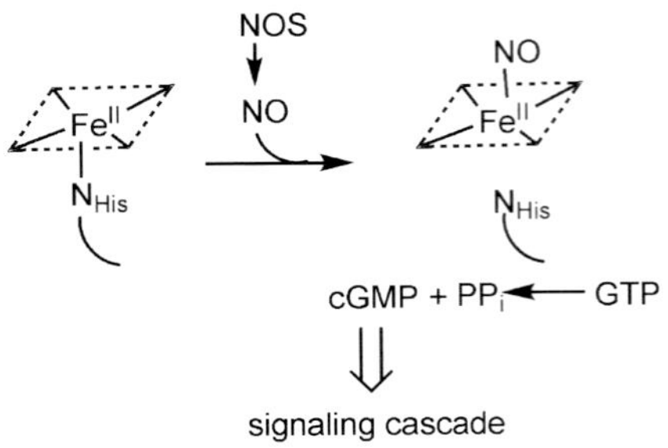

NOS
NO
Fe^{II} N_{His} → NO Fe^{II} N_{His}
cGMP + PP_i ← GTP
⇓
signaling cascade

Scheme 2.1

thetic/chemical research includes the study of the catalytic reduction of NO (or other nitrogen oxides) (mentioned below) and as mentioned above the development of (metallo) reagents as drugs designed to thermally or photochemically release NO [28–38].

In all these aspects, there continues to be widespread efforts to elucidate details of bonding, spectroscopy and structure in penta- and hexacoordinate nitrosyl hemes, including information and insights into Fe–NO versus N–O bond lengths, correlation of ν(Fe–NO) and ν(N–O) frequencies, etc. [15, 63]. A biologically important example worth highlighting is the binding of NO [derived from NO synthase (NOS)] in soluble guanylate cyclase (sGC), which has a signaling function. When sGC is activated by NO binding to a heme, guanosine triphosphate (GTP) is hydrolyzed to cyclic guanosine monophosphate (cGMP) and the latter possesses a well-documented cellular action including blood vessel dilation (Scheme 2.1).

The Enemark and Feltham application of Walsh-type diagrams also predicts that on proceeding from $n=6$ to 7 to 8 for a six-coordinate $\{MNO\}^n$ complex, weakening of the *trans* metal–ligand bond occurs [53]. This is demonstrated by studies on the porphyrinate complexes M(TPP)(L)(NO) (L = 4-methylpiperidine = 4 MePip) [64]. For Mn^{II} ($n=6$), the Mn–NO angle is near-linear (176°) and the Mn–N_{pip} bond length is rather short (2.20 Å). For M = Fe^{II} ($n=7$), bending occurs (∠Fe–N–O = 142°) and the bond to the 4 MePip nitrogen is weakened (Fe–N_{pip} = 2.46 Å). For M = Co^{II} ($n=8$), even greater bending occurs (∠Co–N–O = 128°) and the *trans* ligand labilization occurs to such a large extent that a complex with 4 MePip is not even isolable. Based in part upon such considerations, proposed mechanisms [65, 66] for sGC activation involve binding of the NO signaling molecule to iron, leading to labilization of the *trans* histidine imidazole; this in turn promotes the GTP hydrolysis reaction (*vide supra*). Supporting investigations come from Burstyn and coworkers [67] who studied the activity of sGC prepared by substituting Mn^{II} or Co^{II} into the active site (via techniques affording protein metalloporphyrin substitutions). Addition of NO failed to activate the sGC(Mn^{II}) as the enzyme formed a stable six-coordinate complex without histidine labilization. By contrast, NO activated the cobalt form [i.e. sGC(Co^{II})] even beyond native levels, as a five-coordinate sGC(Co) species formed, consistent with the greater Co–N–O bending expected (*vide supra*) and strong *trans* histidine labilization. Further direct bonding insights were recently obtained by Sage and coworkers [68], who used nuclear resonance vibrational spectroscopy to directly demonstrate weakening of the (porphyrinate) Fe–NO bond strength when in the presence of a *trans* imidazole ligand.

2.2.1.2 **Bridging (η^1-μ_2-) Complexes**

As described above, NO typically bonds to metal ions in an end-on fashion through nitrogen. However, bridging nitrosyl binuclear complexes are in fact quite common. When a NO ligand bridges two metal centers via the N-atom (i.e. in an η^1-μ-NO fashion), three different types of binding modes are possible:

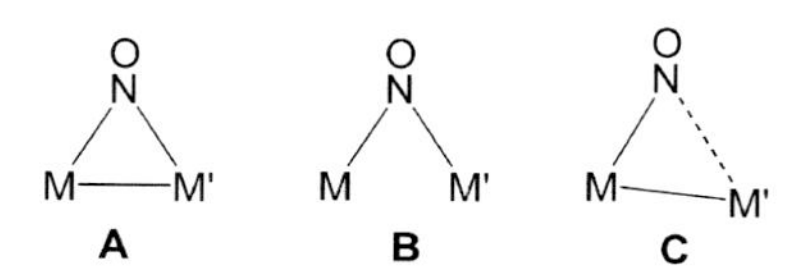

The bridging NO ligand may be symmetric (**A** and **B**) or may be "semi-bridging" (**C**) in which the M–NO–M′ bond is asymmetric. The most widespread case of η^1-μ-NO bimetallic complexes is with structure **A** while a growing number of structures **B** are being reported. Structure **C** is relatively uncommon [69].

Some less common bonding modes are described below.

2.2.1.3 **η^1-μ_3-NO Bridging Complexes**

Several MNO species with triply bridging NO ligands have been structurally characterized by X-ray crystallography. In the trinuclear compound, $Cp_3Mn_3(NO)_4$ (Cp=η^5-cyclopentadienyl), the three manganese ions are found in a triangular array, connected to each other by both doubly and triply bridging nitrosyl ligands [70].

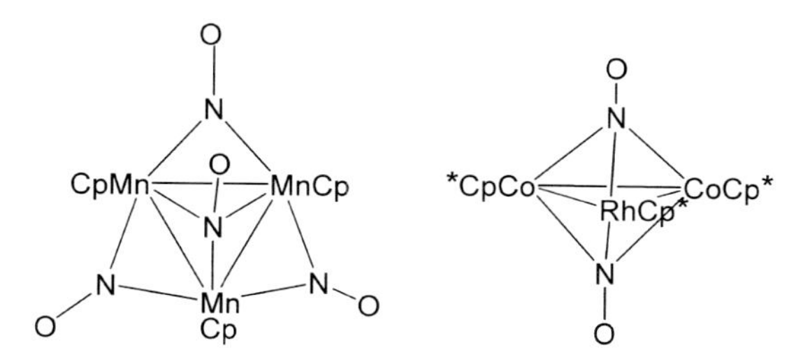

In another case, $Rh(Cp^*)Co_2Cp^*_2(NO)_2$ (Cp*=η^5-pentamethylcyclopentadienyl), two NO ligands cap the triangular plane shaped by two cobalt and one rhodium metal atoms [71].

2.2.1.4 η^2-NO Bridging Complexes

Nitrogen monoxide as a bridging group ligated such that N- *and* O-atoms are bound to both metals, i.e. in an η^2-fashion, is known for only two cases which have been structurally characterized [72, 73]:

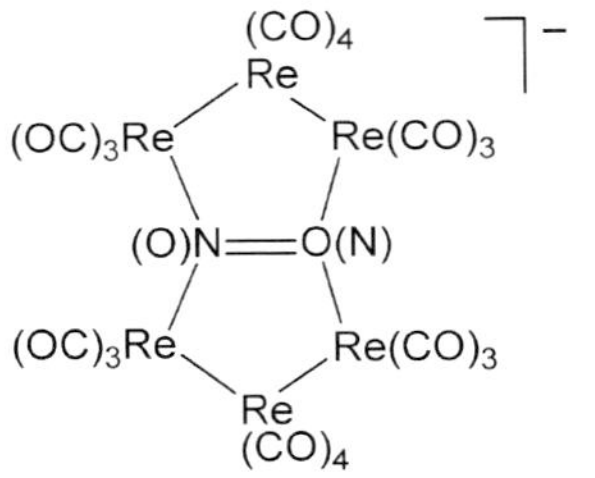

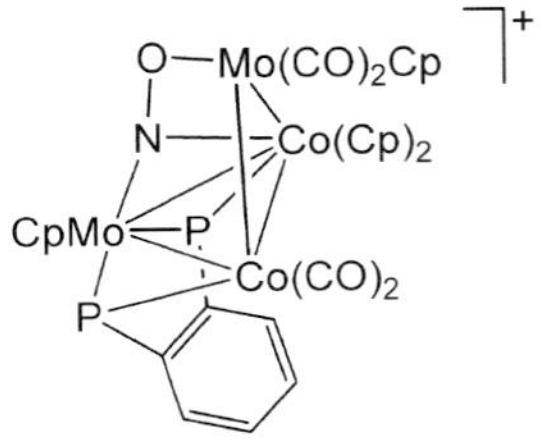

For the molybdenum and cobalt-containing cluster, the metal core is found in a "butterfly" arrangement, with the two molybdenum atoms as the wing tips and the N–Co–Co array as the central core. The bond distance between the oxygen of the bound NO and the molybdenum is 2.158 Å. The O–Mo interaction would seem attributable to the oxophilicity of molybdenum [74].

2.2.1.5 Isonitrosyl and Side-on η^2-NO Complexes

A few NO complexes with η^1-NO coordinated at the oxygen and others with a metal ligated η^2-NO moiety possess the NO bond perpendicular to the metal ligand axis [75]. Laser irradiation of nitroprusside dihydrate ($Na_2[Fe(CN)_5NO]_2 \cdot H_2O$) (**A**) at low temperature produces two different NO species (**B** and **C**) as a mixture:

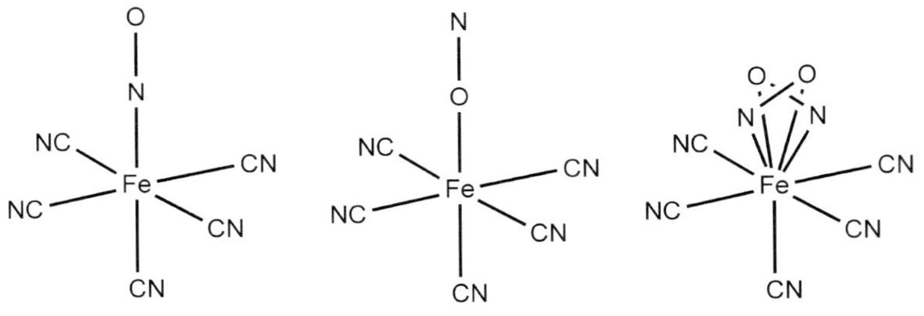

Ground State
A

Metastable State
B

Metastable State
C

The M–NO geometry determined by X-ray crystallographic analysis revealed that those two metastable species possess a linear oxygen-bound isonitrosyl ligand and an η^2-bound NO group, respectively. Such NO linkage isomers, in particular oxygen-bound NO ligands, have also been observed for $\{MNO\}^6$ (M = Fe, Ru, Os) and $\{MNO\}^{10}$ complexes (e.g. M = Ni) [76, 77].

2.2.1.6 Side-on η^2-NO Copper Protein Structures

NO is very reactive towards copper ions, both Cu(I) and Cu(II). Until very recently only two copper nitrosyl X-ray structures were known, one possessing a bridging Cu^{II}–($^{-}$NO)–Cu^{II} structure and formulation [ν(N–O) = 1460 cm^{-1}] [78], the other being a mononuclear copper-nitrosyl complex [(Tp$^{R,R'}$)Cu(NO)] (R = ^{t}Bu, R′ = H):

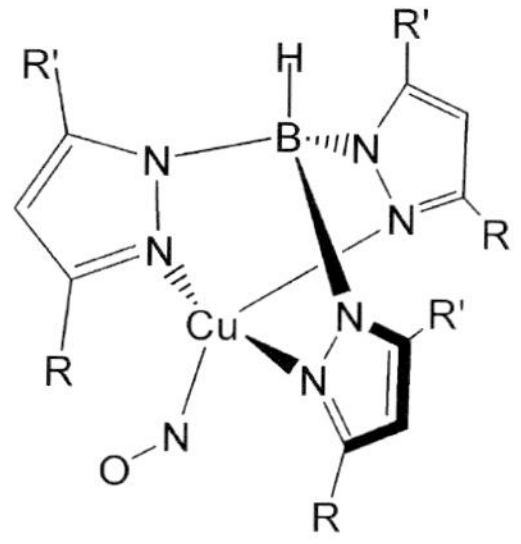

derived from NO (g) reaction with a Cu(I) complex containing a bulky pyrazolylborate ligand; this $\{CuNO\}^{11}$ species is electronic structurally formulated as a Cu^I–($^{\bullet}$NO) species [ν(N–O) = 1712 cm^{-1}; $\angle$Cu–N–O = 163.4°] [54, 79, 80].

Remarkably, crystallographically characterizable side-on η^2-NO copper adducts were recently observed for the protein copper nitrite reductase (Cu-NIR), independently, by two different research groups [81, 82]. For many denitrifying bacteria, it is a copper and not a heme enzyme which effects the NO_2^- (aq) → NO (g) conversion (see above). As such, insight into interactions of NO (g) with the Cu-NIR so-called type 2 Cu-active site (where substrate binding and chemistry occurs) is of great interest. Part of the impetus for the study comes from the fact that Cu^{II} nitrites are normally O-bound, but Cu^I nitrites can be or prefer N-ligation. Thus, either a dissociation and subsequent rebound or "flipping" of an intermediate NO moiety would seem to be eventually required in order to form a copper-nitrosyl species, prior to NO (g) dissociation.

The X-ray structures revealed quite symmetrically side-on bound NO groups, with Cu–N and Cu–O distances ranging between 2.0 and 2.2 Å, depending on the enzyme structure (Fig. 2.2) [81, 82]. The possible occurrence of side-on NO–Cu binding provides a rationale to understand coordination changes or linkage isomerization occurring during the enzyme turnover; thus, structure-based reaction mechanisms were proposed [82] (see, e.g. Fig. 2.2).

The η^2-NO copper protein structures have inspired recent computational studies; Ghosh and coworkers [83] found that $\{CuNO\}^{10}$ as well as $\{CuNO\}^{11}$ species may exhibit quite stable side-on NO bound structures. They also deduced that the $\{CuNO\}^{11}$ moiety is best considered electronically as a Cu^I–($^{\bullet}$NO) moiety, in line with the conclusions reached by Tolman, for his group's synthetic complex (*vide supra*). As described above, Coppens and coworkers [77] observed that the $\{MNO\}^6$ electron count may be important in η^2-NO stabilization and computational studies [83, 84] led to similar conclusions, providing even more

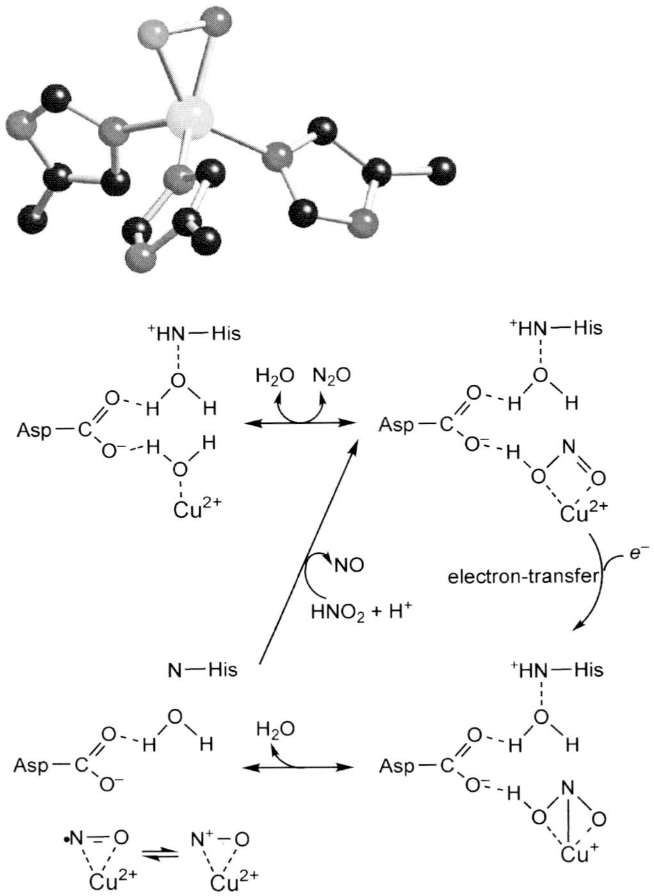

Fig. 2.2 Structural representation of the side-on bound Cu-nitrosyl observed in nitrite reductase (top) [81], and a proposed mechanism of reaction [82] involving binding of nitrite to Cu(II), reduction/isomerization to give an N-bound Cu(I) nitrite, protonation and elimination of water giving the η^2-Cu-nitrosyl, and then NO release from the formally Cu^{II}–($^{\bullet}$NO) moiety.

details such as suggesting that Mn(P)(NO) (P = porphyrinate), $[Fe(P)(NO)]^+$ with no axial ligand, Ru(P)(NO)(Cl) and $[Ru(P)(NO)(L)]^+$ (L = no ligand, imidazole and pyridine) may be primed for such NO ligation.

Thus, in metal-nitrosyl chemistry, it needs to be considered that linkage isomerism, including formation of side-on η^2-NO metal ligation, may occur as a general phenomenon. This may derive from structural/electronic preferences, dynamic behavior or via photochemical initiation. Reactivity patterns and mechanisms, leading from or to M–NO moieties, new or old, should be considered in this light, e.g. as in the Cu-NIR case.

2.2.1.7 Spectroscopic Features of Nitrosyl Metal Complexes

As has been historically well established, infrared (IR) spectroscopy is a simple and powerful means to characterize metal-nitrosyl compounds. As for metal-carbonyl complexes, back-bonding effects are prominent for this strongly π-acidic NO ligand, and the nature of other coligands ligated to the metal in the complex can alter the N–O stretching frequency by more than 100 cm^{-1}. For complexes bearing linear M–N–O moieties, the NO stretching frequency is generally in the range of 1950–1450 cm^{-1}; for those containing bent M–N–O units, it is 1720–1400 cm^{-1}. When NO is bridging two or more metal centers, the frequency may vary from 1650 to 1300 cm^{-1}. The stretching frequency of free NO is 1870 cm^{-1} [12]. As these frequency ranges overlap, clearly it is dangerous to make strong structural conclusions based only upon IR spectroscopy.

While not as convenient, direct monitoring of a nitrosyl group can be carried out by nuclear magnetic resonance (NMR) spectroscopy, since both ^{14}N and ^{15}N have nuclear spins which allow NMR detection. In ^{15}N NMR spectra, the chemical shift observed for bent and bridging M–NO groups typically appears in the range of 300 to 900 ppm, while for linear M–NO moieties, resonances occur from −100 to 200 ppm (referenced to nitromethane) [12]. Isotopically enriched nitrosyl complexes are often prepared to simplify mechanistic investigations or to facilitate the collection of the spectra.

2.2.2 Chemical Reduction of NO and Related Chemistry

Reduction of NO on metal centers is a major theme of this review chapter. As mentioned in the Introduction, such reactivity is relevant to chemical, environmental and biological chemistry, and we here survey these aspects.

2.2.2.1 Chemical Reduction of Metal-bound NO

The reduction, especially catalytic, if possible, of nitrogen oxides is of considerable scientific and technological interest and this has been an active area of research. In particular, conversion of NO generated from internal combustion engines or power plants to more environmentally benign species has become of increased importance. The most practical and convenient method to removing NO might be catalytic reduction using unburnt exhaust gases such as hydrogen, CO and hydrocarbons already contained in the combustion system. For example, NO can be reduced to either N_2O or N_2 with concomitant oxidation of CO:

$$CO + NO \longrightarrow CO_2 + N_2O \quad (1)$$

$$2CO + 2NO \longrightarrow 2CO_2 + N_2 \quad (2)$$

Although both conversions are exergonic, they are kinetically disfavored, having large activation energies. Using heterogeneous catalysts has been an active area of research because it can be considered to be more feasible for practical applications. However, from inorganic chemistry research efforts, many insights have been obtained including the development of a number of well-defined homogeneous catalytic NO reduction systems [85–92].

We start our discussions with reactions of CO with coordinated NO. In early studies, Johnson and Bhaduri demonstrated a stoichiometric CO reduction of iridium-bound NO to give free N_2O while also producing CO_2 [93]:

$$[Ir(NO)_2(PPh_3)_2]^+ + 4CO \longrightarrow [Ir(CO)_3(PPh_3)_2]^+ + CO_2 + N_2O \tag{3}$$

The authors postulated a reaction mechanism involving an O-atom transfer from a coordinated NO to the neighboring CO, followed by an intramolecular attack of the second coordinated NO group on the Ir–N system, invoking a nitrene [94] intermediate. However, follow-up studies employing isotopically labeled NO (^{15}NO) ruled out this nitrene intermediate possibility and provided evidence for dinitrogen dioxide (N_2O_2) as the reactive intermediate which transfers an O-atom to CO [90, 95, 96].

A catalytic NO reduction was effected by the anionic complex $[RhCl_2(CO)_2]^-$ using H_2O, HCl and ethanol [90, 92]. An ^{18}O isotopic-labeling experiment indicated that one of the O-atoms in the CO_2 product is derived from the water in the reaction medium and not directly from the NO reactant [97].

Kubota and coworkers [88, 89] developed a homogeneous NO reduction system, which utilized $PdCl_2$ and $CuCl_2$ (or CuCl) as catalysts in acidic aqueous medium. Their results rather resemble a Wacker chemistry process:

$$PdCl_2 + CO + H_2O \longrightarrow Pd(0) + CO_2 + 2HCl \tag{4}$$

$$Pd(0) + 2NO + 2HCl \longrightarrow PdCl_2 + N_2O + H_2O \tag{5}$$

$$Pd(0) + 2CuCl_2 \longrightarrow PdCl_2 + 2CuCl \tag{6}$$

$$2CuCl + 2NO + 2\,HCl \longrightarrow N_2O + 2CuCl_2 + H_2O \tag{7}$$

Cheng and coworkers [86, 98] took a similar approach, but used $PtCl_4^{2-}$ instead of $PdC1_2$ as catalyst and olefins (instead of CO) as reducing agent, resulting in ketone products plus N_2O (Eq. 8):

$$2NO + RCH{=}CH_2 \longrightarrow N_2O + RC(O)CH_3 \tag{8}$$

The reduction of metal-coordinated NO by a nucleophile/base such as azide, hydroxylamine or hydrazine has been studied with pentacyanonitrosylferrate (nitroprusside, $[Fe(CN)_5(NO)]^{2-}$) [99–103]. Hydrazine treatment of a nitrite complex, $[Fe^{II}(CN)_5(NO_2^-)]^{2-}$ led to nitroprusside formation via acid–base chemistry; this would be followed by attack of a second equivalent of hydrazine at the nitrosyl nitrogen (Scheme 2.2) and this was proposed to form a $[Fe^{II}(CN)_5$-

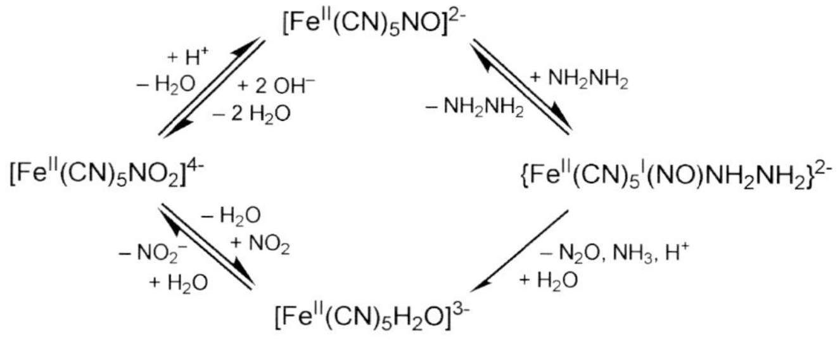

$[Fe^{II}(CN)_5NO]^{2-}$

$+H^+$, $-H_2O$ / $+2\,OH^-$, $-2\,H_2O$

$+NH_2NH_2$ / $-NH_2NH_2$

$[Fe^{II}(CN)_5NO_2]^{4-}$

$\{Fe^{II}(CN)_5{}^{I}(NO)NH_2NH_2\}^{2-}$

$-H_2O$, $+NO_2$ / $-NO_2^-$, $+H_2O$

$-N_2O, NH_3, H^+$, $+H_2O$

$[Fe^{II}(CN)_5H_2O]^{3-}$

Scheme 2.2

$(N(O)NH_2NH_2)]^{2-}$ intermediate. Deprotonation at the bound hydrazine nitrogen followed by tautomerization and N–N bond cleavage results in ammonia generation along with formation of $[Fe(CN)_5N_2O]^{3-}$, which is unstable and releases N_2O.

Using $[Ru(das)_2Cl(NO)]^{2+}$ [das = *o*-phenylenebis(dimethylarsine)] [100] or $[Ru^{II}(NH_3)_5NO]^{3+}$ [101, 102] also led to NO reduction chemistry with added hydrazine, but N_2O was not formed and instead azide (Eq. 9) or dinitrogen (Eq. 10), respectively, were obtained as metal coordinated products.

$$[RuCl(NO)(das)_2]^{2+} + 3\,NH_2NH_2 \longrightarrow [RuClN_3(das)_2] + 2N_2H_5Cl + H_2O \tag{9}$$

$$[Ru^{II}(NH_3)_5(NO)]^{3+} \xrightarrow{NH_2NH_2} [Ru^{II}(NH_3)_5N_2]^{2+} \tag{10}$$

Iron-EDTA and Fe-NTA (where EDTA = ethylenediaminetetraacetic acid and NTA = nitrilotriacetic acid) bind NO strongly but reversibly, formally leading to

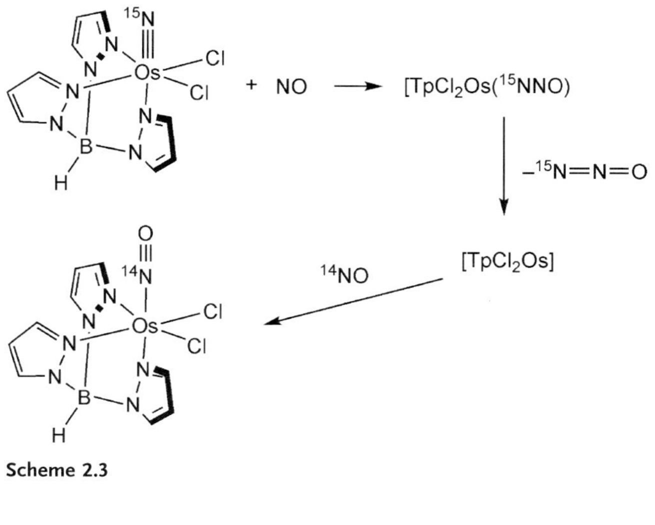

$+ NO \longrightarrow [TpCl_2Os(^{15}NNO)]$

$-{}^{15}N{=}N{=}O$

$[TpCl_2Os]$

^{14}NO

Scheme 2.3

$[Fe^{I}(ligand)(NO^{+})]^{2+}$ species. Reaction of these complexes with sulfite and bisulfite ion yields reduced products N_2O and N_2, along with Fe(III), SO_4^{2-}, $S_2O_6^{2-}$ and $HON(SO_3)_2^{2-}$ [104].

An unusual NO reduction to N_2O that occurs by one-step "N-atom transfer" was demonstrated with a terminal osmium nitride, $TpOs(N)Cl_2$ [Tp = $HBpz_3$, hydrotris(pyrazolyl)borate] [105]. Reaction of this complex with 2 equiv. NO generates N_2O and $\{TpOsCl_2\}$, which is trapped by a second equivalent of NO to form the final $TpOs(NO)Cl_2$ product observed (Scheme 2.3). One can envision "O-atom transfer" from NO to the nitride as a reasonable mechanism for the formation of the nitrosyl complex, but an ^{15}N-labeling experiment ruled out this possibility, instead supporting "N-atom transfer" from the metal nitride to NO to form N_2O. Kinetic data exhibiting bimolecular behavior is also consistent with the rate-determining attack of NO on the nitride ligand.

2.2.2.1.1 **Metal–NO Reduction Accompanied by N–O Cleavage**

Many studies on N–O bond dissociation facilitated by a metal ion have initially been carried out employing cluster compounds, either in solution or the gas phase. In fact, a primary method to generate metal nitride (M–N) compounds is through N–O bond cleavage chemistry. Early examples of low-valent nitride clusters were prepared by using the nitrosonium ion [106]:

$$[Co_6(CO)_{15}]^{2-} + NO^{+} \rightarrow [Co_6N(CO)_{15}]^{-} \quad (11)$$

$$[Rh_6(CO)_{15}]^{2-} + NO^{+} \rightarrow [Rh_6N(CO)_{15}]^{-} \quad (12)$$

Several other nitrido clusters like $[Fe_4N(CO)_{12}]^{-}$ [107], $[Fe_5N(CO)_{14}]^{-}$ [107], $HRu_4N(CO)_{12}$ [108], $H_3Ru_4N(CO)_{11}$ [108], $HOs_4N(CO)_{12}$ [109] and $CoRu_3N(CO)_{12}$ [110] have also been reported to form from clusters containing a coordinated NO ligand.

The synthesis of nitrido clusters by the condensation of monomeric nitrosyl complexes was first reported by Gladfelter and coworkers [111–113]:

$$3\,Fe_3(CO)_{12} + 2[Fe(CO)_3(NO)]^{-} \rightarrow 2[Fe_4N(CO)_{12}]^{-} + CO + 2\,CO_2 + 3\,Fe(CO)_5 \quad (13)$$

Trinuclear molybdenum and tungsten clusters, $Cp_3M_3(O)(\eta^3\text{-}N)(CO)_4$, as well as mixed metal analogs were known to form by N–O cleavage during pyrolysis of $Cp_5M(CO)_2(NO)$ in the presence of the dinuclear complex $Cp_2M_2(CO)_6$ [114]. Gladfelter and coworkers [115] also reported that the deoxygenation of the mixed metal cluster $[FeRu_3(NO)(CO)_{12}]^{-}$ in the presence of CO leads to the reduction of NO and gives an Fe–Ru metal nitride cluster (Eq. 14).

$$[FeRu_3(NO)(CO)_{12}]^{-} + CO \rightarrow [FeRu_3(N)(CO)_{12}]^{-} + CO_2 \quad (14)$$

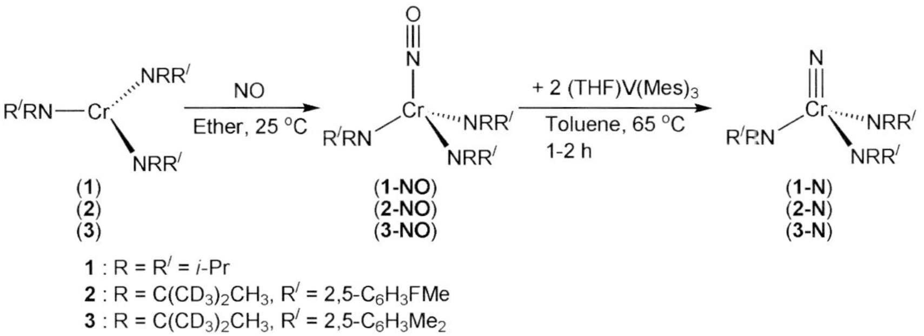
NO
Ether, 25 °C
+ 2 (THF)V(Mes)3
Toluene, 65 °C
1-2 h
(1)
(2)
(3)
(1-NO)
(2-NO)
(3-NO)
(1-N)
(2-N)
(3-N)
1 : R = R' = i-Pr
2 : R = C(CD3)2CH3, R' = 2,5-C6H3FMe
3 : R = C(CD3)2CH3, R' = 2,5-C6H3Me2

Scheme 2.4

Dahl and coworkers [116] reported molybdenum-promoted nitrosyl bond activation leading to the formation of nitrido and/or imido ligand-containing metal clusters. A solution of $Mo_2(\eta^5\text{-}C_5Me_5)_2(CO)_4$ and $Co(CO)_3(NO)$ in tetrahydrofuran, upon photolysis under a slow stream of N_2, gives $(\eta^5\text{-}C_5Me_5)_3Mo_3Co_2(CO)_8(\mu_3\text{-}NH)(\mu_4\text{-}N)$.

Compared to the more extensively studied metal clusters of NO, there are only a handful of examples of monomeric or dimeric nitrosyl complexes undergoing N–O bond dissociation. Cummins and coworkers [117] reported novel *intermolecular* NO reductive cleavage reactions of chromium nitrosyl complexes utilizing a highly reactive low-coordinate V^{III}-triaryl complex $(THF)V(Mes)_3$ ($Mes = 2,4,6\text{-}C_6H_2Me_3$) as the O-atom acceptor (Scheme 2.4).

Legzdins and coworkers [118] reported an unprecedented transformation of NO bound to molybdenum or tungsten in mononitrosyl-diaryl complexes. A solid purple compound $CpW(NO)(o\text{-tolyl})_2$ when exposed to water vapor at ambient temperature gives $CpW(O)(N\text{-}o\text{-tolyl})(o\text{-tolyl})$ with concomitant formation of the di-oxo aryl complex $CpW(O)_2(o\text{-tolyl})$. The latter is exclusively formed when the reactant is exposed to anhydrous O_2 (Scheme 2.5). The mechanism of this conversion is not clear. However, the characteristic chemistry of $CpW(NO)(o\text{-tolyl})_2$ resembles that established for other $CpM(NO)R_2$ complexes such as $CpW(NO)(CH_2SiMe_3)_2$, which indicates that it forms simple 1:1 adducts with both typical Lewis acids and bases. Water as the potential Lewis base is an exception and it is interesting to note that if $CpW(NO)(o\text{-tolyl})_2$ is treated with $^{18}OH_2$, the label is found to be incorporated only into the di-oxo aryl product.

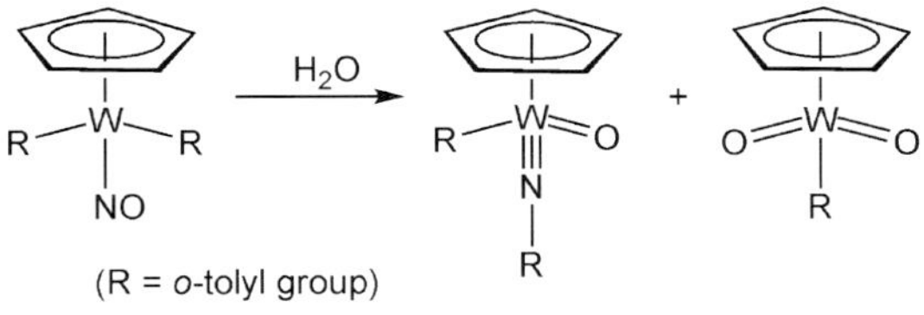
H2O
(R = o-tolyl group)

Scheme 2.5

Thus, it seems that CpW(O)(N-*o*-tolyl)(*o*-tolyl) and $CpW(O)_2$(*o*-tolyl) are formed from CpW(NO)$(o\text{-tolyl})_2$ via independent pathways.

A Mo–NO complex, $[Mo(NO)_2(dttd)]$ ($dttd^{2-}$ = 2,3;8,9-dibenzo-1,4,7,10-tetrathiadecane^{2-}), showed high reactivity towards phosphines, yielding phosphineiminato complexes $[Mo(NO)(NPR_3)(dttd)]$ along with phosphine oxides [119]:

$$Mo(NO)_2(dttd) + 2PR_3 \xrightarrow[(R = Me, Cy, Ph)]{} Mo(NO)(NPR_3)(dttd) + OPR_3 \quad (15)$$

This process seemed to be dominated by the nucleophilicity of PR_3 as was shown by the observation of fast reactions with PMe_3 and PCy_3, but slow transformations with arylphosphines.

2.2.2.2 Electrophilic Attack on Metal-bound NO: HNO (Nitroxyl) Complexes

When the N-atom of an M–NO group is sufficiently electron rich, electrophilic attack may occur, e.g. by protonation. Such reactions are generally observed with bent nitrosyls where the nitrogen is sp^2 hybridized (and thus formally NO^-) and possesses a nitrogen-based electron lone pair. The one-electron reduced form of NO and its protonation product (i.e. nitroxyl (^-NO) or nitrosyl hydride (HNO) [120, 121]) are of considerable current research interest since they appear to possess biological activity distinctive from that of NO [122, 123]. It can act as an enzyme inhibitor and is a possible agent for the treatment of cardiovascular diseases [124]. HNO, as with the isoelectronic O_2 molecule, possesses triplet and singlet states; its known metal complexes are all low-spin d^6 and diamagnetic [121].

The first example of nitrosyl protonation was reported by Roper and coworkers [125]. Reaction of 1 equiv. HCl with $OsCl(CO)(NO)(PPh_3)_2$ resulted in the formation of a stable octahedral Os(II) complex containing a coordinated HNO moiety, $OsCl_2(HNO)(CO)(PPh_3)_2$:

$$OsCl(CO)(NO)L_2 \underset{-\,HCl}{\overset{+\,HCl}{\rightleftharpoons}} OsCl_2(CO)(HNO)L_2 \quad (L = PPh_3) \quad (16)$$

Its formulation was based on elemental analysis and the low energy of the ν(N–O) value observed at 1410 cm^{-1}, which is lowered compared to that known for free HNO (1563 cm^{-1}) [126].

The identity of this complex was subsequently confirmed by X-ray crystallography [127]. The geometry around the osmium ion is octahedral with *trans*-phosphine groups, *cis*-chloro ligands, a carbonyl ligand and the HNO moiety. Structural parameters of the latter (H–N = 0.94 Å, N–O = 1.193 Å and ∠H–N–O = 99°) compare favorably with those of 1.026 Å, 1.211 Å and 108.5° observed for the thermally unstable free molecule [128]. This nitrosyl hydride complex exhibited a characteristic ^{1}H NMR signal at 21.1 ppm, with an observed $^{15}N–^{1}H$ coupling constant value of 75 Hz.

On the other hand, with 2 equiv. HCl, $Os(NO)_2(PPh_3)_2$ forms an adduct which was formulated as $OsCl_2(NHOH)(NO)(PPh_3)_2$ [125]:

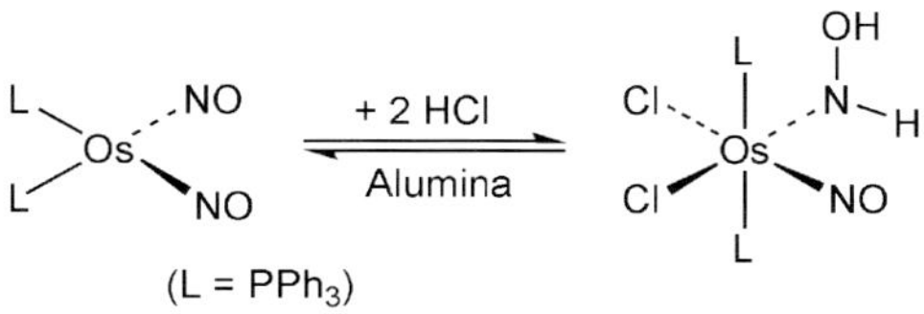

$$L_2Os(NO)_2 \underset{\text{Alumina}}{\overset{+\,2\,HCl}{\rightleftharpoons}} OsCl_2(NHOH)(NO)L_2 \quad (L = PPh_3) \tag{17}$$

the product has a single strong ν(N–O) absorption at 1860 cm^{-1}, a position characteristic of other $OsX_3(NO)(PPh_3)_2$ compounds, along with bands corresponding to ν(NH, OH) at 3310, 3200 and 2600 cm^{-1}. It is interesting to note that the reaction was found to be reversed upon attempted chromatography on alumina and $Os(NO)_2(PPh_3)_2$ was recovered quantitatively.

The complex $IrHCl_2(HNO)(PPh_3)_2$, previously described by Roper, was recently structurally characterized by Hillhouse and coworkers [129]. The iridium atom is in a pseudo-octahedral environment possessing *cis* chlorides and *trans* phosphine ligands. The hydride was not crystallographically located, but it displays a typical *trans* lengthening influence on the chloride ligand. The nitroxyl proton points toward the chloride, while the oxygen points toward the hydride.

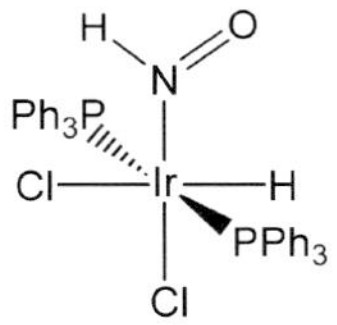

H–N=O, Ph_3P, Cl–Ir–H, PPh_3, Cl

HNO–metal complexes have also been prepared by a number of other synthetic routes, as illustrated in Scheme 2.6 and reviewed recently by Farmer [121].

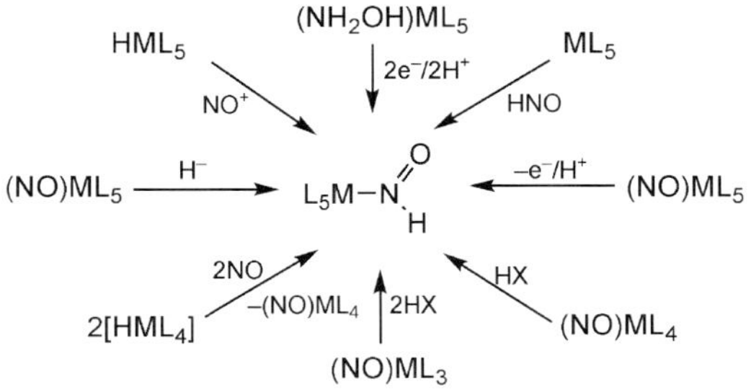

HML_5 →(NO^+) $L_5M–N(H)=O$
$(NH_2OH)ML_5$ →($2e^-/2H^+$) $L_5M–N(H)=O$
ML_5 →(HNO) $L_5M–N(H)=O$
$(NO)ML_5$ →(H^-) $L_5M–N(H)=O$
$(NO)ML_5$ →($-e^-/H^+$) $L_5M–N(H)=O$
$2[HML_4]$ →(2NO, $-(NO)ML_4$) $L_5M–N(H)=O$
$(NO)ML_3$ →(2HX) $L_5M–N(H)=O$
$(NO)ML_4$ →(HX) $L_5M–N(H)=O$

Scheme 2.6

These include insertion of NO into a metal-hydride bond, addition of hydride to a M–NO and redox reactions of NO-related species. Farmer and coworkers [130] recently reported the generation of an unusually stable HNO–metal complex, the first to be formed by direct HNO addition, in this case to a biological heme metal center with a vacant coordination site. HNO, which is short lived in solution, can be generated by decomposition of methylsulfonylhydroxylamine (CH_3SO_2NHOH) in alkaline-buffered solution or by decomposition of Angeli’ salt, $Na_3N_3O_3$, in neutral solutions. In this manner, HNO rapidly reacts with deoxyhemoglobin ($Mb–Fe^{II}$) in neutral to basic solutions, generating this novel Mb–Fe–HNO complex, which has a half-life under anaerobic conditions of more than 6 months.

$Fe^{II}(N)_4(N_{His})$ + HNO ⟶ $(HNO)Fe^{II}(N)_4(N_{His})$

∠ Fe-N-O = 131°

A unique ^1H NMR signal at 14.8 ppm is indicative of the hydrogen of the iron-bound HNO species. Extended X-ray absorption fine structure (EXAFS) and resonance Raman spectroscopic data [131] further support the compound formulation. EXAFS data analysis yielded an Fe–N bond length of 1.82 Å, while the Fe–NO bond angle was determined to be 131°, comparing well with complexes possessing RNO ligands for which X-ray structures are known. Resonance Raman spectra identified an N–O stretch at 1385 cm^{-1} (which was confirmed by ^{15}N labeling), which also corresponds well with those reported for small-molecule HNO complexes [121].

2.2.2.3 Electrocatalytic Reduction of NO

The electrochemical reduction of NO has been studied using metal complexes as homogeneous catalysts. Various reduction products including N_2O, N_2, NH_2OH or NH_3 are observed, depending on the metal employed, the nature of the reducing reagent, the reaction media, etc. For example, Okura [132] reported NO conversion to NH_2OH and N_2H_4, using common iron complexes. The conversion efficiency depended on the catalyst used and diminished in the following order: Fe^{II}phen > aquo-Fe^{II} > Fe^{II}py > Fe^{II}edta > blank. Studies utilizing polypyridyl complexes of ruthenium and osmium have demonstrated the facile electrochemical reduction of bound nitrosyl to coordinated ammonia [133–137]. Electrochemical investigations allowed detection of intermediate products as coordinated ligands and provided support for a mechanism invoking a series of one-electron reductions at NO. In another example, Anson and coworkers [138] used iron-substituted heteropolytungstates, $H_2OFe^{III}XW_{11}O_{39}^{n-}$ (X = Si, Ge, $n = 5$; X = As, P, $n = 4$) as catalysts. The NO-catalyst adduct appeared to utilize the tungsten-oxo core to store reducing electron equivalents temporarily before they were transferred to the iron-coordinated nitrosyl ligand. These researchers made an analogy of the role of iron–tungsten oxo catalyst to the active site of biological nitrite reductase enzymes, where an iron–sulfur cluster provides an electron reservoir and an adjacent iron chlorin (macrocycle) acts to bind the nitrite substrate and effect reduction chemistry.

Electrochemically catalyzed NO reduction by proteins containing redox-active prosthetic groups, such as myoglobin (Mb) and hemoglobin, has been widely explored [139–145]. Although they are not NO reductases (NORs), they were proved to be capable of electrochemically reducing NO. Farmer and coworkers [143, 145] studied and compared the electrocatalytic activity of Mb with a thermophilic cytochrome P450 CYP119, in which these enzymes and didodecyldimethylammonium bromide were cast onto the surface of a pyrclytic graphite electrode. Electrocatalytic NO reduction by CYP119 is very similar to that by Mb, but it is a much more selective catalyst, giving almost exclusively ammonia at least during the initial half-hour of reductive electrolysis chemistry. The catalytic efficiency of NO reduction decreased for CYP119 as compared to Mb, due to both a lower affinity of the protein for NO and a decreased rate of N–N coupling. A recent study using hemin (just the iron porphyrinate portion) as catalyst, instead of whole enzymes, has demonstrated NO reduction to NH_2OH with a selectivity of almost 100% [146]. Thus, heme proteins may be useful NO reduction electrocatalysts, although additional mechanistic work and firm structure–activity relationships need to be deduced.

2.2.2.4 Biological NO Reduction: NORs

Three major types of NORs are known, all involving active site iron ion in catalysis. The coupling of two molecules of NO via reduction and protonation affords N_2O and water, and is a thermodynamically favorable process:

$$2NO + 2H^+ + 2e^- \longrightarrow N_2O + H_2O \quad [E^{\circ\prime}(\text{pH7.0}) = +1.177\ \text{V};\ \Delta G'_0 = -306.3\ \text{kJ mol}^{-1} \tag{18}$$

2.2.2.4.1 Bacterial NORs of the Heme Copper Oxidase (HCO) Type [54, 147]

These enzymes are the most significant physiologically, as they participate in bacterial respiration and denitrification, as mentioned above with respect to the reaction sequence involving Cu-NIR. The enzyme is membrane bound and electron donors pass reducing equivalents through several cytochrome-type (hemes) electron carriers to the active site, a binuclear center with one heme and a non-heme iron referred to as Fe_B (by analogy with HCOs having a copper ion adjacent to a heme; HCOs and NORs are genetically related enzymes [54, 147]). Fe_B is ligated by three histidine imidazole groups and possibly an additional O-atom donor from a glutamic acid protein residue. No X-ray structures are available,

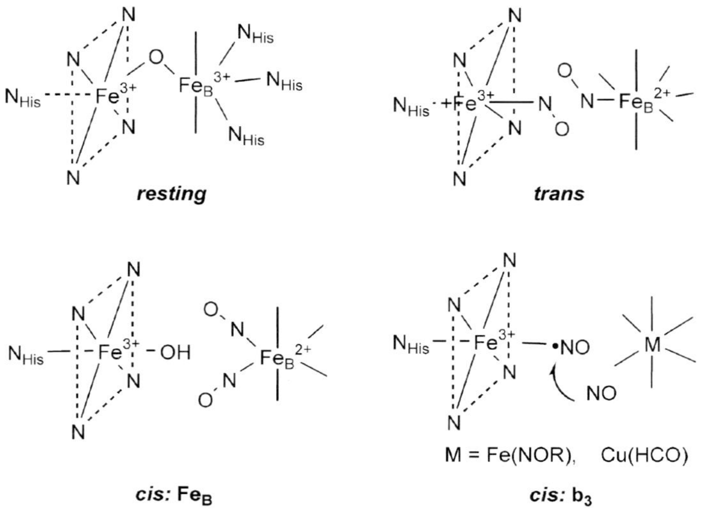

Fig. 2.3 NOR heme/nonheme diiron active site and various structures, including possible reaction intermediates in the coupling of 2 mol equiv. NO to give N_2O. See text for further explanations.

but spectroscopic evidence suggests that the "resting" oxidized diiron(III) enzyme state possesses an oxo bridging ligand (Fig. 2.3) [54, 147, 148]. In the fully reduced enzyme, which may be the active species, the heme is pentacoordinated and bound to the neutral histidine; the latter de-ligates when the iron is in its oxidized Fe(III) state. Various enzyme forms with bound NO or CO (as a surrogate) have been characterized [54].

A number of possible mechanisms for reaction and coupling of two NO molecules to give N_2O have been discussed [54, 147]. The fully reduced enzyme may react with NO, each metal binding one molecule (***trans***, Fig. 2.3), poised for a coupling reaction where each metal ion provides one electron [149]. An alternative mechanism, allowing for the fact that heme–Fe^{II}–NO (i.e. $[FeNC]^7$) species are very stable (perhaps too stable), leads to suggested chemistry occurring at Fe_B, possibly via formation of a dinitrosyl species (***cis:* Fe_B**, Fig. 2.3). In another scenario, a five-coordinated heme–NO complex forms by NO (g) reaction with the heme (Fe^{III} in NOR but Fe^{II} in HCOs, which also effects NO coupling) and a second NO attacks to give a coupled hyponitrite product [i.e. $(N_2O_2)^{2-}$] moiety (***cis:* b_3**, Fig. 2.3), which hydrolyzes to give N_2O and water (with protons provided from the active site environment). See recent references for more details [54, 147, 150, 151].

There is still much to learn concerning the interplay of the two active-site metal ions, such as determination of the order of NO binding or electron transfer to or within the active site and the elucidation of the detailed role and timing of protonation events.

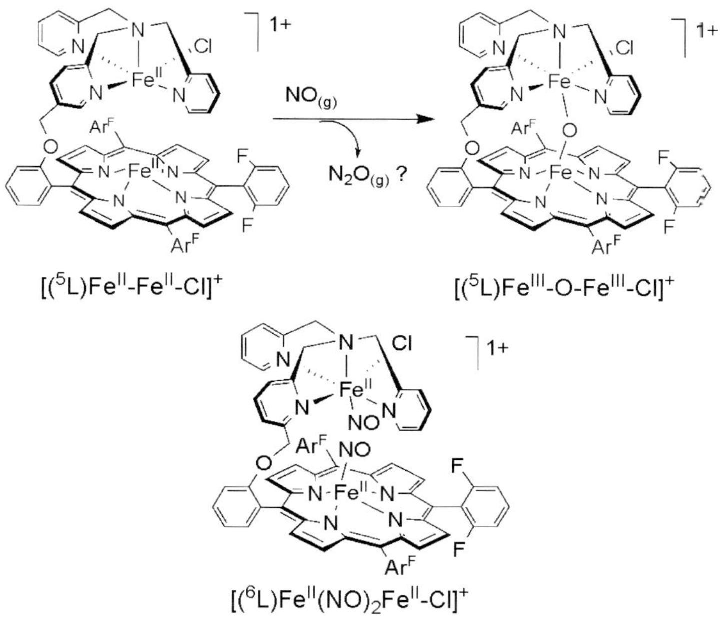

Fig. 2.4 Model compounds for heme/nonheme diiron NOR active site, see text.

2.2.2.4.2 **Models for NORs**

The author's (K.D.K.) research group has been engaged in model coordination chemistry relevant to heme/nonheme diiron NOR active-site chemistry. A number of μ-oxo heme/nonheme diiron(III) complexes have been synthesized and crystallographically characterized, e.g. $[(^5L)Fe^{III}\text{-}O\text{-}Fe^{III}\text{-}Cl]^+$ (Fig. 2.4). In a process relevant to modeling NOR function, a reduced form of the same complex, $[(^5L)Fe^{II}\text{-}Fe^{II}\text{-}Cl]^+$, reacts with NO (g) at low complex concentrations (i.e. below 10 μM) giving back $[(^5L)Fe^{III}\text{-}O\text{-}Fe^{III}\text{-}Cl]^+$, as followed by UV-Vis spectroscopy [152]. This suggests that each iron contributed one electron and 2 mol of NO coupled to give N_2O leaving the O-atom in the μ-oxo product. However, N_2O was not detected; further studies seem warranted. With the related complex $[(^6L)Fe^{II}\text{-}Fe^{II}\text{-}Cl]^+$, reaction with NO (g) afforded a bis-nitrosyl adduct, $[(^6L)Fe^{II}\text{-}(NO)_2\text{-}Fe^{II}\text{-}Cl]^+$ (Fig. 2.4) [153]; however, this is very stable and NO-reductive coupling was not observed under a variety of conditions. Further studies on systems of this type (or others) may help to elucidate structural and mechanistic details concerning NO reductive coupling at a heme/nonheme diiron center.

2.2.2.4.3 **Fungal P450-type NORs**

In fungi, a somewhat different type of NOR activity occurs, at a single heme center, where the proximal axial ligand to the iron is a cysteine thiolate, rendering the system similar to the P450 monooxygenase active-site structure [154]:

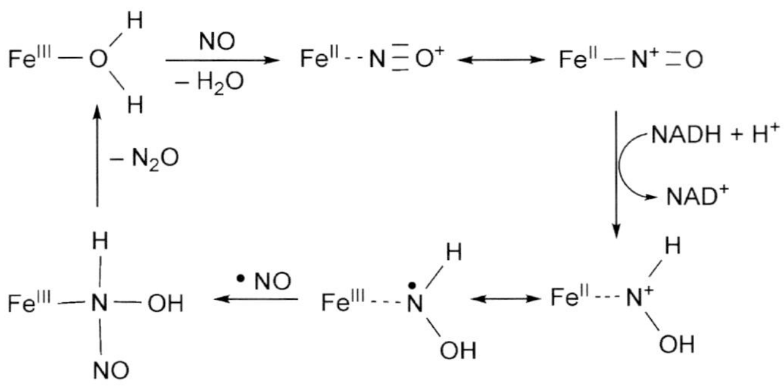

Scheme 2.7

the latter is responsible for O_2 activation and hydrocarbon (or other) substrate oxidation at a heme-thiolate active site. $P450_{NOR}$ is NADH dependent and the currently accepted mechanism of action involves hydride addition to a ferric heme NO moiety (i.e. $\{FeNO\}^6$), giving a nitroxyl-hydride-type intermediate, which with further reaction with another equivalent of NO and protonation would yield N_2O, and return to the resting ferric heme enzyme (Scheme 2.7). An enzyme X-ray crystal structure is available, but many issues remain unresolved, such as understanding the exact role of the thiolate heme ligand and the detailed nature of nitrogen oxide intermediates [155].

2.2.2.4.4 **Flavorubredoxins as Scavenging (S)-NORs**

These bacterial enzymes are relatively more recently described [147, 156, 157], in particular their functions as S-NORs in organisms such as *Escherichia coli*. Recent X-ray crystal structures are available from a number of organisms. The active site of interest consists of a nonheme diiron center which is bridged by carboxylate (aspartate) and either oxo, hydroxo or aquo groups (see diagram). Additional ligands include histidine imidazoles and additional carboxylates (aspartate or glutamate).

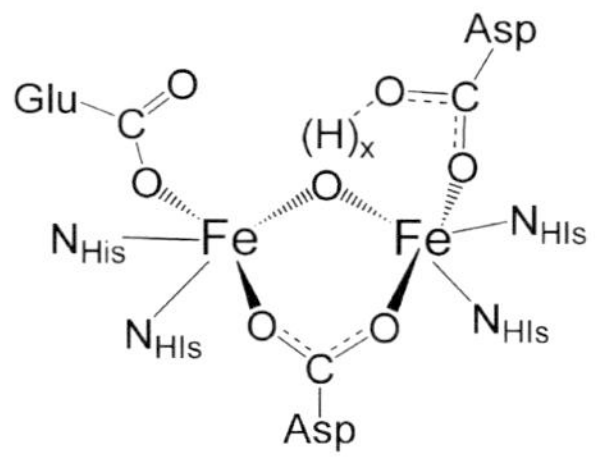

As such, these centers show a great resemblance to other nonheme diiron proteins involved in O_2-activation chemistry. The Fe···Fe distance varies between 2.2 and 2.4 Å, depending on the individual case or state of the enzyme, i.e. "as isolated", reduced $Fe^{II}\cdots Fe^{II}$ or NO-reacted. The course and kinetics of reaction have been determined to be as follows:

$$Fe^{II}Fe^{II} + NO \overset{K_1}{\rightleftharpoons} Fe^{II}(Fe\text{-}NO)$$

$$Fe^{II}(Fe\text{-}NO) + NO \overset{K_2}{\rightleftharpoons} (Fe\text{-}NO)_2$$

$$(Fe\text{-}NO)_2 \xrightarrow{k_{cat}} Fe^{III}Fe^{III} + N_2O$$

Regeneration of the catalytic cycle occurs following reduction of the oxidized diiron center to the active diferrous utilizing electrons coming through the enzyme flavin cofactor.

S-NORs appear to be involved in the bacterial nitrogen oxide metabolism, in this case converting 2 NO to N_2O, protecting the bacterium from "nitrosative" stress [158, 159]. For example, NO may react with superoxide (O_2^-) giving the highly reactive and deleterious nitrating and oxidizing agent peroxynitrite ($ONOO^-$) [5, 6].

2.2.2.5 **Metal Complex-mediated NO Disproportionation**

Metal complex-mediated NO disproportionation is in fact a very common reaction type:

$$3NO + M^{n+} \longrightarrow N_2O + M^{(n+1)+}\text{-}NO_2 \qquad (19)$$

As reduction of NO formally occurs as part of (i.e. one-half of) the process and the chemistry involves interconversion of nitrogen oxide compounds, it is of considerable interest. Thus, we mention a few examples here. The area has been recently reviewed [54].

The copper nitrosyl complex [$(Tp^{tBu,H})Cu(NO)$] was mentioned above [80]. Interestingly, Tolman and coworkers [160] found that with a less-crowded $Tp^{R,R'}$ ligand (R = mesityl, R′ = H or R = R′ = Me), a corresponding Cu–NO species could be detected but ultimately NO disproportionation occurred and a Cu(II) nitrite complex and N_2O were detected as products (Scheme 2.8).

This disproportionation requires N–N coupling chemistry in order to form the N_2O product. Two intermediates are generally considered as plausible precursors to this reaction and subsequent product formation [12]. In form **I** (below), the metal center and corresponding ligand framework accommodate a *cis*-dinitrosyl species. In **II**, a hyponitrite (or hyponitrous) intermediate forms by electrophilic attack of NO onto a previously generated M–NO complex.

O=N–M–N=O (I)   O=N–N(=O)–M (II)

I   II

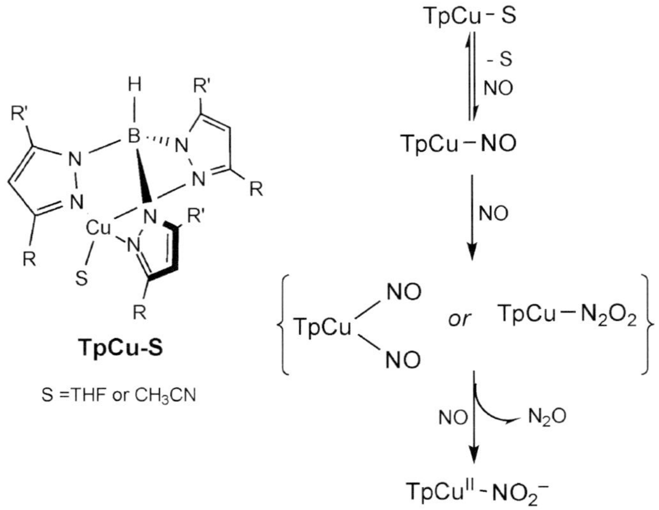

TpCu-S

S = THF or CH_3CN

Scheme 2.8

Following N–N coupling, the next step would be O-atom abstraction (by NO) to yield NO_2 and N_2O; the NO_2 produced often is captured by a reduced metal ion, leading to a metal-nitrite complex.

In addition to the copper complexes mentioned here, nitrogen oxide disproportionation can be mediated by a variety of other metal complexes, including those of cobalt [161], iron [152, 162–164], ruthenium [165–167], manganese [168], rhodium [169] and iridium [93].

2.3 N_2O

2.3.1 Structure and Bonding

N_2O, also known as dinitrogen oxide, is a colorless gas with a sweet odor and taste. The molecule has a linear geometry as predicted with simple valence shell electron pair repulsion (VSEPR) theory. Formal charge considerations suggest that the most important two resonance structures are:

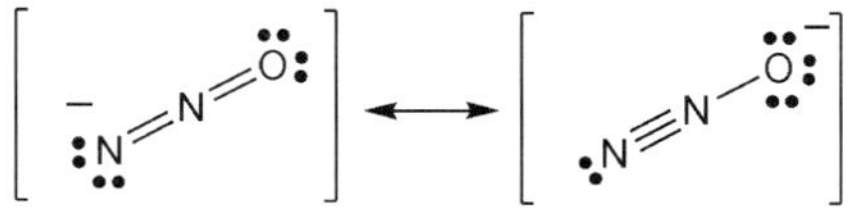

The N–N and N–O bond lengths determined from rotational spectroscopy measurements are 1.128 and 1.184 Å, respectively [170–172]. The N–N bond is slightly longer than the triple bond length in N_2 (1.098 Å); the N–O bond is longer than the typical N=O bond (about 1.14 Å) in NO. Despite the large electronegativity difference between nitrogen and oxygen, the molecule exhibits a relatively low dipole moment. This can be understood by the opposite charge distribution in the two resonance forms.

N_2O, having N and O heteroatoms which may normally serve as good ligands to metal ions, might be expected to have ligating properties somewhat similar to those of dinitrogen or NO. However, N_2O coordination complex adducts reported in the literature are extremely scarce. Actually, only one reasonably well characterized N_2O coordination complex is known up to this day i.e. that reported and studied by Armor and Taube [173–175]. Spectral evidence suggests the existence of $[Ru(NH_3)_5(N_2O)]^{2+}$ which is in equilibrium with N_2O in the reaction of N_2O plus $[Ru(NH_3)_5(H_2O)]^{2+}$. Diamantis and Sparrow soon thereafter isolated this complex in microcrystalline solid forms, and obtained analytical and spectroscopic data consistent with the proposed formulation [176–179].

Two different binding modes of N_2O have been suggested for $[Ru(NH_3)_5(N_2O)]^{2+}$, with terminal N_2O metal binding through either the N- or O-atom [102, 174, 176, 177, 180]. Without any X-ray crystallographic data available, no definite N_2O coordination geometry can be described. Vibrational spectroscopy data and force constant calculations favor bonding through the terminal N-atom. Theoretical considerations have led to suggestions that N_2O would favor Ru-N-N-O binding mode [181, 182]. A recent report by Lehnert and coworkers [183] provides further insights into aspects of the Ru^{II}–N_2O ligation and bonding, following IR and Raman spectroscopic studies of $[Ru(NH_3)_5(N_2O)]^{2+}$, along with normal coordinate analysis and density functional (DFT) calculations. The latter led to two possible structures:

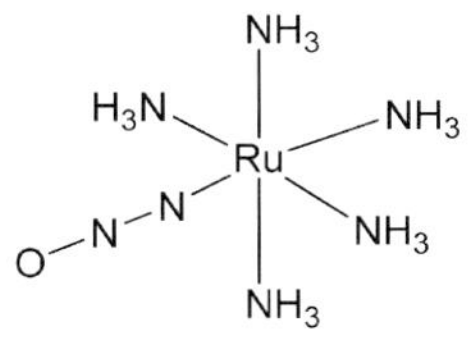

$[Ru(NH_3)_5(NNO)]^{2+}$ 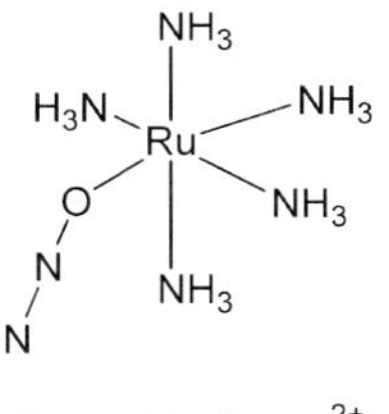 $[Ru(NH_3)_5(ONN)]^{2+}$

When N_2O is bonded through the terminal N-atom, the Ru^{II}-N-N-O unit is linear; however, ruthenium ligation to the O-atom leads to a predicted strongly bent structure with $\angle$Ru-O-N-N = 138°.

The researchers noted that the theoretical model of the end-on terminal binding through the N (rather than O)-atom is more consistent with the experimental spectroscopic data mode. The studies also revealed that Ru^{II}–N_2O bond is dominated by π back-donation, which, however, is weak compared to that found in the corresponding known ruthenium-nitrosyl complex. See further discussions below for possible bridge binding of N_2O in a copper enzyme.

Attempts to isolate M–N_2O complexes have not been successful other than for the ruthenium ion complex above; instead metal oxide, nitride or nitrosyl products are obtained. The existence of a transient N_2O adduct with a coordinatively unsaturated tungsten carbonyl compound, $W(CO)_5$, was observed by examining the IR, Raman and UV-Vis spectra of solid argon or methane matrices at around 20 K [184]. An approximate upper limit for the bond dissociation energy of $W(CO)_5$–N_2O was estimated to be 22 ± 2 kcal mol^{-1} from transient IR spectroscopy [185]. More recently, James and coworkers [186] reported the *in situ* synthesis and NMR characterization of an N_2O complex of a five-coordinate species, $RuCl_2(P\text{-}N)(PPh_3)$ [P-N = [*o*-(*N*,*N*-dimethylamino)phenyl]diphenyl-phosphine], which has a vacant coordination site and is also known to be highly active in that it forms complexes with weak ligands such as H_2S and thiols.

2.3.2
Metal-mediated N_2O Reduction

2.3.2.1 Oxo Transfer Reactions

N_2O is a potentially useful and environmentally friendly oxidant because it is inexpensive and also the only byproduct from an oxo transfer reaction would be dinitrogen. However, N_2O is extremely (kinetically) inert [187] and is one of the least reactive species among nitrogen oxides. The very large activation energy (59 kcal mol^{-1}) [8] for the decomposition of N_2O to $N_2 + O$ makes it very difficult to use N_2O as a practical oxo transfer agent at ambient temperature.

However, there has been progress in metal-mediated N_2O activation. Oxo-transfer from N_2O can lead to formation of metal-oxo complexes or oxidation of ligands contained in the metal complex. In pioneering research of N_2O reactivity, Bottomley and coworkers [188–192] utilized N_2O in the preparation of a series of unusual oxo-bridged clusters with titanium, vanadium and chromium metal ions; M-(N_2O)-M-bridged intermediates were suggested to form.

Groves and coworkers demonstrated that $[Ru^{II}(TMP)(THF)_2]$ (TMP = tetramesitylporphyrinate^{2-}) is efficiently oxidized by N_2O to form a dioxo-Ru(VI) species, $[Ru^{VI}(TMP)(O)_2]$ [193]. A dinitrogen complex, $[(TMP)Ru^{II}N_2(THF)]$ and N_2O-bridged dinuclear complex, [Ru(TMP)-N=N-O-Ru(TMP)], were observed as key intermediates.

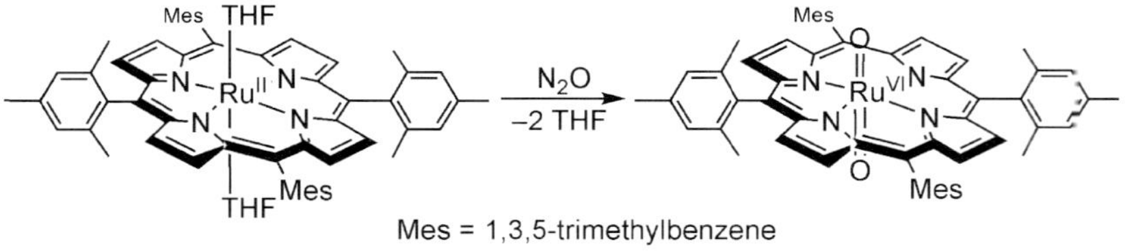

Mes = 1,3,5-trimethylbenzene

N_2O insertion into a metal–ligand bond has been observed with Group IV metallocene complexes (Scheme 2.9) [194]. When reported, these transformations

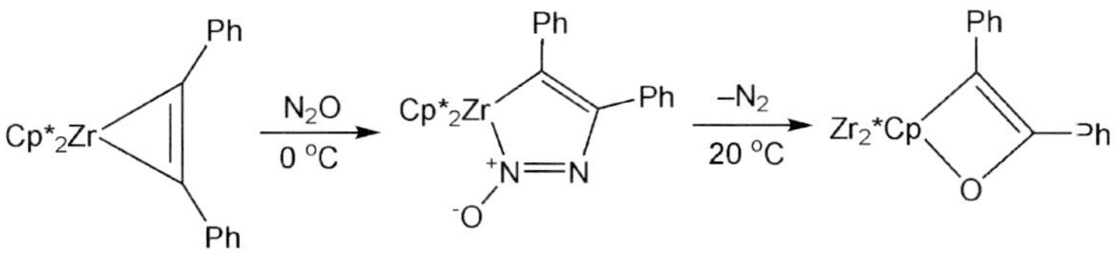

Scheme 2.9

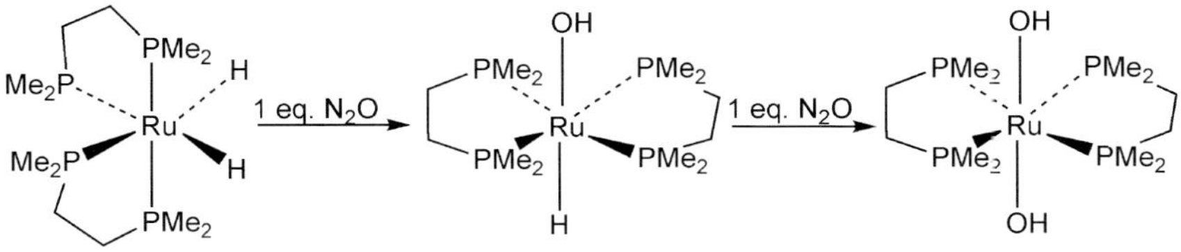

Scheme 2.10

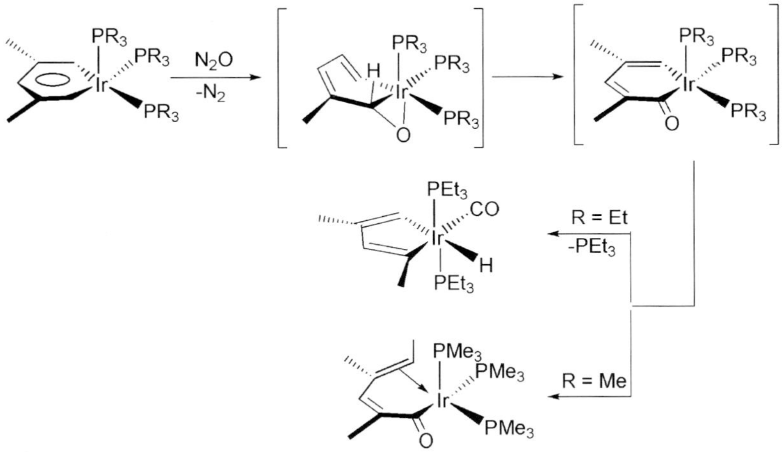

Scheme 2.11

were unprecedented since reactions between N_2O and metal complexes usually lead to extrusion of N_2.

Bergman and coworkers [195, 196] demonstrated that N_2O is capable of mediating O-atom insertion into a late metal–hydrogen bond, in this case to afford a hydroxy-Ru complex (Scheme 2.10).

Bleeke and coworkers [197] found a ligand-based oxidation to occur in the reaction of iridabenzenes with N_2O. The transformations appear to proceed through a metallaepoxide intermediate, although the detailed mechanism of the reaction is not known.

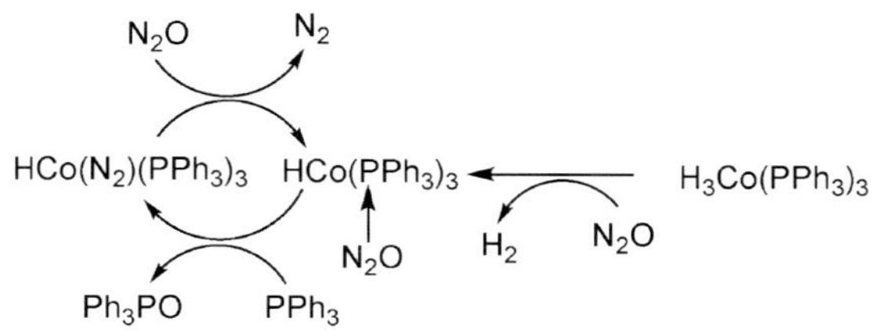

Scheme 2.12

2.3.2.2 **Catalytic Oxo Transfer**

A cobalt complex-mediated catalytic oxidation of triphenylphosphine to triphenylphosphine oxide using N_2O and accompanied by N_2 evolution has been described [198]. The system is proposed to proceed through a $HCo(N_2)(PPh_3)_3$ species as active catalyst (Scheme 2.12).

In the area of heterogeneous catalysis, metal complexes stabilized in zeolite matrix have attracted considerable attention for practical catalytic N_2O decomposition applications. One of the most effective catalysts appear to be iron-containing acidic zeolites which at elevated temperatures decompose N_2O to yield surface-activated iron-oxo species (α-oxygen) which are even capable of oxidizing methane [199–203]. The system can also effect a benzene to phenol oxidation. Physical studies suggested that the active oxidant has a binuclear structure, with two iron atoms bridged by an oxo group. In a related investigation with molybdenum supported on silica, selective oxidation of methane to methanol and formaldehyde was observed [204]. EPR spectroscopic studies suggested that $Mo^{VI}–O^-$ species, formed from the reaction of N_2O with surface Mo(V) ions, effect methane H-atom abstraction reactions; subsequent methyl radical recombination with $Mo^{VI}–O^-$ leads to $Mo^V–^-OCH_3$. This methoxide may decompose to produce HCHO or may react with water to form methanol.

2.3.2.3 **N_2O N–N Bond Cleavage**

An important development in N_2O activation chemistry came with the unprecedented demonstration of metal complex-reductive denitrification of N_2O to yield metal nitride and nitrosyl complexes. The N–N, rather than N–O, bond scission

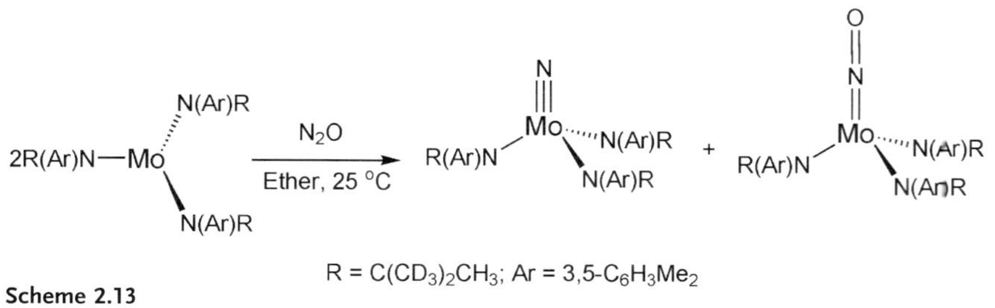

R = $C(CD_3)_2CH_3$; Ar = 3,5-$C_6H_3Me_2$

Scheme 2.13

in N_2O was not known until the recent work of Cummins and coworkers [205]. Exposure of a tri-coordinate molybdenum complex, $Mo(NRAr)_3$, to N_2O in ether resulted in a 1:1 ratio of products, a molybdenum nitride and also a nitrosyl complex (Scheme 2.13). The products were fully characterized by various spectroscopies and mass spectrometry, and their identity was further verified by independent synthesis.

In view of the stronger N–N bond of N_2O compared to its N–O bond (by around 75 kcal mol^{-1}) [206], this reaction is considered to be under kinetic rather than thermodynamic control. N_2O N–N bond cleavage is an important reaction in the generation of stratospheric NO, a process thought to occur by N_2O reaction with excited atomic oxygen (^1D) [207]. Boron cluster ions are also known to effect N_2O N–N bond cleavage [206].

2.3.2.4 **Electrocatalytic Reduction of N_2O to N_2**

Electrocatalytic removal of N_2O from industrial or automobile gas streams emitted to the atmosphere is a technologically important process [208], as is removal of N_2O as a greenhouse gas. As a potentially useful application, it has been demonstrated that a fuel cell can be constructed based on the reduction of N_2O coupled to the oxidation of H_2 [209]. The two-electron reduction of N_2O to N_2 is very favorable from a thermodynamic standpoint (also see below), but the barrier to introduction of the first electron is quite high. For this reason, electrochemical reduction by sequential one-electron steps is unfavorable unless intermediates can be considerably stabilized.

A number of transition metal complexes demonstrate electrocatalytic N_2O reduction activity. Collman and coworkers [210] designed and prepared a series of binary or "face-to-face" metalloporphyrin complexes as catalysts that might be capable of binding N_2O in between the two metal ions. They observed that the electrocatalytic rate of N_2O reduction was sensitive to hydroxide, changes in the functionality of the porphyrin periphery and the presence of proton donors.

Taniguchi and coworkers [211] reported that when using Ni^{II} complexes of macrocyclic polyamines like [15 or 14]ane N_4:

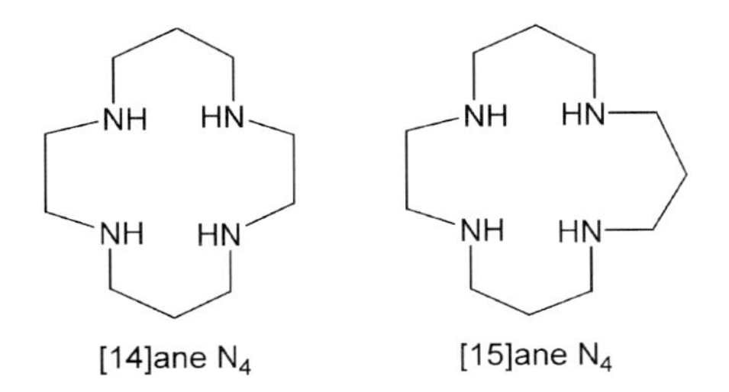

at a mercury electrode, efficient electrocatalytic reduction of N_2O in water occurred.

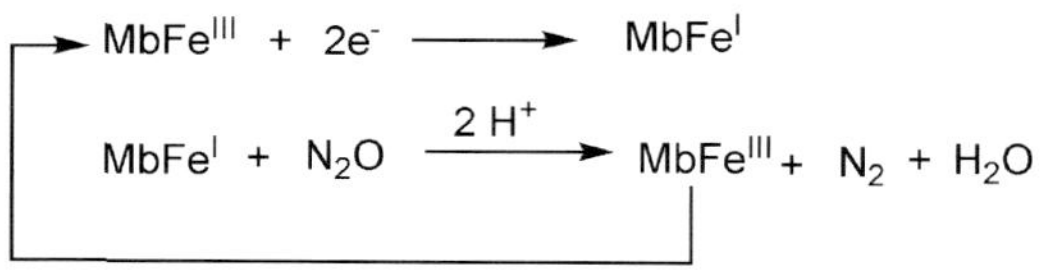

$MbFe^{III} + 2e^- \longrightarrow MbFe^{I}$

$MbFe^{I} + N_2O \xrightarrow{2\,H^+} MbFe^{III} + N_2 + H_2O$

Scheme 2.14

Electrolysis at a nickel electrode (with no catalyst) led to N_2 and H_2 evolution, in contrast to the results with the nickel complex catalysts which gave N_2 in close to 100% yield; no significant amounts of H_2 or NO were detected.

Electrocatalytic N_2O reduction mediated by heme proteins such as Mb or cytochrome P450 monooxygenase has been demonstrated in film-modified electrodes [145, 212]. The authors suggest two-electron redox processes occur with the catalyst shuttling between the Fe(I) and Fe(III) states, e.g. as shown in Scheme 2.14.

2.3.2.5 Biological N_2O Reduction

The copper enzyme N_2O reductase (N_2OR) catalyses the final step in the process of bacterial denitrification (*vide supra*). The two-electron reduction of N_2O to dinitrogen is, as mentioned above, thermodynamically a very downhill reaction (Eq. 20), but N_2O is kinetically inert and a catalyst is necessary to effect this transformation.

$$N_2O + 2\,e^- + 2\,H^+ \rightarrow N_2 + H_2O \, \Delta G^0(\mathrm{pH}\,7) = -340\ \mathrm{kJ\ mol^{-1}} \qquad (20)$$

Recent X-ray structure determinations [213, 214] reveal the so-called Cu_Z catalytic center to be a $Cu_4(S)(His)_7$ (His = histidine amino acid residue, with its imidazole side-chain ligand donor) cluster; the central sulfide (S^{2-}) ion bridges all four copper ions which are arranged in a distorted tetrahedral geometry (Fig. 2.5). Binding of the copper ions to histidines connects the Cu_4S cluster to the protein matrix. Further, a water (or hydroxide) binds to Cu_{IV}, the copper ion which has only one (rather than two) histidine ligands.

The enzyme can exist in a variety of redox states, of which the fully reduced $Cu^{I}_4(S)$ cluster has been shown to be capable of reducing the substrate. Molecular modeling of a high-resolution structure [214] along with computational studies [215, 216] suggest that the N_2O substrate may coordinate by bridging Cu_{IV} and Cu_I in a μ-1,3 fashion (Fig. 2.5). The theoretical work suggests an energetically favorable bent binding (Fig. 2.5) which activates N_2O for favorable Cu(I) d orbital back-donation to N–O π^* orbitals, setting the N_2O molecule up for subsequent electron-transfer reduction. Three possible pathways for N_2O reduction have been suggested (Fig. 2.5) [215]. The binding/activation event may facilitate direct two-electron transfer from two copper ions (path **i**). Current calculations [215] favor reaction **ii**, where the dicopper-bound N_2O moiety gives way to facilitated protonation followed by very low activation barrier reductive cleavage [215].

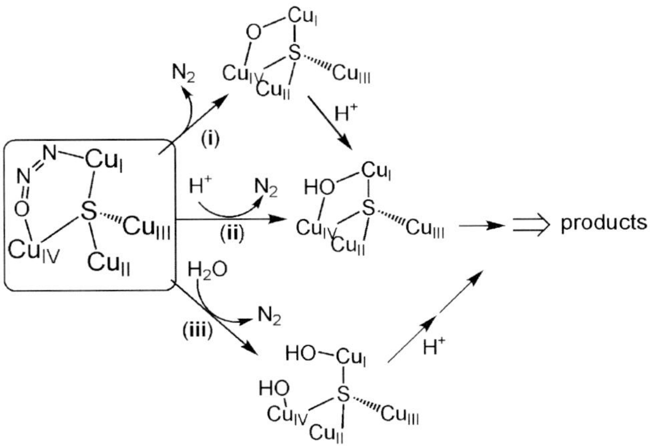

Fig. 2.5 Structure of Cu_Z of N_2OR with its $Cu_4(S)(His)_7$ core and possible pathways (**i**, **ii** and **iii**) for N_2O reduction to N_2. Adapted from Refs. [215, 217].

A reductive cleavage accompanied by hydration and elimination of N_2 is represented by path **iii** [215]. While the active site stores four (or more) electrons with the four copper atoms and reduced sulfur moiety, only two are needed for a stoichiometric reaction; the Cu_4S core provides a reservoir of reducing equivalents which is perhaps key in the chemistry. Once N_2O is reduced, leaving some Cu(II) ions in the cluster, re-reduction of Cu_Z occurs by electron transfer from a neighboring dicopper center (so-called Cu_A), allowing enzyme turnover.

2.4 Summary and Conclusions

We have tried here to overview both basic elements and highlights of the inorganic chemistry of NO and N_2O, especially with respect to the binding of these important nitrogen oxide molecules to transition metal ions as found in coordination and organometallic complexes, along with biologically important centers which process these nitrogen oxide gases. The emphasis has also been placed on reductive chemistry. This is certainly not the only reactivity undergone by NO and N_2O, but it does represent an extensive area of current interest and research activity, due to environmental and bioinorganic considerations.

We hope that readers will realize that while there is a reasonably long history and great deal known about NO and N_2O and their interactions with metal ions, there is still much to be learned. Nitrogen oxide (NO_x) chemistry is in fact quite complex and thus very rich – exciting times lie ahead for further research.

References

1 J. Lelieveld, P.J. Crutzen, *Science* **1994**, *264*, 1759–1761.
2 L.J. Ignarro, in *Nitric Oxide: Principles and Actions*, J. Lancaster, Jr. (Ed.), Academic Press, San Diego, CA, **1996**, p. 111.
3 L.J. Ignarro (Ed.), *Nitric Oxide: Biology and Pathobiology*, Academic Press, San Diego, CA, **2000**.
4 J. Lancaster, Jr. (Ed.), *Nitric Oxide: Principles and Actions*, Academic Press, San Diego, CA, **1996**.
5 R. Radi, *Proc Natl Acad Sci USA* **2004**, *101*, 4003–4008.
6 W.H. Koppenol, *Met Ions Biol Syst* **1999**, *36*, 597–619.
7 W.C. Trogler, *Coord Chem Rev* **1999**, *187*, 303–327.
8 A.V. Leont'ev, O.A. Fomicheva, M.V. Proskurnina, N.S. Zefirov, *Russ Chem Rev* **2001**, *70*, 91–104.
9 W.C. Trogler, *J Chem Educ* **1995**, *72*, 973–976.
10 T.W. Hayton, P. Legzdins, W.B. Sharp, *Chem Rev* **2002**, *102*, 935–991.
11 L. Cheng, G.B. Richter-Addo, in *Porphyrin Handbook 4*, K.M. Kadish, K.M. Simith, R. Guilard (Eds.), Academic Press, San Diego, CA, **2000**, pp. 219–291.
12 G.B. Richter-Addo, P. Legzdins, *Metal Nitrosyls*, Oxford University Press, New York, **1992**.
13 P.C. Ford, B.O. Fernandez, M.D. Lim, *Chem Rev* **2005**, *105*, 2439–2455.
14 P.C. Ford, L.E. Laverman, *Coord Chem Rev* **2005**, *249*, 391–403.
15 M.D. Lim, I.M. Lorkovic, P.C. Ford, *J Inorg Biochem* **2005**, *99*, 151–165.
16 B. D'Autreaux, N.P. Tucker, R. Dixon, S. Spiro, *Nature* **2005**, *437*, 769–772.
17 S. Archer, *FASEB J* **1993**, *7*, 349–360.
18 G. Cirino, *Digest Liver Dis* **2003**, *35*, S2–S8.
19 J.A. Hrabie, L.K. Keefer, *Chem Rev* **2002**, *102*, 1135–1154.
20 L.K. Keefer, *Curr Top Med Chem* **2005**, *5*, 625–634.
21 P.G. Wang, M. Xian, X.P. Tang, X.J. Wu, Z. Wen, T.W. Cai, A.J. Janczuk, *Chem Rev* **2002**, *102*, 1091–1134.
22 M.C. Frost, M.E. Meyerhoff, *J Am Chem Soc* **2004**, *126*, 1348–1349.
23 M.H. Lim, S.J. Lippard, *J Am Chem Soc* **2005**, *127*, 12170–12171.
24 R.C. Smith, A.G. Tennyson, M.H. Lim, S.J. Lippard, *Org Lett* **2005**, *7*, 3573–3575.
25 M.H. Lim, S.J. Lippard, *Inorg Chem* **2004**, *43*, 6366–6370.
26 K. Tsuge, F. DeRosa, M.D. Lim, P.C. Ford, *J Am Chem Soc* **2004**, *126*, 6564–6565.
27 A. Ivanisevic, M.F. Reynolds, J.N. Burstyn, A.B. Ellis, *J Am Chem Soc* **2000**, *122*, 3731–3738.
28 S. Wecksler, A. Mikhailovsky, P.C. Ford, *J Am Chem Soc* **2004**, *126*, 13566–13567.
29 F. Derosa, X.H. Bu, P.C. Ford, *Inorg Chem* **2005**, *44*, 4157–4165.
30 J. Bordini, P.C. Ford, E. Tfouni, *Chem Commun* **2005**, 4169–4171.
31 P.C. Ford, S. Wecksler, *Coord Chem Rev* **2005**, *249*, 1382.
32 L.L. Welbes, A.S. Borovik, *Acc Chem Res* **2005**, *38*, 765–774.
33 J.T. Mitchell-Koch, T.M. Reed, A.S. Borovik, *Angew Chem Int Ed* **2004**, *43*, 2806–2809.
34 R.K. Afshar, A.K. Patra, M.M. Olmstead, P.K. Mascharak, *Inorg Chem* **2004**, *43* 5736–5743.
35 A.K. Patra, M.J. Rose, K.A. Murphy, M.M. Olmstead, P.K. Mascharak, *Inorg Chem* **2004**, *43*, 4487–4495.
36 A.K. Patra, P.K. Mascharak, *Inorg Chem* **2003**, *42*, 7363–7365.
37 K.M. Padden, J.F. Krebs, C.E. MacBeth, R.C. Scarrow, A.S. Borovik, *J Am Chem Soc* **2001**, *123*, 1072–1079.
38 R. Prakash, A.U. Czaja, F.W. Heinemann, D. Sellmann, *J Am Chem Soc* **2005**, *127*, 13758–13759.
39 S.P. Fricker, *Met Ions Biol Syst* **2004**, *41*, 421–480.
40 C.J. Marmion, B. Cameron, C. Mulcahy, S.P. Fricker, *Curr Top Med Chem* **2004**, *4*, 1585–1603.
41 J.S. Stamler, S. Lamas, F.C. Fang, *Cell* **2001**, *106*, 675–683.
42 D.L.H. Williams, *Acc Chem Res* **1999**, *32*, 869–876.
43 M.C. Frost, M.E. Meyerhoff, *J Biomed Mater Res A* **2005**, *72A*, 409–419.

44 B.K. Oh, M.E. Meyerhoff, *J Am Chem Soc* **2003**, *125*, 9552–9553.

45 P.C. Ford, D.A. Wink, D.M. Stanbury, *FEBS Lett* **1993**, *326*, 1–3.

46 J.H. Enemark, R.D. Feltham, *Coord Chem Rev* **1974**, *13*, 339–406.

47 D.M.P. Mingos, D.J. Sherman, *Adv Inorg Chem* **1989**, *34*, 293–377.

48 D. Cornforth, in *Nitric Oxide: Principles and Actions*, J. Lancaster, Jr. (Ed.), Academic Press, San Diego, CA, **1996**, pp. 259–286.

49 M.F. Reynolds, J.N. Burstyn, *Nitric Oxide* **2000**, *4*, 381–399.

50 M. Wolak, R. van Eldik, *Coord Chem Rev* **2002**, *230*, 263–282.

51 J.K.S. Moller, L.H. Skibsted, *Chem Rev* **2002**, *102*, 1167–1178.

52 P.C. Ford, I.M. Lorkovic, *Chem Rev* **2002**, *102*, 993–1017.

53 G.R.A. Wyllie, W.R. Scheidt, *Chem Rev* **2002**, *102*, 1067–1089.

54 I.M. Wasser, S. de Vries, P. Moënne-Loccoz, I. Schröder, K.D. Karlin, *Chem Rev* **2002**, *102*, 1201–1234.

55 J.A. McCleverty, *Chem Rev* **2004**, *104*, 403–418.

56 F.A. Cotton, G. Wilkinson, *Advanced Inorganic Chemistry*, 5th edn, Wiley, New York, **1988**.

57 A.D. Walsh, **1953**, 2266–2288.

58 W.R. Scheidt, M.K. Ellison, *Acc Chem Res* **1999**, *32*, 350–359.

59 P.C. Ford, I.M. Lorkovic, *Chem Rev* **2002**, *102*, 993–1017.

60 J.C. Salerno, in *Nitric Oxide: Principles and Actions*, J. Lancaster, Jr. (Ed.), Academic Press, San Diego, CA, **1996**, pp. 83–110.

61 F.A. Walker, W.R. Montfort, *Adv Inorg Chem* **2001**, *51*, 295–358.

62 B.A. Averill, *Chem Rev* **1996**, *96*, 2951–2964.

63 D.P. Linder, K.R. Rodgers, J. Banister, G.R.A. Wyllie, M.K. Ellison, W.R. Scheidt, *J Am Chem Soc* **2004**, *126*, 14136–14148.

64 M.K. Ellison, W.R. Scheidt, *Inorg Chem* **1998**, *37*, 382–383.

65 T.G. Traylor, V.S. Sharma, *Biochemistry* **1992**, *31*, 2847–2849.

66 Y. Zhao, P.E. Brandish, D.P. Ballou, M.A. Marletta, *Proc Natl Acad Sci USA* **1999**, *96*, 14753–14758.

67 E.A. Dierks, S.Z. Hu, K.M. Vogel, A.E. Yu, T.G. Spiro, J.N. Burstyn, *J Am Chem Soc* **1997**, *119*, 7316–7323

68 W.Q. Zeng, N.J. Silvernail, D.C. Wharton, G.Y. Georgiev, B.M. Leu, W.R. Scheidt, J.Y. Zhao, W. Sturhahn, E.E. Alp, J.T. Sage, *J Am Chem Soc* **2005**, *127*, 11200–11201.

69 J.L. Calderon, F.A. Cotton, B.G. Deboer, N. Martinez, *Chem Commun* **1971**, 1476.

70 R.C. Elder, F.A. Cotton, F.A. Schunn, *J Am Chem Soc* **1967**, *89*, 3645.

71 T. Nakajima, I. Shimizu, K. Kobayashi, Y. Wakatsuki, *Organometallics* **1998**, *17*, 262–269.

72 T. Beringhelli, G. Ciani, G. Dalfonso, H. Molinari, A. Sironi, M. Freni, *Chem Commun* **1984**, 1327–1329.

73 E.P. Kyba, M.C. Kerby, R.P. Kashyap, J.A. Mountzouris, R.E. Davis, *J Am Chem Soc* **1990**, *112*, 905–907.

74 C.P. Gibson, J.S. Huang, L.F. Dahl, *Organometallics* **1986**, *5*, 1676–1681.

75 M.D. Carducci, M.R. Pressprich, P. Coppens, *J Am Chem Soc* **1997**, *119*, 2669–2678.

76 J. Lee, A.Y. Kovalevsky, I.V. Novozhilova, K.A. Bagley, P. Coppens, G.B. Richter-Addo, *J Am Chem Soc* **2004**, *126*, 7180–7181.

77 P. Coppens, I. Novozhilova, A. Kovalevsky, *Chem Rev* **2002**, *102*, 861–883.

78 P.P. Paul, Z. Tyeklár, A. Farooq, K.D. Karlin, S. Liu, J. Zubieta, *J Am Chem Soc* **1990**, *112*, 2430–2432.

79 S. Carrier, C.E. Ruggiero, W.B. Tolman, G.B. Jameson, *J Am Chem Soc* **1992**, *114*, 4408–4410.

80 C.E. Ruggiero, S.M. Carrier, W.E. Antholine, J.W. Whittaker, C.J. Cramer, W.B. Tolman, *J Am Chem Soc* **1993**, *115*, 11285–11298.

81 E.I. Tocheva, F.I. Rosell, A.G. Mauk, M.E.P. Murphy, *Science* **2004**, *304*, 867–870.

82 S.V. Antonyuk, R.W. Strange, G. Sawers, R.R. Eady, S.S. Hasnain, *Proc Natl Acad Sci USA* **2005**, *102*, 12041–12046.

83 I.H. Wasbotten, A. Ghosh, *J Am Chem Soc* **2005**, *127*, 15384–15385.

84 A. Ghosh, *Acc Chem Res* **2005**, *38*, 943–954.

85 J. N. Armor, *Environ Catal* **1994**, *552*, 2–6.

86 K. S. Sun, K. C. Kong, C. H. Cheng, *Inorg Chem* **1991**, *30*, 1998–2004.

87 A. Kudo, M. Steinberg, A. J. Bard, A. Campion, M. A. Fox, T. E. Mallouk, S. E. Webber, J. M. White, *J Catal* **1990**, *125*, 565–567.

88 M. Kubota, K. J. Evans, C. A. Koerntgen, J. C. Marsters, *J Am Chem Soc* **1978**, *100*, 342–343.

89 M. Kubota, K. J. Evans, C. A. Koerntgen, J. C. Marsters, *J Mol Catal* **1980**, *7*, 481–490.

90 C. D. Meyer, R. Eisenberg, *J Am Chem Soc* **1976**, *98*, 1364–1371.

91 R. Eisenberg, C. D. Meyer, *Acc Chem Res* **1975**, *8*, 26–34.

92 J. Reed, R. Eisenberg, *Science* **1974**, *184*, 568–570.

93 B. F. G. Johnson, S. Bhaduri, *Chem Commun* **1973**, 650–651.

94 A. T. McPhail, P. M. Gross, G. R. Knox, C. G. Robertson, G. A. Sim, *J Chem Soc A* **1971**, 205.

95 B. L. Haymore, J. A. Ibers, *J Am Chem Soc* **1974**, *96*, 3325–3327.

96 S. Bhaduri, B. F. G. Johnson, C. J. Savory, J. A. Segal, R. H. Walter, *Chem Commun* **1974**, 809–810.

97 D. E. Hendriksen, R. Eisenberg, *J Am Chem Soc* **1976**, *98*, 4662–4664.

98 C. H. Cheng, K. S. Sun, *Inorg Chem* **1990**, *29*, 2547–2548.

99 S. K. Wolfe, C. Andrade, Jh. Swinehar, *Inorg Chem* **1974**, *13*, 2567–2572.

100 P. G. Douglas, R. D. Feltham, H. G. Metzger, *J Am Chem Soc* **1971**, *93*, 84.

101 F. Bottomley, E. M. R. Kiremire, *Dalton Trans* **1977**, 1125–1131.

102 F. Bottomley, J. R. Crawford, *J Am Chem Soc* **1972**, *94*, 9092–9095.

103 N. E. Katz, M. A. Blesa, J. A. Olabe, P. J. Aymonino, *J Inorg Nucl Chem* **1980**, *42*, 581–585.

104 D. Littlejohn, S. G. Chang, *Ind Eng Chem Res* **1990**, *29*, 10–14.

105 M. R. McCarthy, T. J. Crevier, B. Bennett, A. Dehestani, J. M. Mayer, *J Am Chem Soc* **2000**, *122*, 12391–12392.

106 S. Martinengo, G. Ciani, A. Sironi, B. T. Heaton, J. Mason, *J Am Chem Soc* **1979**, *101*, 7095–7097.

107 M. Tachikawa, J. Stein, E. L. Muetterties, R. G. Teller, M. A. Beno, E. Gebert, J. M. Williams, *J Am Chem Soc* **1980**, *102*, 6648–6649.

108 D. Braga, B. F. G. Johnson, J. Lewis, J. M. Mace, M. McPartlin, J. Puga, W. J. H. Nelson, P. R. Raithby, K. H. Whitmire, *Chem Commun* **1982**, 1081–1083.

109 M. A. Collins, B. F. G. Johnson, J. Lewis, J. M. Mace, J. Morris, M. McPartlin, W. J. H. Nelson, J. Puga, P. R. Raithby, *Chem Commun* **1983**, 689–691.

110 D. E. Fjare, D. G. Keyes, W. L. Gladfelter, *J Organomet Chem* **1983**, *250*, 383–394.

111 D. E. Fjare, W. L. Gladfelter, *J Am Chem Soc* **1981**, *103*, 1572–1574.

112 D. E. Fjare, W. L. Gladfelter, *Inorg Chem* **1981**, *20*, 3533–3539.

113 W. L. Gladfelter, *Adv Organometal Chem* **1985**, *24*, 41–86.

114 N. D. Feasey, S. A. R. Knox, *Chem Commun* **1982**, 1062–1063.

115 D. E. Fjare, W. L. Gladfelter, *J Am Chem Soc* **1984**, *106*, 4799–4810.

116 C. P. Gibson, L. F. Dahl, *Organometallics* **1988**, *7*, 543–552.

117 A. L. Odom, C. C. Cummins, J. D. Protasiewicz, *J Am Chem Soc* **1995**, *117*, 6613–6614.

118 P. Legzdins, S. J. Rettig, K. J. Ross, J. E. Veltheer, *J Am Chem Soc* **1991**, *113*, 4361–4363.

119 D. Sellmann, J. Keller, M. Moll, C. F. Campana, M. Haase, *Inorg Chim Acta* **1988**, *141*, 243–252.

120 M. D. Bartberger, W. Liu, E. Ford, K. M. Miranda, C. Switzer, J. M. Fukuto, P. J. Farmer, D. A. Wink, K. N. Houk, *Proc Natl Acad Sci USA* **2002**, *99*, 10958–10963.

121 P. J. Farmer, F. Sulc, *J Inorg Biochem* **2005**, *99*, 166–184.

122 K. M. Miranda, *Coord Chem Rev* **2005**, *249*, 433–455.

123 J. S. Stamler, D. J. Singel, J. Loscalzo, *Science* **1992**, *258*, 1898–1902.

124 N. Paolocci, W. F. Saavedra, K. M. Miranda, C. Martignani, T. Isoda, J. M.

Hare, M.G. Espey, J.M. Fukuto, M. Feelisch, D.A. Wink, D.A. Kass, *Proc Natl Acad Sci USA* **2001**, *98*, 10463–10468.

125 K.R. Grundy, C.A. Reed, W.R. Roper, *Chem Commun* **1970**, 1501.

126 M.E. Jacox, D.E. Milligan, *J Mol Spectrosc* **1973**, *48*, 536–559.

127 R.D. Wilson, J.A. Ibers, *Inorg Chem* **1979**, *18*, 336–343.

128 F.W. Dalby, *Can J Phys* **1958**, *36*, 1336.

129 R. Melenkivitz, G.L. Hillhouse, *Chem Commun* **2002**, 660–661.

130 F. Sulc, C.E. Immoos, D. Pervitsky, P.J. Farmer, *J Am Chem Soc* **2004**, *126*, 1096–1101.

131 C.E. Immoos, F. Sulc, P.J. Farmer, K. Czarnecki, D.F. Bocian, A. Levina, J.B. Aitken, R.S. Armstrong, P.A. Lay, *J Am Chem Soc* **2005**, *127*, 814–815.

132 K. Ogura, H. Ishikawa, *Faraday Trans I* **1984**, *80*, 2243–2253.

133 F. Bottomley, M. Mukaida, *Dalton Trans* **1982**, 1933–1937.

134 J.N. Armor, M.Z. Hoffman, *Inorg Chem* **1975**, *14*, 444–446.

135 J. Armor, *Inorg Chem* **1973**, *12*, 1959–1961.

136 W.R. Murphy, K. Takeuchi, M.H. Barley, T.J. Meyer, *Inorg Chem* **1986**, *25*, 1041–1053.

137 W.R. Murphy, K.J. Takeuchi, T.J. Meyer, *J Am Chem Soc* **1982**, *104*, 5817–5819.

138 J.E. Toth, F.C. Anson, *J Am Chem Soc* **1989**, *111*, 2444–2451.

139 J.F. Rusling, A.E.F. Nassar, *J Am Chem Soc* **1993**, *115*, 11891–11897.

140 A.E.F. Nassar, W.S. Willis, J.F. Rusling, *Anal Chem* **1995**, *67*, 2386–2392.

141 A.E.F. Nassar, J.M. Bobbitt, J.D. Stuart, J.F. Rusling, *J Am Chem Soc* **1995**, *117*, 10986–10993.

142 Z.Q. Lu, Q.D. Huang, J.F. Rusling, *J Electroanal Chem* **1997**, *423*, 59–66.

143 M. Bayachou, R. Lin, W. Cho, P.J. Farmer, *J Am Chem Soc* **1998**, *120*, 9888–9893.

144 D. Mimica, J.H. Zagal, F. Bedioui, *Electrochem Commun* **2001**, *3*, 435–438.

145 C.E. Immoos, J. Chou, M. Bayachou, E. Blair, J. Greaves, P.J. Farmer, *J Am Chem Soc* **2004**, *126*, 4934–4942.

146 M.T. de Groot, M. Merkx, A.H. Wonders, M.T.M. Koper, *J Am Chem Soc* **2005**, *127*, 7579–7586.

147 W.G. Zumft, *J Inorg Biochem* **2005**, *99*, 194–215.

148 P. Monne-Loccoz, O.-M. H. Richter, H.-W. Huang, I.M. Wasser, R.A. Ghiladi, K.D. Karlin, S. de Vries, *J Am Chem Soc* **2000**, *122*, 9344–9345.

149 P. Moënne-Loccoz, S. de Vries, *J Am Chem Soc* **1998**, *120*, 5147–5152.

150 K.L.C. Grönberg, M.D. Roldán, L. Prior, G. Butland, M.R. Cheesman, D.J. Richardson, S. Spiro, A.J. Thomson, N.J. Watmough, *Biochemistry* **1999**, *38*, 13780–13786.

151 E. Pinakoulaki, S. Gemeinhardt, M. Saraste, C. Varotsis, *J Biol Chem* **2002**, *277*, 23407–23413.

152 T.D. Ju, A.S. Woods, R.J. Cotter, P. Moënne-Loccoz, K.D. Karlin, *Inorg Chim Acta* **2000**, *297*, 362–372.

153 I.M. Wasser, H.W. Huang, P. Moenne-Loccoz, K.D. Karlin, *J Am Chem Soc* **2005**, *127*, 3310–3320.

154 P.R. Ortiz de Montellano (Ed.), *Cytochrome P-450: Structure, Mechanism and Biochemistry*, 2nd edn, Plenum, New York, **1995**.

155 A. Daiber, H. Shoun, V Ullrich, *J Inorg Biochem* **2005**, *99*, 185–193.

156 R. Silaghi-Dumitrescu, D.M. Kurtz, L.G. Ljungdahl, W.N. Lanzilotta, *Biochemistry* **2005**, *44*, 6492–6501.

157 R. Silaghi-Dumitrescu, K.Y. Ng, R. Viswanathan, D.M. Kurtz, *Biochemistry* **2005**, *44*, 3572–3579.

158 D.M. Kurtz, R. Silaghi-Dumitrescu, A. Das, G.N.L. Jameson, L.G. Ljungdahl, B.H. Huynh, *J Inorg Biochem* **2003**, *96*, 174.

159 L.A. Ridnour, D.D. Thomas, D. Mancardi, M.G. Espey, K.M. Miranda, N. Paolocci, M. Feelisch, J. Fukuto, D.A. Wink, *Biol Chem* **2004**, *385*, 1–10.

160 J.L. Schneider, S.M. Carrier, C.E. Ruggiero, J. Young, V.G., W.B. Tolman, *J Am Chem Soc* **1998**, *120*, 11408–11418.

161 P. Gans, *A* **1967**, 943–946.

162 T. Yoshimura, *Inorg Chim Acta* **1984**, *83*, 17–21.

163 M.K. Ellison, C.E. Schulz, W.R. Scheidt, *Inorg Chem* **1999**, *38*, 100–108.

164 K.J. Franz, S.J. Lippard, *J Am Chem Soc* **1999**, *121*, 10504–10512.
165 K.M. Kadish, V.A. Adamian, E. Van Caemelbecke, Z. Tan, P. Tagliatesta, P. Bianco, T. Boschi, G.B. Yi, M.A. Khan, G.B. Richter-Addo, *Inorg Chem* **1996**, *35*, 1343–1348.
166 K.M. Miranda, X. Bu, I. Lorkovic, P.C. Ford, *Inorg Chem* **1997**, *36*, 4838–4848.
167 G.G. Martirosyan, A.S. Azizyan, T.S. Kurtikyan, P.C. Ford, *Chem Commun* **2004**, 1488–1489.
168 K.J. Franz, S.J. Lippard, *J Am Chem Soc* **1998**, *120*, 9034–9040.
169 W.B. Hughes, *Chem Commun* **1969**, 1126.
170 J.L. Griggs, K.N. Rao, L.H. Jones, R.M. Potter, *J Mol Spectrosc* **1968**, *25*, 34.
171 J.H. Callomon, F.R. Creutzberg, *Phil Trans Roy Soc London A* **1974**, *277*, 157–189.
172 M.W. Chase, C.A. Davies, J.R. Downey, D.J. Frurip, R.A. McDonald, A.N. Syverud, *J Phys Chem Ref Data* **1985**, *14*, 19–26.
173 J.N. Armor, H. Taube, *Chem Commun* **1971**, 287–288.
174 J.N. Armor, H. Taube, *J Am Chem Soc* **1970**, *92*, 2560–2561.
175 J.N. Armor, H. Taube, *J Am Chem Soc* **1969**, *91*, 6874–6876.
176 A.A. Diamantis, G.J. Sparrow, M.R. Snow, T.R. Norman, *Aust J Chem* **1975**, *28*, 1231–1244.
177 A.A. Diamantis, G.J. Sparrow, *J Colloid Interf Sci* **1974**, *47*, 455–458.
178 A.A. Diamantis, G.J. Sparrow, *Chem Commun* **1970**, 819–820.
179 A.A. Diamantis, G.J. Sparrow, *Chem Commun* **1969**, 469.
180 F. Bottomley, W.V.F. Brooks, *Inorg Chem* **1977**, *16*, 501–502.
181 D.F.T. Tuan, J.W. Reed, R. Hoffmann, *Theochem J Mol Struct* **1991**, *78*, 111–121.
182 D.F. Tuan, R. Hoffmann, *Inorg Chem* **1985**, *24*, 871–876.
183 F. Paulat, T. Kuschel, C. Nather, V.K.K. Praneeth, O. Sander, N. Lehnert, *Inorg Chem* **2004**, *43*, 6979–6994.
184 M.J. Almond, A.J. Downs, R.N. Perutz, *Inorg Chem* **1985**, *24*, 275–281.
185 P.L. Bogdan, J.R. Wells, E. Weitz, *J Am Chem Soc* **1991**, *113*, 1294–1299.
186 C.B. Pamplin, E.S.F. Ma, N. Safari, S.J. Rettig, B.R. James, *J Am Chem Soc* **2001**, *123*, 8596–8597.
187 R.G.S. Banks, R.J. Henderson, J.M. Pratt, *Inorg Phys Theor* **1968**, 2886–2889.
188 F. Bottomley, H.H. Brintzinger, *Chem Commun* **1978**, 234–235.
189 F. Bottomley, D.F. Drummond, D.E. Paez, P.S. White, *Chem Commun* **1986**, 1752–1753.
190 F. Bottomley, P.S. White, *Chem Commun* **1981**, 28–29.
191 F. Bottomley, G.O. Egharevba, I.J.B. Lin, P.S. White, *Organometallics* **1985** *4*, 550–553.
192 F. Bottomley, D.E. Paez, P.S. White, *J Am Chem Soc* **1982**, *104*, 5651–5657.
193 J.T. Groves, J.S. Roman, *J Am Chem Soc* **1995**, *117*, 5594–5595.
194 G.A. Vaughan, G.L. Hillhouse, A.L. Rheingold, *J Am Chem Soc* **1990**, *112*, 7994–8001.
195 A.W. Kaplan, R.G. Bergman, *Organometallics* **1998**, *17*, 5072–5085.
196 A.W. Kaplan, R.G. Bergman, *Organometallics* **1997**, *16*, 1106–1108.
197 J.R. Bleeke, R. Behm, *J Am Chem Soc* **1997**, *119*, 8503–8511.
198 A. Yamamoto, S. Kitazume, L.S. Pu, S. Ikeda, *J Am Chem Soc* **1971**, *93*, 371.
199 P.P. Notte, *Top Catal* **2000**, *13*, 387–394.
200 D.P. Ivanov, M.A. Rodkin, K.A. Dubkov, A.S. Kharitonov, G.I. Panov, *Kinet Catal* **2000**, *41*, 771–775.
201 N.S. Ovanesyan, A.A. Shteinman, K.A. Dubkov, V.I. Sobolev, G.I. Panov, *Kinet Catal* **1998**, *39*, 792–797.
202 G.I. Panov, A.K. Uriarte, M.A. Rodkin, V.I. Sobolev, *Catal Today* **1998**, *41*, 365–385.
203 K.A. Dubkov, V.I. Sobolev, G.I. Panov, *Kinet Catal* **1998**, *39*, 72–79.
204 H.F. Liu, R.S. Liu, K.Y. Liew, R.E. Johnson, J.H. Lunsford, *J Am Chem Soc* **1984**, *106*, 4117–4121.
205 C.E. Laplaza, A.L. Odom, W.M. Davis, C.C. Cummins, J.D. Protasiewicz, *J Am Chem Soc* **1995**, *117*, 4999–5000.

206 P.A. Hintz, M.B. Sowa, S.A. Ruatta, S.L. Anderson, *J Chem Phys* **1991**, *94*, 6446–6458.

207 J.C. Kramlich, W.P. Linak, *Prog Energy Combust Sci* **1994**, *20*, 149–202.

208 G. Centi, L. Dall'Olio, S. Perathoner, *Appl Catal A* **2000**, *194*, 79–88.

209 N. Furuya, H. Yoshiba, *J Electroanal Chem* **1991**, *303*, 271–275.

210 J.P. Collman, M. Marrocco, C.M. Elliott, M. Lher, *J Electroanal Chem* **1981**, *124*, 113–131.

211 I. Taniguchi, T. Shimpuku, K. Yamashita, H. Ohtaki, *Chem Commun* **1990**, 915–917.

212 M. Bayachou, L. Elkbir, P.J. Farmer, *Inorg Chem* **2000**, *39*, 289–293.

213 K. Brown, K. Djinovic-Carugo, T. Haltia, I. Cabrito, M. Saraste, J.J.G. Moura, I. Moura, M. Tegoni, C. Cambillau, *J Biol Chem* **2000**, *275*, 41133–41136.

214 T. Haltia, K. Brown, M. Tegoni, C. Cambillau, M. Saraste, K. Mattila, K. Djinovic-Carugo, *Biochem J* **2003**, *369*, 77–88.

215 P. Chen, S.I. Gorelsky, S. Ghosh, E.I. Solomon, *Angew Chem Int Ed* **2004**, *43*, 4132–4140.

216 S.I. Gorelsky, S. Ghosh, E.I. Solomon, *J Am Chem Soc* **2006**, *128*, 278–290.

217 M. Sommerhalter, R.L. Lieberman, A.C. Rosenzweig, *Inorg Chem* **2005**, *44*, 770–778.

3
Bio-organometallic Approaches to Nitrogen Fixation Chemistry

Jonas C. Peters and Mark P. Mehn

3.1
Introduction – The N_2 Fixation Challenge

Biological and industrial nitrogen fixation are the processes by which one of the most inert molecules, N_2, is transformed into a bioavailable nitrogen source (e.g. NH_3) that can be incorporated into all nitrogen-containing biomolecules [1]. As such, nitrogen fixation is essential to sustaining life on this planet, and has attracted intense scrutiny among biological and chemical communities for decades [2–4]. The mechanism by which nitrogenase enzymes promote the biological reduction of nitrogen under ambient conditions remains an unsolved and fascinating problem [4–6]. Nature's solution to fixing nitrogen stands in sharp contrast to that which humanity has adopted, i.e. the Haber-Bosch process for ammonia synthesis that is carried out at approximately 200 atm and 500 °C. This Herculean feat of chemical engineering provides, in net, about 50% of the nitrogen atoms that wind up in all human beings on this planet [7]. From a chemist's perspective, however, this remains a brute force and high-energy solution to the nitrogen fixation problem, especially given the knowledge that nature can fix N_2 under ambient conditions.

Chemists began to become more keenly aware of the apparent shortcomings of Haber-Bosch chemistry around 1965, when Allen and Senoff reported their landmark discovery that N_2 could coordinate as a ligand to a transition metal as they established for the complex ($[(NH_3)_5Ru(N_2)]^{2+}$) [8, 9]. It was already known that metal-rich enzymes in nature could catalyze nitrogen fixation under ambient conditions and therefore it could be reasonably postulated that synthetic metal catalysts might also be able to catalyze nitrogen reduction under suitable conditions. After all, the reduction of plenty of related substrates (e.g. acetylene and ethylene) could already be mediated by synthetic metal catalysts. An exciting surge of research activity followed the Allen and Senoff discovery, and within just a decade profound progress had been made towards the ultimate goal of viable synthetic catalysts for nitrogen fixation under ambient conditions. Literally dozens of dinitrogen complexes were prepared using various transition

Activation of Small Molecules. Edited by William B. Tolman

ISBN: 3-527-31312-5

metals [2]. This established an unexpectedly rich chemistry for N_2 as a ligand throughout much of the d block. Additionally, Shilov and his collaborators (of the former Soviet Union) reported the exciting discovery that certain aqueous transition metal mixtures containing, for example, molybdenum precursors mixed with $Mg(OH)_2$, could in fact catalyze N_2 reduction to hydrazine and ammonia in the presence of reducing agents such as sodium amalgam [3, 10]. In a complementary fashion, Chatt and his team in England discovered a family of low-valent, phosphine-supported molybdenum and tungsten complexes that could mediate the stoichiometric reduction of N_2 to NH_3 [2]. These systems were better suited to detailed solution studies and solid-state characterization than those of Shilov. The work of Chatt's research group, particularly their successful isolation and identification of many partially reduced N_xH_y complexes that might be formed *en route* to N_2 reduction at molybdenum and tungsten centers, led to their proposal of the now famous "Chatt cycle" for N_2 reduction, a scheme that to this day occupies a central role in N_2 research. Indeed, it was eventually demonstrated (in 1985) that one of these systems (based on tungsten) could be rendered modestly catalytic by using a potentiostat to reduce the tungsten center with electrons once the nitrogen molecule had been protonated and released as either N_2H_4 or NH_3, triggering the uptake of another N_2 molecule for further reduction [11]. By any account, these and the many related discoveries that were reported in the years that followed the Allen and Senoff discovery helped to define a truly remarkable period for research in inorganic and organometallic chemistry.

Perhaps not surprisingly – there had been so much work done by so many talented chemists in such a short period of time – the "principle of initial optimization" [12] took hold in the synthetic N_2 reduction field until the 1990s. While many systems were further developed that helped to establish what was, in rough outline, relatively well understood by then, it proved difficult to move towards a truly catalytic system. This is not to say there was not forward progress, but the pace of this progress slowed considerably until the 1990s. Fortunately for those now working in this area, a renaissance of interest and activity in the development of synthetic nitrogen-fixing catalysts occurred. Much of this activity can perhaps be traced to the report by Rees and coworkers in 1992 on the structure of the active site of the MoFe nitrogenase enzyme [6]. This picture, more than any other piece of data, was uniquely able to engender creative mechanistic proposals. This was good news for the synthetic community, because molecular complexes were needed more than ever to test the chemical viability of certain assumptions and proposals, and to provide the necessary databank of spectroscopic signatures associated with the diverse intermediate species of envisioned catalytic cycles.

In this chapter, we focus predominantly on recent experimental and theoretical work that specifically pertains to modeling aspects of biological nitrogen fixation. As numerous thorough and recent reviews have already been written on this topic [2, 4, 6, 13–15], and as there are reviews that also cover inorganic nitrogen activation chemistry with a more broadly defined scope [15, 16], we will emphasize experimental systems that appear to be functionally most related to intermediates of biological nitrogen fixation. To further limit the scope of this chapter, only molybdenum and iron model systems are discussed in detail. A few comments that directly pertain to biological nitrogen fixation, and the trials and tribulations involved in studying this complicated enzymatic reaction, are first in order.

3.2
Biological N_2 Reduction

3.2.1
General Comments

Biological nitrogen reduction is an unusually difficult biocatalytic transformation to study because the nitrogenase apparatus needs to be partially loaded with electron (and presumably proton) equivalents before substrate uptake can occur [5]. The exact number of electrons (and protons) transferred at the earliest stage of dinitrogen reduction remains unclear, but is likely to be as many as three or four [5]. Moreover, while the isolated MoFe protein of MoFe nitrogenase exhibits nitrogenase activity *in vitro*, the presence of various cofactors (e.g. $MgCl_2$, ATP, and hexose kinase) in large molar excess relative to the protein itself, an excess of the obligatory Fe protein and NADH as an electron source are necessary to attain activity. This situation makes the identification of highly reduced intermediate states a daunting task. Nonetheless, a suite of biochemical, spectroscopic and crystallographic studies has revealed a great deal about nitrogenase enzymes. Some of the most recent spectroscopic data available will be discussed below.

All nitrogenases feature a cofactor that is generally regarded as the site of N_2 binding and reduction. For the purposes of this chapter, we assume that the transformations converting nitrogen to ammonia occur at the cofactor. The nitrogenase cofactors known are comprised of molybdenum and iron, vanadium and iron, or iron alone. The MoFe nitrogenases are generally presumed to be the most efficient [17, 18], but some degree of cautious interpretation is prudent with respect to this fundamental assumption. Establishing the exact stoichiometry for activity of a nitrogenase enzyme is difficult. FeFe and VFe nitrogenases remain difficult to purify, and the measurement of the catalytic stoichiometries is invariably dependent on the exact conditions of the experiment.

$$N_2 + 8\,H^+ + 8e^- + 16\,MgATP \rightarrow 2\,NH_3 + H_2 + 16\,MgADP + 16\,P_i$$

The molybdenum-containing nitrogenase was the first to be identified and is the most thoroughly studied enzyme of the three known nitrogenase types. The solid-state crystal structure of the MoFe protein, obtained from *Azetobacter vinelandii*, has been solved in both a dithionite-reduced and an oxidized form. The enzyme consists of two components, one that couples ATP hydrolysis to electron transfer (the Fe protein) and a second (the MoFe protein) where the reduction of dinitrogen to ammonia occurs. This component is an $\alpha_2\beta_2$ tetramer that contains two unique metal clusters: a P-cluster, thought to be involved in the electron transfer pathway, and an FeMo cofactor (FeMoco), generally regarded as the site of N_2 binding and reduction. Despite abundant structural, spectroscopic and biochemical data now available, we still do not definitively know which metal(s) initially bind(s) dinitrogen, what metal(s) oxidation state(s) promotes binding nor which reduced nitrogen-based intermediates are generated (e.g. N^{3-}, NH^{2-}, N_2H, N_2H_3, etc.) *en route* to ammonia formation.

3.2.2
Structural Data

Detailed structural knowledge of the FeMoco provided an invaluable piece of data in considering the mechanism of biological nitrogen fixation. The high-resolution FeMo protein crystal structure recently published by Rees' group provided the best structural data to date regarding the FeMoco, in what is presumably its $S=3/2$ dithionite-reduced state (Fig. 3.1) [19]. The cofactor contains seven iron centers and a single molybdenum center that are held together by nine bridging sulfides ($MoFe_7S_9$) and a single interstitial light X-atom that has been postulated to be a nitride (N^{3-}). This assignment remains problematic.

On the basis of recent electron-nuclear double resonance (ENDOR) studies, Hoffman and Seefeldt suggested that the interstitial X-atom is not exchanged during catalysis [20, 21]. This would seem to suggest that the X-atom, even if it is a nitride, is not derived from N_2 during turnover. Most recently, Hoffman and Seefeldt have gone further in their analysis and have proposed that the central light atom is *not* an N-atom [21]. They noted a loss of ^{14}N electron

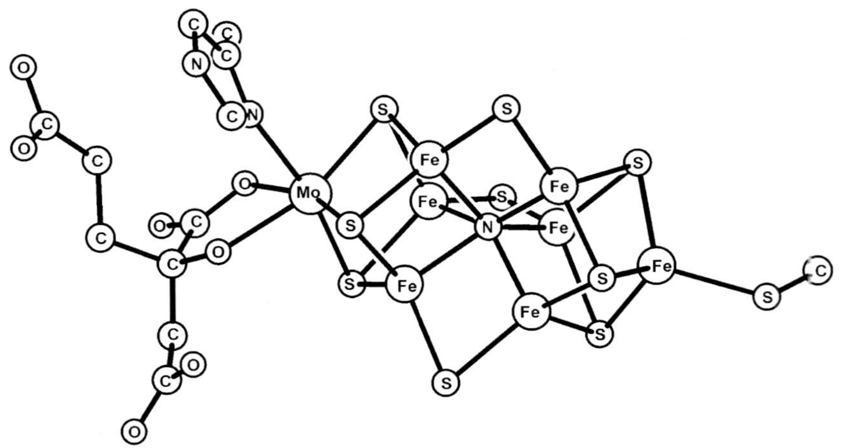

Fig. 3.1 Structural representation of the FeMoco at 1.16-Å resolution (1M1N.pdb).

spin-echo envelope modulation (ESEEM) and ^{14}N ENDOR in the resting-state FeMo protein after the FeMoco had been extracted into *N*-methylformamide, and also a loss of ^{15}N ENDOR signals in an R-70 Gly MoFe protein upon extraction of the FeMoco. These observations, according to their reasoning, show that the $^{14/15}N$ ENDOR signals observed from the resting-state MoFe protein all arise from protein-bound N nuclei, not from the cofactor itself. If one assumes that the interstitial light atom is coupled to the electron-spin system, with a coupling constant that exceeds around 0.1 MHz, then the spectroscopic methods employed would be expected to reveal the interaction. That no such interaction is observed is highly consistent with the absence of a central N-atom, leaving an O-, C- or B-atom as the next most likely candidates. The correct assignment of the central X-atom no doubt remains an issue for future clarification.

The cofactor is attached to its surrounding protein matrix by only two linkages: a histidine residue (His442) that is covalently tethered to the axially positioned molybdenum center and a cysteine residue (Cys275) that is covalently attached to the opposite iron center (Fig. 3.1). The geometry of each central iron atom and the axial iron nucleus is best described as pseudotetrahedral. The axial molybdenum center is six-coordinate and features a κ^2-homocitrate ligand, in addition to its His442 and three μ-sulfide linkages. Homocitrate is required for enzymatic activity and even replacing it with citrate will shut down nitrogenase activity [4]. A number of residues participate in weak interactions that may aid in controlling the activity of the FeMoco (e.g. His195).

3.2.3 Assigning the FeMoco Oxidation States

Knowledge of the relevant valence states within the cofactor is mechanistically important. The Hoffman and Münck groups have independently proposed valence assignments for the inorganic portion of the cofactor in its $S=3/2$ state using Q-band ENDOR and Mössbauer spectroscopies [22, 23]. While a two-electron discrepancy exists in their originally proposed assignments ($Mo^{4+}/Fe^{3+}/6Fe^{2+}$ versus $Mo^{4+}/3Fe^{3+}/4Fe^{2+}$), it is known that during turnover conditions at least three or four Fe protein cycles occur. Each Fe protein cycle involves an association of the Fe protein with MoFe protein which results in electron transfer from the Fe to the MoFe protein with concomitant hydrolysis of MgATP. Each cycle is followed by a dissociation and re-reduction of the Fe protein, dissociation of MgADP, and binding of ATP to poise the enzyme for another cycle [4–6]. Presumably three or four reduction events occur *prior* to N_2 uptake and further reduction. Very little information is available regarding the specific properties of the more reduced states of the cofactor, because directly studying these states has been experimentally so difficult. To a large extent, educated suppositions have been the dominant vehicle for conversations about key mechanistic aspects of biological nitrogen fixation.

3.3 Biomimetic Systems that Model Structure and Function

3.3.1 General Comments

Well-defined transition metal complexes that are amenable to detailed solution studies provide a critical tool to test whether certain mechanistic proposals are chemically feasible, at least with respect to fundamental reaction steps under a specific set of conditions. Systems that are possibly biomimetic can also provide spectroscopic insights that motivate more biologically relevant experiments by exposing key spectroscopic signatures indicative of certain intermediates. The most common *biomimetic* modeling approach applied to nitrogenase has involved the synthesis of well-defined mono- and/or dinuclear complexes in which one or two metal sites are supported by a set of donor ligand auxiliaries. While such complexes represent an obvious departure from the cofactor itself, the approach greatly simplifies spectroscopic studies and allows local structure–function relationships to be more clearly identified because variables can be altered in a systematic fashion via chemical synthesis.

3.3.2 Mononuclear Molybdenum Systems of Biomimetic Interest

As noted above, much of the early biomimetic work along these lines focused on Mo-N_2 (and related W-N_2) complexes due to the identification of molybdenum in the active site of FeMo nitrogenase [4]. Biochemical experiments had also suggested that nitrogenase activity was highly sensitive to alteration at or near the molybdenum site. For instance, simply changing the homocitrate ligand to citrate shuts down nitrogenase activity [4]. These data provided impetus for the development of functional model systems featuring molybdenum. Numerous molybdenum model systems were developed that established N_2 binding and activation, and that promoted its functionalization and even its direct cleavage [2, 13, 24–27]. In certain instances, nitrogen could be successfully protonated and reduced to hydrazine and ammonia using molybdenum and tungsten coordination complexes [2]. Some tungsten systems were elucidated that could reduce nitrogen electrocatalytically when conditions were appropriate [11]. Most recently, it has been shown that nitrogen fixation can be mediated catalytically using an organic proton source (lutidinium), an inorganic reductant [decamethylchromocene (Cp^*_2Cr)] and a sterically encumbered triamidoamine molybdenum scaffold [28, 29]. The early work of Chatt and his colleagues regarding low-valent, phosphine-supported molybdenum systems that mediated the conversion of N_2 to ammonia laid the foundation for many of these Mo-centered studies (Fig. 3.2). Because so many well-defined Mo-N_2H_x species were isolated and characterized by Chatt's group, and because a hypothetical but relatively

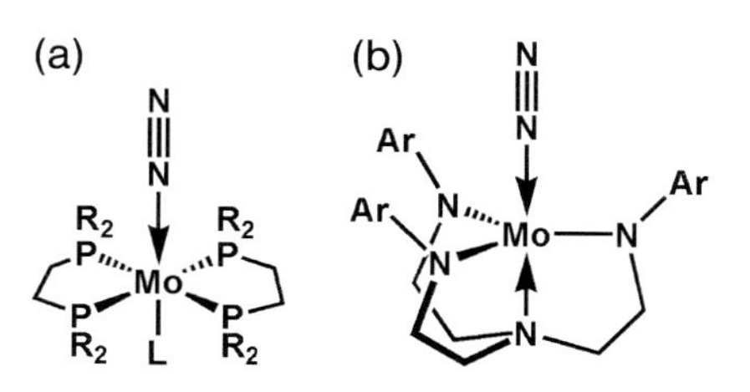

Fig. 3.2 (a) Chatt- and (b) Schrock-type molybdenum systems for N_2 reduction.

simple cycle was forwarded for N_2 reduction at a single metal center, synthetic molybdenum systems dominated the modeling arena for many years.

3.3.2.1 The Originally Proposed "Chatt Cycle"

Chatt and coworkers devised their original scheme for dinitrogen reduction in the 1970s. This work has been reviewed numerous times [2, 30, 31] and the cycle has been outlined in two recent reviews [15, 32]. Some important points and clarifications, however, are worth underscoring. Figure 3.3 illustrates a general (and *hypothetical*) Chatt-type cycle for dinitrogen reduction, constituting a heterolytic reduction process that shuttles between Mo(0) and Mo(III). Essential characteristics of the scheme shown here are: (i) dinitrogen adduct formation by a π-basic $L'P_4Mo(0)$ scaffold, (ii) conversion by formal H-atom transfer to a Mo(I) N_2H (diazenido) species, (iii) further reduction by a second H-atom transfer step to a terminally bonded Mo(II)-N_2H_2 (hydrazido) species and (iv) generation of NH_3 and a terminally bonded Mo(III)-(N) (nitride) species by a third H-atom transfer to the β-N-atom [which initially generates a Mo(I)-N $\leftarrow NH_3$ (hydrazidium) species that loses NH_3]. Continuing along this cycle, the terminal nitride complex is then transformed to a second NH_3 equivalent by successive H-atom transfers via intermediates including (v) Mo(II)-NH (imide), (vi) Mo(I)-NH_2 (amide) and (vii) Mo(0)-NH_3 (amine). An alternative pathway generates a Mo(I)-N_2H_3 complex by transfer of the third H-atom to the α-N-atom of the Mo(II)-N_2H_2 (hydrazido) species. This secondary cycle bypasses the terminally bonded nitride species by releasing ammonia from Mo(I)-N_2H_3 by H-atom transfer to its β-N-atom, which concomitantly generates the Mo(II)-NH (imide). If H-atom transfer instead occurs to the α-N-atom the hydrazine adduct Mo(0)-N_2H_4 is generated, accounting for the alternative byproduct of protonation, hydrazine. The secondary pathway has been suggested to be more prevalent in related tungsten phosphine systems that have been studied [2], and also in the $Cp^*MMe_3(N_2)$ (M = Mo, W) systems discussed below [33–39].

It is often assumed that certain features of this original Chatt cycle have been explicitly demonstrated, that some of the intermediates invoked have been isolated or at least spectroscopically observed and that some of the elementary transformations of the sequence have been demonstrated [15, 32]. While examples of these phosphine-supported molybdenum and tungsten N_2H_x and NH_x complexes have been thoroughly characterized (except perhaps an N_2H species,

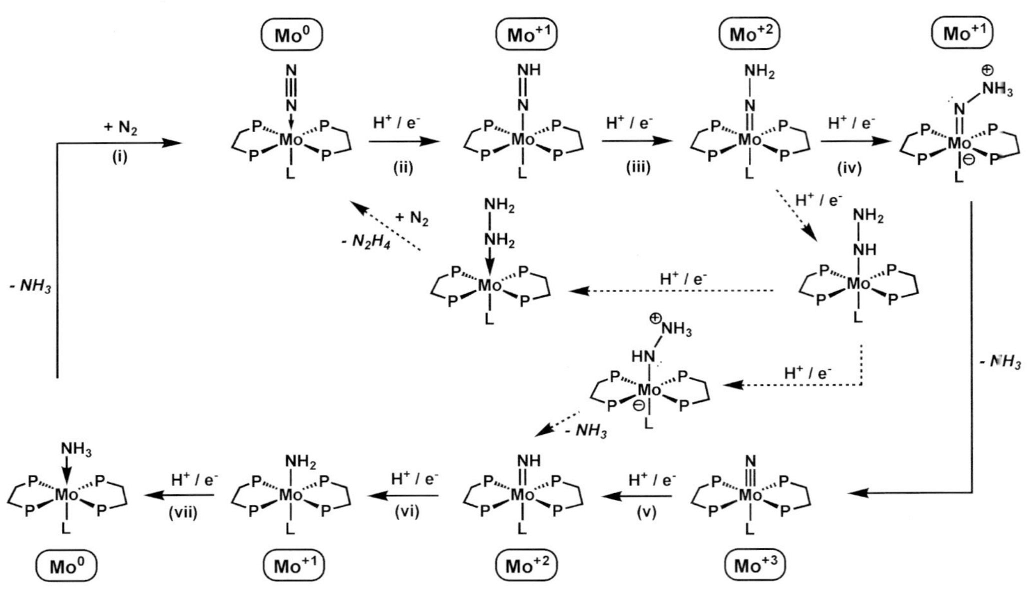

Fig. 3.3 The original Chatt cycle.

see below), in virtually *no case* is an example of such a species known in the oxidation state that is invoked by the original Chatt cycle. Some specific notes are worth providing for added clarification.

(i) Phosphine-supported octahedral Mo(0) and W(0) complexes are well known to bind N_2 due to their high π-basicity.

(ii) Nitrides of the general formula $LP_4M{\equiv}N$ (M = Mo, W) are known in the +4 oxidation state [40], but *not* the +3 oxidation state. This is presumably because a d^2 configuration places two electrons in a nonbonding d_{xy} orbital and leads to a stable configuration, whereas a d^3 configuration places a single unpaired electron in a much higher-lying $d_{xz,yz}$ set, a Jahn-Teller unstable configuration that decreases the $M{\equiv}N$ π-bond order and should lead to strong electronic destabilization.

(iii) Whereas the Chatt cycle shown in Fig. 3.3 invokes $LP_4Mo(I)$-N_2H and $LP_4Mo(I)$-NH_2 as intermediate species, examples of these species have only been reported in higher oxidation states. Moreover, while there are spectroscopic and electrochemical data for LP_4M-N_2H (M = Mo, W) diazenido species that are consistent with the assignments of such complexes [11, 41], such structures were never substantiated crystallographically. Indeed, it is possible that the data available are alternatively consistent with other isomeric forms (e.g. species where the hydride is coordinated to the metal center rather than the β-N-atom of ligated N_2).

(iv) Even for hydrazido species of the general formula $LP_4M{=}N$-NH_2, of which there are numerous thoroughly characterized examples that have been prepared by direct protonation of N_2, the molybdenum and tungsten centers are in the +4 oxidation state, not the +2 oxidation state [2, 42, 43].

(v) A similar situation exists for imide and hydrazidium species of the type $LP_4M{=}NH$ [40] and $LP_4M{\equiv}N{\leftarrow}NH_3$ [24, 44]. Well-defined examples of these types of species have been characterized, but not in the +2 and +1 oxidation states that are invoked in the hypothetical cycle [2].

Given this situation, the general scheme shown in Fig. 3.3, regardless of its relevance to biological nitrogen fixation, lacks synthetic precedent both with respect to its catalytic feasibility, and with respect to the nature of the specific intermediates and transformations implicated. Nonetheless, certain low-valent tungsten and, to a lesser extent, molybdenum systems of these general types can release significant quantities of NH_3 upon addition of strong acid. An early system that provided the best-reported NH_3 product yield was *cis*-$W(N_2)_2(PMe_2Ph)_4$. The stoichiometry of the reaction, as it was originally proposed [2, 45], is shown below. This stoichiometry is probably misleading because the PMe_2Ph ligands most likely undergo oxidation under the acidic alcohol conditions used. Given these numerous caveats, neither the operative redox chemistry nor the mechanism by which N_2 is stoichiometrically transformed into ammonia at these low-valent synthetic molybdenum and tungsten systems is well defined. Any connection to the synthetic intermediates shown in the Chatt cycle in Fig. 3.3 is therefore tenuous.

$$\begin{aligned} &cis\text{-}W(N_2)_2(PMe_2Ph)_4 + 8\,H_2SO_4 \\ &\quad \rightarrow N_2 + 2\,NH_3 + \{W(VI)\text{byproducts}\}^{4+} + 4[HPMe_2Ph][HSO_4] + 4\,HSO_4^- \end{aligned}$$

3.3.2.2 **An Electrocatalytic Reduction Cycle using Low-valent Tungsten**

While synthetic precedent for a catalytic NH_3 production cycle using the original Chatt-type systems was never realized using chemical reductants, Pickett and Talarmin established that a cycle could be demonstrated electrochemically [11]. More specifically, they were able to demonstrate that by first supplying a proton source (tosic acid) to *trans*-$(dppe)_2W(N_2)_2$ to generate $[(dppe)_2(OTs)W{=}N\text{-}NH_2][OTs]$, they could subsequently reduce $[(dppe)_2(OTs)W{=}N\text{-}NH_2][OTs]$ at a mercury-pool cathode (–2.6 V versus Fc/Fc^+) to reform $(dppe)_2W(N_2)_2$ in good yield along with substantial amounts of ammonia and a small amount of hydrazine. This manual cycle could be repeated several times and the authors were able to speculate on the presence of several intermediates. The most interesting intermediate they proposed to electrochemically observe was the diazenido species $[(dppe)_2(THF)W(N_2H)]^+$. Unfortunately, the overall stoichiometry of the transformation (shown below) was difficult to establish because H_2, one of the requisite byproducts, could not be detected. There was also evidence that radical chemistry involving the tetrahydrofuran (THF) solvent was operative during the reductive cycle.

$$16\,e^- + 8[(dppe)_2(OTs)W{=}N\text{-}NH_2]^+ + 9\,N_2 \rightarrow 2\,NH_3 + 8(dppe)_2W(N_2)_2 + 8\,TsO^- + 5\,H_2$$

3.3.2.3 A Mo(III)-mediated Catalytic N_2 Reduction System

It seems unlikely that very low-valent molybdenum species could play a mechanistic role during biological nitrogen fixation. A molybdenum center in the +4 or –3 oxidation state during turnover conditions is more plausible, and schemes for N_2 reduction at a single molybdenum center proceeding from such a complex have been postulated and probed experimentally [30]. The simplest and perhaps most plausible scenario with respect to the cofactor itself is to suggest that the homocitrate ligand that binds molybdenum in the resting state of the enzyme might be hemilabile, dechelating after electron loading (Fig. 3.4) [46]. Such a step would expose a coordination site at molybdenum, perhaps in the formal +3 state, for N_2 binding and subsequent reduction (Fig. 3.4, top cycle).

Schrock's group has spent a great deal of experimental effort trying to demonstrate the chemical feasibility of such a cycle at Mo and has most recently focused its attention on triamidoamine $[RN_3N]Mo$ platforms (Fig. 3.2) to specifically test whether catalytic N_2 fixation might proceed from a well-defined $[RN_3N]Mo(III)\text{-}N_2$ species. They have in fact reported one such system that is

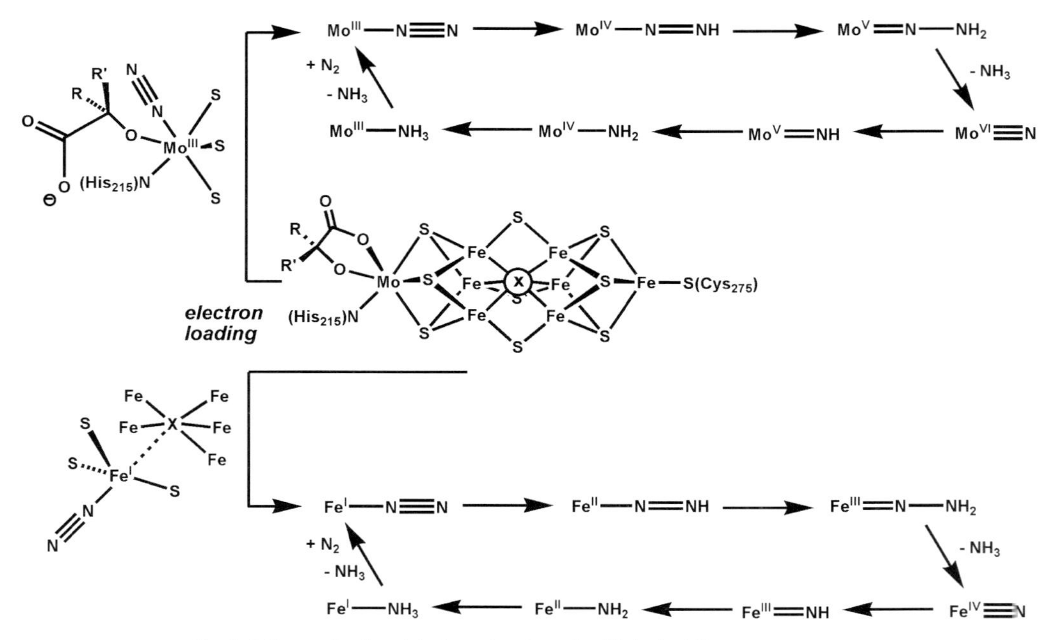

Fig. 3.4 Conceptually related mechanisms for biological N_2 fixation mediated by a single molybdenum or iron site.

capable of such a transformation, albeit with modest catalytic turnover (up to seven turnovers) and slow rates [28, 47]. This system is particularly fascinating because so many of the intermediates that are relevant to the proposed catalytic cycle can be independently generated and thoroughly characterized [29, 47, 48]. In this regard the Schrock system is distinct from all previously developed molybdenum N_2 systems.

The $[RN_3N]^{3-}$ auxiliary ligand (Fig. 3.5) is typically abbreviated as $[HIPTN_3N]^{3-}$. As the HIPT nomenclature indicates, the extremely encumbering hexaisopropylterphenyl substituents effectively preclude the formation of less reactive bimetallic $[RN_3N]Mo\text{-}N{=}N\text{-}Mo[RN_3N]$ species that would otherwise be a thermodynamic sink. Schrock and Yandulov were able to demonstrate catalytic turnover using the isolable $[HIPTN_3N]Mo\text{-}N_2$ precursor with [2,6-lutidinium]$^+$ {LutH} as an acid source with a compatible $[BAr'_4]^-$ noncoordinating counter anion [where $Ar' = 3,5\text{-}(CF_3)_2C_6H_3$] and Cp^*_2Cr as a stoichiometric reductant. These data have been carefully described [28, 29, 47, 49]. In the most favorable conditions, a 66% yield of NH_3 can be realized – the yield being based upon the addition of 36 equiv. Cp^*_2Cr (and 48 equiv. acid) relative to the starting molybdenum precursor. The catalysis is slow. It takes approximately 6 h to realize just four turnovers and only 8 equiv. of NH_3 has been generated in a single catalytic run. Competitive degradation reactions presumably poison the system over time and the possible role(s) of these degradation products, in addition to the Cp^*_2Cr reductant itself, with respect to the catalysis remains unclear.

The judicious choice of the nonpolar solvent heptane for the reaction appears to have been critical to the success of the Schrock system. The {LutH}$\{BAr'_4\}$ acid is poorly soluble in heptane and, as such, is only sparingly solubilized in the reaction slurry. This fact, along with a slow rate of addition of the Cp^*_2Cr reductant, circumvents to a large extent the direct reaction between Cp^*_2Cr and {LutH}$\{BAr'_4\}$ to generate hydrogen. It appears that the kinetics of electron transfer between Cp^*_2Cr and the various molybdenum species present during the reaction cycle are more favorable than those for the background H_2 evolution reaction that would be operative if Cp^*_2Cr and {LutH}$\{BAr'_4\}$ were to be

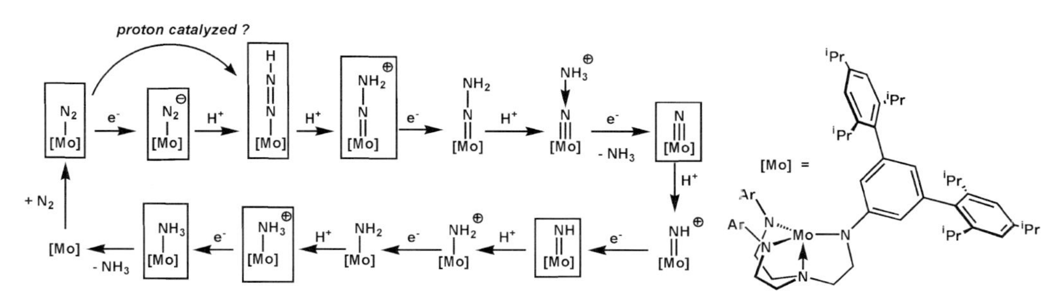

Fig. 3.5 Proposed intermediates and steps in the reduction of dinitrogen at a $[HIPTN_3N]Mo$ center. Characterized intermediates are enclosed in boxes.

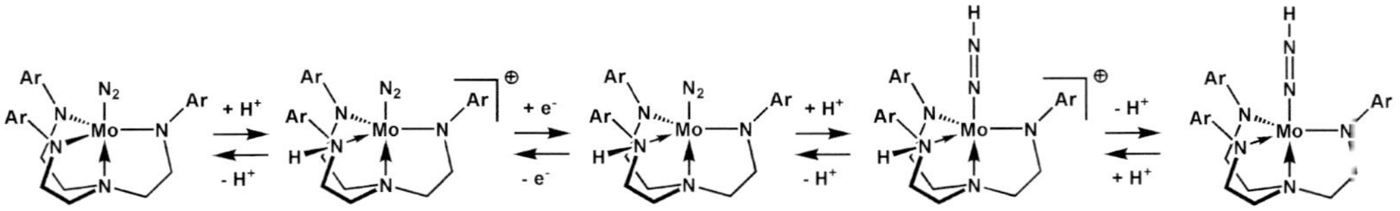

Fig. 3.6 Schrock's postulated proton-catalyzed reductive protonation.

mixed without any molybdenum species present. Control experiments including the quantification of the amount of background H_2 produced during the catalytic cycle and the exploration of different solvent systems would certainly be of interest for this remarkable catalytic system.

Several molybdenum complexes that are invoked as intermediates of the N_2 fixation cycle can be used as the initial catalyst precursors, instead of the starting N_2 complex. That these species provide similar overall results with respect to yields of reduced N_2 is consistent with the cycle proposed. Therefore, they are either (i) species generated along the N_2 fixation path or possibly (ii) "pre-catalysts" that can be chemically converted to the relevant active catalyst.

The complete catalytic cycle proposed by Schrock is illustrated in Fig. 3.5. Eight of the intermediate species in this cycle can be prepared and thoroughly characterized, and these are enclosed in boxes. Perhaps the most synthetically interesting and unusual species in the cycle shown is the parent diazenido species [$HIPTN_3N$]Mo(N_2H). This complex is prepared from [$HIPTN_3N$]Mo-N_2 by the addition of 1.0 equiv. [2,6-lutidinium][BAr'_4] and 2.0 equiv. Cp^*_2Cr, and is the only example of such a species to have been structurally characterized. Schrock has suggested the possibility that the first formal H-atom transfer step proceeds via a "proton-catalyzed" reductive protonation (Fig. 3.6). In other words, initial protonation at the $[HIPTN_3N]^{3-}$ ligand shifts the reduction potential of the complex anodically, allowing electron transfer to proceed more readily, and this electron transfer is then followed by protonation at N_2 and subsequent deprotonation at the $[HIPTN_3N]^{3-}$ ligand. From a mechanistic perspective, each of the isolable and presumed intermediate species can serve as a pre-catalyst

3.3.2.4 A $Cp^*MMe_3(N_2)$ Model System (M = Mo, W)

Prior to their emphasis of [RN_3N]Mo systems for N_2 fixation, Schrock and coworkers had extensively studied $Cp^*MMe_3(N_2)$ systems (M = Mo, W) that mediated the reduction of N_2 to ammonia in nearly quantitative fashion (but not catalytically) under certain conditions [33–39]. A possible mechanistic difference between systems of this latter type and the now better known [RN_3N] Mo systems concerns the terminal nitride intermediate. Ammonia release was presumed to proceed without generation of a terminal nitride species in the $Cp^*MMe_3(N_2)$ systems, as illustrated in Fig. 3.7. Note the similarity between this proposed pathway and the secondary Chatt pathway illustrated in Fig. 3.3. A detailed mechanistic study by Norton regarding the protonation of the species

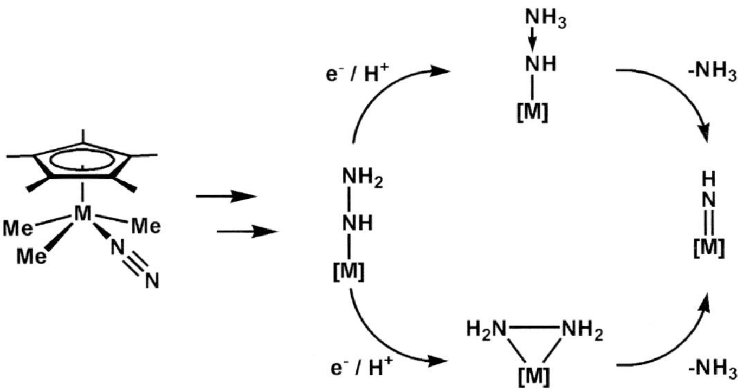

Fig. 3.7 Schrock's $Cp^*MMe_3(N_2)$ systems appear to avoid a terminally bonded nitride intermediate *en route* to NH_3 release.

$Cp^*WMe_3NNH_2$ established the β-N-atom to be the kinetic site of protonation to provide $\{Cp^*Me_3W{=}N \leftarrow NH_3\}^+$, although another isomeric form of this latter species, $\{Cp^*Me_3W(\eta^2\text{-}NH\text{-}NH_2)\}^+$, is also accessible [50].

3.3.2.5 **Bimetallic Molybdenum Systems that Cleave N_2**

One possible pathway for dinitrogen fixation in biology (and also industry) concerns the formal scission of dinitrogen to metal-bound nitride followed by the transfer of proton and electron equivalents to release ammonia. Whether and how this might happen at the cofactor of nitrogenase is unclear, but a number of transition metal complexes are now known that can mediate the formal reductive cleavage of dinitrogen [51]. These reactions typically require the formation of bimetallic dinitrogen adduct complexes and subsequent addition of further reducing equivalents from a secondary source, such as sodium or potassium graphite [52], a silane or other main group hydride [53–56], or even hydrogen itself [57]. The simplest dinitrogen cleavage that has been elucidated concerns Cummins' molybdenum tris(amido) complex $Mo(N[R]Ar)_3$ (Fig. 3.8) [26, 27]. Like the related $[RN_3N]Mo$ system of Schrock, $Mo(N[R]Ar)_3$ is able to bind dinitrogen at room temperature, although in this case reversibly and with an equilibrium constant that lies far to the side of the unbound species. In fact, at room temperature in ethereal solution only the unbound orange species $Mo(N[R]Ar)_3$ can be observed by nuclear magnetic resonance (NMR) spectroscopy. However, when such a solution is cooled to –35 °C over a period of several days, the dinuclear dinitrogen-bridged species $\{(N[R]Ar)_3Mo\}_2(\mu\text{-}N_2)$ is formed quantitatively. Upon warming, this complex breaks apart to generate two equivalents of the terminally bonded nitride $(N[R]Ar)_3Mo(N)$. Plausible symmetry arguments, in addition to supporting density functional theory (DFT) calculations, suggest that the cleavage reaction proceeds via a zig-zag transition state

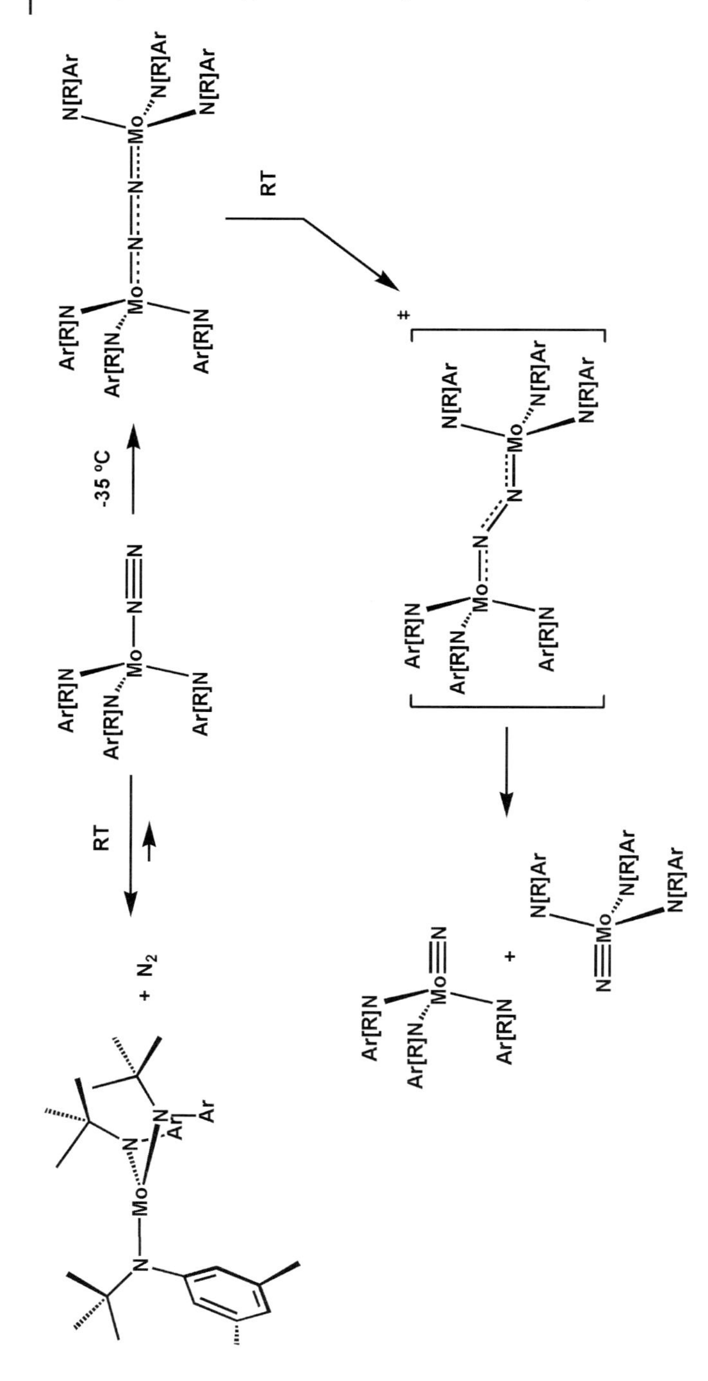

Fig. 3.8 Cummins' original dinitrogen cleavage reaction mediated by $Mo(N[R]Ar)_3$.

(Fig. 3.8) [27, 58]. While the terminal dinitrogen adduct species $(N[R]Ar)_3Mo(N_2)$ is not readily detected, it can be trapped by one-electron reductants such as Na/Hg or Ti(III) [59]. Indeed, the cleavage reaction can be facilitated rapidly at ambient temperature if even catalytic amounts of certain reductants (Na/Hg, BH_4^-) [59, 60] are added to THF solutions of $Mo(N[R]Ar)_3$. The protonation chemistry of these N_2 adduct species has not been reported, but competitive aminolysis of the auxiliary amido ligands would likely compete with or dominate productive protonation at the coordinated N_2 ligand.

3.3.2.6 Sulfur-supported Mo-N_2 Complexes

Despite the sulfur-rich environment of the nitrogenase cofactor, sulfur-supported transition metal complexes that provide well-defined dinitrogen adduct species are relatively uncommon. This statement is certainly true of molybdenum and iron systems. Indeed, we are aware of no examples of well-defined dinitrogen adducts of iron in which one of the supporting donor atoms is sulfur. For molybdenum the literature provides a limited number of examples, although relatively little is known regarding the utility of sulfur donors to facilitate dinitrogen reduction. One interesting exception concerns Yoshida's *trans*-$[Mo(N_2)_2(Me_8[16]\text{-ane-}S_4)]$ [61]. This complex has been crystallographically characterized and is reported to display reactivity patterns reminiscent of the *trans*-$Mo(N_2)_2P_4$ and *trans*-$W(N_2)_2P_4$ systems of Chatt and Hidai [43, 62]. Its preparation (Fig. 3.9) proceeded from sodium-mercury amalgam reduction of *trans*-$[Mo(Br)_2(Me_8[16]\text{-ane-}S_4)]$ in THF solution. Curiously, the cyclic voltammogram of *trans*-$[Mo(N_2)_2(Me_8[16]\text{-ane-}S_4)]$ displays an irreversible one-electron oxidation potential that is cathodically shifted relative to related phosphine systems. These data imply that the all sulfur macrocycle is more electron-releasing than the structurally related tetrakis phosphine donor sets (e.g. *trans*-$Mo(N_2)_2(R_2PCH_2CH_2PR_2)_2$ systems (R=Ph [63], Et [43]). While this is only a crude approximation given the irreversibility of the $Mo^{I/0}$ redox couple, comparison of the ν(N–N) stretching frequencies for these types of complexes also supports the idea that the $[Mo(Me_8[16]\text{-ane-}S_4)]$ unit is a better π-base than a related $Mo(R_2PCH_2CH_2PR_2)_2$ unit. In accord with this model, a β-N-atom can be doubly methylated by treatment of $[Mo(N_2)_2(Me_8[16]\text{-ane-}S_4)]$ with MeBr to generate

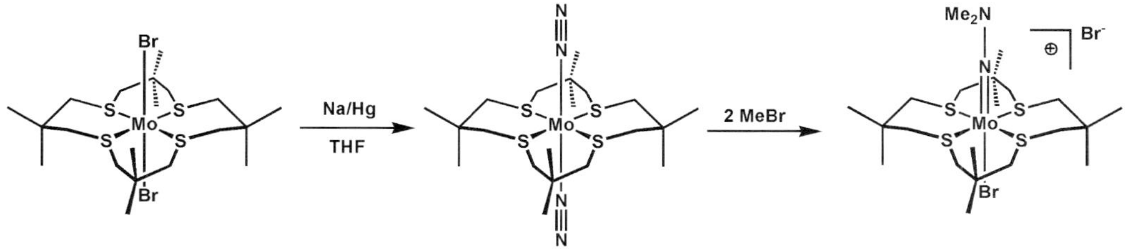

Br
S
Mo
S
Br
Na/Hg
THF
N
N
2 MeBr
Me_2N
N
Br^-

Fig. 3.9 Synthesis and reactivity of Yoshida's $[Mo(N_2)_2(Me_8[16]\text{-ane-}S_4)]$ system.

the hydrazido species $[Mo(NNMe_2)(Br)(Me_8[16]\text{-ane-}S_4)][Br]$ [61]. Methylation of the corresponding *trans*-$Mo(N_2)_2(Ph_2PCH_2CH_2PPh_2)_2$ complex only provides monomethylation to generate *trans*-$Mo(N_2Me)(Br)(Ph_2PCH_2CH_2PPh_2)_2$ [25].

3.3.3
Considering Mechanisms Involving Multiple and Single Iron Sites for N_2 Reduction

3.3.3.1 General Comments

The model studies discussed above, collectively, establish the chemical feasibility of a nitrogen reduction scheme to generate ammonia mediated by a single molybdenum center. At this stage it is of interest to alternatively consider iron-mediated N_2 reduction schemes, as these schemes have garnered a great deal of speculative attention with respect to the mode by which biology catalyzes this reaction. This attention has not only come from the experimental communities, but also from the many theorists that have employed *ab initio* and DFT methods to probe aspects of N_2 fixation. There are two limiting scenarios to consider for an iron-based N_2 fixation mechanism: one that invokes two or more iron centers intimately involved in the N_2 reduction process, and one that focuses on a single iron site mediating each of the key chemical steps *en route* to ammonia generation. A third scenario, one in which a single iron site initiates the N_2 reduction process, but at some later stage samples a bi- or multimetallic N_xH_y species, should also be considered.

The multiple-iron-site scenario has received the most attention during the past 15 years. The reasons for this situation seem clear. The FeMoco features seven iron centers, six of them in the center of the cluster. Prior to the elucidation of a central light atom in the cluster, each of these central iron centers was thought to be unusually low coordinate and, hence, likely reactive towards N_2. Also, the oxidation states known to be available to synthetic tetrahedral iron complexes have been typically limited to Fe^{2+} and Fe^{3+}. It has therefore been widely assumed that a single iron site would not likely support the various N_xH_y intermediates or the requisite oxidation states (at least four) and transformations that would need to be invoked *en route* to NH_3 production from N_2. This situation sharply contrasts with single-site molybdenum schemes, where a rich redox chemistry at molybdenum at first glance appears far more reasonable and for which numerous examples of intermediate N_xH_y complexes have been reported. Given this dichotomy and nature's employment of an unusual inorganic cluster featuring so many iron centers, many have perhaps reasonably assumed a scheme that invokes multiple iron centers. This point is especially evident in several theoretical DFT studies that have been undertaken [64–69].

3.3.3.2 Theoretical Studies that Invoke Iron-mediated Mechanisms

We felt it worthwhile to provide some level of theoretical overview given the number of DFT studies that have appeared recently and here limit the discussion to theoretical studies that have included an interstitial light atom in consideration of the FeMoco structure. Three of the most recent and comprehensive studies that have been reported in this regard come from the respective research groups of Nørskov, Blöchl and Ahlrichs [70–76]. Figures 3.10–3.12 depict, using chemical line representations with some degree of our own interpretation, the overall mechanisms for N_2 reduction that are postulated by these three DFT studies.

A few cautionary notes are worth highlighting prior to discussing the results of these efforts. Many of the fundamental conclusions of these DFT studies rest upon a number of critical assumptions that need to be made at the outset of any set of calculations. For instance, the number of electrons assigned to the cofactor, and therefore its net charge, invariably depends on how the central light atom is assigned. For simplicity, and based upon the availability of some experimental data (e.g. the Mössbauer work of Münck and coworkers) [22], most authors have recently opted to regard the cofactor unit as neutral and with a central N-atom providing the empirical formula $[MoFe_7S_9N]^0$ [22]. While the resting state FeMoco is thought to feature seven high-spin iron centers antiferromagnetically coupled to one another [22], spin is treated somewhat differently in each study. Additionally, each study examines the net transfer of H-atoms to the N_2 substrate. Blöchl, Nørskov and Ahlrichs have treated the successive proton and electron transfers as correlated events (i.e. PCET), in other words as net H-atom transfer steps. To consider the relative exo- or endothermicity of every H-atom transfer step, each of their studies has also needed to provide a basis for how the energies of "free" protons and electrons (or H-atoms) are handled within the protein matrix. Finally, in these three studies, and virtually every theoretical study that has been undertaken, the cofactor has been truncated with respect to the peripheral coordination environment. The homocitrate and histidine units bound to molybdenum have been approximated, as has the cysteine unit that ties the opposite iron center to the surrounding protein. Each of these simplifications is likely to impact the energies of various intermediates, and the barriers encountered along a given reaction coordinate. More critical, however, is that each DFT study by necessity removes the protein matrix surrounding the cofactor – a matrix that provides substantial asymmetry about the cofactor core that, through H-bonding interactions, may lower or raise the energy of certain intermediates and reaction barriers preferentially.

3.3.3.2.1 **Comparing Several Proposed Mechanisms**

In broad outline, the three mechanisms that can be distilled out of the theoretical studies referred to above differ in several key ways. Blöchl's work proposes an N_2 reduction model that invokes two iron centers as participants in N_2 coordination and reduction, whereas the Nørskov model invokes N_2 reduction mediated by a single iron site. Both of these models converge on one important point. Each assigns a hemi-labile role to the central light atom that frees a site(s) at the reactive iron center(s). A similar role for the central atom has been independently suggested by our group at Caltech to maintain tetra-coordination at iron upon N_2 binding and reduction, or possibly to sample a five-coordinate trigonal bipyramidal geometry [77, 78]. Blöchl's study goes further and also invokes dissociation of one of the central bridged sulfides. In all of these models, the central atom, assumed to be a nitride by both Blöchl and Nørskov, is *not* implicated as an intermediate species generated by N_2 reduction. Rather, it plays an important structural role. By contrast, the DFT study by Ahlrichs and coworkers ascribes an active role to the central N-atom, explicitly suggesting that it is an intermediate generated by loss of NH_3 from N_2. It is then further protonated to generate a second ammonia equivalent.

Each study underscores the energetically expensive, hence chemically challenging nature of the initial N_2 reduction step that generates N_2H. Both Nørskov and Blöchl propose initial axial N_2 binding, indicated by intermediates **B** in Fig. 3.10 and **C** in Fig. 3.11. In the Nørskov model, N_2 binding is accommodated by slippage of the participating iron center away from the central N-atom (Fe–N = 2.12 Å), the specific coordinates of which are depicted in Fig. 3.13, such that it obtains an approximately five-coordinate trigonal bypyramidal geometry. The participating iron center also shifts slightly away from the central sulfide unit that links it to the opposite iron center of the trigonal prism. Closer inspection of the minimized structure shown in Fig. 3.13 also reveals that the central N-atom has slipped considerably away from another iron center (Fe–N = 2.18 Å) at the stage of N_2 coordination. In the Blöchl model, N_2 binding is concurrent with displacement of a protonated sulfide linkage such that in the first intermediate of N_2 binding all iron centers remain pseudotetrahedral. Blöchl's study calculates intermediate **C** of Fig. 3.11 to be of slightly lower energy than bridged N_2 adduct **D**. However, the productive N_2 reduction path is proposed to proceed via bridged N_2H intermediate **E**, and hence an equilibrium distribution of structures **C** and **D** is proposed, with **D** being more likely suited to the first N_2 protonation step.

N_2 adducts of iron are known for both four- and five-coordinate geometries from the experimental literature [31, 77, 79–86], providing structural precedent for species such as intermediate **B** of Fig. 3.10 and **C** of Fig. 3.11. However, the structure of intermediate **D** in Fig. 3.11, which is depicted explicitly in Fig. 3.14, is to our knowledge without precedent in the literature. In fact, we are unaware of any precedent for a bimetallic N_2 adduct of a transition metal system that obtains a geometry that is grossly similar to the geometry of intermediate **D**. In general, an N_2 ligand is known to bridge two transition metals in one of two ways, either by forming a linear M-N-N-M linkage or coordinating in a side-on fashion such that each metal center symmetrically interacts with each N-atom [15]. There are a few exceptions to this statement. One of the more fascinating examples is an asymmetrically bridged ditantulum N_2 adduct [87], but the structure suggested for intermediate **D** of Fig. 3.11 and provided explicitly in Fig. 3.14 would be unique, at least to the best of our knowledge.

In any case, once N_2H has been generated, coordinated either axially or in a bridging fashion, subsequent H-atom transfers proceed more readily. In the Nørskov scheme, the second H-atom transferred to N_2 is delivered to the μ-N-atom providing a terminally bonded η^1-N_2H_2 adduct **D** (Fig. 3.10). While such a

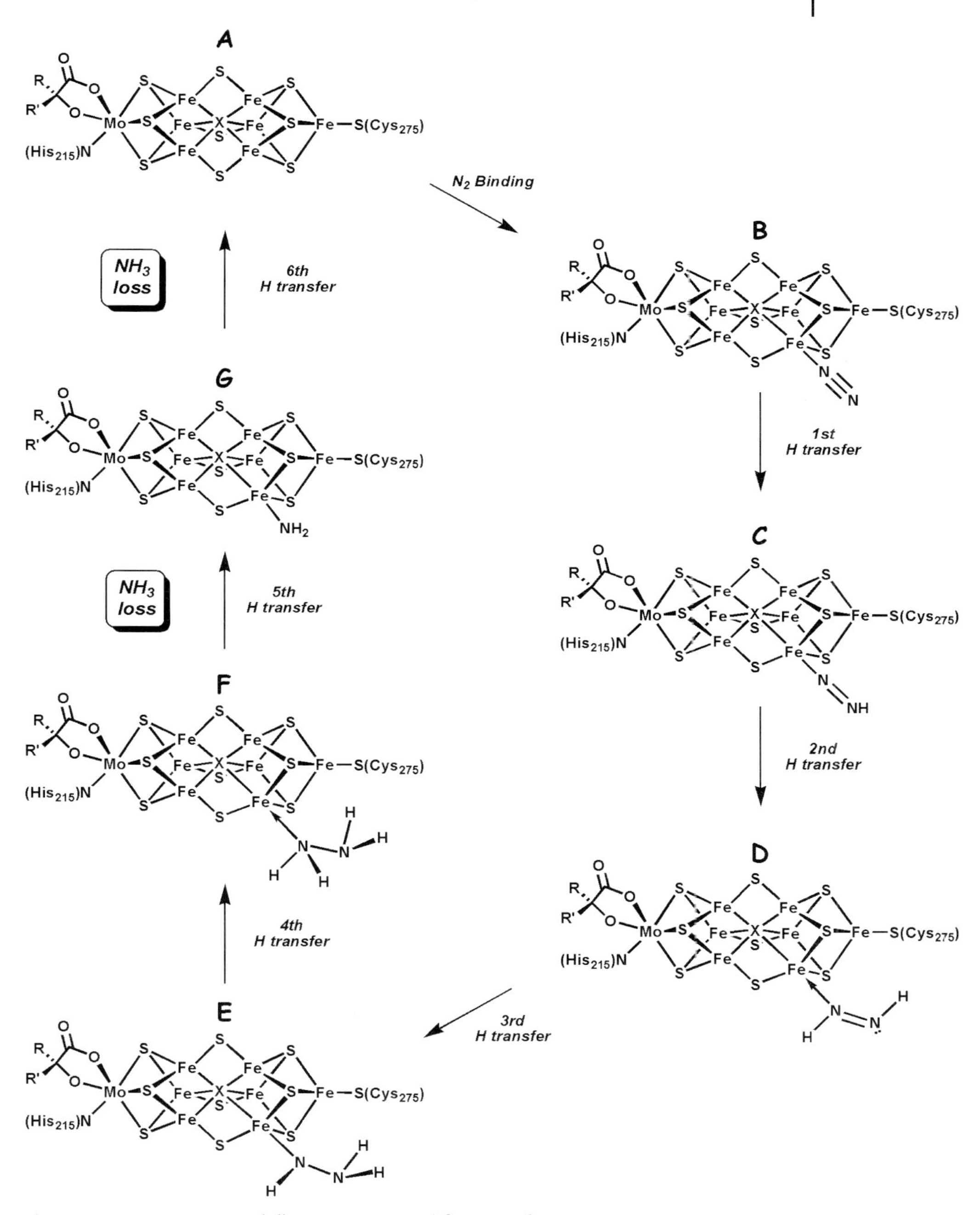

Fig. 3.10 Diiron site proposal illustrating essential features of the lowest energy path forwarded by Nørskov and coworkers.

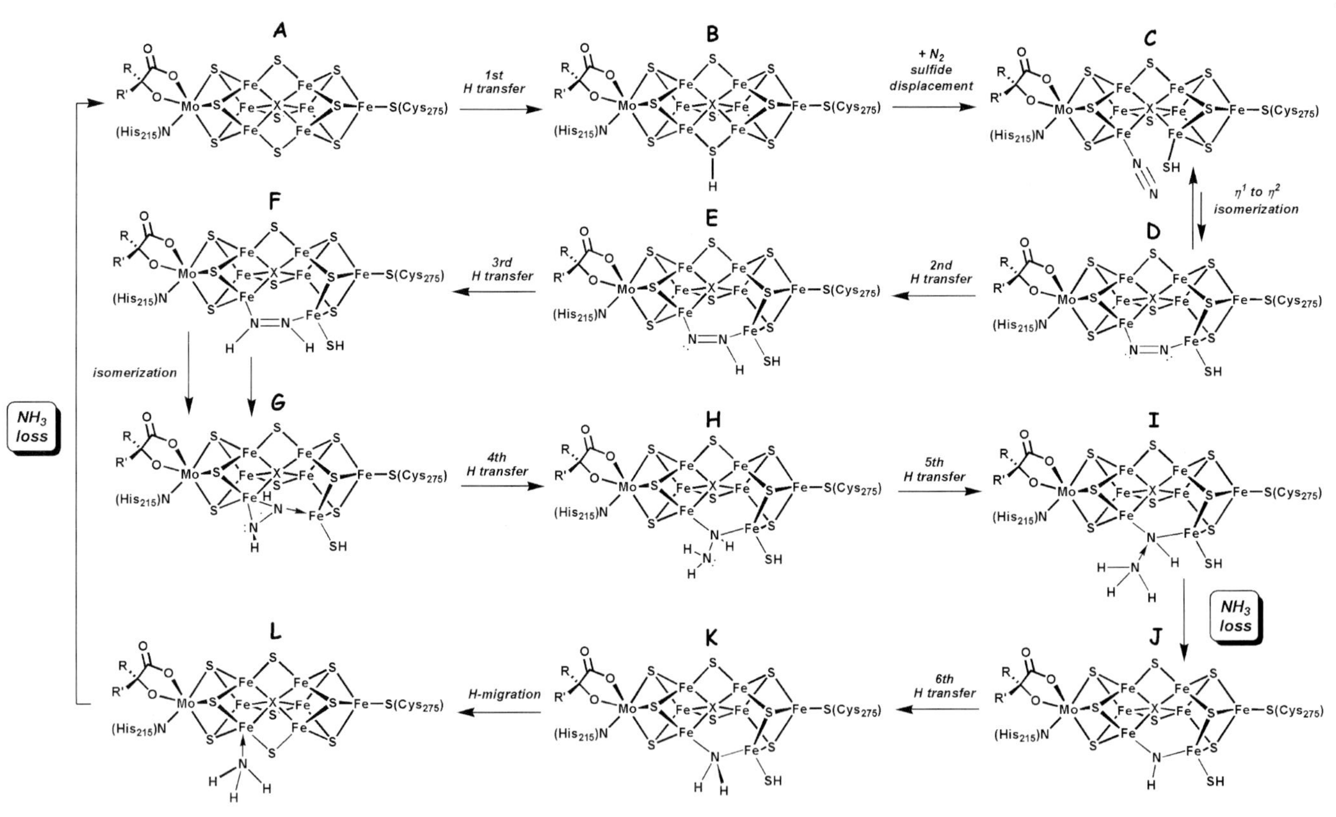

Fig. 3.11 Single iron site proposal illustrating essential features of the lowest energy path forwarded by Blöchl and coworkers.

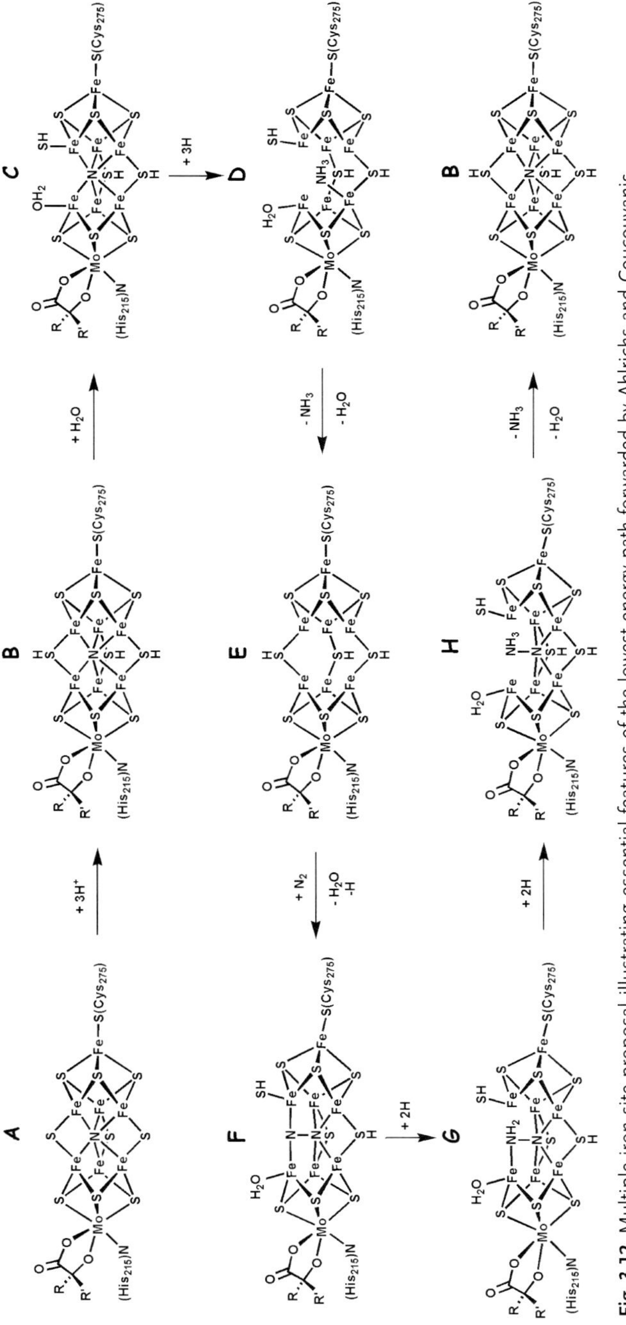

Fig. 3.12 Multiple iron site proposal illustrating essential features of the lowest energy path forwarded by Ahlrichs and Coucouvanis.

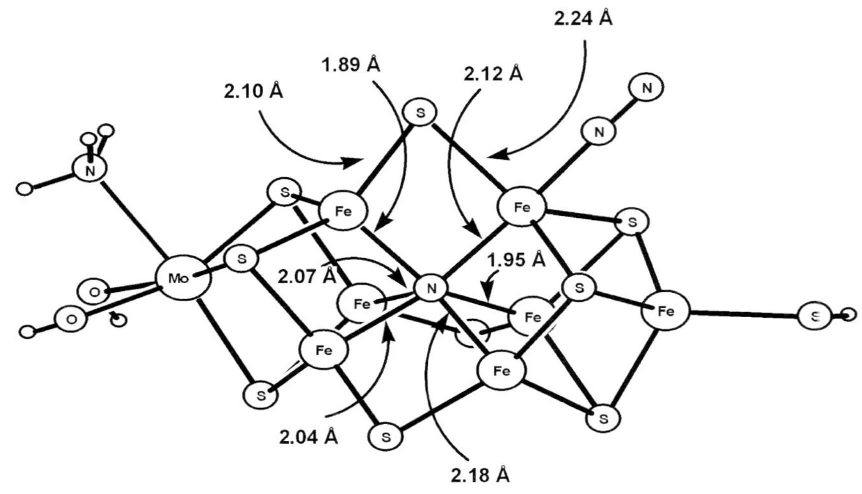

Fig. 3.13 Structural representation of the N_2 adduct of FeMoco obtained from minimized coordinates from the Nørskov study (intermediate **B** in Fig. 3.10). Arrows indicate bonds to which the distances provided refer.

species is synthetically unknown for iron, it is certainly a plausible intermediate and Hillhouse has provided precedent for exactly this type of species in a well-defined tungsten complex [88]. The third H-atom transferred generates the N_2H_3 adduct **E** containing a doubly protonated β-N-atom. The next intermediate in the cycle is that of the hydrazine adduct **F**. This species is further reduced to release the first equivalent of NH_3, which generates the terminally bonded amide (NH_2) species **G**. Species **G** is an H-atom transfer away from release of the second NH_3 equivalent and restarting the cycle. In the Blöchl scheme (Fig. 3.11), the symmetrically bridged diimide intermediate **F** is generated first from **E**, but this species isomerizes to an asymmetrically coordinate adduct **G** prior to being protonated at the terminal nitrogen position. This step sets up the generation and subsequent loss of the first NH_3 equivalent and leads to the structurally unusual diiron μ-NH intermediate **J**. While unusual, our group at Caltech has provided synthetic precedent for a diiron μ-NH species very recently [89]. Another H-atom transfer generates the μ-NH_2 species **K**, and proton transfer from the terminal SH to the bridging amide generates the second NH_3 equivalent and also re-establishes the μ-sulfide linkage. This final proton transfer might also be base mediated.

As noted above, the mechanistic scenario proposed by Ahlrichs and Coucouvanis (Fig. 3.12) invokes protonation at a central nitride atom and generation of ammonia from it. The mechanistic hypothesis therefore rests on the assignment of the central light atom as an N-atom. The cycle is set up by protonating each of the central μ-sulfide linkages of the starting cluster **A** to generate **B**, followed by opening of the cluster at one sulfide hinge by its displacement from

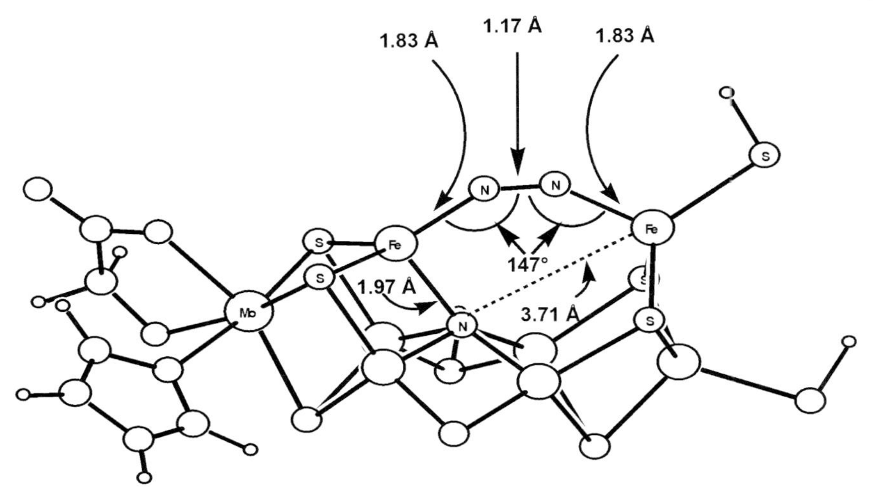

Fig. 3.14 Structural representation of intermediate C from Fig. 3.11 obtained from minimized coordinates from the Blöchl study.

one central iron center by coordination of H_2O. This generates **C**, which is now sufficiently opened-up to allow for protonation at the central atom generating the first NH_3 equivalent (**D**) and, upon its release as well as release of the water ligand, a cofactor cluster **E** which is devoid of a central light atom. Indeed, **E** represents the originally proposed cofactor structure. Water again displaces one sulfide linkage to open the cluster and N_2 coordination in the center of the cluster affords **F**. This N_2 adduct, which from a structural viewpoint is totally unprecedented, is poised for protonation at the terminal N-atom, and this process (**G** and **H**) generates a second ammonia equivalent and re-inserts the nitride subunit into the cofactor.

3.3.3.3 Synthetic Efforts to Model N_2 Reduction by Multiple Iron Sites

Although a great deal of effort has been put into the development of iron centers in sulfur-rich coordination spheres, as noted previously no examples of an iron center are known for which N_2 and an S-donor ligand are present. Sellmann and coworkers postulated some years ago an opening of the cofactor cluster that leads to a diiron N_2 reduction scheme [90]. His group thus designed a number of ligands that utilize thioethers and thiophenolates to try to model bimetallic N_2 activation schemes (Fig. 3.15) [90, 91]. Perhaps the most intriguing report was of a ("N_HS_4")Fe complex that supports the binding of CO, NH_3, N_2H_4 and N_2H_2. Oxidation of the ("N_HS_4")Fe(CO) complex with hydrazine led to the diiron diazene adduct shown in Fig. 3.15 [92]. This adduct is stabilized by a number of factors, including H-bonding interactions between the diazene protons and the ligand sulfides. While the coordination chemistry of Sellmann's

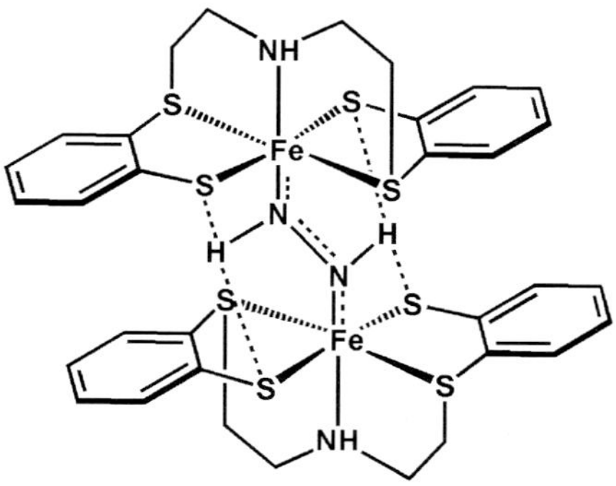

Fig. 3.15 Sellmann's iron diazene adduct.

sulfur-rich iron systems is fascinating, the reduction of nitrogenous substrates was never realized.

3.3.3.4 Nitrogenase-related Transformations at Cluster Models

To gain insight into possible polymetallic iron mechanisms for substrate reduction significant synthetic effort has also been placed into replicating the structure of the FeMoco (both prior to and following the observation of the interstitial atom). This synthetically challenging task builds upon a large body of work pertaining to the synthesis of iron sulfur cluster model compounds [14, 93, 94]. A recent review has been published that covers the assembly and physical properties of these complex metal sulfur clusters [93]. The high-spin nature of the iron centers, combined with the sheer complexity of the nitrogenase cofactor, makes realizing its synthesis an especially difficult goal. Although the exact cofactor structure has yet to be synthetically duplicated, there are a number of biomimetic FeMo clusters that capture several of its key structural parameters.

Coucouvanis' group has carried out several studies describing the synthesis of cuboidal $[MFe_3S_4]^{n+}$ (M = Mo, n = 3; M = V, n = 2) units and have subsequently shown that these complexes will react with certain nitrogenase substrates (Fig. 3.16) [95]. These moieties can, for example, catalytically reduce *cis*-dimethyldiazene in the presence of a reducing agent (e.g. $CoCp_2$) and proton source (2,6-lutidinium chloride). Methyl amine production ceases when substitutionally inert ligands block the metal coordination sphere. This implies that a coordinatively unsaturated heterometallic center is important to this reactivity. These complexes also carry out the reduction of acetylene to ethylene and ethane. Interestingly, $[Fe_4S_4Cl_4]^{2-}$, though capable of slowly reducing acetylene, is ineffective in the reduction of *cis*-dimethyldiazene. These studies appear to emphasize the importance of the heteroatom in the reaction with nitrogenase substrates. Yet none of these clusters are known to bind and reduce dinitrogen. These observations emphasize the importance of the protein environment in carrying out these transformations and open the possibility of different portions of the cofactor carrying out different tasks depending upon the state of N_2 reduction.

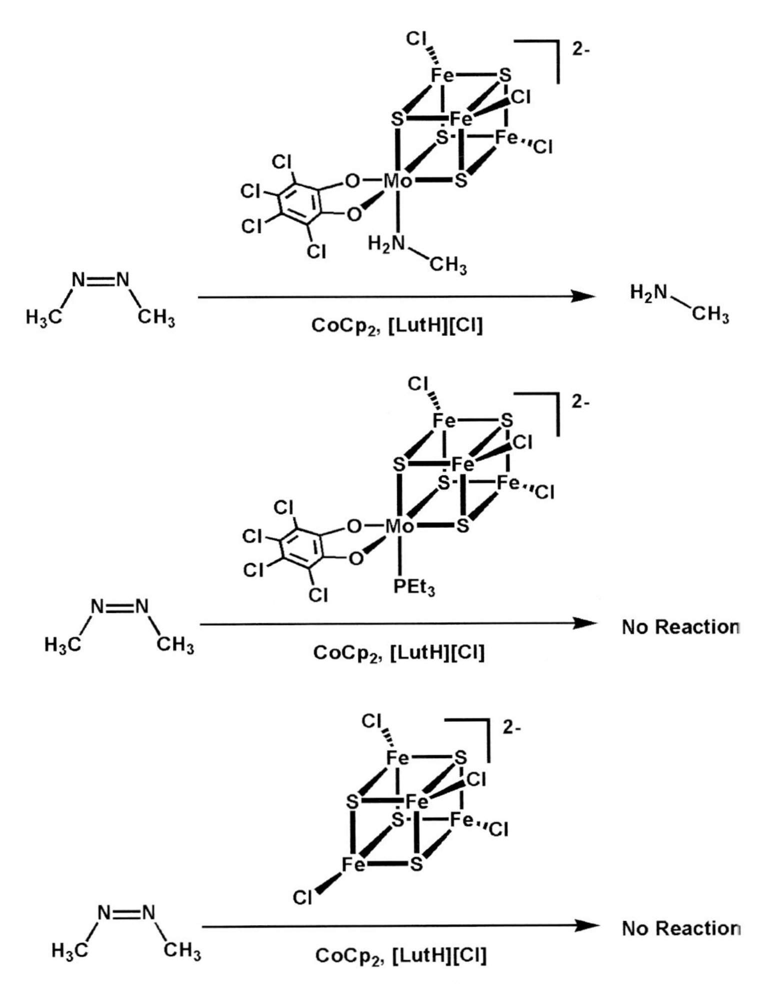

Fig. 3.16 Reactivity of Coucouvanis' heterometallic cubanes towards 1,2-dimethyldiazene with added reductant and protons.

These model complexes do not include an interstitial atom nor do they capture the geometry of the central iron core of the cofactor which recent calculations point to as vital for nitrogenase activity [73–76]. One possibility is that the central atom is a nitride; if this is so, then the coordination geometry is unprecedented in small-molecule model chemistry of iron nitrides. Holm and co-workers have recently begun to examine the coordination chemistry of iron nitrides to see if a nitride which bridges six iron centers can be obtained [96]. Literature precedent shows ample evidence for linear Fe–N–Fe bridges for octahedrally coordinated iron [97–100] and, more recently, a bent Fe–N–Fe bridge for pseudotetrahedral iron centers from our laboratory at Caltech (*vide infra*) [89,

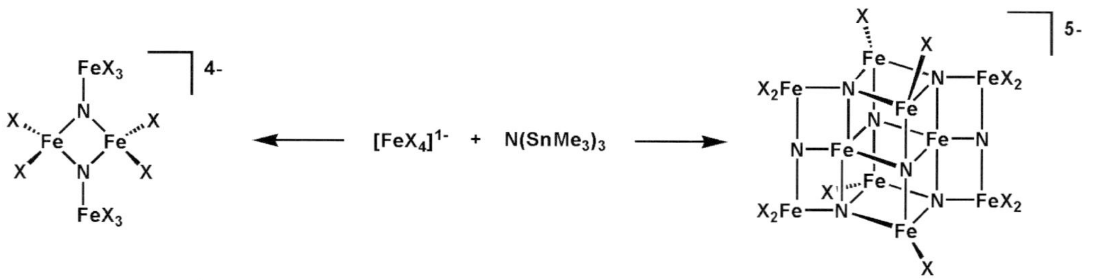

$[FeX_4]^{1-}$ + $N(SnMe_3)_3$

Fig. 3.17 Holm's nitride-bridged clusters.

101]. There is a report of a nitride bridging as many as four iron carbonyl clusters [102]. Holm's elegant work utilizes $N(SnMe_3)_3$ as a nitride transfer agent and examines the reactivity of simple iron halides with this reagent (Fig. 3.17) [96]. Variation of the stoichiometry of the reagents has resulted in the preparation of two different cluster geometries. The first of these is a tetranuclear assembly of high-spin Fe(III) atoms in which each nitride bridges three iron centers. The other form is a larger 10 iron cluster in which there are formally nine Fe(III) centers and a single Fe(IV) center. There are two types of nitrides in this cluster: μ_3-nitrides and μ_4-nitrides. It is hoped that future studies will reveal a similar bridging order to that proposed for the FeMoco and that the coordination chemistry will be complementary to the existing transformations of heterometallic metal–iron–sulfur clusters, thereby allowing the assembly of clusters that are structurally similar to the FeMoco.

Another recently emerging area of synthesis is that of imidoiron clusters (Fig. 3.18) [103–105]. Lee and coworkers have developed a series of weak field iron clusters supported by imide ligands displaying a wide range of cluster nuclearities (2–4) and mid- to high-valent iron oxidation states. Of particular inter-

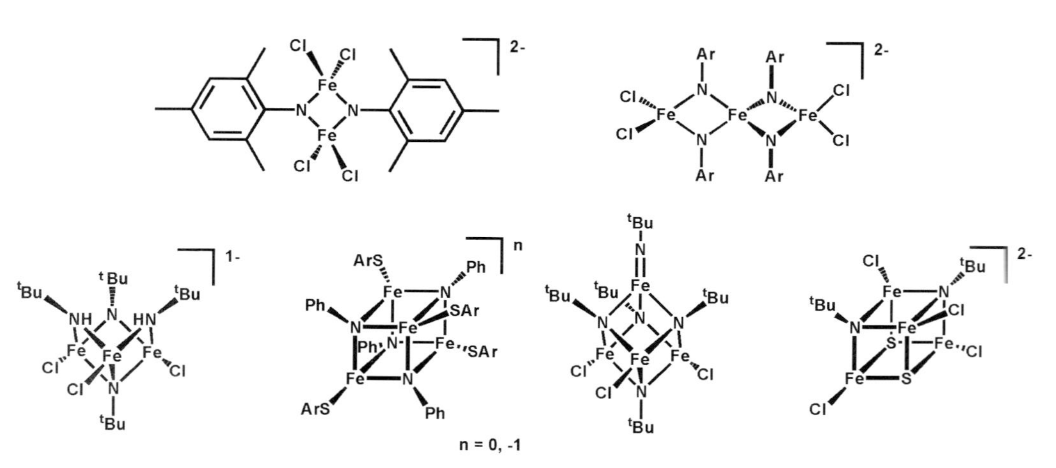

Fig. 3.18 Selected examples of imido-bridged clusters.

est are the monoanionic ($Fe^{II}Fe_3^{III}$) and neutral (Fe_4^{III}) cubanes, and the methodologies for the sequential installation of arylthiolates and bridging imides. In one instance, a cluster with a terminal iron imide was isolated in very low yield (1–2%) [104]. This cluster likely has a single Fe(IV) center and three Fe(III) sites. Spectroscopic measurements indicate that the valencies are localized in this cluster. The chemistry of these imido complexes correlates quite well with principles already established for iron–sulfur clusters and a few results have been reported of mixing the two methodologies to obtain complexes with both sulfide bridges and imido bridges [94].

3.3.3.5 Considering N_2 Fixation Involving a Scheme Single Iron Site

By analogy to the Chatt- and Schrock-type systems that have already been discussed (Section 3.3.2), it is possible that a single iron site is responsible for mediating the initial, and perhaps even all, of the steps leading to dinitrogen reduction in biology (Fig. 3.4, bottom cycle) [70, 78]. The conceptual outline for such a mechanism is appealing by its analogy to the mode of O_2 reduction by cytochrome P450 and related biocatalytic systems. P450-type systems are thought to bind and heterolytically reduce O_2 at a single iron site, after which reductive loss of water generates a high-valent Fe(IV)-oxo intermediate [i.e. $PFe^{II}O_2 \rightarrow PFe^{III}O_2^- \rightarrow PFe^{III}O_2H \rightarrow (P^{+\bullet})Fe^{IV}{=}O + H_2O$; P = porphyrin] [106]. The iron-mediated fixation scheme proposed in Fig. 3.4 (bottom cycle) is conceptually related, but in this case heterolytic release of ammonia would generate a high-valent iron nitride (e.g. $Fe^{I}N_2 + 3H^+ + 3e^- \rightarrow Fe^{IV}{\equiv}N + NH_3$). Alternative scenarios might lead to bridged Fe–N–Fe and Fe–NH–Fe species after release of the first NH_3 equivalent.

According to the single-iron-site cycle shown in Fig. 3.4, initial electron loading (constituting three or four Fe protein cycles) would render the cofactor sufficiently electron-rich to bind N_2. Substitution of the interstitial X-atom linkage, whatever its specific identity, by N_2 might coincide with slippage of the iron center into and perhaps through the S_3 plane provided by its three sulfide (or possibly protonated SH) linkages. This motion would serve to redirect the iron center's d orbitals to the outer face of the cluster to engage in a favorable Fe (d) $\rightarrow$ N_2 (π^*) back-bonding interaction. Bonds between the interstitial X-atom and the remaining five iron centers might contract to help maintain the cluster's overall structural integrity. It is also conceivable that an Fe–S linkage, rather than the Fe–X linkage, is broken during electron-loading conditions. The key point is that reduction of the cluster is likely to labilize some iron–ligand interactions that expose a site for N_2 uptake, possibly giving rise to a four-coordinate iron geometry or expansion to a five-coordinate species to a five-coordinate species upon substrate binding.

Once N_2 uptake occurs, the Fe-N_2 intermediate would be set up for further reduction. Regardless of the iron center's formal oxidation state at the *initial stage* of N_2 binding, it can be presumed that subsequent reduction to lower-valent iron (e.g. Fe^{I}-N_2) becomes more accessible due to the π-acidic character of N_2.

An illustrative example comes from the Cummins group who observed that the complex (^{t}Bu(Ar)N)$_3$Mo exhibits a $Mo^{III/II}$ reduction potential at –2.9 V relative to ferrocene in THF. However, upon binding N_2 [i.e. (^{t}Bu(Ar)N)$_3$Mo-N_2], its $Mo^{III/II}$ potential is shifted by 1.2 V (!) to –1.7 V [59].

Experimental data from the synthetic modeling community to support the chemical feasibility for a single iron site N_2 fixation scheme are sparse by comparison to the molybdenum data that are available. This is in part because less attention has been devoted to biomimetic iron systems than to those for molybdenum. Biochemical experiments have nevertheless motivated various iron-mediated mechanistic postulates. These iron-focused proposals derive some circumstantial support from site-directed mutagenesis studies at residues far removed from the molybdenum center that demonstrate a dramatic impact on substrate binding and reduction. Also, N_2 substrate analog studies by Seefeldt and coworkers [68] using unsaturated hydrocarbon substrates have been used to argue that nitrogen is likely fixed at an iron center. Most recently, the Hoffman and Seefeldt team has proposed the direct spectroscopic detection of cofactor bound intermediates of nitrogen fixation, and their analysis allows them to suggest one or perhaps two of the central iron centers act as the active metal sites [107, 108].

3.3.3.6 Model Studies that May be Relevant to N_2 Fixation Involving a Single Iron Site

3.3.3.6.1 Fe(0)-N_2 Complexes and NH_3 versus N_2H_4 Production

To probe the possibility of an N_2 fixation cycle mediated by a single iron site, it is instructive to consider several model studies. Leigh and coworkers demonstrated in the early 1990s that iron N_2 complexes supported by chelating bis-(phosphine) ligands (e.g. $Me_2PCH_2CH_2PMe_2$) could be protonated in THF to release small amounts of ammonia (Fig. 3.19) [31]. The yields they obtained were invariably low (0.12 equiv. per iron or less), and were moreover highly dependent on the choice of phosphine ligand, acid, solvent and the presence of co-additives such as $MgCl_2$. Leigh proposed an Fe^0-N_2 species to be active towards protonation, although this suggestion seems unlikely based upon related work by Komiya and coworkers [83]. The latter team was able to isolate and structurally characterize $(Et_2PCH_2CH_2PEt_2)_2Fe^0$-$N_2$, and to show unequivocally that this latter species does not react with H^+ sources to afford NH_3 or N_2H_4.

Along a similar line of investigation, Zubieta and George later reported that iron N_2 complexes supported by tetradentate $N(CH_2CH_2PPh_2)_3$ ligands will release hydrazine (around 0.1 equiv. per iron) and trace ammonia upon the addition of HBr in CH_2Cl_2 [85]. Similarly, Tyler and coworkers recently adapted Leigh's Fe-N_2 systems to aqueous media by appending OH groups to the chelating phosphine ligands [109, 110]. In broad terms, their work has shown that the generation of N_2 adduct species, and the N_2 reduction chemistry that follows, proceeds similarly in water and in CH_2Cl_2. Unfortunately, the reaction

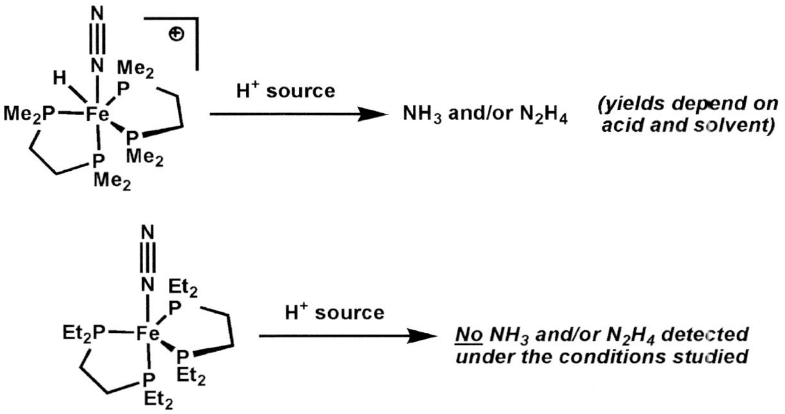

Fig. 3.19 Leigh's phosphine-supported iron system that established iron-mediated N_2 protonation and Komiya's N_2 complex.

yields of all of these systems are too low to warrant thorough mechanistic studies and no intermediates of N_2 reduction have been observed. This situation provides a sharp contrast to the various molybdenum (and tungsten) systems known to mediate N_2 functionalization, for which a great deal of mechanistic information is now available. Nonetheless, these phosphine-supported iron systems provide direct evidence that synthetic iron complexes can mediate the generation of NH_3 (and/or N_2H_4), albeit thus far inefficiently and by what remain ill-defined routes.

3.3.3.6.2 Low-coordinate Iron Model Systems

If iron-based mechanistic models are to be better evaluated, mechanistically well-defined, sterically protected iron systems that enable the careful study of specific reaction steps/intermediates possibly relevant to nitrogen fixation are clearly needed. In this context, Holland and coworkers have reported methodical studies that involve three-coordinate, β-diketiminate (abbreviated as *nacnac*) supported iron scaffolds (Fig. 3.20) [111, 112]. These systems have been shown to support N_2 at one of their three coordination sites and such low-coordination geometries, in addition to the [BP_3]Fe systems described below, lend credence to the general notion that low-coordinate iron centers within the nitrogenase cofactor might be able to activate N_2. Holland's dinitrogen complex (*nacnac*)FeNNFe·(*nacnac*) exhibits short Fe–N bonds (1.77–1.78 Å) and elongated N–N bond lengths [1.182(5) Å]. This species exhibits an ^{15}N-sensitive feature in the resonance Raman spectrum at 1778 cm^{-1} (60 cm^{-1} downshift upon ^{15}N substitution). Furthermore, the bound N_2 complex can also be reduced with sodium or potassium metal. The two-electron reduction shows changes in both N–N bond length (lengthens to 1.23–1.24 Å) as well as vibrational features with high N–N stretching modes

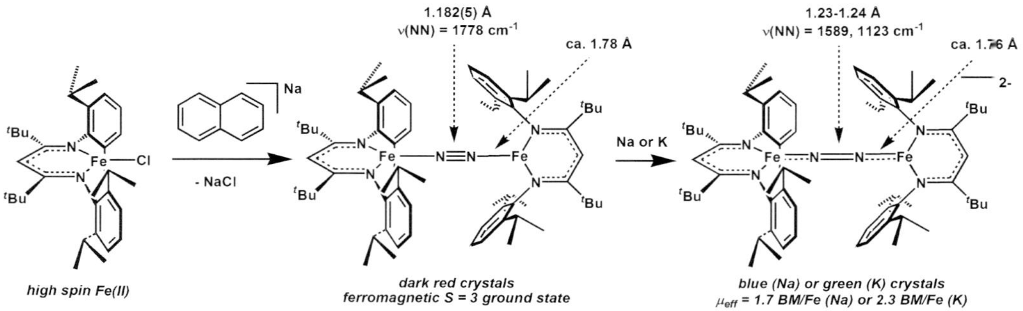

Fig. 3.20 Holland's three-coordinate iron N_2 system.

(1589 and 1123 cm^{-1}). Computational studies on these iron complexes suggest that the weakening of the N–N bond arises from cooperative back-bonding into the π^* orbitals of the N_2 fragment. However, the reactivity of these *nacnac* complexes with CO, PR_3 and benzene is governed by the loss of N_2, and in no case has ammonia or hydrazine production been demonstrated. It appears that despite the reduced bond order of N_2, these complexes serve as Fe(I) sources.

The dinitrogen adduct (*nacnac*)FeNNFe(*nacnac*) also reacts with elemental sulfur to yield a sulfide-bridged diiron species. The low-coordinate sulfide-bridged iron complex forms adducts with nitrogen donors such as ammonia or 1,1-dimethylhydrazine. In addition, the sulfide-bridged species reacts with phenylhydrazine to give a mixed valent (II/III) phenylhydrazido species, (*nacnac*)Fe(μ-1,2-$PhNNH_2$)(μ-S)Fe(*nacnac*). This process requires 1.4 equiv. phenylhydrazine and is believed to proceed with N–N bond cleavage due to the observation of ammonia and aniline byproducts [113].

Our own laboratory has examined tris(phosphino)borate (BP_3) supported Fe-N_x systems [77, 79, 89, 101, 114–117] of various types and has prepared pseudotetrahedral examples of $[L_3Fe\text{-}N_2]^{n-}$ and $[L_3Fe\text{-}N_2\text{-}FeL_3]^{n-}$ species. These are the first structurally authenticated N_2 adducts of four-coordinate iron. In addition to N_2, these scaffolds accommodate a wide range of N_x-type ligands at the fourth binding site, including diazenido (N_2R^-), amido (NR_2^-), imido (NR^{2-}) and even nitride (N^{3-}) functionalities. To stabilize this range of ligands the iron center must be able to span a remarkable range of formal oxidation states. Furthermore, reactions that have been elucidated using these scaffolds have exposed an unexpectedly rich suite of iron redox chemistry. Well-defined examples of one-, two- and even three-electron redox changes at a single iron site are possible by chemically modifying or substituting the N_x functionality. The wide range of redox chemistry spanned by these $BP_3Fe\text{-}N_x$ systems is thus far unique and we are unaware of any other iron systems that support an equally rich range of N_x functionalities. This discovery may have ramifications with respect to the consideration of single-site biological N_2 fixation schemes.

Our group has been able to establish straightforward access to pseudotetrahedral $[PhBP_3^{iPr}]Fe(\mu\text{-}N_2)Fe[PhBP_3^{iPr}]$, $\{[PhBP_3^{iPr}]Fe(\mu\text{-}N_2)Fe[PhBP_3^{iPr}]\}\{Na(THF)_3\}$

and {[$PhBP^{iPr}{}_3$]$Fe(N_2)$}$_2$($Mg(THF)_4$) species by judicious choice of the reducing conditions with [$PhBP_3^{iPr}$]Fe-X (X=Cl, I) precursors (Fig. 3.21). For instance, exposure of [$PhBP_3^{iPr}$]Fe-Cl to a stoichiometric equivalent of Na/Hg amalgam affords the diiron(I) complex [$PhBP_3^{iPr}$]Fe(μ-N_2)Fe[$PhBP_3^{iPr}$]. This complex features an N–N bond length of 1.138(6) Å [79] and has an $S=3$ ground state electronic configuration that is believed to arise from two ferromagnetically coupled $S=3/2$ Fe(I) centers. The presence of excess Na/Hg reducing agent results in the formation of the structurally related, but one electron reduced, diiron N_2 adduct {[$PhBP_3^{iPr}$]Fe(μ-N_2)Fe[$PhBP_3^{iPr}$]}{$Na(THF)_5$} [N–N 1.171(4) Å]. This species is formally an Fe(0)/Fe(I) mixed-valence species and the observed electron paramagnetic resonance (EPR) features are consistent with an $S=5/2$ ground state [79]. Interestingly, reduction of the halide precursors by Mg^0 results in the formation of a dark red complex that features two formally anionic [$PhBP_3^{iPr}$]$Fe(N_2)^-$ units sandwiching a $Mg(THF)_4$ dication: {[$PhBP_3^{iPr}$]$Fe(N_2)$}$_2$($Mg(THF)_4$). Structurally related dinitrogen complexes are also known for molybdenum [79]. This latter species has been reported to react with MeOTf to provide the diazenide species [$PhBP_3^{iPr}$]Fe-N=NMe [79]. This reaction establishes that electrophilic attack at the β-N-atom of ligated N_2 is possible at iron and such a step models, albeit crudely, the first protonation step shown in the iron cycle of Fig. 3.4.

A key feature of the hypothetical Fe-N_2 reduction cycle shown in Fig. 3.4 is the assumption that a single iron site can accommodate ligands as electronically distinct as π-acidic N_2 and π-basic nitride (N^{3-}) or imide (NH^{2-}). The [$PhBP_3^{iPr}$]Fe platform our group has popularized has been used to successfully demonstrate the feasibility of this idea. For example, a highly unusual terminal nitride species, [$PhBP_3^{iPr}$]$Fe^{IV}\equiv N$, can be generated and thoroughly characterized using this system. This terminal nitride ligand is installed using the N-atom transfer agent Li(dbabh) (dbabh=2,3:5,6-dibenzo-7-aza bicyclo[2.2.1]hepta-2,5-diene) to form an amide that releases anthracene to deliver the N-atom to iron (Fig. 3.22) [77]. This type of N-atom transfer reactivity for Li(dbabh) was first reported by Cummins [118]. For the iron system of interest here, metathesis of Li(dbabh) with [$PhBP_3^{iPr}$]$Fe^{II}Cl$ provides the high-spin Fe(II) amide [$PhBP_3^{iPr}$]Fe(dbabh) cleanly at –35 °C. Subsequent warming of a sample of this solution to 0 °C cleanly produces a stoichiometric equivalent of anthracene and the diamagnetic nitride product. The distinctive ^{15}N NMR resonance for [$PhBP_3^{iPr}$]$Fe^{IV}\equiv N$ at 952 ppm is most indicative of the terminal nitride moiety. The complex exhibits an IR active mode at 1034 cm^{-1} that shifts to 1007 cm^{-1} upon ^{15}N substitution, as expected from the theoretically predicted shift (28 cm^{-1}) based upon an *FeN* harmonic oscillator approximation. [$PhBP_3^{iPr}$]$Fe^{IV}\equiv N$ is diamagnetic and is predicted to feature the electronic configuration $(3d_{xy})^2(3d_{x2-y2})^2(3d_z^2)^0(3d_{yz})^0(3d_{xz})^0$ based upon DFT calculations [77].

Attempts to obtain single crystals of [$PhBP_3^{iPr}$]$Fe^{IV}\equiv N$ have been thwarted by its propensity to undergo bimolecular nitride coupling to form [$PhBP_3^{iPr}$]Fe-(μ-N_2)Fe[$PhBP_3^{iPr}$]. This bimolecular coupling reaction can be attenuated by storing relatively dilute solutions of [$PhBP_3^{iPr}$]$Fe^{IV}\equiv N$ at temperatures below 0 °C. Reactivity studies can therefore be executed and it has been observed that NH_3

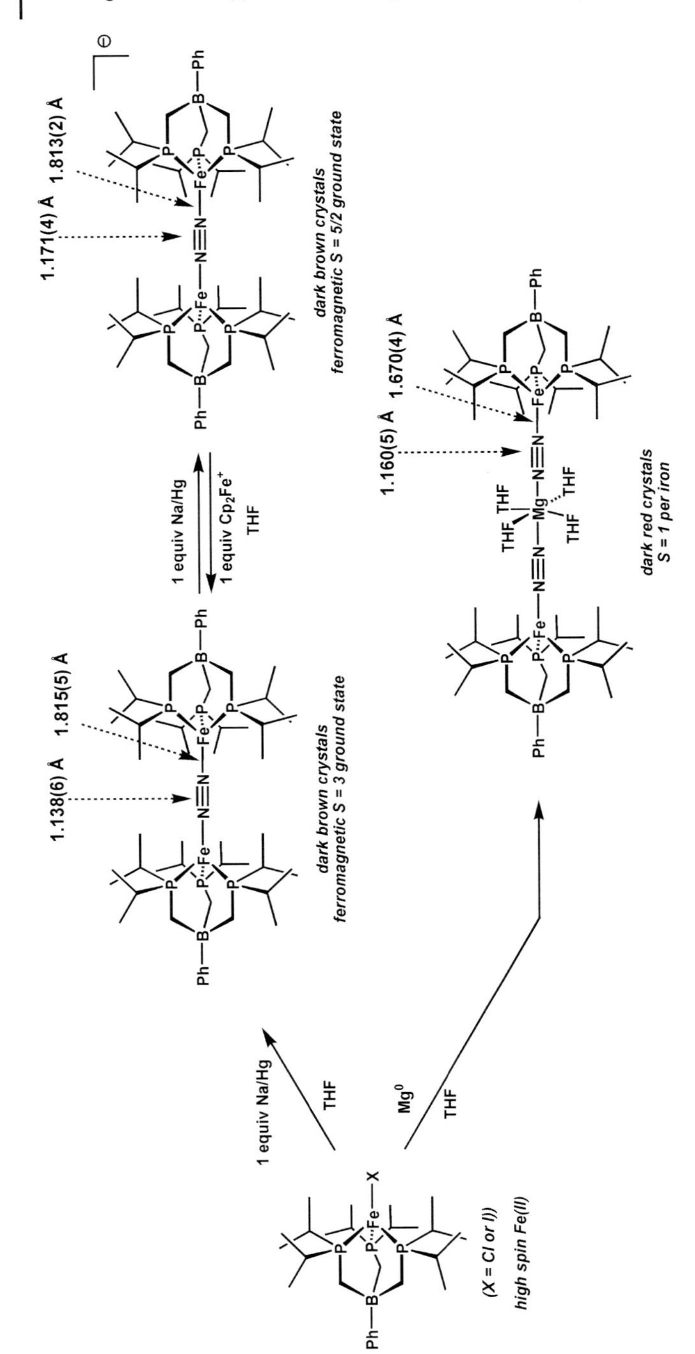

Fig. 3.21 Synthesis of 4-coordinate Fe-N_2 complexes.

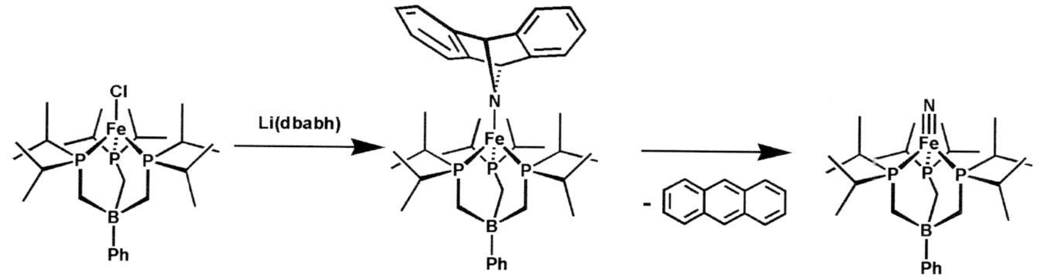

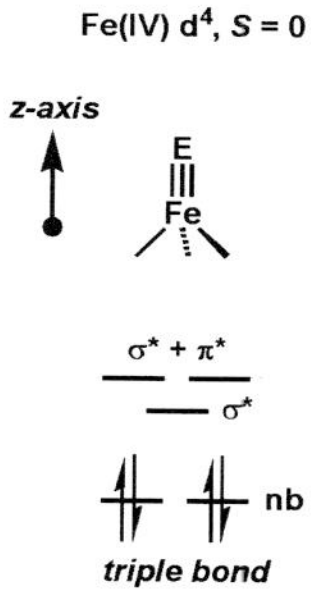

Fig. 3.22 Synthesis of the terminal nitride species $[PhBP_3^{iPr}]Fe^{IV}\equiv N$.

release from $[PhBP_3^{iPr}]Fe^{IV}\equiv N$ can be effected by its treatment with 3 equiv. $[LutH][BPh_4]$ and 3 equiv. $CoCp_2$ (around 45% yield of NH_3). To estimate the Fe–N bond distance in $[PhBP_3^{iPr}]Fe^{IV}\equiv N$, X-ray absorption spectroscopy (XAS) studies have been carried out, and show both a large pre-edge feature consistent with the Fe(IV) oxidation state assignment and EXAFS features including a scatterer at 1.54 Å that has been tentatively assigned as the $Fe\equiv N$ bond distance [119]. The coupling of the two terminal nitrides to form dinitrogen is a fascinating six-electron redox process mediated by two iron centers.

Reductive release of ammonia from $[PhBP_3^{iPr}]Fe^{IV}\equiv N$ may proceed via the steps shown in Fig. 3.4. Because the one-electron reduction of $[PhBP_3^{iPr}]Fe^{IV}\equiv N$ is difficult to realize, it is most likely the case that protonation of $[PhBP_3^{iPr}]Fe(N)$ to initially generate $[PhBP_3^{iPr}]Fe^{IV}(NH)^+$ precedes electron transfer from $CoCp_2$ possibly to generate $[PhBP_3^{iPr}]Fe^{III}(NH)$. This species may then be further reduced according to the scheme shown below.

This scheme invokes unusual imides of iron in the oxidation states +2, +3 and +4. While efforts in our laboratories are ongoing with respect to the generation and characterization of the directly relevant parent imide derivatives, our group has achieved the preparation of well-defined $[BP_3]Fe(NR)$ species in the Fe(II) and Fe(III) oxidation states [79, 101, 117]. We have most recently also realized the preparation of a related system, $[PhB(CH_2P^tBu_2)_2(pyrazole)]Fe(NR)$ [120], that supports imide functionalities in the Fe(III) and Fe(IV) oxidation states. Remarkably the Fe–N bond distance changes only negligibly across the three iron oxidation states (Fe–N ranges between 1.63 and 1.65 Å) for this series of pseudotetrahedral imide complexes. This fact is consistent with the d orbital splitting diagram that we have proposed for low-spin, four-coordinate iron imides of approximate

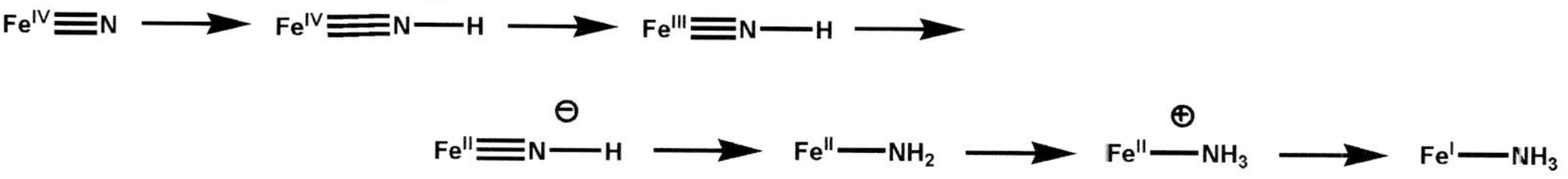

$Fe^{IV}\equiv N \longrightarrow [Fe^{IV}\equiv N{-}H]^{+} \longrightarrow Fe^{III}\equiv N{-}H \longrightarrow [Fe^{II}\equiv N{-}H]^{-} \longrightarrow Fe^{II}{-}NH_2 \longrightarrow [Fe^{II}{-}NH_3]^{+} \longrightarrow Fe^{I}{-}NH_3$

Fig. 3.23 Possible mode of reduction of the terminal nitride $[PhBP_3^{iPr}]Fe(N)$.

3-fold symmetry, whereby two high-lying antibonding orbitals ($3d_{xz}$, $3d_{yz}$) with both σ- and π-antibonding contributions are empty and three lower lying, largely nonbonding orbitals ($3d_{xy}$, $3d_{x2-y2}$, $3d_z^2$) house either six, five or four electrons for the Fe(II), Fe(III) and Fe(IV) imides, respectively [79, 101, 104, 117, 120]. This electronic arrangement gives rise to their corresponding singlet, doublet and triplet ground states (Fig. 3.24) [79, 101, 117, 120].

As noted previously, it is possible that a single iron site initiates the N_2 reduction process, but that bi- or polymetallic N_xH_y intermediates are formed at later stages. While Sellmann's work provides examples of some such complexes for octahedral iron, scant precedent exists for such species with respect to low-coordinate iron, and this is especially true for bridged N, NH and NH_2 ligands. Our laboratory has therefore pursued the systematic preparation of such species using the $[BP_3]Fe$ platforms and has found that several unique species can indeed be prepared [89, 101].

The Na/Hg amalgam reduction of the Fe(II) azide complex $\{[PhBP_3]Fe\}_2(\mu$-1,3-$N_3)$, itself a ferromagnetically coupled dimer in the solid state, gives rise to the dinuclear nitride-bridged anion $\{[PhBP_3]Fe(\mu\text{-}N)Fe[PhBP_3]\}^-$. This complex is a distinct example of a bimetallic Fe–N–Fe species in that it features (i) pseudotetrahedral iron centers, (ii) a severely bent Fe–(μ-N)–Fe linkage (135°), and (iii) two relatively low-valent Fe(II) centers. The complex also exhibits rather short Fe–N bond distances (around 1.70 Å). Because the sample is diamagnetic at room temperature, it is possible to obtain its ^{15}N NMR spectrum ($\delta = 801$ ppm versus NH_3). Cyclic voltammetry for this species establishes a well-

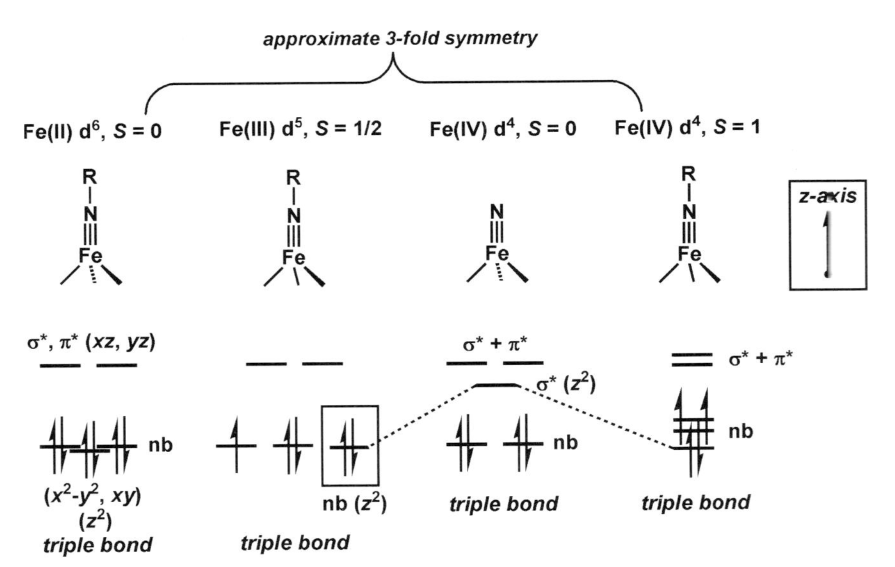

Fig. 3.24 Qualitative d orbital splitting diagrams for pseudotetrahedral $L_3Fe(N)$ and $L_3Fe(NR)$ complexes prepared by our group at Caltech.

defined $Fe^{III}Fe^{II}/Fe^{II}Fe^{II}$ redox process and chemical oxidation has provided the crystallographically characterized $[PhBP_3]Fe^{III}(\mu\text{-N})Fe^{II}[PhBP_3]$ species. Interestingly, ammonia release from $\{[PhBP_3]Fe(\mu\text{-N})Fe[PhBP_3]\}^-$ can be cleanly effected by its treatment with three equivalents of HCl:

$$\{[PhBP_3]Fe(\mu\text{-N})Fe[PhBP_3]\}Na + 3\,HCl \rightarrow NH_3 + 2[PhBP_3]FeCl + NaCl$$

The nitride-bridged Fe^{III}–N–Fe^{II} and $\{Fe^{II}\text{–N–}Fe^{II}\}^-$ complexes have been shown to activate hydrogen under ambient conditions to provide their respective μ-NH/μ-H products $\{[PhBP_3]Fe^{III}(\mu\text{-NH})(\mu\text{-H})Fe^{II}[PhBP_3]\}$ and $\{[PhBP_3]Fe^{II}(\mu\text{-NH})(\mu\text{-H})Fe^{II}[PhBP_3]\}^-$ [89]. These unusual diiron -NH species can be electrochemically and chemically interconverted. The anionic species is diamagnetic and consists of two low-spin, five-coordinate Fe(II) centers. Its diamagnetic nature allows it to be thoroughly characterized by multinuclear NMR spectroscopy. Signature resonances include the μ-NH (18.5 ppm, ^{1}H NMR; 406 ppm $\{^1J_{N-H}=64$ Hz$\}$, ^{15}N NMR) and the μ-H (–22.4 ppm, ^{1}H NMR) functionalities. The low-spin character of each iron center is maintained in the neutral $Fe^{III}Fe^{II}$ analog, which exhibits a characteristic $S=1/2$ EPR signal with superhyperfine coupling to six phosphorus nuclei. Synthetic complexes containing bridging parent imides and hydrides have become especially interesting as spectroscopic models in light of the recent ENDOR detection of several possible intermediates of proton uptake and N_2 reduction at the FeMoco of nitrogenase [121].

3.4 Concluding Remarks

Describing the biochemistry of nitrogenases and capturing the process of fixing nitrogen has proven a difficult and exciting challenge, and much progress has been made during the past 15 years. In recent years, Seefeldt and Hoffman have gathered evidence for a number of possible intermediates utilizing partially reduced forms of dinitrogen (shunt pathways) and in some cases enzyme mutants. The relationship between the postulated intermediates and the *native* catalytic cycle is not yet, at least in all instances, readily apparent. The seminal observation of an interstitial atom in the center of the nitrogenase cofactor by the Rees group has lent new life to both theoretical and synthetic models for nitrogen fixation. It has spurred not only an exciting debate over the identity but also the structural and possibly catalytic role of the interstitial atom (X). This observation has also led to new work in the synthesis of iron-imide and iron-nitride clusters. The three scenarios that have been advanced involve: (i) dinitrogen reduction centered at a molybdenum center (Figs. 3.3, 3.4, top cycle and 3.5), (ii) dinitrogen reduction at a four- or five-coordinate iron center (Figs. 3.4, bottom cycle and 3.10), or (iii) a multimetallic mechanism in which some or all of the reduced N_xH_y species are stabilized by two or more of the metal centers within the FeMoco (Figs. 3.11 and 3.12). The synthetic community has provided a number of monometallic complexes supporting a wide range of N_xH_y ligands. Particularly fascinating is Schrock's work using a $[HIPTN_3N]Mo$ center to elegantly demonstrate (i) that a Mo(III) center can facilitate N_2 reduction to ammonia, and (ii) that a number of the intermediates proposed for the catalytic pathway can be isolated and independently studied. Schrock's studies have provided a much needed experimental basis for many of the species postulated (but never before observed) for a *true* Chatt-type N_2 fixation cycle. Mirroring this molybdenum work is a growing interest in synthetic iron systems similarly capable of mediating dinitrogen reduction. Most recently, our own group and that of Holland's have independently established that low-coordinate iron complexes can activate dinitrogen, although much work remains to realize viable iron-based catalysts. The $[BP_3]Fe$ iron systems provide a basis to further consider single-site mechanisms for iron-mediated nitrogen reduction, and hint at the possibility of a viable Chatt-type cycle that would proceed via an Fe(I) to Fe(IV) redox loop. In particular, the ability of these iron systems to support both π-acidic (N_2) and π-basic ligands (N^{3-}) establishes the limiting structure types one would require for such an iron-based mechanism to be chemically feasible.

There does not yet appear to be consensus with respect to the mode(s) by which Nature fixes dinitrogen, implying that much work remains to gain deeper insight into this fascinating problem.

Acknowledgments

We are grateful to the NIH for their generous support: GM-070757 (J. C. P.), post-doctoral fellowship GM-072291 (M. P. M.).

We also express thanks to Dr. Johannes Kaestner and Profressor Peter Blöchl for providing the coordinate file to generate Fig. 3.14.

References

1 Smil, V. *Enriching the Earth*, MIT Press, Cambridge, MA, **2001**.
2 Chatt, J., Dilworth, J. R., Richards, R. L. *Chem Rev* **1978**, *78*, 589.
3 Bazhenova, T. A., Shilov, A. E. *Coord Chem Rev* **1995**, *144*, 69.
4 Burgess, B. K., Lowe, D. J. *Chem Rev* **1996**, *96*, 2983.
5 Thorneley, R. N. F., Lowe, D. J. In *Molybdenum Enzymes*, Spiro, T. G. (Ed.), Wiley, New York, **1985**, vol. 7, p. 221.
6 Howard, J. B., Rees, D. C. *Chem Rev* **1996**, *96*, 2965.
7 Smil, V. *Sci Am* **1997**, *277*, 76.
8 Allen, A. D., Bottomley, F., Harris, R. O., Reinsalu, V. P., Senoff, C. V. *J Am Chem Soc* **1967**, *89*, 5595.
9 Allen, A. D., Senoff, C. V. *Chem Commun* **1965**, 621.
10 Shilov, A. E., Denisov, N. T., Efimov, N. O., Shuvalov, N., Shuvalov, N. I., Shilova, A. K. *Nature* **1971**, *231*, 460.
11 Pickett, C. J., Talarmin, J. *Nature* **1985**, *317*, 652.
12 Bercaw, J. E. *Personal communication*. The principle of initial optimization is a playful phrase often invoked in the chemistry department at Caltech to describe the all too frequently encountered observa-

tion that an initial catalytic system discovered is quite often the best system that is ever discovered, despite years of hard labor by graduate students and postdoctoral associates to try and devise a better system through optimization, mechanistic understanding, etc.

13 Hidai, M., Mizobe, Y. *Chem Rev* **1995**, *95*, 1115.
14 Lee, S.C., Holm, R.H. *Proc Natl Acad Sci USA* **2003**, *100*, 12522.
15 MacKay, B.A., Fryzuk, M.D. *Chem Rev* **2004**, *104*, 385.
16 Gambarotta, S. *J Organomet Chem* **1995**, *500*, 117.
17 Eady, R.R. *Chem Rev* **1996**, *96*, 3013.
18 Schrock, R.R. *Philos Trans Royal Soc A Math Phys Eng Sci* **2005**, *363*, 959.
19 Einsle, O., Tezcan, F.A., Andrade, S.L.A., Schmid, B., Yoshida, M., Howard, J.B., Rees, D.C. *Science* **2002**, *297*, 1696.
20 Lee, H.-I., Benton, P.M.C., Laryukhin, M., Igarashi, R.Y., Dean, D.R., Seefeldt, L.C., Hoffman, B.M. *J Am Chem Soc* **2003**, *125*, 5604.
21 Yang, T.-C., Maeser, N.K., Laryukhin, M., Lee, H.-I., Dean, D.R., Seefeldt, L.C., Hoffman, B.M. *J Am Chem Soc* **2005**, *127*, 12804.
22 Vrajmasu, V., Munck, E., Bominaar, E.L. *Inorg Chem* **2003**, *42*, 5974.
23 Lee, H.-I., Hales, B.J., Hoffman, B.M. *J Am Chem soc* **1997**, *119*, 11395.
24 Chatt, J., Pearman, A.J., Richards, R.L. *Dalton Trans* **1977**, 1852.
25 Chatt, J., Diamantis, A.A., Heath, G.A., Hooper, N.E., Leigh, G.J. *Dalton Trans* **1977**, 688.
26 Laplaza, C.E., Cummins, C.C. *Science* **1995**, *268*, 861.
27 Laplaza, C.E., Johnson, M.J.A., Peters, J.C., Odom, A.L., Kim, E., Cummins, C.C., George, G.N., Pickering, I.J. *J Am Chem Soc* **1996**, *118*, 8623.
28 Yandulov, D.V., Schrock, R.R. *Science* **2003**, *301*, 76.
29 Yandulov, D.V., Schrock, R.R. *Inorg Chem* **2005**, *44*, 1103.
30 Pickett, C.J. *J Biol Inorg Chem* **1996**, *1*, 601.
31 Leigh, G.J. *Acc Chem Res* **1992**, *25*, 177.
32 Barriere, F. *Coord Chem Rev* **2003**, *236*, 71.
33 Glassman, T.E., Vale, M.G., Schrock, R.R. *Organometallics* **1991** *10*, 4046.
34 Glassman, T.E., Vale, M.G., Schrock, R.R. *J Am Chem Soc* **1992**, *114*, 8098.
35 O'Regan, M.B., Liu, A.H., Finch, W.C., Schrock, R.R., Davis, W.M. *J Am Chem Soc* **1990**, *112*, 4331.
36 Schrock, R.R., Glassman, T.E., Vale, M.G. *J Am Chem Soc* **1991**, *113*, 725.
37 Schrock, R.R., Glassman, T.E., Vale, M.G., Kol, M. *J Am Chem Soc* **1993**, *115*, 1760.
38 Schrock, R.R., Kolodziej, R.M., Liu, A.H., Davis, W.M., Vale, M.G. *J Am Chem Soc* **1990**, *112*, 4338.
39 Schrock, R.R., Liu, A.H., O'Regan, M.B., Finch, W.C., Payack, J.F. *Inorg Chem* **1988**, *27*, 3574.
40 Dilworth, J.R., Dahlstrom, P.L., Hyde, J.R., Zubieta, J. *Inorg Chim Acta* **1983**, *71*, 21.
41 Chatt, J., Pearman, A.J., Richards, R.L. *Dalton Trans* **1976**, 1520.
42 Chatt, J., Heath, G.A., Richards, R.L. *Dalton Trans* **1974**, 2074.
43 Chatt, J., Hussain, W., Leigh, G.J., Neukomm, H., Pickett, C.J., Rankin, D.A. *Chem Commun* **1980**, 1024.
44 Chatt, J., Richards, R.L. *J Less-Common Met* **1977**, *54*, 477.
45 Chatt, J., Pearman, A.J., Richards, R.L. *Nature* **1975**, *253*, 39.
46 Hughes, D.L., Ibrahim, S.K., Pickett, C.J., Querne, G., Laouenan, A., Talarmin, J., Queiros, A., Fonseca, A. *Polyhedron* **1994**, *13*, 3341.
47 Yandulov, D.V., Schrock, R.R. *J Am Chem Soc* **2002**, *124*, 6252.
48 Ritleng, V., Yandulov, D.V., Weare, W.W., Schrock, R.R., Hock, A.S., Davis, W.M. *J Am Chem Soc* **2004**, *126*, 6150.
49 Yandulov, D.V., Schrock, R.R. *Can J Chem* **2005**, *83*, 341.
50 Wagenknecht, P.S., Norton, J.R. *J Am Chem Soc* **1995**, *117*, 1841.
51 Shaver, M.P., Fryzuk, M.D *Adv Synth Catal* **2003**, *345*, 1061.
52 Clentsmith, G.K.B., Bates, V.M.E., Hitchcock, P.B., Cloke, F.G.N. *J Am Chem Soc* **1999**, *121*, 10444

53 MacKay, B.A., Patrick, B.O., Fryzuk, M.D. *Organometallics* **2005**, *24*, 3836.
54 MacKay, B.A., Johnson, S.A., Patrick, B.O., Fryzuk, M.D. *Can J Chem* **2005**, *83*, 315.
55 Fryzuk, M.D., MacKay, B.A., Patrick, B.O. *J Am Chem Soc* **2003**, *125*, 3234.
56 Fryzuk, M.D., MacKay, B.A., Johnson, S.A., Patrick, B.O. *Angew Chem Int Ed* **2002**, *41*, 3709.
57 Pool, J.A., Lobkovsky, E., Chirik, P.J. *Nature* **2004**, *427*, 527.
58 Cui, Q., Musaev, D.G., Svensson, M., Sieber, S., Morokuma, K. *J Am Chem Soc* **1995**, *117*, 12366.
59 Peters, J.C., Cherry, J.-P.F., Thomas, J.C., Baraldo, L., Mindiola, D.J., Davis, W.M., Cummins, C.C. *J Am Chem Soc* **1999**, *121*, 10053.
60 Tsai, Y.-C., Johnson, M.J.A., Mindiola, D.J., Cummins, C.C., Klooster, W.T., Koetzle, T.F. *J Am Chem Soc* **1999**, *121*, 10426.
61 Yoshida, T., Adachi, T., Kaminaka, M., Ueda, T., Higuchi, T. *J Am Chem Soc* **1988**, *110*, 4872.
62 Seino, H., Mizobe, Y., Hidai, M. *Chem Rec* **2001**, *1*, 349.
63 Al-Salih, T.I., Pickett, C.J. *Dalton Trans* **1985**, 1255.
64 Lovell, T., Liu, T.Q., Case, D.A., Noodleman, L. *J Am Chem Soc* **2003**, *125*, 8377.
65 Lovell, T., Li, J., Case, D.A., Noodleman, L. *J Biol Inorg Chem* **2002**, *7*, 735.
66 Lovell, T., Li, J., Case, D.A., Noodleman, L. *J Am Chem Soc* **2002**, *124*, 4546.
67 Lovell, T., Li, J., Liu, T.Q., Case, D.A., Noodleman, L. *J Am Chem Soc* **2001**, *123*, 12392.
68 Seefeldt, L.C., Dance, I.G., Dean, D.R. *Biochemistry* **2004**, *43*, 1401.
69 Dance, I.G. *Austr J Chem* **1994**, *47*, 979.
70 Hinnemann, B., Nørskov, J.K. *J Am Chem Soc* **2003**, *125*, 1466.
71 Schimpl, J., Petrilli, H.M., Blöchl, P.E. *J Am Chem Soc* **2003**, *125*, 15772.
72 Kästner, J., Blöchl, P.E. *Inorg Chem* **2005**, *44*, 4568.
73 Kästner, J., Hemmen, S., Blöchl, P.E. *J Chem Phys* **2005**, *123*, 074306/1.
74 Kästner, J., Blöchl, P.E. *ChemPhysChem* **2005**, *6*, 1724.
75 Huniar, U., Ahlrichs, R., Coucouvanis, D. *J Am Chem Soc* **2004**, *126*, 2588.
76 Coucouvanis, D., Han, J.H., Ahlrichs, R., Nava, P., Huniar, U. *J Inorg Biochem* **2003**, *96*, 19.
77 Betley, T.A., Peters, J.C. *J Am Chem Soc* **2004**, *126*, 6252.
78 MacBeth, C.E., Harkins, S.B., Peters, J.C. *Can J Chem* **2005**, *83*, 332.
79 Betley, T.A., Peters, J.C. *J Am Chem Soc* **2003**, *125*, 10782.
80 Hughes, D.L., Jimenez-Tenorio, M., Leigh, G.J., Rowley, A.T. *Dalton Trans* **1993**, 3151.
81 Hills, A., Hughes, D.L., Jimenez-Tenorio, M., Leigh, G.J., Rowley, A.T. *Dalton Trans* **1993**, 3041.
82 Hughes, D.L., Leigh, G.J., Jimenez-Tenorio, M., Rowley, A.T. *Dalton Trans* **1993**, 75.
83 Komiya, S., Akita, M., Yoza, A., Kasuga, N., Fukuoka, A., Kai, Y. *Chem Commun* **1993**, 787.
84 Hirano, M., Akita, M., Morikita, T., Kubo, H., Fukuoka, A., Komiya, S. *Dalton Trans* **1997**, 3453.
85 George, T.A., Rose, D.J., Chang, Y.D., Chen, Q., Zubieta, J. *Inorg Chem* **1995**, *34*, 1295.
86 Field, L.D., Messerle, B.A., Smernik, R.J. *Inorg Chem* **1997**, *36*, 5984.
87 Fryzuk, M.D., Johnson, S.A., Rettig, S.J. *J Am Chem Soc* **1998**, *120*, 11024.
88 Smith, M.R., Cheng, T.Y., Hillhouse, G.L. *J Am Chem Soc* **1993**, *115*, 8638.
89 Brown, S.D., Mehn, M.P., Peters, J.C. *J Am Chem Soc* **2005**, *127*, 13146.
90 Sellmann, D., Sutter, J. *Acc Chem Res* **1997**, *30*, 460.
91 Sellmann, D., Utz, J., Blum, N., Heinemann, F.W. *Coord Chem Rev* **1999**, *192*, 607.
92 Sellmann, D., Soglowek, W., Knoch, F., Ritter, G., Dengler, J. *Inorg Chem* **1992**, *31*, 3711.
93 Rao, P.V., Holm, R.H. *Chem Rev* **2004**, *104*, 527.
94 Lee, S.C., Holm, R.H. *Chem Rev* **2004**, *104*, 1135.
95 Malinak, S.M., Coucouvanis, D. *Prog Inorg Chem* **2001**, *49*, 599.

96 Bennett, M.V., Stoian, S., Bominaar, E.L., Munck, E., Holm, R.H. *J Am Chem Soc* **2005**, *127*, 12378.

97 Summerville, D.A., Cohen, I.A. *J Am Chem Soc* **1976**, *98*, 1747.

98 Donzello, M.P., Ercolani, C., Kadish, K.M., Ou, Z., Russo, U. *Inorg Chem* **1998**, *37*, 3682.

99 Donzello, M.P., Ercolani, C., Russo, U., Chiesi-Villa, A., Rizzoli, C. *Inorg Chem* **2001**, *40*, 2963.

100 Li, M., Shang, M., Ehlinger, N., Schulz, C.E., Scheidt, W.R. *Inorg Chem* **2000**, *39*, 580.

101 Brown, S.D., Peters, J.C. *J Am Chem Soc* **2005**, *127*, 1913.

102 Fjare, D.E., Gladfelter, W.L. *Inorg Chem* **1981**, *20*, 3533.

103 Verma, A.K., Lee, S.C. *J Am Chem Soc* **1999**, *121*, 10838.

104 Verma, A.K., Nazif, T.N., Achim, C., Lee, S.C. *J Am Chem Soc* **2000**, *122*, 11013.

105 Duncan, J.S., Nazif, T.M., Verma, A.K., Lee, S.C. *Inorg Chem* **2003**, *42*, 1211.

106 Sono, M., Roach, M.P., Coulter, E.D., Dawson, J.H. *Chem Rev* **1996**, *96*, 2841.

107 Barney, B.M., Yang, T.-C., Igarashi, R.Y., Dos Santos, P.C., Laryukhin, M., Lee, H.-I., Hoffman, B.M., Dean, D.R., Seefeldt, L.C. *J Am Chem Soc* **2005**, *127*, 14960.

108 Barney, B.M., Laryukhin, M., Igarashi, R.Y., Lee, H.-I., Dos Santos, P.C., Yang, T.-C., Hoffman, B.M., Dean, D.R., Seefeldt, L.C. *Biochemistry* **2005**, *44*, 8030.

109 Gilbertson, J.D., Szymczak, N.K., Tyler, D.R. *Inorg Chem* **2004**, *43*, 3341.

110 Gilbertson, J.D., Szymczak, N.K., Tyler, D.R. *J Am Chem Soc* **2005**, *127*, 10184.

111 Smith, J.M., Lachicotte, R.J., Holland, P.L. *J Am Chem Soc* **2003**, *125*, 15752.

112 Smith, J.M., Lachicotte, R.J., Pittard, K.A., Cundari, T.R., Lukat-Rodgers, G., Rodgers, K.R., Holland, P.L. *J Am Chem Soc* **2001**, *123*, 9222.

113 Vela, J., Stoian, S., Flaschenriem, C.J., Munck, E., Holland, P.L. *J Am Chem Soc* **2004**, *126*, 4522.

114 Betley, T.A., Peters, J.C. *Inorg Chem* **2003**, *42*, 5074.

115 Jenkins, D.M., Betley, T.A., Peters, J.C. *J Am Chem Soc* **2002**, *124*, 11238.

116 Brown, S.D., Peters, J.C. *J Am Chem Soc* **2004**, *126*, 4538.

117 Brown, S.D., Betley, T.A., Peters, J.C. *J Am Chem Soc* **2003**, *125*, 322.

118 Mindiola, D.J., Cummins, C.C. *Angew Chem Int Ed* **1998**, *37*, 945.

119 Rohde, J.U., Betley, T.A., Peters, J.C., Que, L. In preparation.

120 Thomas, C.M., Mankad, N.P., Peters, J.C. *J Am Chem Soc* **2005**, *128*, 4956.

121 Igarashi, R.Y., Laryukhin, M., Dos Santos, P.C., Lee, H.-I., Dean, D.R., Seefeldt, L.C., Hoffman, B.M. *J Am Chem Soc* **2005**, *127*, 6231.

4
The Activation of Dihydrogen

Jesse W. Tye, Marcetta Y. Darensbourg, and Michael B. Hall

4.1
Introduction

Dihydrogen, consisting of two protons and two electrons, is the simplest of all stable molecules. Its apparent simplicity belies its rich and complex chemistry; a chemistry that keeps a host of scientists busy studying its reactions.

4.1.1
Why Activate H_2?

Dihydrogen has the potential to act as a "clean" alternative to fossil fuels [1]. The oxidation of dihydrogen, either electrochemically or via combustion, leads only to the production of water. One of the major drawbacks of solar, hydroelectric and wind power is that periods of peak energy production do not necessarily coincide with periods of peak energy consumption. Solar, hydroelectric and wind power, however, could be used to electrochemically generate H_2 that can be stored and later burned to produce thermal power or converted back to H^+ and e^- to produce electrical power [2].

In many industrially important reactions, such as hydroformylation and hydrogenation, dihydrogen gas serves as a reducing agent and/or hydrogen atom (H-atom) source. Even small improvements in the efficiency of these reactions translate into large monetary savings.

This review will focus on the homogeneous catalysis of H–H bond cleavage and formation by discrete transition metal complexes and enzymes. These topics have been the subject of a number of excellent reviews [3–21]. This review will not discuss in detail the electrochemical H^+ reduction or H_2 oxidation [22–24], or the heterogeneous activation of H_2 by extended systems [25–29].

Activation of Small Molecules. Edited by William B. Tolman

ISBN: 3-527-31312-5

4.1.2
Why is it so Difficult to Activate H_2?

The H_2 molecule is, in fact, so stable that it was used as an "inert" gas in early air-free chemistry. The following physical properties of H_2 combine to make it a very unreactive molecule: (i) the H–H bond is remarkably strong, (ii) the H_2 molecule is completely nonpolar, (iii) the frontier molecular orbitals of H_2 do not permit most direct, concerted reactions between dihydrogen and other nonmetals, and (iv) the H_2 molecule is a very poor acid.

The amount of energy required for homolytic cleavage of the H–H bond ($H_2 \rightarrow 2H$) is +103.25(1) kcal mol^{-1} [30]. As shown in Table 4.1, this value places the H–H bond among the strongest single bonds [31]. Since most H–X bonds will generally be weaker than the H–H bond, there is often little or no thermodynamic driving force for the cleavage of the H–H bond.

The polarity of the reacting molecules often enhances the rates of chemical reactions. Scheme 4.1 contrasts the proton transfer reaction between the very polar reactants $MeNH_3^+$ and $EtNH_2$ and the reaction between nonpolar reactants

Table 4.1 Bond dissociation energies for H–X bonds [30, 31]

Bond type	Average bond dissociation energy (kcal mol^{-1}) a)
H–F	135(1)
H–O	109.60(4)
H–H	103.25(1)
H–Cl	102.3(1)
H–C	98.3(8)
H–N	92(2)

a) Error in last digit is given parenthetically.

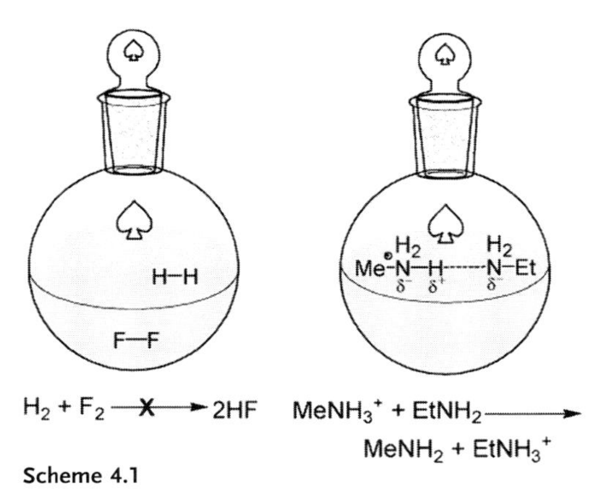

Scheme 4.1

H_2 and F_2 to produce HF. For the reaction between $MeNH_3^+$ and $EtNH_2$, the partially positively charged ammonium H-atoms of $MeNH_3^+$ are attracted to the negatively charged nitrogen center of $EtNH_2$. The polarity of the reactants therefore helps to organize them spatially for effective proton transfer between $MeNH_3^+$ and $EtNH_2$. No such strong intermolecular forces exist for the reaction between H_2 and F_2. Since the H_2 molecule is completely nonpolar, it is a poor target for attack by either electrophiles or nucleophiles, resulting in large activation energies for H_2 activation. Even when direct reaction with H_2 is thermodynamically feasible, the rates of reactions with H_2 are often extremely slow.

The analysis of the frontier molecular orbitals of dihydrogen also suggests that direct, concerted reactions between H_2 and most nonmetals should have high activation energies. In frontier molecular orbital theory [32], a high activation energy is predicted when the symmetry of the highest occupied molecular orbital (HOMO) of one reactant does not match the symmetry of the lowest unoccupied molecular orbital (LUMO) of the second reactant and *vice versa*. An illustrative example is the concerted addition of dihydrogen to ethylene to yield ethane ($C_2H_4 + H_2 \rightarrow C_2H_6$). Although the hydrogenation of ethylene is a thermodynamically favorable process (ΔH_{298} around -32 kcal mol^{-1}) [33], mixtures of the two gases are stable indefinitely in the absence of an appropriate catalyst. The implied high activation energy for the reaction $C_2H_4 + H_2 \rightarrow C_2H_6$ may be explained in the context of frontier molecular orbital theory [34]. The frontier molecular orbitals of H_2 are H–H bonding (HOMO) and H–H antibonding (LUMO), and the frontier molecular orbitals of ethylene are C–C π bonding (HOMO) and C–C π antibonding (LUMO). As shown in Fig. 4.1, the frontier molecular orbitals of H_2 and ethylene are inappropriate for a direct concerted reaction between these species to yield ethane. In the reaction of ethylene with dihydrogen, the HOMO of ethylene is not a match for the LUMO of dihydrogen (likewise, the HOMO of H_2 does not match the LUMO of ethylene). The direct reaction of H_2 with C_2H_4 is therefore expected to have a large activation barrier.

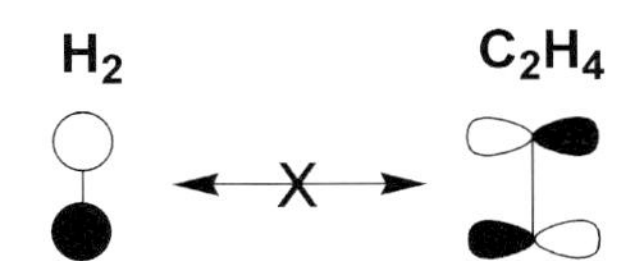

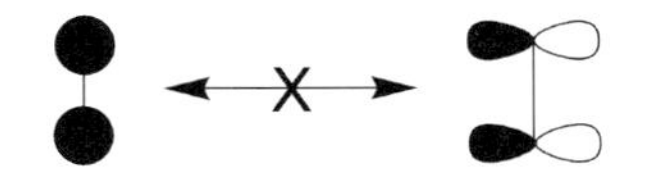

Fig. 4.1 The shapes of the frontier molecular orbitals of H_2 and C_2H_4 are inappropriate for a direct, concerted reaction. The interactions between the LUMOs of H_2 and C_2H_4 have no effect on the energy since they are interactions between empty orbitals. The interaction between the filled HOMOs of H_2 and C_2H_4 is inherently destabilizing, since the interaction is between two filled orbitals.

Table 4.2 Proton dissociation constants for several compounds in organic solvents

Acid	pK_a	Solvent	Reference
HH	49	THF	35
Ph_3CH	44	THF	35
Cyclohexane-OH	38	THF	35
Ph_2PH	35	THF	35
CH_3COOH	22.3	CH_3CN	36
$CH_3(C_6H_4)SO_3H$	8.0	CH_3CN	37
CF_3SO_3H	2.6	CH_3CN	37

The uncatalyzed heterolytic cleavage of the H–H bond ($H_2 \rightarrow H^+ + H^-$) is also difficult. The strength of the H–H bond and its lack of polarity contribute to the poor kinetic and thermodynamic acidity of H_2. The pK_a values for a series of mono-protic "acids", dissolved in either tetrahydrofuran (THF) or acetonitrile solvent, are given in Table 4.2. Dihydrogen with an estimated pK_a of 49 in THF solvent is among the weakest acids.

4.2 Structure and Bonding of Metal-bound H-Atoms

4.2.1 Why can Metal Centers React Directly with H_2, while most Nonmetals Cannot?

Metal centers have low-energy d orbitals. The nodal character and energy of the d orbitals allows a transition metal center to react directly with H_2 in a concerted reaction with a low activation barrier. As shown in Fig. 4.2, for the reaction between H_2 and the ML_5 center, the symmetry of the LUMO of the metal complex matches the symmetry of the HOMO of H_2 and a filled orbital of the metal center matches the symmetry of the LUMO of H_2. In the transition state for H_2 binding to the metal center, there is a synergistic flow of electron density from H_2 to the metal center and from the metal center to the H_2 ligands. A low energy transition state is therefore expected for addition of H_2 to the ML_5 fragment. (Contrast the reaction of ML_5, given in Fig. 4.2, with that of H_2 and C_2H_4 given in Fig. 4.1.)

The synergistic flow mentioned above has two parts: dihydrogen to metal donation [$H_2(\sigma) \rightarrow M$], which consists of the transfer of electron density from the H–H bonding orbital to an empty orbital on the transition metal center, and metal to dihydrogen back-donation [$M \rightarrow H_2(\sigma^*)$], which consists of the transfer of electron density from a filled orbital on the metal center to the H–H antibonding orbital of dihydrogen. Both $H_2 \rightarrow M$ donations and $M \rightarrow H_2$ back-donation weaken the H–H bond and if strong enough will lead to the eventual cleav-

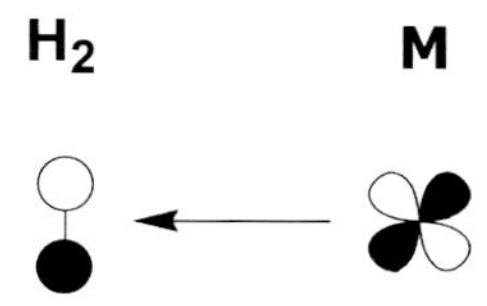

empty H_2 orbital + filled M orbital

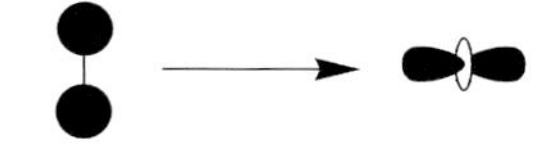

filled H_2 orbital + empty M orbital
low energy interactions
predict low energy TS

Fig. 4.2 The shapes of the frontier molecular orbitals of H_2 and M (left) are appropriate for a direct, concerted reaction. In the transition state for M–(η^2-H_2) bond formation, there is a synergistic flow of electron density from the HOMO of H_2 into the LUMO of M and from an occupied orbital of ML_5 to the LUMO of H_2.

age of the H–H bond. Interestingly, the Dewar-Chatt-Duncanson (DCD) bonding model [38–43] that was originally devised to explain the binding of olefins to transition metal centers consists of similar electron density transfers. Thus, it is somewhat surprising that metal-bound H_2 complexes were unknown until 1984 [4].

4.2.2 Seminal Work: The Discovery of Metal-bound H_2 Complexes

Since the H_2 molecule has a very strong H–H bond and possesses no nonbonding electrons or "lone pairs", it was generally believed that an intact dihydrogen molecule could not act as a ligand to a transition metal center. In the early 1980s, however, Kubas and coworkers reported evidence for the existence of a transition metal complex containing side-on bound H_2 as a ligand to a tungsten center [44]. They showed that the reaction of the coordinatively unsaturated tungsten complex $(P^iPr_3)_2W(CO)_3$ (see "Nomenclature" at end of chapter) with H_2 gas resulted in the formation of a complex of the form $(\eta^2\text{-}H_2)(P^iPr_3)_2W\text{-}(CO)_3$ (η^2 reflects the fact that both H-atoms of the intact H_2 molecule are bound to the tungsten center) (Scheme 4.2).

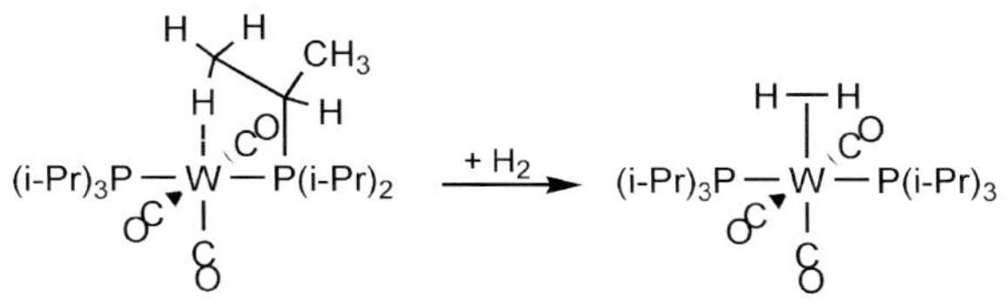

Scheme 4.2

The molecular structure of $(\eta^2\text{-}H_2)(P^iPr_3)_2W(CO)_3$ as determined by single-crystal neutron diffraction studies shows that the binding of H_2 to the tungsten has led to an increase in the H–H distance from 0.74 (free H_2) to 0.84 Å. The observed H–H distance suggests that coordination to the tungsten center has weakened the H–H bond. The metal–ligand distances and ligand–metal–ligand angles of the six-coordinate $(\eta^2\text{-}H_2)(P^iPr_3)_2W(CO)_3$ complex are similar to those observed in the molecular structure of the five-coordinate $(P^iPr_3)_2W(CO)_3$ derived from X-ray diffraction. The similarity between these two structures suggest that the complex is more accurately described as a six-coordinate W^0 $\eta^2\text{-}H_2$ complex and not a seven-coordinate W^{II} bis-hydride complex.

4.2.3
What are the Possible Consequences when H_2 Approaches a Coordinatively Unsaturated Transition Metal Center?

When a molecule of H_2 approaches a coordinatively unsaturated [45] transition metal center, there are at least four possible outcomes (Scheme 4.3). (i) There is no reaction whatsoever. (ii) An essentially intact H_2 molecule is bound to the transition metal center. (iii) The H–H bond is homolytically cleaved, resulting in a bis-hydride complex. (iv) The H–H bond is heterolytically cleaved, resulting in the formation of a metal hydride with loss of H^+. Note that throughout this chapter, $(\eta^2\text{-}H_2)$ indicates a dihydrogen ligand, $(H)_n$ ($n=1, 2, \ldots$) denotes n classical hydride ligands and H_x ($x=1, 2, \ldots$) indicates an unspecified structural form.

There is often no reaction whatsoever between a coordinatively unsaturated metal complex and H_2. As Morris said: "The first step in the activation of dihydrogen is the formation of a transition state, intermediate or product structure in which an essentially intact H_2 molecule is bound side-on to the metal center" [46]. In order for H_2 to bind to a coordinatively unsaturated metal complex, it must displace any intermolecular and/or intramolecular interactions that are stabilizing the "vacant" site. The formation of an $\eta^2\text{-}H_2$ complex, even transiently, is often a difficult task since intact H_2 is generally the most weakly binding ligand for a given complex [47].

Once dihydrogen is bound to the metal center, the steric and electronic properties of the metal center and ligand set determine whether the complex will

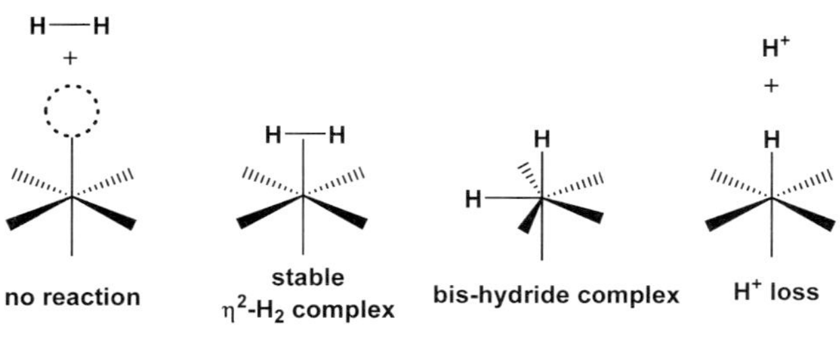

Scheme 4.3

exist as a stable η^2-H_2 complex, or whether homolytic or heterolytic cleavage of the H–H bond will result. Homolytic cleavage of the H–H bond to generate the corresponding bis-hydride complex is favored for electron-rich metal centers and for metal centers with less sterically demanding ligands. Heterolytic cleavage of the H–H bond to generate the corresponding hydride complex and release H^+ is favored for electron-poor metal centers.

The reactivity of dihydrogen can be explained qualitatively by the DCD bonding model. This model partitions metal–H_2 bonding into two parts: $H_2(\sigma) \rightarrow M$ donation and $M \rightarrow H_2(\sigma^*)$ back-donation. When attached to an electron-poor metal center, $H_2(\sigma) \rightarrow M$ donation dominates, leading to a weakening of the H–H bond and the depletion of electron density on the H-atoms. Therefore, the bonding of dihydrogen to an electron-poor metal center increases its acidity relative to free H_2. When attached to an electron-rich metal center, $M \rightarrow H_2(\sigma^*)$ back-donation dominates, leading to a weakening of the H–H bond and the build-up of electron density on the H-atoms, ultimately leading to bis-hydride formation.

Hall and coworkers have utilized *ab initio* electronic structure calculations to examine the factors that lead to the structures of hydride and η^2-H_2 complexes [48–51]. For 114 polyhydride species, Bayse and Hall found that the most stable structure maximizes use of the n d and $(n+1)$ s orbitals on the metal in the formation of the M–H bonds [51]. Lin and Hall examined factors which lead to oxidative addition to form a bis-hydride or association of an intact molecule of H_2 to form the corresponding η^2-H_2 complex. They concluded that a bis-hydride is preferred over an η^2-H_2 complex when twice the ionization enthalpy of an electron in the M–H bond is greater than sum of the ionization enthalpies of an electron in the H–H bond and one in the metal d orbital [49].

Lin and Hall have also examined the periodic trends for the formation of stable η^2-H_2 complexes [48]. They concluded that for neutral phosphine complexes [i.e. $M(PH_3)_x$] a diagonal line passing through ruthenium and iridium divides the periodic table into bis-hydride complexes (left) from the η^2-H_2 complexes (right). For monocationic complexes, this line shifts to between technetium/ruthenium and osmium/iridium. They find that the stability of η^2-H_2 complex relative to the corresponding bis-hydride complex is directly related to the number and type of π-accepting ligands. Successive replacement of PH_3 by CO shifts the dividing line to the left of the periodic table.

Morris has shown that the product of the reaction of a coordinatively unsaturated metal complex with H_2 correlates quantitatively with the electron density available at the H_2 binding site as determined by the ν(N–N) stretching frequency and redox potential of the corresponding η^1-N_2 complex [46, 52, 53]. For η^1-N_2 complexes with $\nu(\text{N–N}) < 2050\ \text{cm}^{-1}$ and $E_{1/2} < 0.5$ V, the reaction with H_2 led to H–H cleavage and only the dihydride form was observed. For η^1-N_2 complexes with $\nu(\text{N–N}) = 2050$–$2200\ \text{cm}^{-1}$ and $E_{1/2} = 0.5$–2.0 V, the reaction with H_2 led to stable η^2-H_2 complexes. The η^2-H_2 complexes became increasingly acidic as ν(N–N) and $E_{1/2}$ approached their threshold values of 2200 cm^{-1} and 2.0 V. For η^1-N_2 complexes with $\nu(\text{N–N}) > 2200\ \text{cm}^{-1}$ and $E_{1/2} > 2.0$ V, H_2 did not react at all or was found to bind only transiently. It is sometimes found that seem-

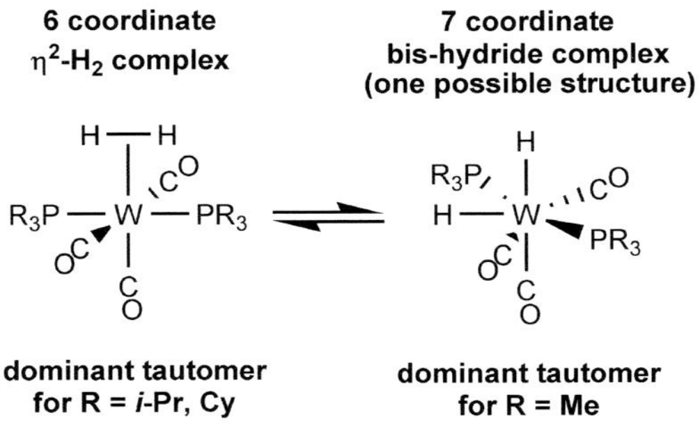

6 coordinate
η^2-H_2 complex

dominant tautomer for R = *i*-Pr, Cy

7 coordinate
bis-hydride complex
(one possible structure)

dominant tautomer for R = Me

Scheme 4.4

ingly subtle changes of the ancillary ligands have a profound effect on the relative stabilities of the η^2-H_2 and bis-hydride forms [54].

The steric bulk of the ligand set is also important in determining whether a given complex will exist in the η^2-H_2 form or the corresponding bis-hydride form. The higher formal oxidation state and higher coordination number of the bis-hydride form leads to shorter metal–ligand distances and smaller ligand–metal–ligand angles than in the corresponding η^2-H_2 complex. The presence of bulky ligands therefore disfavors the bis-hydride form.

The work of Kubas, Heinekey and their respective coworkers illustrates the importance of the steric properties of the ligand set in determining the relative energies of the bis-hydride and η^2-H_2 forms (Scheme 4.4). Although the PCy_3, P^iPr_3 and PMe_3 ligands are similar in their electron-donating ability, the PCy_3 and P^iPr_3 ligands are much larger. For the PCy_3- and P^iPr_3-containing complex, the six-coordinate $(\eta^2\text{-}H_2)W(CO)_3(PR_3)_2$ form is in dynamic equilibrium with the seven-coordinate $(H)_2W(CO)_3(PR_3)_2$ form with the η^2-H_2 form being the dominant tautomer [55]. The corresponding PMe_3-containing complex exists exclusively in the bis-hydride tautomer [56].

4.2.4
Elongated η^2-H_2 Complexes

Complexes with H–H distances between 1.0 and 1.5 Å are often referred to as "elongated" or "stretched" dihydrogen complexes [57]. The properties of these complexes make it difficult to justify their classification as either "true" dihydrogen complexes ($d_{H-H} < 1.0$ Å) or *cis*-dihydride complexes ($d_{H-H} > 1.5$ Å). Other authors further categorize these complexes as "true" elongated dihydrogen complexes ($d_{H-H} < 1.0$–1.3 Å) and compressed bis-hydrides ($d_{H-H} < 1.3$–1.5 Å) [46]

There is no consensus on the nature of elongated dihydrogen complexes and how they should best be described. While electronic structure calculations generally provide accurate H–H distances for true η^2-H_2 complexes and true *cis*-dihydride complexes, they often fail to replicate the H–H distances observed in the molecular structures determined by single-crystal neutron diffraction for elongated dihydrogen complexes. Instead, these computations often predict that these complexes should exist as either a true η^2-H_2 complex, a true *cis*-hydride complex or an equilibrium mixture of the two species.

Lluch, Lledos and coworkers have shown that the potential energy surface for elongation or contraction of the H–H distance is remarkably flat in the case of elongated dihydrogen complexes [58–62]. Heinekey and coworkers state, "...the description of ... [certain elongated dihydrogen complexes]... as a dihydrogen or dihydride complex loses its significance, and it is more appropriate to describe it as a complex containing two H atoms moving freely in a wide region of the coordination sphere of the metal" [20]. They argue that the elongated dihydrogen complexes cannot, in general, be described by a single static structure.

The H–H distance in elongated dihydrogen complexes, as determined by the value of J_{HD}, is often dramatically affected by small changes in temperature [63–65]. This result provided strong experimental evidence that the potential energy surface is flat with respect to changes in the H–H distance.

Hall and coworkers used density functional theory calculations to assign the inelastic neutron scattering derived vibrational spectrum of the elongated dihydrogen complex, (Tp*)Rh(H)$_2$(η^2-H_2) [66]. They conclude that the H–H distance derived from neutron diffraction for the (Tp*)Rh(H)$_2$(η^2-H_2) complex may in fact correspond to the average of the H–H distances of the tetrahydride and bishydride/η^2-H_2 species.

4.2.5 Experimental Gauges of the H–H Interaction and Degree of Activation

Ever since the discovery of the first η^2-H_2 complex by Kubas, there has been the lingering question: after coordination to a metal center, what remains of the H–H bond? A series of experimental methods for answering this question are presented below.

There are several experimental tools available for the determination of the H–H distance and the degree of the H–H bonding interaction. Neutron diffraction studies provide an accurate measure of the H–H distance. The measurement of the spin-lattice proton relaxation time, T_1, for an η^2-H_2 complex or the proton–deuteron coupling constant, J_{HD}, for the corresponding isotopically substituted η^2-HD complex via ^{1}H nuclear magnetic resonance (NMR) spectroscopy provides a quantitative measure of the H–H distance. The frequency of the ν(H–H) stretching band, as determined by Raman or infrared (IR) spectroscopy of η^2-H_2 complexes provides semiquantitative information about the strength of the H–H interaction.

4.2.5.1 Neutron Diffraction

The two main methods for the determination of the three-dimensional (3-D) structures of molecules are single crystal X-ray diffraction and single-crystal neutron diffraction studies [67]. Single-crystal X-ray diffraction generally requires smaller crystals and is widely available. It is the method of choice for determining the 3-D structure of most molecules. The constituent atoms of the crystal are located by the way in which their electron clouds scatter X-ray radiation. Since H-atoms have only a single electron, they are difficult to locate accurately using X-ray diffraction. Neutrons are scattered by the atomic nuclei and H-atoms have the largest neutron cross-section of all elements. Thus, single-crystal neutron diffraction is the method of choice for determining the positions of H-atoms.

4.2.5.2 ^{1}H NMR Studies: HD Coupling

The magnitude of the NMR coupling constant between atoms A and B, J_{AB}, is related to the spatial orientation of those atoms. Non-interacting atoms have J_{AB} values at or near 0 Hz. The largest values of J_{AB} occur when atoms A and B are connected by a direct chemical bond.

The J_{HD} coupling constant provides a measure of the H–D interaction in η^2-HD complexes. The observed J_{HD} coupling constant for HD gas is 43.2 Hz [68]. The coordination of HD to a transition metal center leads to weakening of the H–D bond, an increase in the H–D distance and a decrease in the J_{HD} coupling constant. Morris, Heinekey and their respective coworkers have shown that the value of J_{HD} for a given η^2-HD complex is linearly related to the H–H distance of the corresponding η^2-H_2 complex as determined by neutron diffraction [69]. The J_{HD} coupling constant ranges from the maximum value of 43.2 Hz for free HD to 0–5 Hz for *cis*-hydride–deuteride complexes [46].

4.2.5.3 ^{1}H NMR Studies: Proton Relaxation Time (Measurements)

It has been observed that the proton spin lattice relaxation time, T_1, measured for the H-atoms of the η^2-H_2 ligand is extraordinarily short (tens of milliseconds) when compared to dihydride complexes (hundreds of milliseconds). Crabtree and coworkers have shown that in general the value of T_1 is related to the inverse of the sixth power of the H–H distance [70–73]. Unfortunately the interpretation of T_1 can be complicated by other factors, such as the magnetogyric ratio of the metal nucleus and the proximity of other ligands to the η^2-H_2 ligand [74, 75].

4.2.5.4 IR and Raman Spectral Studies: ν(H–H) Measurements

The analysis of the IR or Raman spectra of the H–H stretching band of η^2-H_2 complexes can give qualitative and semiquantitative information about the strength of the H–H interaction, but suffers from several drawbacks. For IR spectroscopy, the ν(H–H) band is often weak and/or obscured by the IR bands of co-ligands. Since both $H_2 \rightarrow M$ donation and $M \rightarrow H_2$ back-donation weaken the H–H bond, the ν(H–H) stretch does not correlate exactly with the electron

Table 4.3 IR data for several η^2-H_2 complexes

Complex	ν(H–H)	Reference
$Tp^*RuH(\eta^2\text{-}H_2)$ (THT)	2250	78
$Tp^*RuH(\eta^2\text{-}H_2)_2$	2361	78
$CpNb(CO)_3(\eta^2\text{-}H_2)$	2600	79
$CpV(CO)_3(\eta^2\text{-}H_2)$	2642	79
$W(CO)_3(PCy_3)_2(\eta^2\text{-}H_2)$	2690	80
$W(CO)_3(P^iPr_3)_2(\eta^2\text{-}H_2)$	2695	80
$W(CO)_5(\eta^2\text{-}H_2)$	2711	81
$Cr(CO)_5(\eta^2\text{-}H_2)$	3030	81, 82

density provided by the metal center. The H–H stretch becomes a H–M–H bend as the H–H distance increases [60, 76]. In other words, the ν(H–H) band observed in the spectra of an η^2-H_2 complex is not a pure H–H stretch; it consists of a complex mixture of H–H, M–H stretching and H–M–H bending motions. The nature of the metal center and the coligands exerts a large influence on the H–M–H bend and the value of the "ν(H–H)" band in the spectra of these complexes.

The coordination of H_2 to a metal center leads to weakening of the H–H bond as evidenced by the ν(H–H) stretching frequency. The ν(H–H) stretching frequency of free H_2 as measured by gas-phase Raman spectroscopy is 4161 cm^{-1} [77]. As shown in Table 4.3, the ν(H–H) stretching frequencies of η^2-H_2 complexes are significantly lower than the ν(H–H) stretch of free H_2. The comparison of $Tp^*RuH(\eta^2\text{-}H_2)_2$ and $Tp^*RuH(\eta^2\text{-}H_2)$(THT) or $W(CO)_5(\eta^2\text{-}H_2)$ and $W(CO)_3(PR_3)_2(\eta^2\text{-}H_2)$ shows that the presence of better donor ligands in the complex lowers the ν(H–H) stretching frequency. The comparison of $CpM(CO)_3(\eta^2\text{-}H_2)$ (M = Nb, V) or $M(CO)_5(\eta^2\text{-}H_2)$ (M = Cr, Mo, W) shows that the heaviest congener has the lowest ν(H–H) stretching frequency. This trend is expected because the heavier congeners have more stable higher oxidation states are more capable of donating electron density to the σ^* orbital of H_2.

4.3 Intramolecular H-Atom Exchange

Metal bound H-atoms undergo a number of unique intermolecular and intramolecular exchange processes. In the words of Greg Kubas: "Transition metal complexes containing η^2-H_2 ligands and hydride ligands are unquestionably the most dynamic ligand systems known" [4].

In general, the rapid exchange of metal-bound H-atoms benefits from the nature of the 1s valence orbital of hydrogen. The spherical shape of the 1s orbital allows H-atom exchange to occur by associative mechanisms (with low barriers) in which H–H bond formation coincides with metal–hydrogen bond breaking.

Theoretical studies of the reductive elimination reactions of Pd^{II} and Pt^{II} bis-hydride, bis-alkyl and *cis*-hydride alkyl complexes illustrate the important role of the shape of 1s valence orbital of hydrogen on the rates of these reactions. It had been observed experimentally that CH_4 loss from $Pt(H)(CH_3)(PPh_3)_2$ is quite facile even at $-25\,^\circ C$ [83], while the $Pt(CH_3)_2(PPh_3)_2$ complex is stable against CH_3CH_3 loss up to $237\,^\circ C$ [84]. Low and Goddard, utilizing electronic structure calculations, demonstrated that the observed reactivity (or lack thereof) was a result of the barrier to reductive elimination and not the thermodynamics of the reaction [85–90]. They argue that while the sp^3 hybridized orbital of CH_3 requires nearly complete scission of the M–C bond prior to C–C bond formation for $Pt(CH_3)_2(PPh_3)_2$ the spherical nature of the 1s orbital of H allows for the formation of the C–H bond to begin without complete cleavage of M–C and M–H bonds of $Pt(H)(CH_3)_2(PPh_3)_2$

4.3.1 Rotation of η^2-H_2 Ligands

In many true η^2-H_2 complexes (H–H distance < 1.0 Å), the H-atoms of the η^2-H_2 ligand rapidly interconvert on the NMR timescale, even at very low temperatures. The putative exchange process involves rotation of the η^2-H_2 ligand about the M-H_2 axis (Scheme 4.5). The electronic portion of the barrier to H_2 rotation is directly related to the change in $M \rightarrow H_2$ back-donation between the minimum and transition state structures [17]. The H_2 ligand's orientation is largely dictated by $M \rightarrow H_2$ back-donation, since the $H_2 \rightarrow M$ portion of η^2-H_2 bonding is relatively unaffected by the orientation of the H_2 ligand with respect to the M-H_2 axis.

Complexes of low-spin d^6 metals have the d_{xy}, d_{xz} and d_{yz} orbitals fully occupied. In η^2-H_2 complexes of low-spin d^6 metals, the d_{xz}, d_{yz} or a linear combination of the d_{xz} and d_{yz} orbitals can effectively overlap with the σ^* orbital of H_2. Therefore, the orientation of the η^2-H_2 ligand is largely dictated by the π-donating or π-accepting abilities of the ligands bound *cis* to the η^2-H_2 ligand. For complexes in which the *cis* ligands are similar to one another, the change in $M \rightarrow H_2$ back-donation and the barrier to H_2 rotation is expected to be relatively small. For complexes in which the *cis* ligands are very different from one another, the change in $M \rightarrow H_2$ back-donation and the barrier to H_2 rotation is expected to be relatively large. For example, the $(\eta^2\text{-}H_2)W(CO)_3(PCy_3)_2$ complex, in which the d orbitals responsible for $M \rightarrow H_2$ back-donation are all filled and

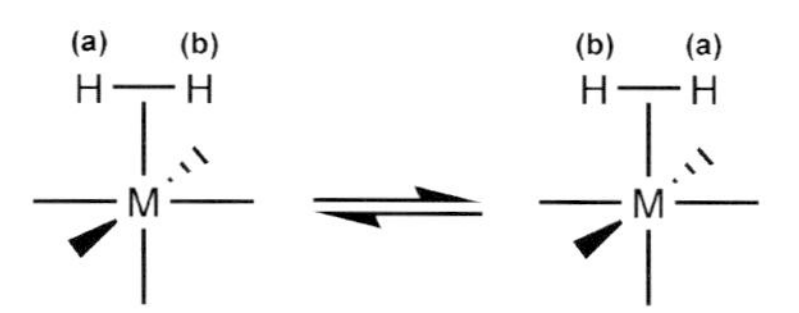

Scheme 4.5

have reasonably similar energies, generally have low barriers to η^2-H_2 rotation ($\Delta G^{\neq} = 2.2$ kcal mol^{-1}) [41].

For complexes that have only one filled orbital that is capable of $M \rightarrow H_2$ back-donation or in which the filled d orbitals have drastically different energies, the barrier to η^2-H_2 rotation can be quite large. The d^2 complexes, $[CpTa(CO)(\eta^2\text{-}H_2)]^{1+}$ and $[Cp'Nb(CNR)(\eta^2\text{-}H_2)]^{1+}$ $[Cp' = C_5H_4Si(CH_3)_3]$, have been shown experimentally to have particularly large barriers to η^2-H_2 rotation (around 10 kcal mol^{-1}) [91–93]. In these complexes, the low-energy orientation of the η^2-H_2 ligand generally corresponds to that which optimizes $M \rightarrow H_2$ back-donation between the sole filled d orbital on the metal center and the η^2-H_2 ligand. In the transition state for η^2-H_2 rotation there can be complete loss of $M \rightarrow H_2$ back-donation.

Interactions between the bound η^2-H_2 ligand and the co-ligands can also affect the barrier to H_2 rotation by making some orientations of the η^2-H_2 ligand significantly more stable others. In general, these interactions may be electrostatic or may result from orbital overlap between the metal-bound η^2-H_2 ligand and one of the *cis* coligands. Large barriers for η^2-H_2 rotation are often observed when an η^2-H_2 ligand is bound *cis* to a hydride ligand [94]. This effect has been examined computationally for a series of octahedral Ir^{III} complexes [95, 96]. The computed barriers to η^2-H_2 rotation for three of these complexes are shown in Scheme 4.6. The leftmost and rightmost complexes in Scheme 4.6 have either two chloride ligands or two hydride ligands *cis* to the η^2-H_2 ligand and have low computed barriers to η^2-H_2 rotation. The center complex in Scheme 4.6 has one chloride ligand and one hydride ligand *cis* to the η^2-H_2 ligand, and has a significantly larger barrier. If the interaction were purely electrostatic, one would expect the value of the computed barrier to η^2-H_2 rotation for the center structure to be intermediate between that of the other two structures. Albinati and coworkers [95] and Eckert and coworkers [96] show the stabilizing interaction observed for the central structure in Scheme 4.6 arises from donation of electron density from the M–H σ bond into the η^2-H_2 σ^* bond leading to a weak bonding interaction between the hydride ligand and the nearby hydrogen of the η^2-H_2 ligand. In our view this effect is related to the partial formation of a three-center, four-electron bond, i.e. contribution from an H_3^- resonance structure.

Eckert and coworkers have used inelastic neutron scattering (INS) experiments to measure transitions between the rotational energy levels of the bound H_2 ligand, which allows the determination of the barrier to rotation of the η^2-H_2 ligand [97–99] and the H–H distance [100]. One difficulty in INS experi-

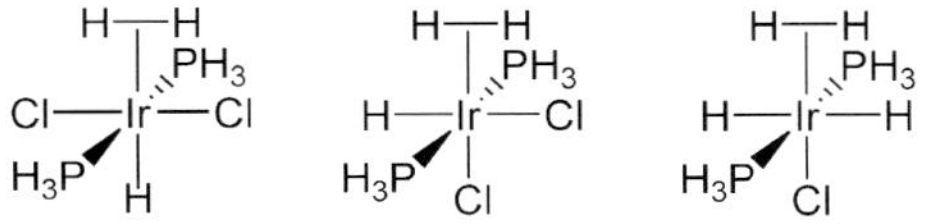

	Cl, Cl *cis*	H, Cl *cis*	H, H *cis*
$\Delta G^{\ddagger}$	2.2	6.5	2.3

Scheme 4.6

ments is that the large number of H-atoms present in the coligands can obscure the transitions due to the η^2-H_2 ligands. This problem is often remedied by measuring the difference spectrum between the η^2-H_2 complex and the corresponding η^2-D_2 complex or by synthesizing a complex which contains perdeuterated coligands.

4.3.2
H_2/H^- Exchange

There is often rapid exchange between η^2-H_2 ligands and neighboring hydrides as evidenced by NMR experiments. Even at very low temperatures, a single "hydride" resonance is observed in the ^1H NMR spectra of most of these complexes. In addition, reaction of the protio form of many of these complexes with D_2 gas leads to the incorporation of deuterium into the complex and the formation of HD.

In one mechanism, the H_2/H^- exchange process can be considered an example of internal heterolytic cleavage of the η^2-H_2 ligand (Fig. 4.3). The protonation of hydride ligands by an external acid and deprotonation of η^2-H_2 by an external base are common processes. In the H_2/H^- exchange process η^2-H_2 acts as an internal acid and the hydride ligand acts as an internal base. In this process, the η^2-H_2 ligand transfers H^+ to the neighboring hydride ligand to afford a new hydride and a new η^2-H_2 ligand. The transition state structures for this type of rearrangement often feature a linear or nearly linear orientation of the three H-atoms, such that an H_3^- ligand appears bound to the metal center in the transition state. For example, low-temperature protonation of d^0 pentahydride com-

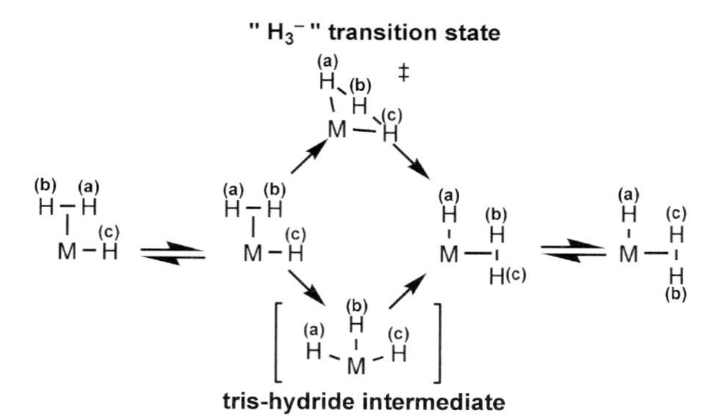

Fig. 4.3 Possible mechanisms of H-atom exchange for a complex containing both an η^2-H_2 ligand and a single hydride ligand. In the top mechanism, there is direct exchange of "H^+" between the η^2-H_2 and H^- ligands. In the bottom mechanism, the η^2-H_2 ligand oxidatively adds to the metal center to yield a tris-hydride species. Reductive elimination of hydrides (b) and (c) yields a "new" hydride/η^2-H_2 complex. Either mechanism requires η^2-H_2 rotation for complete H-atom exchange.

plexes of the form $[Cp^*M(H)_5(PR_3)]^{1+}$ (M = Mo, W) leads to thermally unstable "hexahydride" species [101]. These complexes display a single resonance in the hydride region of their ^1H NMR spectra. Bayse et al., utilizing electronic structure calculations, demonstrated that the lowest energy structure for these complexes corresponds to a dihydrogen/tetrahydride complex of the form $[Cp^*M(H)_4(\eta^2H_2)(PR_3)]^{1+}$. Both η^2-H_2 rotation and H_2/H^- exchange pathways were computed to have low activation energies. The structure of the transition state for H-atom exchange between the η^2-H_2 ligand and its neighboring hydride ligands resembles a linear H_3 molecule bound to the molybdenum center. Recently, Webster et al. reported experimental and theoretical evidence for a similar exchange mechanism in Cp^*OsH_4L (L = PPh_3, $AsPh_3$, PCy_3) [102].

An alternative mechanism for H_2/H^- exchange requires oxidative addition of the η^2-H_2 ligand (also shown in Fig. 4.3). The $IrX(H)_2(\eta^2$-$H_2)(P(i\text{-}Pr)_3)_2$ (X = Cl, Br, I) complexes undergo rapid H-atom exchange [103–105]. Li et al. conclude on the basis of density functional theory calculations that the most likely mechanism involves oxidative addition of the η^2-H_2 ligand to form a transient tetrahydride complex, followed by pairwise reductive elimination of hydride ligands [106].

4.3.3 Hydride–Hydride Exchange

There is often rapid exchange between neighboring hydride ligands as evidenced by NMR experiments. Even at very low temperatures, a single hydride resonance is observed in the ^1H NMR spectra of most of these complexes. The most common mechanism proposed for H-atom exchange in polyhydride complexes involves the formation of transient η^2-H_2 ligands (Scheme 4.7). In this mechanism, a pair of neighboring hydride ligands interact to form an η^2-H_2 ligand. The η^2-H_2 ligand can undergo rotation about the M–H_2 axis before converting back to two hydride ligands. When three or more hydride ligands are found to be exchanging, it is generally assumed that rapid pair-wise exchange between each neighboring set is involved.

Although pairwise exchange of hydride ligands via formation of transient η^2-H_2 is the most commonly invoked mechanism to explain hydride–hydride exchange in polyhydride complexes, other low-energy pathways may exist. In examining possible hydride exchange processes for pentahydride complex $CpOs(H)_5$, Bayse et al. concluded that the hydride ligands of this complex exchange three at a time via a trigonal twist mechanism [107].

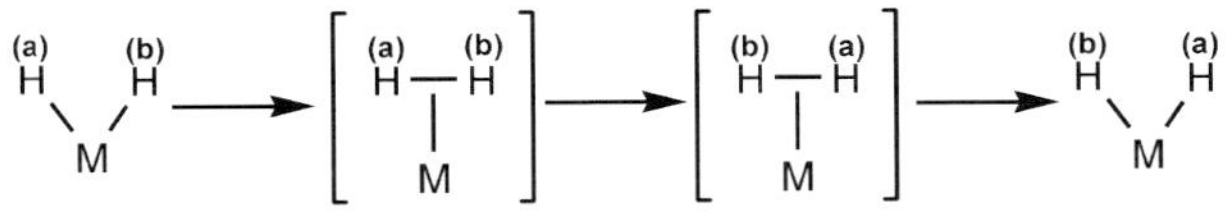

(a) H (b) H bonded to M → [(a) H–H (b) bonded to M] → [(b) H–H (a) bonded to M] → (b) H (a) H bonded to M

Scheme 4.7

4.4 Nonclassical H-Bonds

4.4.1 Hydride Ligands as Nonclassical H-bond Acceptors

The hydride ligand can act as a nonclassical H-bond acceptor (Scheme 4.8). The noncovalent interactions between H-bond donors and hydride ligands are often referred to as protonic–hydridic bonding [108] interactions. Morris, Crabtree and their respective coworkers reported examples of intramolecular protonic–hydridic interactions, which produce H–H distances of 1.7–1.8 Å [109, 110]. It remained difficult for some time to prove definitively that the protonic–hydridic interaction is a stabilizing interaction and not simply a consequence of the steric constraints imposed by the ligand set. Crabtree and coworkers [111–114] cocrystallized indole and $Re(H)_5(PPh_3)_3$. The solid-state structure demonstrated the presence of an intermolecular nonclassical H-bond between the NH hydrogen of indole and a hydride ligand of $Re(H)_5(PPh_3)_3$.

4.4.2 η^2-H_2 as a Nonclassical H-Bond Donor

The donation of electron density from dihydrogen to the metal center depletes the electron density of the η^2-H_2 ligand. The η^2-H_2 ligand, therefore, can act as a nonclassical H-bond donor (Scheme 4.9). The η^2-H_2 ligand may act as an intramolecular H-bond donor to hydrides, halides and other negatively charged ligands as well as common H-bond donors such as amines and alcohols. Neutron diffraction studies often find that the η^2-H_2 ligand is oriented to place its H-atoms as close as possible to *cis*-hydride ligands. In fact, the preference for the H-atoms of the η^2-H_2 ligand to be coplanar with a *cis*-hydride ligand is so common that it has been termed the "*cis* effect of hydrides" [94]. The η^2-H_2 ligand is

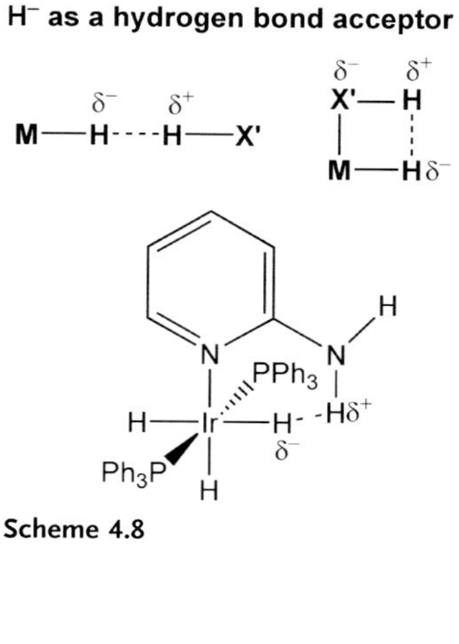

H^- as a hydrogen bond acceptor

Scheme 4.8

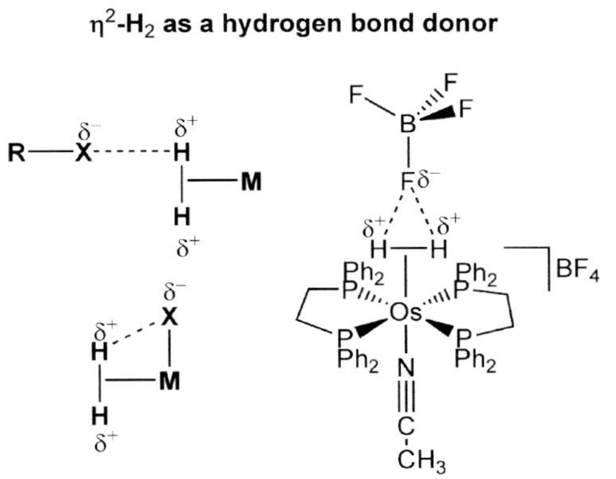

η^2-H_2 as a hydrogen bond donor

Scheme 4.9

also capable of forming intermolecular H-bonds. Morris and coworkers have noted that one of the fluorine atoms of the BF_4 or PF_6 counterions is consistently oriented toward the η^2-H_2 ligand in the solid-state structures of a series of cationic η^2-H_2 complexes [115, 116].

4.5 Reactivity of Metal-bound H-Atoms

4.5.1 How Does the Reactivity of Metal-bound H-Atoms Compare to that of Free H_2?

The binding of dihydrogen to a transition metal center can greatly change its reactivity (Scheme 4.10). The H–H bond of an η^2-H_2 complex is weaker than the H–H bond in free dihydrogen, making cleavage of the H–H more thermodynamically favorable. An η^2-H_2 complex is therefore generally more thermodynamically acidic than free H_2. When coordinated to an electrophilic transition metal center, in which $H_2 \rightarrow M$ donation dominates, the hydrogens atoms of the

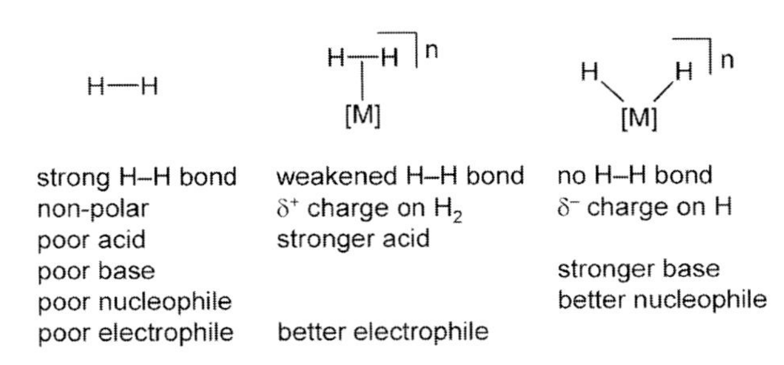

Scheme 4.10

η^2-H_2 ligand become positively charged, and therefore more kinetically acidic than free H_2. When coordinated to an electron-rich transition metal center, in which $M \rightarrow H_2$ donation dominates, dihydrogen may oxidatively add to the metal center to generate bis-hydride complex. The resulting hydride ligands may be either hydridic or protonic in nature.

4.5.2
Metal-Monohydride Species – "Hydride Ligands can be Acidic!"

Before venturing into the more complicated case of polyhydrides, there is one point that should be made about transition metal hydrides in general. The term hydride is sometimes misleading. To many chemists, this term implies an H^- ionically bound to metal. This term may be used with relative inpunity for main group hydrides such as NaH (Na^+H^-). When applied to transition metal complexes, however, this term is sometimes a misnomer. In terms of their reactivity, transition metal hydrides can function as hydride (H^-) donors, proton (H^+) donors or be quite covalent in their bonding.

Morris and coworkers have measured the proton dissociation constants for a series of mono-protic species in THF solution (pK_a^{THF}) and/or estimated these values from pK_a measurements in other solvents. The selected pK_a^{THF} values given in Table 4.4 span 24 pK_a units. The most acidic of these complexes, $[(\eta^5\text{-}C_5Me_5)_2OsH]^{1+}[OTf]^{1-}$, is significantly more acidic than $[HNEt_3]^{1+}[BPh_4]^{1-}$. Dubois and coworkers have determined the pK_a values for a number of nickel, palladium and platinum complexes [117–127].

Norton and coworkers examined the thermodynamic and kinetic acidity of a series of transition metal complexes [128–133]. They found that the rates of H^+ transfer from transition metal hydrides can be quite slow when compared to organic and mineral acids of similar pK_a's [134]. Walker et al. attribute the slow rates of H^+ transfer from hydride ligands to the following factors (illustrated in

Table 4.4 Proton dissociation constants for a series of compounds in THF solvent

Compound	pK_a (THF)
HH	49 a)
$(\eta^5\text{-}C_5H_5)W\mathbf{H}(PMe_3)(CO)_2$	32 a)
$(\eta^5\text{-}C_5Me_5)Fe\mathbf{H}(CO)_2$	31 ± 4
$[Pt\mathbf{H}(dmpe)_2]^{1+}[PF_6]^{1-}$	21 a)
$[Ni\mathbf{H}(dmpe)_2]^{1+}[PF_6]^{1-}$	18 a)
$(\eta^5\text{-}C_5H_5)Mo\mathbf{H}(CO)_3$	17 ± 1
$[Pt\mathbf{H}(dppe)_2]^{1+}[PF_6]^{1-}$	16 a)
$[\mathbf{H}NEt_3]^{1+}[BPh_4]^{1-}$	12.5
$[(\eta^5\text{-}C_5Me_5)_2Os\mathbf{H}]^{1+}[OTf]^{1-}$	6

a) Estimated from pK_a in another solvent.

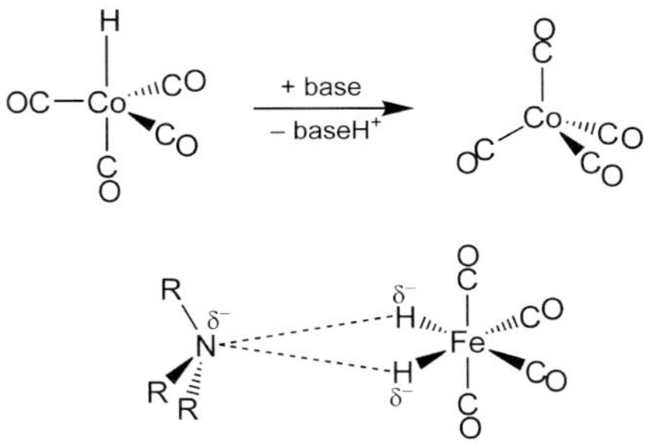

Scheme 4.11

Scheme 4.11). (i) Bases generally have a partial negative charge on the atom that is to accept H^+. This partial negative charge is repelled by the anionic nature of the hydride ligand. (ii) The acid and conjugate base may have very different orientation of the ligand set. (iii) The steric bulk of the coligands may not allow for a close approach to the hydride ligand.

4.5.3
Increased Acidity of η^2-H_2

Dihydrogen gas is a very poor acid (pK_a of H_2 dissolved in THF = 49). The binding of dihydrogen to a transition metal complex to form a η^2-H_2 complex can greatly increase both its thermodynamic and kinetic acidity [8, 16]. The *thermodynamic acidity* of an given acid in a given solvent is related to the change in free energy (ΔG) upon proton loss in that solvent. The pK_a of an acid, which is measured when the acid and its conjugate base have reached equilibrium, is a measure of that acid's thermodynamic acidity. *Kinetic acidity* deals with the rate of proton transfer. Acids with a high kinetic acidity lose H^+ at a fast rate.

The binding of dihydrogen to a metal to form an (η^2-H_2)[M] can increase the thermodynamic acidity of the H_2 molecule relative to free H_2 in at least two ways: (i) the binding of H_2 to a metal to form an η^2-H_2 complex weakens the H–H bond and (ii) the metal fragment acts as a built-in hydride acceptor following deprotonation: $(\eta^2\text{-}H_2)M + B \rightarrow BH^+ + MH^-$. The binding of dihydrogen to a metal to form an (η^2-H_2)[M] can increase the kinetic acidity of the H_2 molecule relative to free H_2. Molecular hydrogen is completely nonpolar. Dihydrogen to metal [$H_2(\sigma) \rightarrow M$] donation induces a partial positive charge on the H-atoms, making these atoms more kinetically accessible to nucleophiles (bases), and even without a permanent positive charge the metal-bound H_2 is more polarizable and can use the metal to shift electron density such that upon the approach of a base one of the H-atoms can become more positively charged.

The binding of dihydrogen to a metal center can lead to highly acidic η^2-H_2 complexes. The most acidic of these complexes are generated by protonation of a hydride complex by strong acid. A few of these complexes have been prepared

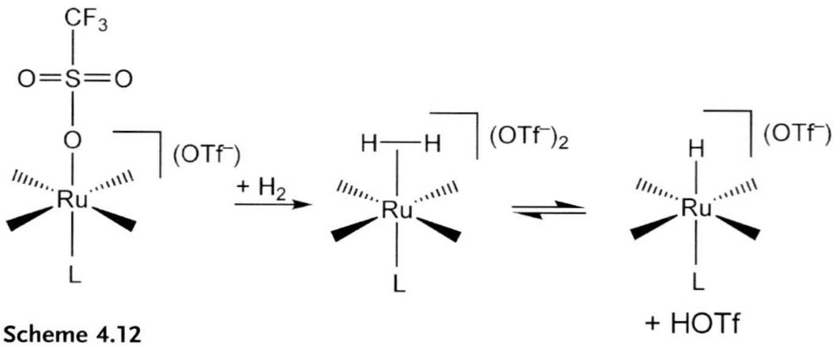

Scheme 4.12

using dihydrogen gas. Morris, Jagirdar and their respective coworkers have demonstrated that the coordination of dihydrogen to a Ru^{II} center produces an η^2-H_2 complex that is as acidic as triflic acid (CF_3SO_3H) (Scheme 4.12) [135, 136].

Many η^2-H_2 complexes are in dynamic equilibrium with the corresponding bis-hydride complexes. The η^2-H_2 complex and corresponding bis-hydride complex generally may differ in terms of thermodynamic and kinetic acidity. The minor tautomer is the stronger thermodynamic acid (lowest pK_a) [137, 138]. Deprotonation of either the η^2-H_2 complex or bis-hydride complex generally leads to the same monohydride complex. As shown in Scheme 4.13, the minor tautomer is necessarily less stable in terms of its free energy (ΔG). Therefore, the difference in free energy between the protonated and deprotonated forms is smaller for the minor tautomer, and hence the pK_a of the minor tautomer is lower than that of the major tautomer. The H-atoms of an η^2-H_2 complex generally have a partial positive charge, while those of a bis-hydride complex generally have a partial negative charge. Thus, the η^2-H_2 complex generally has a higher kinetic acidity than the corresponding bis-hydride complex.

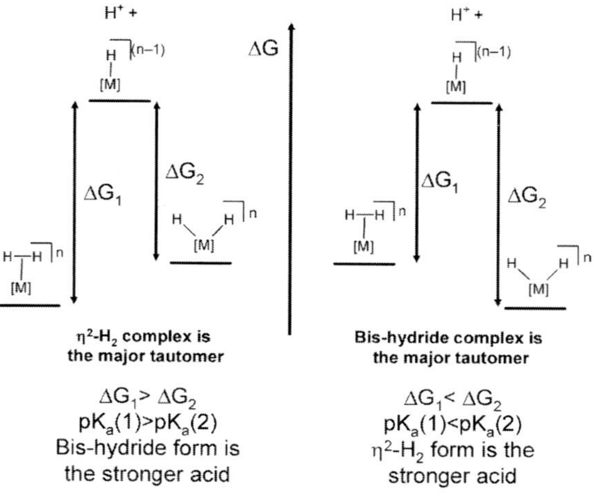

Scheme 4.13

4.5.4
Seminal Work: Intramolecular Heterolytic Cleavage of H_2

Morris [139], Crabtree [140] and their respective coworkers have investigated the binding of dihydrogen to transition metal centers which contain "built-in" basic functionalities near the H_2 binding site (as shown in Scheme 4.14). Using these complexes, they observed the intramolecular heterolytic splitting of dihydrogen.

Nonclassical H-bonding has an important role in the intramolecular heterolytic cleavage of H_2 in these complexes (as illustrated in Scheme 4.15). H-bonding between the internal base and the η^2-H_2 ligand stabilizes the η^2-H_2 complex with respect to H_2 loss, and orients the reactant atoms for efficient H^+ transfer. The stabilization of the η^2-H_2 form with respect to H_2 loss is quite important since the most acidic η^2-H_2 complexes often contain the most labile H_2 ligands. Once H^+ transfer has occurred, a nonclassical H-bond between the resulting hydride ligand and protonated base adds to the stability of the deprotonated form.

4.6
Recent Advances in the Activation of Dihydrogen by Synthetic Complexes

4.6.1
H_2 Uptake by a Pt–Re Cluster

Adams and Captain demonstrated that the heterometallic cluster $[Pt(P^tBu_3)]_3$-$[Re(CO)_3]_2$ reacts with three equivalents of H_2 at 25 °C to yield $(\mu_2$-$H)_6$-$[Pt(P^tBu_3)]_3[Re(CO)_3]_2$ [141]. The spatial arrangement of the metal center and the coligands are nearly identical for the two structures. The authors state that [this system provides] "...the first example of the addition of three equivalents of H_2 by an intact molecular metal–cluster complex" [141].

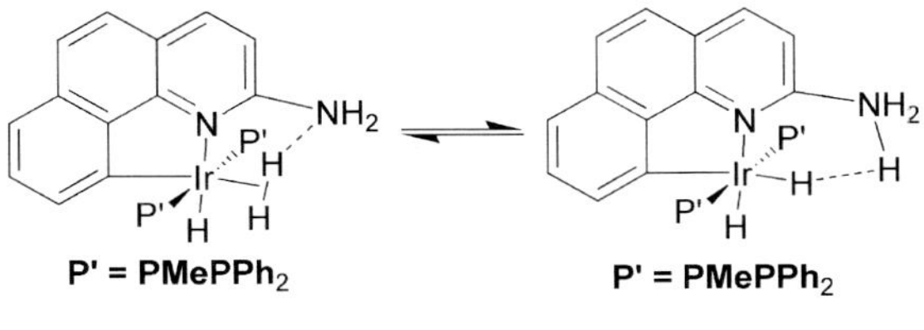

P' = $PMePPh_2$ P' = $PMePPh_2$

Scheme 4.14

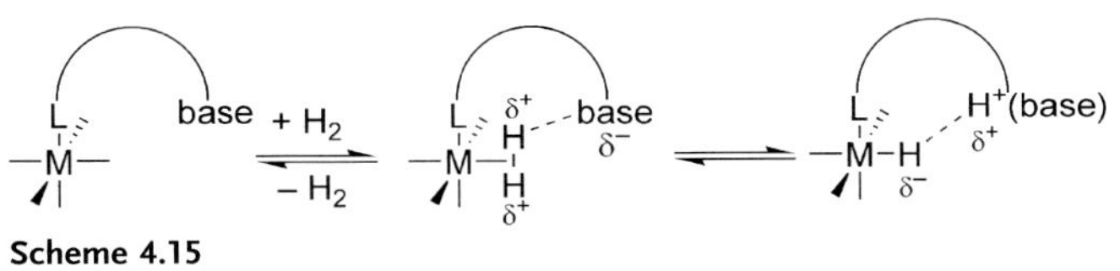

Scheme 4.15

4.6.2
H_2 Binding to Ir^{III} Initiates Conversion of CF_3 to CO

The addition of a strong mineral acid to complexes with fluoroalkyl ligands sometimes leads to cleavage of a C–F bond [142–144]. Eisenberg and coworkers recently demonstrated that the binding of dihydrogen to the electrophilic Ir^{III} center of $[Ir(CF_3)(CO)(dppe)(DIB)][B^{ArF}]_2$ leads to the facile cleavage of a C–F bond of the CF_3 ligand [145]. The proposed mechanism involved deprotonation of an iridium-bound η^2-H_2 ligand and loss of HF.

4.6.3
Encapsulation of H_2 in C_{60}

Komatsu and coworkers developed a new process for the insertion of small molecules into fullerenes [146]. In this process, referred to as "molecular surgery", the fullerene is opened, a small molecule is inserted and the fullerene is reclosed. Using this procedure, they were able to synthesize C_{60} with a single molecule of H_2 trapped inside ($H_2@C_{60}$ in the parlance of fullerene chemistry).

4.6.4
Conversion of Biomass to H_2

The confectionary and agricultural industries produce large amounts of biomass, which contains unutilized energy. Cortright and coworkers have demonstrated the effective conversion of alcohols and sugars derived from biomass to H_2 using 3 wt% Pt/Al_2O_3 as a catalyst [147]. The H_2/CO_x ($x=1$, 2) can be used directly or converted into small molecular weight hydrocarbons.

4.6.5
First Group 5 η^2-H_2 Complex

Chaudret and coworkers reported the synthesis of the first η^2-H_2 complex of a Group 5 metal [148]. The reaction of $Cp''_2Ta(H)(CO)$ ($Cp''=C_5H_5$ or $C_5H_4{}^tBu$) reacts with HBF_4 at –78 °C to yield the corresponding η^2-H_2 complex, $[Cp''_2Ta(\eta^2\text{-}H_2)(CO)]^{1+}$ as the major product. This complex is shown to possess an η^2-H_2 ligand by both X-ray diffraction and measurement of the J_{HD} coupling constant of the corresponding η^2-HD complex. As might be expected for a d^2 complex the barrier to η^2-H_2 rotation is fairly large (9.6 kcal mol^{-1}).

4.7
Enzymatically Catalyzed Dihydrogen Oxidation and Proton Reduction

Two major classes of H_2 activating and/or producing enzymes are the hydrogenases (H_2ases) [149–151] and nitrogenases (N_2ases) [152, 153]. The H_2ase enzymes are known to catalyze dihydrogen oxidation, proton reduction, dihydrogen detection and dihydrogen utilization in cells. The N_2ase enzymes catalyze the reduction of molecular nitrogen to ammonia (nitrogen fixation). For reasons that are not completely clear, the N_2ase enzymes couple H^+ reduction to the N_2 reduction process, i.e. $N_2 + 8\,H^+ + 8\,e \rightarrow 2\,NH_3 + H_2$.

4.7.1 General Information about H_2ase Enzymes

The H_2ase enzymes may be broadly classified by specifying the transition metal content of their active sites. The three main classes of H_2ase enzymes are the nickel–iron ([NiFe]) [154–156], iron–iron ([FeFe]) [157, 158] and the so called iron–sulfur cluster-free H_2ases, which until recently were thought to be "metal-free" [159–162]. The [NiFe] enzymes are primarily utilized for hydrogen oxidation, while the [FeFe] enzymes are primarily utilized for proton reduction. The iron–sulfur cluster-free H_2ases are H_2-utilizing enzymes, which activate dihydrogen for use in catabolic processes within the cell, but do not catalyze H^+ reduction or H_2 oxidation. Certain organisms also contain H_2-sensing H_2ases, which regulate H_2 oxidation and/or H^+ reduction in these organisms [163, 164].

4.7.1.1 [NiFe]H_2ase

Single-crystal X-ray diffraction studies have defined the basic framework of the active site of the [NiFe]H_2ases as consisting of an iron center and a nickel center bridged by cysteinate sulfur atoms (Scheme 4.16) [165–169]. The nickel center is further coordinated by terminal cysteinate sulfur ligands, and the iron center is further coordinated by two CN^- ligands and one CO ligand. The nature of a third ligand, which bridges the two metal centers in certain redox states of the enzyme, is currently a matter of some contention [170, 171]. A gas access channel [172] and a series of ferrodoxin-like iron–sulfur clusters extend from the enzyme active site to the protein surface.

The use of IR and electron paramagnetic resonance (EPR) spectroscopies has demonstrated the existence of at least seven different forms of the Ni–Fe center. EPR studies on the enzyme active site have identified three $S=1/2$, EPR-active states designated as Ni-A, Ni-B and Ni-C [173–180]. IR studies on the enzyme

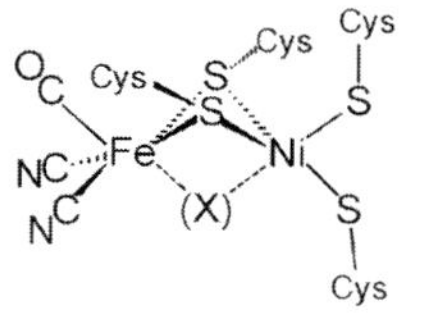

(X) = O^{2-}, OH^-, HO_2^-, H^-, H_2, CO, or nothing

Scheme 4.16

identified four EPR-silent states which have been designated as Ni-SU, Ni-SI_I, Ni-SI_{II} and Ni-SR (alternatively known as Ni-R).

The structures of the various species, their roles in the catalytic cycle and the details of their interconversions are not fully understood. The Ni-A and Ni-B forms correspond to over-oxidized species, which are not active in the catalytic cycle for H_2 oxidation. Both Ni-A and Ni-B may be re-activated by reduction, although the rate of reactivation is markedly slower for Ni-A. Species designated as Ni-C and Ni-R are believed to be intermediates in the oxidation of H_2. Other species designated as Ni-SU, Ni-SI_I and Ni-SI_{II} are presumed to be intermediates in the re-activation of the over-oxidized forms of the enzyme, and one of the Ni-SI species may play a role in the catalytic cycle [166, 181–187].

The nickel center is the putative site of H_2 activation, as it appears that the iron center remains low-spin Fe^{II}, while the nickel center takes the electrons and passes protons to nearby bases. In the active cycle, nickel changes from Ni^{III} to Ni^{II}, and finally to Ni^{I}, which has not been observed because of its rapid electron transfer. All of the observed EPR-active species appear to be Ni^{III}, while the remaining, observed species appear to be Ni^{II}. Interestingly, the Ni^{II} forms were initially assumed to be low-spin, but nickel L-edge soft X-ray spectroscopy and density functional theory (DFT) calculations suggest that the enzyme structure (i.e. the twisted ligand arrangement about the nickel center) results in high-spin Ni^{II} in the EPR-silent forms [188, 189].

The addition of CO gas to preparations of the active enzyme leads to essentially no change in the ν(C–O) band (around 2 cm^{-1}) of the $Fe(CO)(CN)_2$ unit and the appearance of a new ν(C–O) band at 2055 cm^{-1} in the IR spectrum [186]. The CO-inhibited form of the enzyme that is produced is incapable of catalyzing H^+ reduction or H_2 oxidation. The molecular structure derived from single-crystal X-ray diffraction studies of the CO-inhibited form of the [NiFe]H_2ase enzyme from *Desulvibrio vulgaris* (Miyazaki) clearly shows that the exogenous CO ligand is bound to the Ni center, but predicts Ni–CO angles that range from 136.2 to 160.9° [190].

The synthesis of small-molecule models of the active site of [NiFe]H_2ase has proven quite difficult [191]. A variety of synthetic active site models complexes are presented in Scheme 4.17. The majority of [NiFe]H_2ase model complexes, synthesized to date, are mononuclear models that attempt to model either the Fe site or the Ni site. Liaw and coworkers synthesized the mononuclear iron complex $[Fe(CO)_2(CN)_2(\eta^2\text{-}S_2COCH_2CH_3)]^{1-}$ and found it to be an excellent model of the coordination environment of the Fe center of [NiFe]H_2ase [192]. Darensbourg and coworkers showed that the IR spectrum of the $[CpFe(CO)(CN)_2]^{1-}$ complex is similar to the IR spectra of the oxidized forms of [NiFe]H_2ase [193]. Sellmann and coworkers designed a mononuclear nickel complex capable of catalyzing D_2/H_2O exchange [194]. Very recently, Tatsumi and coworkers developed the $[(\mu\text{-}S(CH_2)_3S)[Fe(CO)_2(CN)_2][Ni(\eta^2\text{-}S_2CNR_2)]^{1-}$ complex as excellent models of the composition of the [NiFe]H_2ase active site [195]. The Sellmann laboratory has also published a series of trinuclear ($FeNi_2$) complexes which (ignoring the extra Ni center) serve as excellent models

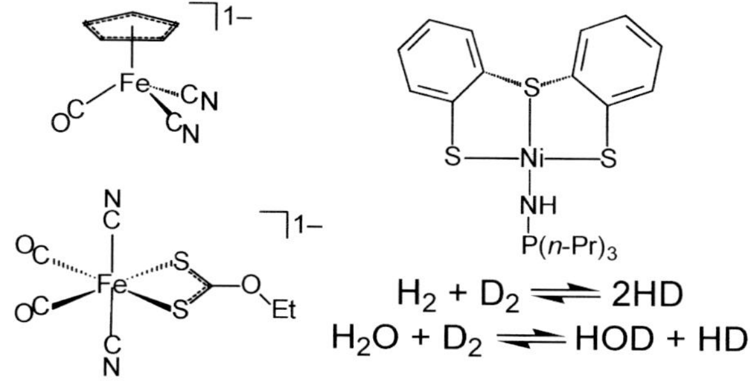

Scheme 4.17

of the structural and compositional models of the [NiFe]H_2ase active site [196, 197].

Computational studies on the [NiFe] active site have been performed by a number of workers [198–214]. Theory supports high-spin nickel [189] and suggests that the sulfur atoms of the terminal cysteine ligands act as bases in the heterolytic cleavage of dihydrogen.

4.7.1.2 [FeFe]H_2ase

Single-crystal X-ray diffraction studies have defined the basic framework of the active site of the [FeFe]H_2ases as consisting of two iron centers bridged by a novel dithiolate ($^-SCH_2XCH_2S^-$; X=CH_2, NH, O) linker (Scheme 4.18) [157, 215–218]. Each of the iron centers is further coordinated by one terminal CO ligand and one terminal CN^- ligand. The proximal iron center is further coordinated by a sulfur atom from a protein-bound cysteinate ligand, which bridges the FeFe active site and a nearby ferrodoxin-like [Fe_4S_4] cluster. (The two iron centers are commonly designated as proximal and distal by noting their spatial relation to the nearby [Fe_4S_4] cluster and protein backbone.) An additional CO ligand either bridges the two iron centers or is terminally coordinated to the distal iron center, depending on the redox state of the diiron center. A gas access

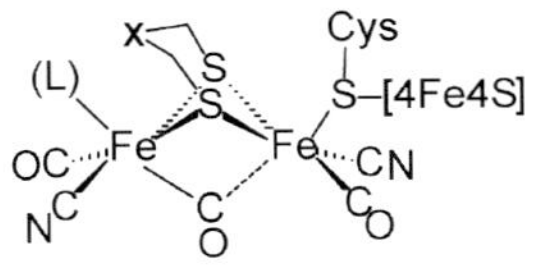

(L) = H_2O, OH^-, HO_2^-,
H^-, H_2, CO, or nothing

Scheme 4.18

channel [219] and a series of ferrodoxin-like iron–sulfur clusters extend from the enzyme active site to the protein surface.

The use of IR [218, 220–222] and EPR [223–226] spectroscopies have demonstrated the existence of at least four different forms of the [FeFe]–[Fe_4S_4] active site. EPR studies on the enzyme active site have identified two $S=1/2$, EPR-active states designated as H_{ox} (g=2.06) and H_{ox} (g=2.10). IR studies on the enzyme identified two EPR-silent states which have been designated as H_{ox}^{air} and H_{red}. The H_{ox}^{air} form corresponds to over-oxidized species, which is not an active catalyst for H^+ reduction or H_2 oxidation. The H_{ox}^{air} form may be re-activated either electrochemically or via the introduction of chemical reductants. In the presence of low potential reductants, the active site undergoes a one-electron reduction. The added electron is found to initially localize on the [Fe_4S_4] portion of the 6Fe active site, generating a species designated as H_{ox}(g=2.06). A conformational change of the protein superstructure is believed to initiate the transfer of the electron from the [Fe_4S_4] cluster to the [FeFe] cluster, yielding a species designated as H_{ox}(g=2.10). The addition of high potential reductants leads to a second one-electron reduction to yield a species designated as H_{red}.

The distal iron center is the putative site of H_2 activation. The addition of CO gas to preparations of the H_{ox}(g=2.10) form of the enzyme derived from *Desulfovibrio desulfuricans* Hildenborough (DdH) or *Clostridium pasteurianum* I (CpI) leads to an inhibited form of the enzyme that is incapable of catalyzing H^+ reduction or H_2 oxidation and the appearance of an additional ν(C–O) band. The molecular structure derived from single-crystal X-ray diffraction studies of the CO-inhibited form of the [FeFe]H_2ase enzyme shows that the distal iron center is coordinated by an additional CO ligand.

The synthesis of small-molecule models of the active site of [FeFe]H_2ase have contributed to a better understanding of its basic structure and the catalytic mechanism of H_2 oxidation and H^+ reduction [191]. A variety of synthetic active site models complexes are presented in Scheme 4.19. Simple dithiolate bridged diiron clusters have been found to act as structural and functional models of the active site of [FeFe]H_2ase. Complexes of the form [(μ-SCH_2XCH_2S)-[$Fe(CO)_2(CN)]_2]^{2-}$ (X=CH_2, NH, O) are excellent models of the composition of the [FeFe]H_2ase active site [227–231]. Darensbourg argued for a formal Fe^IFe^I redox state assignment for the H_{red} form of [FeFe]H_2ase based on the similarity of the ν(C–O) bands of $[(\mu\text{-}SCH_2CH_2CH_2S)[Fe(CO)_2(CN)]_2]^{2-}$ and

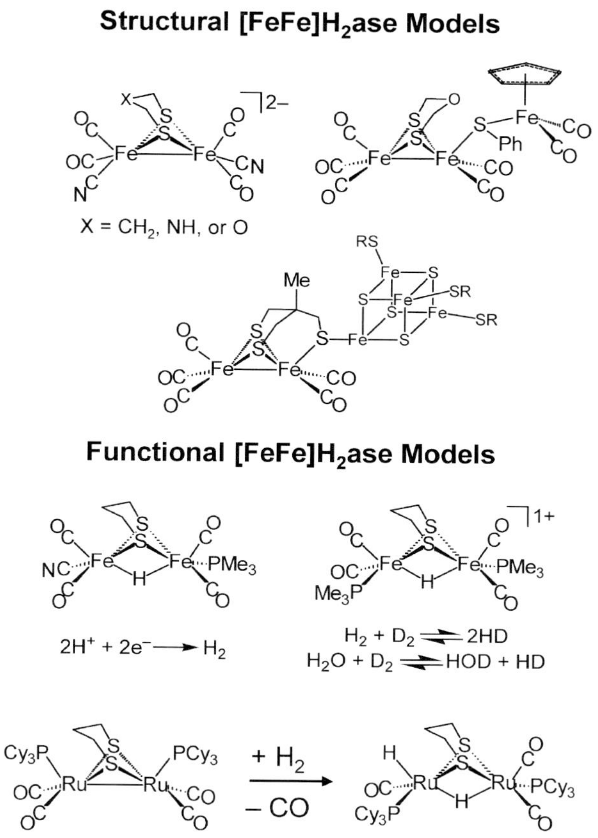

Structural [FeFe]H_2ase Models

X = CH_2, NH, or O

Functional [FeFe]H_2ase Models

$2H^+ + 2e^- \longrightarrow H_2$

$H_2 + D_2 \rightleftharpoons 2HD$

$H_2O + D_2 \rightleftharpoons HOD + HD$

$\xrightarrow[-\,CO]{+\,H_2}$

Scheme 4.19

those of the enzyme active site [232]. Pickett and coworkers thoroughly explored the chemistry of first-generation Fe_2S_3 models, which feature coordination of a pendant thioether sulfur (RSR) to one of the iron centers [233–238]. Complexes synthesized by Pickett, Song and their respective coworkers feature coordination of an external iron-bound thiolate ligand to one of the iron centers of the $Fe_2S_2(CO)_5$ core [231, 239]. A range of simple dithiolate bridged diiron complexes have been shown to act as solution electrocatalysts for H_2 production [240–247]. Darensbourg and coworkers have demonstrated that the $[(\mu\text{-}S(CH_2)_3S)[Fe(CO)_2(PMe_3)]_2]^{1+}$ complex catalyzes isotopic exchange in H_2/D_2 and D_2/H_2O mixtures, under photolytic, CO loss conditions [248–250].

Cao and Hall utilized the experimentally determined ν(C–O) and ν(C–N) stretching frequencies for the $(\mu\text{-}S(CH_2)_3S)[Fe(CO)_3]_2$ and $[(\mu\text{-}S(CH_2)_3S)\text{-}[Fe(CO)_2(CN)]_2]^{2-}$ complexes in order to calibrate their computationally derived ν(C–O) and ν(C–N) stretching frequencies. Using this method, they were able to show that the H_{ox}^{air}, H_{ox} and H_{red} forms of the [FeFe]H_2ase enzymes must corre-

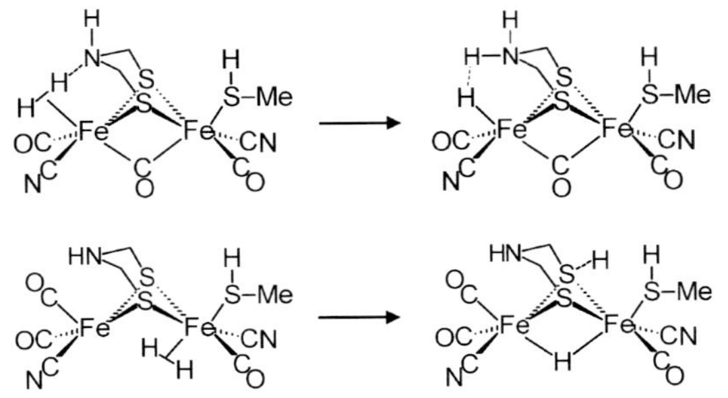

Fig. 4.4 Heterolytic cleavage of H_2 using small-molecule computational models of the $[FeFe]H_2$ase active site. Some workers favor heterolytic cleavage of H_2 bound to the distal iron center and utilizing the central nitrogen of the S-to-S linker as an internal base [251–254] (a). Other workers favor heterolytic cleavage of H_2 bound to the proximal iron and utilizing a bridging thiolate sulfur atom as an internal base [255–258] (b).

spond to the $Fe^{II}Fe^{II}$, $Fe^{I}Fe^{II}$ and $Fe^{I}Fe^{I}$ formal oxidation states of the FeFe cluster, rather than the higher formal oxidation states initially suggested [251].

DFT calculations have been applied by several research groups to give a better understanding of the molecular details of H_2 oxidation and H^+ reduction at the [FeFe] active site. The various proposed mechanisms differ mainly in the prospective location of H_2 binding to the [FeFe] cluster (as shown in Fig. 4.4). Nicolet and coworkers assigned the central atoms of the S-to-S linker as NH and identified a possible H-bonded proton-transfer pathway leading from these atoms to the protein surface [218]. Using DFT calculations, Hall and coworkers established that a bridgehead N-atom provides a kinetically and thermodynamically favorable route for the heterolytic cleavage of H_2 bound at the distal iron [251, 252]. Hu [253, 254] later reported similar results. De Gioia [255–258] and coworkers found an alternative kinetically and thermodynamically favorable route for the heterolytic cleavage of bound H_2. In mechanism of De Gioia, the active site rearranges from the structure observed in the X-ray structures of the enzyme. Dihydrogen then binds to the proximal iron in the area "between" the two iron atoms. A proton is transferred from the bound η^2-H_2 to a μ-S atom of the dithiolate cofactor to afford heterolytic cleavage.

4.7.2
H_2 Production by N_2ase

4.7.2.1 General Information about N_2ase Enzymes

Nitrogenase enzymes catalyze the reduction of molecular nitrogen to ammonia. The idealized stoichiometry for this reaction is $N_2 + 8e^- + 16\,MgATP + 8H^+ \; 2NH_3 + H_2 + 16\,MgADP + 16\,P_i$. For reasons that are not completely clear, the N_2ase enzymes couple the reduction of dinitrogen to ammonia to the reduction of protons to produce dihydrogen.

4.7.2.2 Molybdenum–Iron-containing N_2ase

The molybdenum-containing enzyme Mo-N_2ase is the most well-studied, although all-iron, vanadium-containing and tungsten-containing enzymes are also known to exist. The Mo-N_2ase enzyme is composed of two proteins, which are referred to as the iron protein and the molybdenum–iron protein [259–267]. The iron protein contains a single [Fe_4S_4] cluster, which mediates electron transfer to the FeMo-containing protein. The FeMo protein contains two 8Fe–7S clusters referred to as P-clusters and two 1Mo7Fe–9S clusters referred to as FeMo cofactors (FeMocos).

The FeMoco is the putative active site for dinitrogen and proton reduction. The iron and molybdenum centers are extensively bridged by sulfide ligands. The arrangement of the metal centers in FeMoco may be described as a bicapped trigonal prism, where the trigonal prism is formed by six iron atoms and capped on opposite sides by an iron and a molybdenum center (Scheme 4.20). An interstitial atom, believed to be nitrogen, is found occupy the center of the trigonal prism. The FeMoco is attached to the protein by coordination of a protein-bound cysteinate ligand to the capping iron center and coordination of a protein bound histidine ligand to the molybdenum center. The molybdenum center is further coordinated by a homocitrate ligand.

The mechanism of hydrogen production at FeMoco is not well understood. Only recently has direct evidence for the presence of a hydride ligand bound to FeMoco become available [268]. On the basis of thermodynamic arguments, Alberty contends that the highly reduced state of FeMoco required for N_2 reduction leads to the incidental production of H_2. Others have suggested that the reductive elimination of H_2 is necessary to produce a more reduced form of the FeMoco center and an open coordination site for N_2 binding [269].

4.8 Conclusions

The looming shortage of fossil fuels is a global problem. We are in immediate need of alternative energy sources. In theory, the generation of dihydrogen, and its use as a source of thermal and electrochemical energy, is a potential contri-

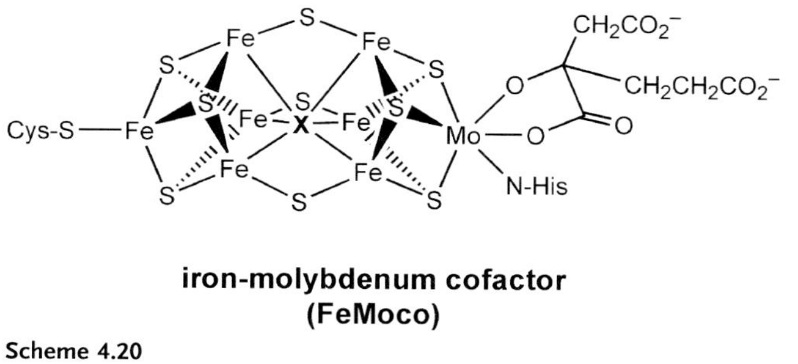

iron-molybdenum cofactor (FeMoco)

Scheme 4.20

bution to the solution of this problem. In practice, the implication of this so-called hydrogen economy has many immediate problems. The processes of H^+ reduction and H_2 oxidation are most readily accomplished by the noble metals, platinum and palladium, but these metals are expensive and in short supply. The storage of dihydrogen is a major problem for other applications, such as hydrogen-powered automobiles.

The new model complexes and other synthetic material inspired by the H_2ase and N_2ase enzymes hold out the hope of replacing platinum- and palladium-based H^+ reduction/H_2 oxidation catalysts with those constructed from the base metals iron and nickel [180]. The design of these new catalysts will require the cooperation of scientists from the fields of biology/biochemistry, synthetic organometallic chemistry and theoretical chemistry.

Acknowledgments

The authors acknowledge financial support from the National Science Foundation (CHE-0111629 to MYD and CHE-9800184/CHE-0518047 to MBH) and The Welch Foundation (A-0924 to MYD and A-0648 to MBH).

Abbreviations

B^{ArF} = $[B(3,5\text{-}(CF_3)_2C_6H_3)_4]^{1-}$
Cy = cyclohexyl = $\text{-}CH(CH_2)_5$
Cp = cyclopentadienyl = $[C_5H_5]^{1-}$
Cp* = pentamethyl-cyclopentadienyl = $[C_5(CH_3)_5]^{1-}$
^{i}Pr = isopropyl group = $\text{-}CH(CH_3)_2$
Tp = tris(pyrazolyl)hydroborate
Tp* = tris(3,5-dimethylpyrazolyl)hydroborate
η^x = x atoms of a ligand are attached to a metal
μ^x = a ligand atom bridges x metal centers
THT = tetrahydrothiofuran
THF = tetrahydrofuran

References

1 S.S. Penner, *Energy* **2006**, *31*, 33–43.
2 S.A. Sherif, F. Barbir, T. N. Veziroglu, *Solar Energy* **2005**, *78*, 647–660.
3 A. Dedieu (Ed.), *Transition Metal Hydrides*. VCH, New York, **1992**.
4 G.J. Kubas (Ed.), *Metal Dihydrogen and σ-Bond Complexes: Structure, Theory and Reactivity*, Kluwer/Plenum, Dordrecht, **2001**.
5 M. Peruzzini, R. Poli (Eds.), *Recent Advances in Hydride Chemistry*, Elsevier Amsterdam, **2001**.
6 G.J. Kubas, *J Less-Common Met* **1991**, *173*, 475–484.

7 P. Espinet, M.A. Esteruelas, L.A. Oro, J.L. Serrano, E. Sola, *Coord Chem Rev* **1992**, *117*, 215–274.
8 P.G. Jessop, R.H. Morris, *Coord Chem Rev* **1992**, *121*, 155–284.
9 D.M. Heinekey, W.J. Oldham, Jr., *Chem Rev* **1993**, *93*, 913–926.
10 R. Kuhlman, *Coord Chem Rev* **1997**, *167*, 205–232.
11 R.H. Crabtree, *J Organomet Chem* **1998**, *557*, 111–115.
12 R.H. Crabtree, O. Eisenstein, G. Sini, E. Peris, *J Organomet Chem* **1998**, *567*, 7–11.
13 M.A. Esteruelas, L.A. Oro, *Chem Rev* **1998**, *98*, 577–588.
14 S. Sabo-Etienne, B. Chaudret, *Coord Chem Rev* **1998**, *178–180*, 381–407.
15 S. Sabo-Etienne, B. Chaudret, *Chem Rev* **1998**, *98*, 2077–2091.
16 G. Jia, C.-P. Lau, *Coord Chem Rev* **1999**, *190–192*, 83–108.
17 F. Maseras, A. Lledos, E. Clot, O. Eisenstein, *Chem Rev* **2000**, *100*, 601–636.
18 M.Y. Darensbourg, E.J. Lyon, J.J. Smee, *Coord Chem Rev* **2000**, *206/207*, 533–561.
19 G.S. McGrady, G. Guilera, *Chem Soc Rev* **2003**, *32*, 383–392.
20 D.M. Heinekey, A. Lledos, J.M. Lluch, *Chem Soc Rev* **2004**, *33*, 175–182.
21 V.I. Bakhmutov, *Eur J Inorg Chem* **2005**, 245–255.
22 A. Heller, *Acc Chem Res* **1981**, *14*, 154–162.
23 W. Kreuter, H. Hofmann, *Int J Hydrogen Energy* **1998**, *23*, 661–666.
24 B. Losiewicz, A. Budniok, E. Rowinski, E. Lagiewka, A. Lasia, *Int J Hydrogen Energy* **2003**, *29*, 145–157.
25 H.F. Schaefer III, *Acc Chem Res* **1977**, *10*, 287–293.
26 T.R. Lee, G.M. Whitesides, *Acc Chem Res* **1992**, *25*, 266–272.
27 H. Uchida, *Int J Hydrogen Energy* **1999**, *24*, 861–869.
28 G.-J. Kroes, A. Gross, E.-J. Baerends, M. Scheffler, D.A. McCormack, *Acc Chem Res* **2002**, *35*, 193–200.
29 H. Jacobsen, *Angew Chem Int Ed* **2004**, *43*, 1912–1914.
30 X.-M. Zhang, J.W. Bruno, E. Enyinnaya, *J Org Chem* **1998**, *63*, 4671–4678.
31 B. Darwent, *Bond Dissociation Energies in Simple Molecules (NSRDS-NBS 31)*, National Bureau of Standards, Washington, DC, **1970**.
32 Arguments based on Woodward and Hoffmann's ideas of the conservation of orbital symmetry give similar results.
33 G.B. Kistiakowsky, H. Romeyn, Jr., J.R. Ruhoff, H.A. Smith, W.E. Vaughan, *J Am Chem Soc* **1935**, *57*, 55–75.
34 M.S. Gordon, T.N. Truong, J.A. Pople, *Chem Phys Lett* **1986**, *130*, 245–248.
35 K. Abdur-Rashid, T.P. Fong, B. Greaves, D.G. Gusev, J.G. Hinman, S.E. Landau, A.J. Lough, R.H. Morris, Jr., *Am Chem Soc* **2000**, *122*, 9155–9171.
36 K. Izutsu, *Acid–Base Dissociation Constants in Dipolar Aprotic Solvents*, Blackwell Scientific, Oxford, **1990**.
37 I.M. Kolthoff, M.K. Chantooni, Jr., *J Chem Eng Data* **1999**, *44*, 124–129.
38 G.J. Kubas, *J Organomet Chem* **2001**, *635*, 37–68.
39 J.O. Noell, P.J. Hay, *J Am Chem Soc* **1982**, *104*, 4578–4584.
40 P.J. Hay, *J Am Chem Soc* **1987**, *109*, 705–710.
41 J. Eckart, G.J. Kubas, J.H. Hall, P.J. Hay, C.M. Boyle, *J Am Chem Soc* **1990**, *112*, 2324–2332.
42 G.R. Haynes, R.L. Martin, P.J. Hay, *J Am Chem Soc* **1992**, *114*, 28–36.
43 Z. Lin, M.B. Hall, *J Am Chem Soc* **1992**, *114*, 2928–2932.
44 G.J. Kubas, R.R. Ryan, B.I. Swanson, P.J. Vergamini, H.J. Wasserman, *J Am Chem Soc* **1984**, *106*, 451–452.
45 A transition metal complex is termed coordinatively unsaturated when it contains less ligands than normally observed for complexes of the metal's formal oxidation state. These complexes are often discussed in terms of containing one or more open sites or vacant coordination sites on the transition metal center. Real coordinatively unsaturated complexes generally only exist under very special conditions. In general, "coordinatively unsaturated" complexes are actually stabilized by weak intermolecular forces, such as interactions with the solvent and/or weak intramolecular forces between the ligand and the "open" site.
46 R.H. Morris, *Can J Chem* **1996**, *74*, 1907–1915.

47 A. A. Gonzalez, K. Zhang, S. L. Mukerjee, C. D. Hoff, G. R. K. Khalsa, G. J. Kubas, in *Bonding Energetics in Organometallic Compounds*, T. J. Marks (Ed.), American Chemical Society, Washington, DC, **1990**, pp. 133–147.
48 Z. Lin, M. B. Hall, *J Am Chem Soc* **1992**, *114*, 6102–6108.
49 Z. Lin, M. B. Hall, *Inorg Chem* **1992**, *31*, 4262–4265.
50 Z. Lin, M. B. Hall, *Coord Chem Rev* **1994**, *135/136*, 845–879.
51 C. A. Bayse, M. B. Hall, *J Am Chem Soc* **1999**, *121*, 1348–1358.
52 R. H. Morris, K. A. Earl, R. I. Luck, N. J. Lazarowych, A. Sella, *Inorg Chem* **1987**, *26*, 2674–2683.
53 R. H. Morris, *Inorg Chem* **1992**, *31*, 1471–1478.
54 V. Pons, S. L. J. Conway, M. L. H. Green, J. C. Green, B. J. Herbert, D. M. Heinekey, *Inorg Chem* **2004**, *43*, 3475–3483.
55 G. J. Kubas, C. J. Unkefer, B. I. Swanson, E. Fukushima, *J Am Chem Soc* **1986**, *108*, 7000–7009.
56 D. M. Heinekey, J. K. Law, S. M. Schultz, *J Am Chem Soc* **2001**, *123*, 12728–12729.
57 D. Michos, X. L. Luo, J. A. K. Howard, R. H. Crabtree, *Inorg Chem* **1992**, *31*, 3914–3916.
58 R. Gelabert, M. Moreno, J. M. Lluch, A. Lledos, *J Am Chem Soc* **1997**, *119*, 9840–9847.
59 R. Gelabert, M. Moreno, J. M. Lluch, A. Lledos, *J Am Chem Soc* **1998**, *120*, 8168–8176.
60 L. Torres, R. Gelabert, M. Moreno, J. M. Lluch, *J Phys Chem A* **2000**, *104*, 7898–7905.
61 G. Barea, M. A. Esteruelas, A. Lledos, A. M. Lopez, E. Onate, J. I. Tolosa, *Organometallics* **1998**, *17*, 4065–4076.
62 G. Barea, M. A. Esteruelas, A. Lledos, A. M. Lopez, E. Onate, J. I. Tolosa, *Inorg Chem* **1998**, *37*, 5033–5035.
63 W. T. Klooster, T. F. Koetzle, G. Jia, T. P. Fong, R. H. Morris, A. Albinati, *J Am Chem Soc* **1994**, *116*, 7677–7681.
64 J. K. Law, H. Mellows, D. M. Heinekey, *J Am Chem Soc*, **2002**, *124*, 1024–1030.
65 V. Pons, D. M. Heinekey, *J Am Chem Soc*, **2003**, *125*, 8428–8429.
66 J. Eckert, C. E. Webster, M. B. Hall, A. Albinati, L. M. Venanzi, *Inorg Chim Acta* **2002**, *330*, 240–249.
67 J. L. Finney, *Acta Crystallogr B* **1995**, *B51*, 447–467.
68 B. D. N. Rao, L. R. Anders, *Phys Rev* **1965**, *140*, 112–117.
69 T. A. Luther, D. M. Heinekey, *Inorg Chem* **1998**, *37*, 127–132.
70 R. H. Crabtree, M. Lavin, *Chem Commun* **1985**, 1661–1662.
71 D. G. Hamilton, R. H. Crabtree, *J Am Chem Soc* **1988**, *110*, 4126–4133.
72 R. H. Crabtree, *Acc Chem Res* **1990**, *23* 95–101.
73 W. Yao, J. W. Faller, R. H. Crabtree, *Inorg Chim Acta* **1997**, *259*, 71–76.
74 P. J. Desrosiers, L. Cai, Z. Lin, R. Richards, J. Halpern, *J Am Chem Soc* **1991**, *113*, 4173–4184.
75 M. T. Bautista, K. A. Earl, P. A. Maltby, R. H. Morris, C. T. Schweitzer, A. Sella, *J Am Chem Soc* **1988**, *110*, 7031–7036.
76 R. Gelabert, M. Moreno, J. M. Lluch, A. Lledos, *Chem Phys* **1999**, *241*, 155–166.
77 G. Tejeda, J. M. Fernandez, S. Montero, D. Blume, J. P. Toennies, *Phys Rev Lett* **2004**, *92*, 223401/1–223401/4.
78 B. Moreno, S. Sabo-Etienne, B. Chaudret, A. Rodriguez, F. Jalon, S. Trofimenko, *J Am Chem Soc* **1995**, *117*, 7441.
79 M. W. George, M. T. Haward, P. A. Hamley, C. Hughes, F. P. A. Johnson, V. K. Popov, M. Poliakoff, *J Am Chem Soc* **1993**, *115*, 2286.
80 G. J. Kubas, C. J. Unkefer, B. I. Swanson, E. Fukushima, *J Am Chem Soc* **1986**, *108*, 7000.
81 R. K. Upmacis, M. Poliakoff, J. J. Turner, *J Am Chem Soc* **1986**, *108*, 3645.
82 R. L. Sweany, A. Moroz, *J Am Chem Soc* **1989**, *111*, 3577.
83 L. Abis, A. Sen, J. Halpern, *J Am Chem Soc* **1978**, *100*, 2915–2916.
84 J. Chatt, B. L. Shaw, *J Chem Soc* **1959**, 705–716.
85 L. M. Rendina, R. J. Puddephatt, *Chem Rev* **1997**, *97*, 1735–1754.
86 J. M. Wisner, T. J. Bartczak, J. A. Ibers J. J. Low, W. A. Goddard III, *J Am Chem Soc* **1986**, *108*, 347–348.

87 J.J. Low, W.A. Goddard III, *J Am Chem Soc* **1984**, *106*, 8321–8322.

88 J.J. Low, W.A. Goddard III, *J Am Chem Soc* **1984**, *106*, 6928–6937.

89 J.J. Low, W.A. Goddard III, *Organometallics* **1986**, *5*, 609–622.

90 J.J. Low, W.A. Goddard III, *J Am Chem Soc* **1986**, *108*, 6115–6128.

91 S. Sabo-Etienne, B. Chaudret, H.B. el Makarim, J.-C. Barthelat, J.-P. Daudey, S. Ulrich, H.H. Limbach, C. Moïse, *J Am Chem Soc* **1995**, *117*, 11602–11603.

92 F.A. Jalon, A. Otero, B.R. Manzano, E. Villaseñor, B. Chaudret, *J Am Chem Soc* **1995**, *117*, 10123–10124.

93 A. Antiñolo, F. Carillo-Hermosilla, M. Fajardo, S. Garcia-Yuste, A. Otero, S. Camanyes, F. Maseras, M. Moreno, A. Lledórs, J.M. Lluch, *J Am Chem Soc* **1997**, *119*, 6107–6114.

94 L.S. Van der Sluys, J. Eckert, O. Eisenstein, J.H. Hall, J.C. Huffman, S.A. Jackson, T.F. Koetzle, G.J. Kubas, P.J. Vergamini, K.G. Caulton, *J Am Chem Soc* **1990**, *112*, 4831–4841.

95 A. Albinati, V.I. Bakhmutov, K.G. Caulton, E. Clot, J. Eckert, O. Eisenstein, D.G. Gusev, V.V. Grushin, B.E. Hauger, W.T. Klooster, T.F. Koetzle, R.K. McMullan, T.J. O'Loughlin, M. Pelissier, J.S. Ricci, M.P. Sigalas, A.B. Vymenits, *J Am Chem Soc* **1993**, *115*, 7300–7312.

96 J. Eckert, C.M. Jensen, G. Jones, E. Clot, O. Eisenstein, *J Am Chem Soc* **1993**, *115*, 11056–11057.

97 J. Eckert, *Spectrochim Acta A* **1992**, *48A*, 271–283.

98 J. Eckert, *Spectrochim Acta A* **1992**, *48A*, 363–378.

99 J. Eckert, G.J. Kubas, A.J. Dianoux, *J Chem Phys* **1988**, *88*, 466–468.

100 E. Clot, J. Eckert, *J Am Chem Soc* **1999**, *121*, 8855–8863.

101 C.A. Bayse, M.B. Hall, B. Pleune, R. Poli, *Organometallics* **1998**, *17*, 4309–4315.

102 C.E. Webster, C.L. Gross, D.M. Young, G.S. Girolami, A.J. Schultz, M.B. Hall, J. Eckert, *J Am Chem Soc* **2005**, *127*, 15091–15101.

103 M. Mediati, G.N. Tachibana, C.M. Jensen, *Inorg Chem* **1990**, *29*, 3–5.

104 M. Mediati, G.N. Tachibana, C.M. Jensen, *Inorg Chem* **1992**, *31*, 1827–1832.

105 J. Eckert, C.M. Jensen, T.F. Koetzle, T.L. Husebo, J. Nicol, P. Wu, *J Am Chem Soc* **1995**, *117*, 7271–7272.

106 S. Li, M.B. Hall, J. Eckert, C.M. Jensen, A. Albinati, *J Am Chem Soc* **2000**, *122*, 2903–2910.

107 C.A. Bayse, M. Couty, M.B. Hall, *J Am Chem Soc* **1996**, *118*, 8916–8919.

108 Protonic–hydridic interactions are also referred as "dihydrogen bonding".

109 A.J. Lough, S. Park, R. Ramachandran, R.H. Morris, *J Am Chem Soc* **1994**, *116*, 8356–8357.

110 J.C. Lee, Jr., E. Peris, A.L. Rheingold, R.H. Crabtree, *J Am Chem Soc* **1994**, *116*, 11014–11019.

111 J. Wessel, J.C. Lee, Jr., E. Peris, G.P.A. Yap, J.B. Fortin, J.S. Ricci, G. Sini, A. Albinati, T.F. Koetzle, C. Eisenstein, A.L. Rheingold, R.H. Crabtree, *Angew Chem Int Ed* **1995**, *34*, 2507–2509.

112 B.P. Patel, W. Yao, G.P.A. Yap, A.L. Rheingold, R.H. Crabtree, *Chem Commun* **1996**, 991–992.

113 B.P. Patel, J. Wessel, W.B. Yao, J.C. Lee, E. Peris, T.F. Koetzle, G.P.A. Yap, J.B. Fortin, J.S. Ricci, G. Sini, A. Albinati, O. Eisenstein, A.L. Rheingold, R.H. Crabtree, *New J Chem* **1997**, *21*, 413–421.

114 G. Sini, O. Eisenstein, W. Yao, R.H. Crabtree, *Inorg Chim Acta* **1998**, *280*, 26–29.

115 M. Schlaf, A.J. Lough, P.A. Maltby, R.H. Morris, *Organometallics* **1996**, *15*, 2270–2278.

116 P.A. Maltby, M. Schlaf, M. Steinbeck, A.J. Lough, R.H. Morris, W.T. Klooster, T.F. Koetzle, R.C. Srivastava, *J Am Chem Soc* **1996**, *118*, 5396–5407.

117 D.E. Berning, B.C. Noll, D.L. DuBois, *J Am Chem Soc* **1999**, *121*, 11432–11447.

118 D.E. Berning, A. Miedaner, C.J. Curtis, B.C. Noll, M.C.R. DuBois, D.L. DuBois, *Organometallics* **2001**, *20*, 1832–1839.

119 C.J. Curtis, A. Miedaner, W.W. Ellis, D.L. DuBois, *J Am Chem Soc* **2002**, *124*, 1918–1925.

120 W.W. Ellis, A. Miedaner, C.J. Curtis, D.H. Gibson, D.L. DuBois, *J Am Chem Soc* **2002**, *124*, 1926–1932.

121 A.J. Price, R. Ciancanelli, B.C. Noll, C.J. Curtis, D.L. DuBois, M.R. DuBois, *Organometallics* **2002**, *21*, 4833–4839.

122 C.J. Curtis, A. Miedaner, R. Ciancanelli, W.W. Ellis, B.C. Noll, M.R. DuBois, D.L. DuBois, *Inorg Chem* **2003**, *42*, 216–227.

123 C.J. Curtis, A. Miedaner, J.W. Raebiger, D.L. DuBois, *Organometallics* **2004**, *23*, 511–516.

124 A. Miedaner, J.W. Raebiger, C.J. Curtis, S.M. Miller, D.L. DuBois, *Organometallics* **2004**, *23*, 2670–2679.

125 J.W. Raebiger, A. Miedaner, C.J. Curtis, S.M. Miller, O.P. Anderson, D.L. DuBois, *J Am Chem Soc* **2004**, *126*, 5502–5514.

126 R.M. Henry, R.K. Shoemaker, R.H. Newell, G.M. Jacobsen, D.L. DuBois, M.R. DuBois, *Organometallics* **2005**, *24*, 2481–2491.

127 J.W. Raebiger, D.L. DuBois, *Organometallics* **2005**, *24*, 110–118.

128 R.F. Jordan, J.R. Norton, *J Am Chem Soc* **1982**, *104*, 1255–1263.

129 E.J. Moore, J.M. Sullivan, J.R. Norton, *J Am Chem Soc* **1986**, *108*, 2257–2263.

130 R.T. Edidin, J.M. Sullivan, J.R. Norton, *J Am Chem Soc* **1987**, *109*, 3945–3953.

131 S.S. Kristjansdottir, A.E. Moody, R.T. Weberg, J.R. Norton, *Organometallics* **1988**, *7*, 1983–1987.

132 R.T. Weberg, J.R. Norton, *J Am Chem Soc* **1990**, *112*, 1105–1108.

133 S.S. Kristjansdottir, A.J. Loendorf, J.R. Norton, *Inorg Chem* **1991**, *30*, 4470–4471.

134 H.W. Walker, R.G Pearson, P.C. Ford, *J Am Chem Soc* **1983**, *105*, 1179–1186.

135 T.P. Fong, A.J. Lough, R.H. Morris, A. Mezzetti, E. Rocchini, P. Rigo, *Dalton Trans* **1998**, 2111–2114.

136 C.M. Nagaraja, M. Nethaji, B.R. Jagirdar, *Inorg Chem* **2005**, *44*, 4145–4147.

137 G. Jia, A.J. Lough, R.H. Morris, *Organometallics* **1992**, *11*, 161–171.

138 M.S. Chinn, D.M. Heinekey, *J Am Chem Soc* **1987**, *109*, 5865–5867.

139 M. Schlaf, A.J. Lough, R.H. Morris, *Organometallics* **1996**, *15*, 4423–4436.

140 D.-H. Lee, B.P. Patel, E. Clot, O. Eisenstein, R.H. Crabtree, *Chem Commun* **1999**, 297–298.

141 R.D. Adams, B. Captain, *Angew Chem Int Ed* **2005**, *44*, 2531–2533.

142 P.J. Brothers, W.R. Roper, *Chem Rev* **1988**, *88*, 1293–1326.

143 D. Huang, K.G. Caulton, *J Am Chem Soc* **1997**, *119*, 3185–3186.

144 D. Huang, P.R. Koren, K. Folting, E.R. Davidson, K.G. Caulton, *J Am Chem Soc* **2000**, *122*, 8916–8931.

145 P.J. Albietz, Jr., J.F. Houlis, R. Eisenberg, *Inorg Chem* **2002**, *41*, 2001–2003.

146 K. Komatsu, M. Murata, Y. Murata, *Science* **2005**, *307*, 238–240.

147 R.D. Cortright, R.R. Davda, J.A. Dumesic, *Nature* **2002**, *418*, 964–967.

148 S. Sabo-Etienne, V. Rodriguez, B. Donnadieu, B. Chaudret, H.A. el Makarim, J.C. Barthelat, S. Ulrich, H.-H. Limbach, C. Moïse, *New J Chem* **2001**, *25*, 55–62.

149 M.W.W. Adams, *Biochim Biophys Acta* **1990**, *1020*, 115–145.

150 S.P.J. Albracht, *Biochim Biophys Acta* **1994**, *1188*, 167–204.

151 M. Frey, *Struct Bonding* **1998**, *90*, 97–126.

152 J. Kim, D.C. Rees, *Biochemistry* **1994**, *33*, 389–397.

153 B.A. MacKay, M.D. Fryzuk, *Chem Rev* **2004**, *104*, 385–401.

154 A.E. Przbyla, J. Robbins, N. Menon, H.D. Peck, Jr., *FEMS Microbiol Rev* **1992**, *8*, 10913–10915.

155 E. Garcin, X. Vernede, E.C. Hatchikian, A. Volbeda, M. Frey, J.C. Fontecilla-Camps, *Structure* **1999**, *7*, 557–566.

156 M. Teixeira, I. Moura, A.V. Xavier, B.H. Huynh, D.V. DerVartanian, H.D. Peck, Jr., J. LeGall, J.J.G. Moura, *J Biol Chem* **1985**, *260*, 8942–8950.

157 Y. Nicolet, B.J. Lemon, J.C. Fontecilla-Camps, J.W. Peters, *Trends Biochem Sci* **2000**, *25*, 138–143.

158 J.W. Peters, *Curr Opin Struct Biol* **1999**, *9*, 670–676.

159 E.J. Lyon, S. Shima, G. Buurman, S. Chowdhuri, A. Batschauer, K. Stein-

bach, R. K. Thauer, *Eur J Biochem* **2004**, *271*, 195–204.

160 A. Berkessel, R. K. Thauer, *Angew Chem Int Ed* **1995**, *34*, 2247–2250.

161 R. K. Thauer, A. R. Klein, G. C. Hartmann, *Chem Rev* **1996**, *96*, 3031–3042.

162 E. J. Lyon, S. Shima, R. Boecher, R. K. Thauer, F.-W. Grevels, E. Bill, W. Roseboom, S. P. J. Albracht, *J Am Chem Soc* **2004**, *126*, 14239–14248.

163 P. M. Vignais, B. Dimon, N. A. Zorin, M. Tomiyama, A. Colbeau, *J Bacteriol* **2000**, *182*, 5997–6004.

164 M. Bernhard, T. Buhrke, B. Bleijlevens, A. L. De Lacey, V. M. Fernandez, S. P. J. Albracht, B. Friedrich, *J Biol Chem* **2001**, *276*, 15592–15597.

165 A. Volbeda, M.-H. Charon, C. Piras, E. C. Hatchikian, M. Frey, J. C. Fontecilla-Camps, *Nature* **1995**, *373*, 580–587.

166 A. Volbeda, E. Garcia, C. Piras, A. L. De Lacey, V. M. Fernández, E. C. Hatchikian, M. Frey, J. C. Fontecilla-Camps, *J Am Chem Soc* **1996**, *118*, 12989–12996.

167 Y. Higuchi, T. Yagi, N. Yasuoka, *Structure* **1997**, *5*, 1671–1680.

168 Y. Higuchi, H. Ogata, K. Miki, N. Yasuoka, T. Yagi, *Structure* **1999**, *7*, 549–556.

169 A. Volbeda, Y. Montet, X. Vernede, E. C. Hatchikian, J. C. Fontecilla-Camps, *Int J Hydrogen Res* **2002**, *27*, 1449–1461.

170 S. E. Lamle, S. P. J. Albracht, F. A. Armstrong, *J Am Chem Soc* **2005**, *127*, 6595–6604.

171 A. Volbeda, L. Martin, C. Cavazza, M. Matho, B. W. Faber, W. Roseboom, S. P. J. Albracht, E. Garcin, M. Rousset, J. C. Fontecilla-Camps, *J Biol Inorg Chem* **2005**, *10*, 239–249.

172 Y. Montet, P. Amara, A. Volbeda, X. Vernede, E. C. Hatchikian, M. J. Field, M. Frey, J. C. Fontecilla-Camps, *Nat Struct Biol* **1997**, *4*, 523–526.

173 S. P. Albracht, *J Biochem Soc Trans* **1985**, *13*, 582–585.

174 V. M. Fernandez, E. C. Hatchikian, R. Cammack, *Biochim Biophys Acta* **1985**, *832*, 69–79.

175 R. Cammack, V. M. Fernandez, K. Schneider, *Biochimie* **1986**, *68*, 85–91.

176 V. M. Fernandez, E. C. Hatchikian, D. S. Patil, R. Cammack, *Biochim Biophys Acta* **1986**, *883*, 145–154.

177 M. J. Maroney, M. A. Pressler, S. A. Mirza, J. P. Whitehead, R. J. Gurbiel, B. M. Hoffman, in *Mechanistic Bioinorganic Chemistry*, H. H. Thorp, V. L. Pecoraro (Eds.), American Chemical Society, Washington, DC, **1995**, vol. 246, pp. 21–60.

178 F. Dole, A. Fournel, V. Magro, E. C. Hatchikian, P. Bertrand, B. Guigliarelli, *Biochemistry* **1997**, *36*, 7847–7854.

179 B. Bleijlevens, B. W. Faber, S. P. J. Albracht, *J Biol Inorg Chem* **2001**, *6*, 763–769.

180 K. A. Vincent, J. A. Cracknell, A. Parkin, F. A. Armstrong, *Dalton Trans* **2005**, 3397–3403.

181 Y. Berlier, G. D. Fauque, J. LeGall, E. S. Choi, H. D. Peck, Jr., P. A. Lespinat, *Biochem Biophys Res Commun* **1987**, *146*, 147–153.

182 K. A. Bagley, C. J. Van Garderen, M. Chen, E. C. Duin, S. P. J. Albracht, W. H. Woodruff, *Biochemistry* **1994**, *33*, 9229–9236.

183 K. A. Bagley, E. C. Duin, W. Roseboom, S. P. J. Albracht, W. H. Woodruff, *Biochemistry* **1995**, *34*, 5527–5535.

184 T. M. van der Spek, A. F. Aredsen, R. P. Happe, S. Yun, K. A. Bagley, D. J. Stufkens, W. R. Hagen, S. P. J. Albracht, *Eur J Biochem* **1996**, *237*, 629–634.

185 A. L. De Lacey, E. C. Hatchikian, A. Volbeda, M. Frey, J. C. Fontecilla-Camps, V. M. Fernandez, *J Am Chem Soc* **1997**, *119*, 7181–7189.

186 A. L. de Lacey, C. Stadler, V. M. Fernandez, E. C. Hatchikian, H.-J. Fan, S. Li, M. B. Hall, *J Biol Inorg Chem* **2002**, *7*, 318–326.

187 B. Bleijlevens, F. A. van Broekhuizen, A. L. de Lacey, W. Roseboom, V. M. Fernandez, S. P. J. Albracht, *J Biol Inorg Chem* **2004**, *9*, 743–752.

188 H. Wang, C. Y. Ralston, D. S. Patil, R. M. Jones, W. Gu, M. Verhagen, M. Adams, P. Ge, C. Riordan, C. A. Marganian, P. Mascharak, J. Kovacs, C. G. Miller, T. J. Collins, S. Brooker, P. D. Croucher, K. Wang, E. I. Stiefel, S. P.

Cramer, *J Am Chem Soc* **2000**, *122*, 10544–10552.

189 H.-J. Fan, M.B. Hall, *J Am Chem Soc* **2002**, *124*, 394–395.

190 H. Ogata, Y. Mizoguchi, N. Mizuno, K. Miki, S.-I. Adachi, N. Yasuoka, T. Yagi, O. Yamauchi, S. Hirota, Y. Higuchi, *J Am Chem Soc* **2002**, *124*, 11628–11635.

191 I.P. Georgakaki, M.Y. Darensbourg, in *Comprehensive Coordination Chemistry II*, J.A. McCleverty, T.J. Meyer (Eds.), Elsevier, New York, **2004**, vol. 8, pp. 549–568.

192 W.-F. Liaw, J.-H. Lee, H.-B. Gau, C.-H. Chen, S.-J. Jung, C.-H. Hung, W.-Y. Chen, C.-H. Hu, G.-H. Lee, *J Am Chem Soc* **2002**, *124*, 1680–1688.

193 C.-H. Lai, W.-Z. Lee, M.L. Miller, J.H. Reibenspies, D.J. Darensbourg, M.Y. Darensbourg, *J Am Chem Soc* **1998**, *120*, 10103–10114.

194 D. Sellmann, F. Geipel, M. Moll, *Angew Chem Int Ed* **2000**, *39*, 561–563.

195 Z. Li, Y. Ohki, K. Tatsumi, *J Am Chem Soc* **2005**, *127*, 8950–8951.

196 D. Sellmann, F. Lauderbach, F. Geipel, F.W. Heinemann, M. Moll, *Angew Chem Int Ed* **2004**, *43*, 3141–3144.

197 D. Sellmann, F. Lauderbach, F.W. Heinemann, *Eur J Inorg Chem* **2005**, 371–377.

198 L. De Gioia, P. Fantucci, B. Guigliarelli, P. Bertrand, *Inorg Chem* **1999**, *38*, 2658–2662.

199 L. De Gioia, P. Fantucci, B. Guigliarelli, P. Bertrand, *Int J Quantum Chem* **1999**, *73*, 187–195.

200 M. Bruschi, L. De Gioia, G. Zampella, M. Reiher, P. Fantucci, M. Stein, *J Biol Inorg Chem* **2004**, *9*, 873–884.

201 S. Niu, L.M. Thomson, M.B. Hall, *J Am Chem Soc* **1999**, *121*, 4000–4007.

202 H.-J. Fan, M.B. Hall, *J Biol Inorg Chem* **2001**, *6*, 467–473.

203 S. Li, M.B. Hall, *Inorg Chem* **2001**, *40*, 18–24.

204 S. Niu, M.B. Hall, *Inorg Chem* **2001**, *40*, 6201–6203.

205 A.L. DeLacey, C. Stadler, V.M. Fernandez, E.C. Hatchikian, H.-J. Fan, S. Li, M.B. Hall, *J Biol Inorg Chem* **2002**, *7*, 318–326.

206 M. Stein, E. van Lenthe, E.J. Baerends, W. Lubitz, *J Am Chem Soc* **2001**, *123*, 5839–5840.

207 M. Stein, W. Lubitz, *Curr Opin Chem Biol* **2002**, *6*, 243–249.

208 S. Foerster, M. Stein, M. Brecht, H. Ogata, Y. Higuchi, W. Lubitz, *J Am Chem Soc* **2003**, *125*, 83–93.

209 M. Stein, W. Lubitz, *J Inorg Biochem* **2004**, *98*, 862–877.

210 M. Gastel, M. Stein, M. Brecht, O. Schroeder, F. Lendzian, R. Bittl, H. Ogata, Y. Higuchi, W. Lubitz, *J Biol Inorg Chem* **2006**, *11*, 41–51.

211 M. Pavlov, P.E.M. Siegbahn, M.R.A. Blomberg, R.H. Crabtree, *J Am Chem Soc* **1998**, *129*, 548–555.

212 M. Pavlov, M.R.A. Blomberg, P.E.M. Siegbahn, *Int J Quant Chem* **1999**, *73*, 197–207.

213 P.E.M. Siegbahn, M.R.A. Blomberg, M.W. Pavlov, R.H. Crabtree, *J Biol Inorg Chem* **2001**, *6*, 460–466.

214 P.E.M. Siegbahn, *Adv Inorg Chem* **2004**, *56*, 101–125.

215 J.W. Peters, W.N. Lanzilotta, B.J. Lemon, L.C. Seefeldt, *Science* **1998**, *282*, 1853–1858.

216 B.J. Lemon, J.W. Peters, *Biochemistry* **1999**, *39*, 12969–12973.

217 Y. Nicolet, C. Piras, P. Legrand, E.C. Hatchikian, J.C. Fontecilla-Camps, *Structure* **1999**, *7*, 13–23.

218 Y. Nicolet, A.L. De Lacey, X. Vernede, V.M. Fernandez, E.C. Hatchikian, J.C. Fontecilla-Camps, *J Am Chem Soc* **2001**, *123*, 1596–1601.

219 Y. Nicolet, C. Cavazza, J.C. Fontecilla-Camps, *J Inorg Biochem* **2002**, *91*, 1–8.

220 A.T. Pierik, M. Hulstein, W.R. Hagen, S.P.J. Albracht, *Eur J Biochem* **1998**, *258*, 572–578.

221 A.L. De Lacey, C. Stadler, C. Cavazza, E.C. Hatchikian, V.M. Fernandez, *J Am Chem Soc* **2000**, *122*, 11232–11233.

222 Z. Chen, B.J. Lemon, S. Huang, D.J. Swartz, J.W. Peters, K.A. Bagley, *Biochemistry* **2002**, *41*, 2036–2043.

223 M.W.W. Adams, L.E. Mortenson, *J Biol Chem* **1984**, *259*, 7045–7055.

224 M.W.W. Adams, *J Biol Chem* **1987**, *262*, 15054–15061.

225 I.C. Zambrano, A.T. Kowal, L.E. Mortenson, M.W.W. Adams, M.K. Johnson, *J Biol Chem* **1989**, *264*, 20974–20983.
226 B. Bennett, B.J. Lemon, J.W. Peters, *Biochemistry* **2000**, *39*, 7455–7560.
227 E.J. Lyon, I.P. Georgakaki, J.H. Reibenspies, M.Y. Darensbourg, *Angew Chem Int Ed* **1999**, *38*, 3178–3180.
228 A. Le Cloirec, S.C. Davies, D.J. Evans, D.L. Hughes, C.J. Pickett, S.P. Best, S. Borg, *Chem Commun* **1999**, 2285–2286.
229 M. Schmidt, S.M. Contakes, T.B. Rauchfuss, *J Am Chem Soc* **1999**, *121*, 9736–9737.
230 H. Li, T.B. Rauchfuss, *J Am Chem Soc* **2002**, *124*, 726–727.
231 L.-C. Song, Z.-Y. Yang, H.-Z. Bian, Q.-M. Hu, *Organometallics* **2004**, *23*, 3082–3084.
232 C.Y. Popescu, E. Muenck, *J Am Chem Soc* **1999**, *121*, 7877–7884.
233 M. Razavet, S.C. Davies, D.L. Hughes, C.J. Pickett, *Chem Commun* **2001**, 847–848.
234 S.J. George, Z. Cui, M. Razavet, C.J. Pickett, *Chem Eur J* **2002**, *8*, 4037–4046.
235 M. Razavet, S.J. Borg, S.J. George, S.P. Best, S.A. Fairhurst, C.J. Pickett, *Chem Commun* **2002**, 700–701.
236 M. Razavet, S.C. Davies, D.L. Hughes, J.E. Barclay, D.J. Evans, S.A. Fairhurst, X. Liu, C.J. Pickett, *Dalton Trans* **2003**, 586–595.
237 X. Yang, M. Razavet, X.-B. Wang, C.J. Pickett, L.-S. Wang, *J Phys Chem A* **2003**, *107*, 4612–4618.
238 G. Zampella, M. Bruschi, P. Fantucci, M. Razavet, C.J. Pickett, L. De Gioia, *Chem Eur J* **2005**, *11*, 509–520.
239 C. Tard, X. Liu, S.K. Ibrahim, M. Bruschi, L. De Gioia, S.C. Davies, X. Yang, L.-S. Wang, G. Sawers, C.J. Pickett, *Nature* **2005**, *433*, 610–613.
240 D. Chong, I.P. Georgakaki, R. Mejia-Rodriguez, J. Sanabria-Chinchilla, M.P. Soriaga, M.Y. Darensbourg, *Dalton Trans* **2003**, *21*, 4158–4163.
241 R. Mejia-Rodriguez, D. Chong, J.H. Reibenspies, M.P. Soriaga, M.Y. Darensbourg, *J Am Chem Soc* **2004**, *126*, 12004–12014.
242 F. Gloaguen, J.D. Lawrence, T.B. Rauchfuss, M. Benard, M.-M. Rohmer, *Inorg Chem* **2002**, *41*, 6573–6582.
243 S. Ott, M. Kritikos, B. Åkermark, L. Sun, R. Lomoth, *Angew Chem Int Ed* **2004**, *43*, 1006–1009.
244 J.-F. Capon, F. Gloaguen, P. Schollhammer, J. Talarmin, *J Electroanal Chem* **2004**, *566*, 241–247.
245 T. Liu, M. Wang, Z. Shi, H. Cui, W. Dong, J. Chen, B. Äkermark, L. Sun, *Chem Eur J* **2004**, *10*, 4474–4479.
246 S.J. Borg, T. Behrsing, S P. Best, M. Razavet, X. Liu, C.J. Pickett, *J Am Chem Soc* **2004**, *43*, 5635–5644.
247 J.W. Tye, J. Lee, H.-W. Wang, R. Mejia-Rodriguez, J.H. Reibenspies, M.B. Hall, M.Y. Darensbourg, *Inorg Chem* **2005**, *44*, 5550–5552.
248 X. Zhao, I.P. Georgakaki, M.L. Miller, J.C. Yarbrough, M.Y. Darensbourg, *J Am Chem Soc* **2001**, *123*, 9710–9711.
249 X. Zhao, I.P. Georgakaki, M.L. Miller, R. Mejia-Rodriguez, C.-Y. Chiang, M.Y. Darensbourg, *Inorg Chem* **2002**, *41*, 3917–3928.
250 J.W. Tye, M.B. Hall, I.P. Georgakaki, M.Y. Darensbourg, *Adv Inorg Chem* **2004**, *56*, 1–26.
251 Z. Cao, M.B. Hall, *J Am Chem Soc* **2001**, *123*, 3734–3742.
252 H.-J. Fan, M.B. Hall, *J Am Chem Soc* **2001**, *123*, 3828–3829.
253 Z.-P. Liu, P. Hu, *J Am Chem Soc* **2002**, *124*, 5175–5182.
254 Z.-P. Liu, P. Hu, *J Chem Phys* **2002**, *117*, 8177–8180.
255 M. Bruschi, P. Fantucci, L. De Gioia, *Inorg Chem* **2002**, *41*, 1421–1429.
256 M. Bruschi, P. Fantucci, L. De Gioia, *Inorg Chem* **2003**, *42*, 4773–4781.
257 G. Zampella, C. Greco, P. Fantucci, L. De Gioia, *Inorg Chem* **2006**, *45*, 4109–4118.
258 M. Bruschi, G. Zampella, P. Fantucci, L. De Gioia, *Coord Chem Rev* **2005**, *249*, 1620–1640.
259 J. Kim, D.C. Rees, *Science* **1992**, *257*, 1677–1682.
260 J. Kim, D.C. Rees, *Nature* **1992**, *360*, 553–560.
261 M.K. Chan, J. Kim, D.C. Rees, *Science* **1993**, *260*, 792–794.

262 J. Kim, D. Woo, D.C. Rees, *Biochemistry* **1993**, *32*, 7104–7115.
263 J.T. Bolin, N. Campobasso, S.W. Muchmore, T.V. Morgan, L.E. Mortenson, in *Molybdenum Enzymes, Cofactors and Model Systems*, E.I. Stiefel, D. Coucouvanis, W.E. Newton (Eds.), American Chemical Society, Washington, DC, **1993**, pp. 186–195.
264 J.W. Peters, M.H.B. Stowell, S.M. Soltis, M.G. Finnegan, M.K. Johnson, D.C. Rees, *Biochemistry* **1997**, *36*, 1181–1187.
265 S.M. Mayer, D.M. Lawson, C.A. Gormal, S.M. Roe, B.E. Smith, *J Mol Biol* **1999**, *292*, 871–891.
266 O. Einsle, F.A. Tezcan, S.L.A. Andrade, B. Schmid, M. Yoshida, J.B. Howard, D.C. Rees, *Science* **2002**, *297*, 1696–1700.
267 J.C. Peters, M.P. Mehn, see Chapter 3 in this book.
268 R.Y. Igarashi, M. Laryukhin, P.C. Dos Santos, H.-I. Lee, D.R. Dean, L.C. Seefeldt, B.M. Hoffman, *J Am Chem Soc* **2005**, *127*, 6231–6241.
269 S. Ogo, B. Kure, H. Nakai, Y. Watanabe, S. Fukuzumi, *Appl Organomet Chem* **2004**, *18*, 589–594.

5
Molecular Oxygen Binding and Activation: Oxidation Catalysis

Candace N. Cornell and Matthew S. Sigman

5.1
Introduction

The cost of using common oxidation reagents is rising due to the need to dispose of stoichiometric quantities of metal waste. For this reason, a strong emphasis has been put on reducing the environmental impact of oxidations in fine and commodity chemical syntheses [1]. Molecular oxygen (O_2) has been targeted as an ideal oxidant because of its abundance, atom efficiency and benign byproducts (H_2O_2 and/or H_2O). As researchers incorporate the advances in inorganic/organometallic oxidation chemistry with a more detailed understanding of biological aerobic oxidations, chemoselective metal-centered oxygen activation and subsequent substrate functionalization has become a realistic goal.

This chapter highlights many of the recent advances in homogeneous aerobic oxidation catalysis. As a primer, aerobic oxidations, in both biological and organometallic chemistry, have been classified as either "oxygenase" or "oxidase" reaction types based on whether oxygen atoms from O_2 are incorporated into the product [2]. In oxygenase systems, either one or both of the oxygen atoms are incorporated into the substrate (O-atom transfer) (Scheme 5.1 A and B) [3, 4]. In oxidase systems, O_2 acts as an electron/proton acceptor (the oxidant) where either H_2O or H_2O_2 is the inorganic coproduct (Scheme 5.1 C) [5, 6]. Oxidase systems increase the scope of aerobic oxidation catalysis, in that the oxidation of the organic substrate does not include the transfer of oxygen and any oxidation reaction can potentially be coupled to an environmentally friendly dioxygen reduction.

The oxidase/oxygenase reactivity distinction will be used throughout the chapter. Molecular oxygen may also react via an autoxidation, single-electron mechanism, but the nonmetal-centered reactivity of dioxygen will not be discussed in this chapter. Rather, the focus shall be on the advancement of chemoselective reactions through modifications of the metal-centered catalyst system.

To achieve successful aerobic oxidation catalysis, the redox potential of the metal catalyst must be tuned for both substrate oxidation and turnover. A high

Activation of Small Molecules. Edited by William B. Tolman

ISBN: 3-527-31312-5

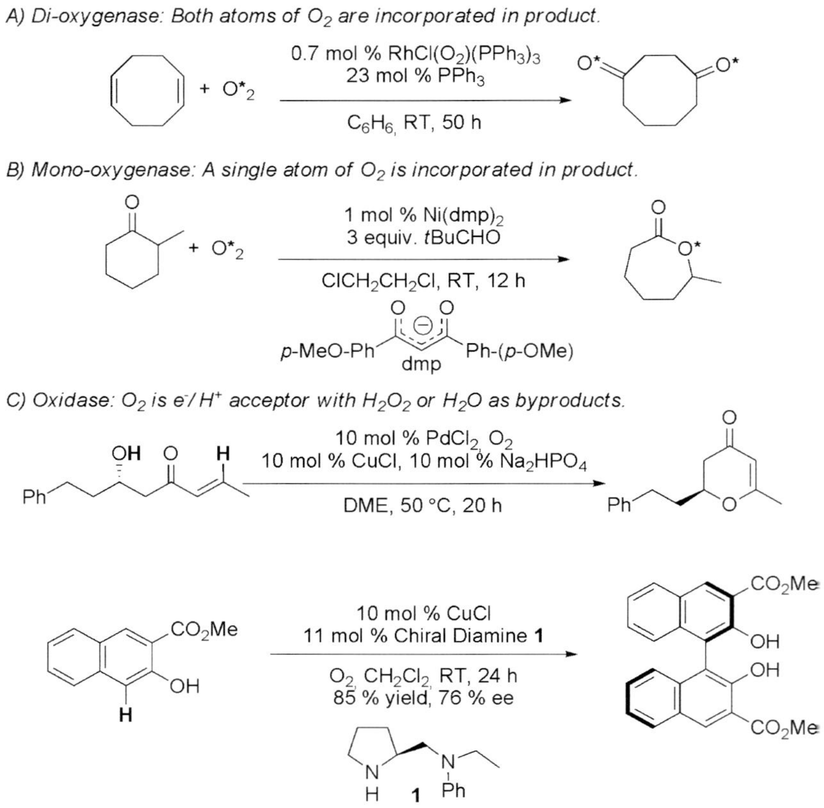

A) Di-oxygenase: Both atoms of O_2 are incorporated in product.

+ O^*_2 — 0.7 mol % $RhCl(O_2)(PPh_3)_3$, 23 mol % PPh_3; C_6H_6, RT, 50 h

B) Mono-oxygenase: A single atom of O_2 is incorporated in product.

+ O^*_2 — 1 mol % $Ni(dmp)_2$, 3 equiv. *t*BuCHO; $ClCH_2CH_2Cl$, RT, 12 h

p-MeO-Ph, Ph-(*p*-OMe), dmp

C) Oxidase: O_2 is e^-/H^+ acceptor with H_2O_2 or H_2O as byproducts.

10 mol % $PdCl_2$, O_2, 10 mol % CuCl, 10 mol % Na_2HPO_4; DME, 50 °C, 20 h

10 mol % CuCl, 11 mol % Chiral Diamine **1**; O_2, CH_2Cl_2, RT, 24 h, 85 % yield, 76 % ee

Scheme 5.1 Examples of organometallic aerobic oxidation catalytic systems in which O_2 acts as either the oxygen source in oxygenase-type activity (A or B) or an electron/proton acceptor in oxidase-type activity (C).

catalyst oxidation potential is required for substrate oxidation (an electrophilic catalyst), but must also be moderated to provide a chemoselective oxidation. In other words, an extremely "hot" oxidant may react at more than one site in the molecule. Once the substrate is oxidized, direct O_2 turnover of the catalyst requires a low oxidation potential where the catalyst is nucleophilic enough to react with O_2. Activation of O_2 by the metal must also be facile to avoid competing catalyst decomposition pathways. In order to achieve these somewhat conflicting goals many types of modifications have been utilized, including (i) additive coreductants (Section 5.2), (ii) coupling more than one redox cycle to substrate oxidation (Section 5.2.2) and (iii) ligand-supported catalysis for direct metal–O_2 oxidation systems (Section 5.3). A comprehensive examination of all reactions in each of these areas is neither possible due to the vast nature of the topic nor necessary thanks to the many recently published reviews [2]. Rather, selected examples which highlight recent developments and the mechanistic complexity associated with aerobic oxidation catalysis are presented.

5.2 Additive Coreductants

Ideally, a metal catalyst should be able to perform a chemoselective substrate oxidation followed by direct reoxidation with molecular oxygen. Due to the extreme difference in required reactivity of the metal catalyst for each of these processes, cocatalysts are frequently necessary. Two distinct strategies will be discussed: (i) modification of the metal-based oxidant with a coreductant allowing oxygenase activity (Section 5.2.1) or (ii) inclusion of another catalyst which is directly oxidized by O_2 and is responsible for reoxidation of the substrate catalyst (Section 5.2.2). The following sections describe specific examples of each approach.

5.2.1 Aldehydes

Epoxidation reactions have been widely utilized for over 100 years with peracids, peroxides and, more recently, metal catalysts [7]. However, direct metal-catalyzed aerobic epoxidations are rare and generally require an aldehyde coreductant. In this case, the metal is proposed to catalyze radical formation (**A–C**, Scheme 5.2) followed by O_2 insertion to form acyl peroxide **D**. Metal-catalyzed aerobic oxidation of aldehydes to peracids has previously been observed [8]. With the formation of species **D**, either an outer-sphere path similar to a peracid-type oxidation occurs (**Path 1**) or an inner-sphere metal-catalyzed oxidation in which the metal-based oxidant and substrate interact during oxygen transfer (**Path 2** or **3**). Mukaiyama and coworkers were the first to report an aerobic epoxidation of olefins catalyzed by transition metals using either a primary alcohol or an aldehyde as coreductants [9]. The role of the metal was probed through parallel studies of peracid and metal-catalyzed epoxidations of **2** which yielded different stereochemical outcomes. Therefore, a metal-centered mechanism for olefin epoxidation was proposed which implicates an oxygenase system, **Path 2** or **3** (Table 5.1) [10].

Further evidence for a metal-mediated process is the observation of asymmetric induction when using a chiral ligand on the catalyst. The chiral complex must be closely associated with the olefin during oxidation to induce an enantioselective outcome. Several olefins were epoxidized using Jacobsen's catalyst, **3** [11], under Mukaiyama's conditions. Good enantioselection was observed (Fig. 5.1) [12], indicating that the rate of metal-centered epoxidation is greater than the rate of peracid epoxidation. However, under standard Jacobsen epoxidation conditions using bleach as an oxidant and lower catalyst loading (2 mol%), a significantly improved outcome is achieved (72% yield and 98% *ee*) [11]. Thus,

Table 5.1 Stereochemical evidence for metal-centered epoxidation

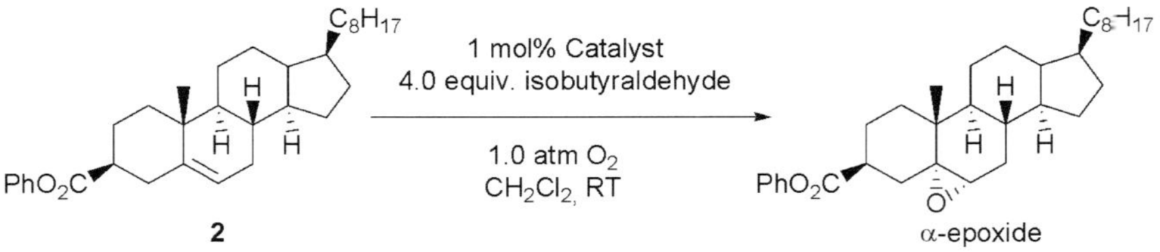

1 mol% Catalyst, 4.0 equiv. isobutyraldehyde, 1.0 atm O_2, CH_2Cl_2, RT; **2** → α-epoxide

Entry	Catalyst	*α*-Epoxide	*β*-Epoxide
1	$Ni(dmp)_2$	31	69
2	$Mn(acac)_2$	20	80
3	*m*CPBA	71	29

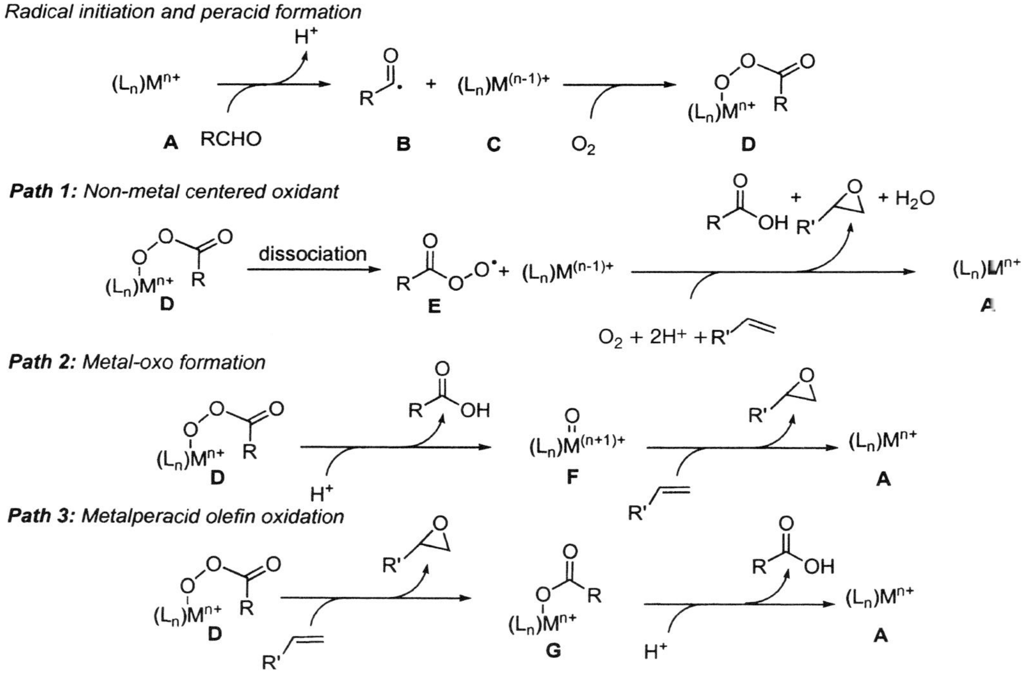

Scheme 5.2 Proposed mechanistic paths for aerobic olefin epoxidation with aldehyde coreductant.

while the use of aerobic conditions in a metal-centered asymmetric epoxidation is possible, the inherent background processes still remain problematic.

The stereoselective epoxidation examples in Table 5.1 and Fig. 5.1 are rare examples where the metal-mediated process (**Path 2** or **3**) clearly competes with

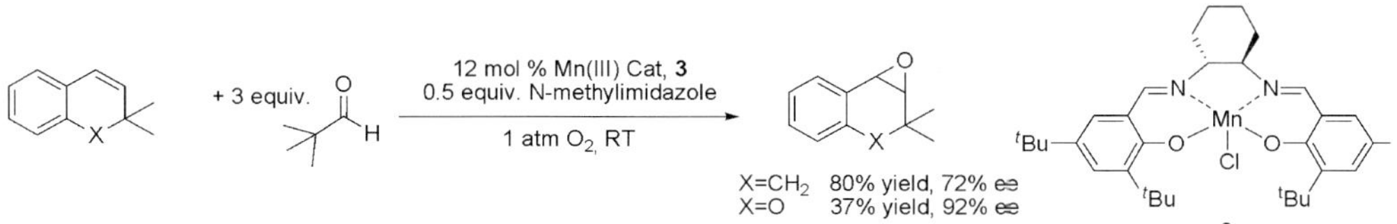

Fig. 5.1 Asymmetric Mn(salen) oxidation utilizing O_2 and additive aldehyde supports a metal-centered oxidation.

background autoxidation [13]. As noted in Scheme 5.2, the epoxidation mechanism using additive aldehydes has been proposed to proceed through either a metal-oxo **F** (**Path 2**) or a metal-peroxo **D** (**Path 3**). Evaluating these possibilities, Nolte and coworkers were unable to detect a Ni-oxo species via electron paramagnetic resonance (EPR). However, they observed two species attributed to an acyl radical (**B**) when metal and aldehyde are combined under an inert atmosphere and a second radical species, upon O_2 exposure, proposed to be a metal acylperoxy radical (similar to **D**) (Scheme 5.2) [13d]. In contrast, Talsi's later studies in 2001 provided the first spectroscopic data supporting a monomeric Mn(IV)-oxo intermediate formed when **3** is combined with pivaldehyde or isobutyraldehyde and allowed to react with O_2 [14]. A metal-bound acylperoxide was not observed in these studies. Additionally, Talsi examined Mn(salen)/(*N*-Me-Imd)/aldehyde/O_2 used in Mukaiyama's asymmetric aerobic epoxidation, to probe potential intermediates. Again, a Mn(IV)(salen)oxo was observed and allowed to react with dihydronaphthalene leading to a racemic product (25% yield). Based on this result, a more complex interaction/mechanism is implicated and it clearly shows the challenges in deciphering or predicting pathways in Mukaiyama-type epoxidation reactions.

Mukaiyama's conditions have also been used in other aerobic oxidation reactions of substrates including thiols (Table 5.2, entries 1–4, 10 and 11), alkanes (entries 8, 12 and 14) and alcohols (entries 9 and 13), as well as reactions involving lactone formation via a Baeyer-Villiger oxidation (entries 5–7) and oxidative decarboxylation (entry 16) [15–17]. While nickel, iron and cobalt all selectively oxidize thiols to sulfoxides, Co(II) is the most active (entries 1–4) [15b]. Of particular synthetic interest, the chemoselective and diastereoselective aerobic oxidation of the complex sulfide, exomethylenecepham (entries 10 and 11), was observed with no overoxidation to the sulfone or oxidation of the olefin [16a]. The diverse substrate scope in entries 1–9 suggest iron and nickel species tend to have similar reactivity with substrates, but cobalt behaves differently. For example, both iron and nickel displayed similar reactivity in Baeyer-Villiger oxidations, with cobalt being much less active (entries 5–7), yet the opposite trend was observed for sulfide oxidation (entries 1–4) [15]. Lastly, illustrating the broad potential scope of Mukaiyama-type oxidations, alcohol oxidation (entries 9 and 13) and oxidative decarbonylation (entry 15) reactions, which are oxidase systems, have also been reported [16b, 17b].

Table 5.2 Aerobic metal-catalyzed organic substrate oxidations in the presence of aldehyde

Entry [Reference]	Substrate	Catalyst	Catalyst (mol%)	Aldehyde	Time (h)	Conversion (%)	Yield (%)	Product
1 [15b]	$(CH_3)_2S$	$Co(AAEMA)_2$	0.8	IVA	6	55	52	$(CH_3)_2SO$
2 [15b]	$(CH_3)_2S$	$Co(AAEMA)_2$	0.8	IVA	20	100	90	$(CH_3)_2SO_2$
3 [15b]	$(CH_3)_2S$	$Ni(AAEMA)_2$	0.8	IVA	20	5	5	$(CH_3)_2SO$
4 [15b]	$(CH_3)_2S$	$Fe(AAEMA)_3$	0.8	IVA	20	15	15	$(CH_3)_2SO$
5 [15b]	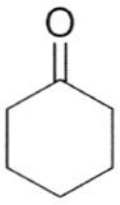	$Co(AAEMA)_2$	0.8	IVA	36	2	2	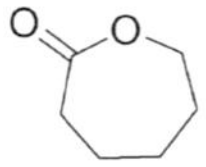
6 [15b]		$Ni(AAEMA)_2$	0.8	IVA	36	75	75	
7 [15b]		$Fe(AAEMA)_3$	0.8	IVA	36	75	75	
8 [15b]	Ethylbenzene	$Co(AAEMA)_2$	0.8	IVA	40	10	10	acetophenone
9 [15b]	cyclohexanol	$Co(AAEMA)_2$	0.8	IVA	48	56	52	cyclohexanone
10 [16a]	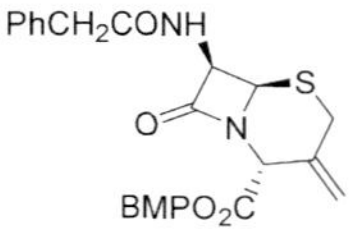 $PhCH_2CONH$ … $BMPO_2C$	$Co(acac)_3$	10	IBA	3	–	23α,49β	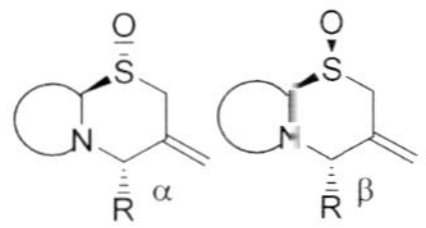 α, β
11 [16a]		$CoCl_2$	10	IBA	22	<45	41 (68:32)	
12 [16b]	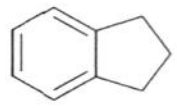	Co(TPP-OMe)	2.5	IBA	3	–	86	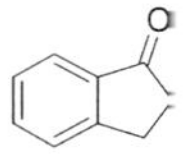
13 [16b]	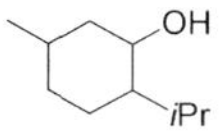 OH, iPr	Co(TPP-OMe)	2.5	IBA	3	–	76	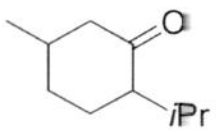 iPr
14 [16b]		Co(TPP-OMe)	2.5	IBA	3	–	26 (44:56)	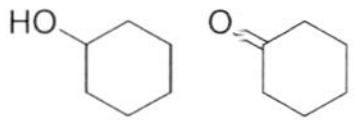 HO, O
15 [17b]	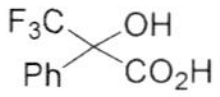 F_3C, OH, Ph, CO_2H	4	5	PivA	–	>95	87	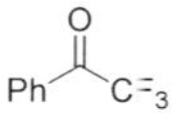 Ph, CF_3

IVA = isovaleraldehyde, IBA = isobutyraldehyde, PivA = pivaldehyde.

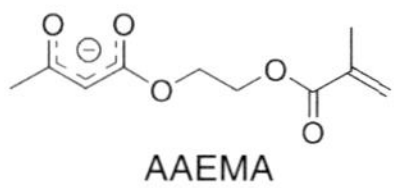

AAEMA
2-(acetoacetoxy)methyl methacrylate

TPP-OMe
tetra-(*para*-methoxyphenyl) porphyrin

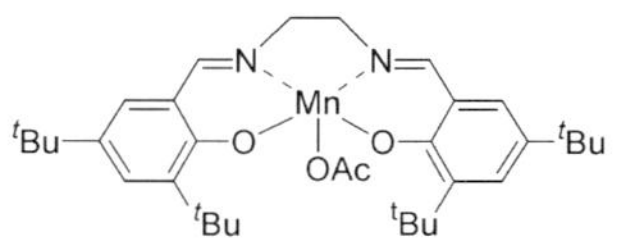

Mn(III)(Salen)OAc

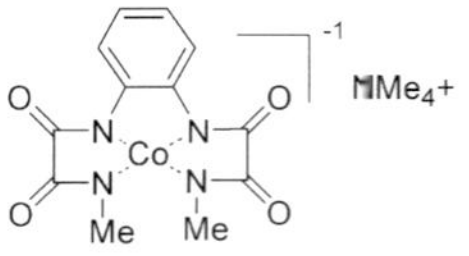

4 (NMe_4^+)

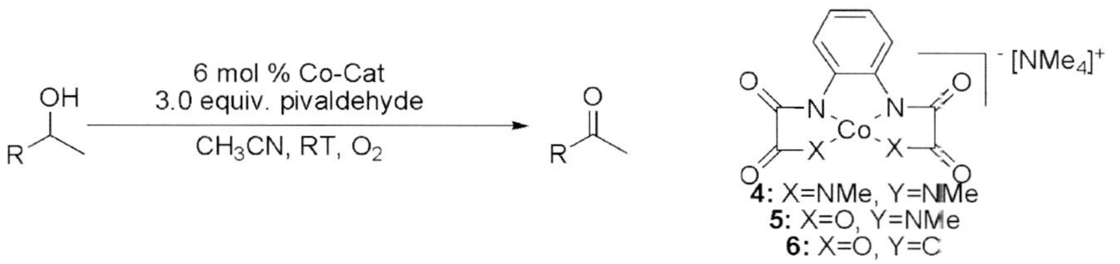

Fig. 5.2 Cobalt-catalyzed alcohol oxidation with pivaldehyde additive.

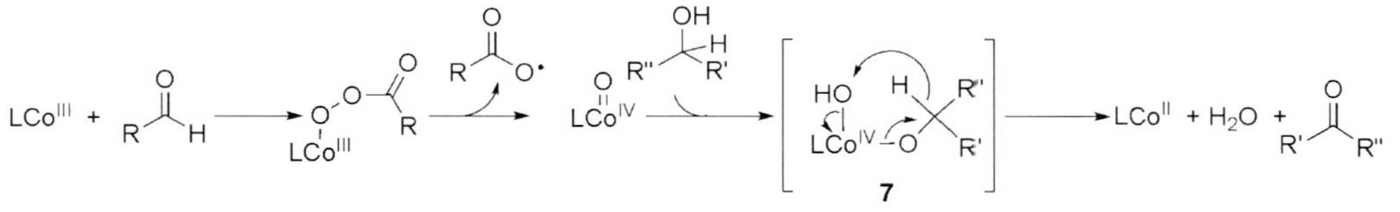

Scheme 5.3 Pedro's proposed Co(III) Mukaiyama alcohol oxidation mechanism.

In a step towards predicting and systematically modifying catalysts for Mukaiyama-type oxidations, Pedro and coworkers conducted a mechanistic study of a secondary alcohol oxidation [17a]. Using square-planar Co(III)-oxamidate and/or oxamate catalysts in the presence of pivaldehyde, kinetic analysis showed a correlation between catalytic activity and the electron donor capacity of the ligand ($t_{1/2}$ **4** > $t_{1/2}$ **5** > $t_{1/2}$ **6**) (Fig. 5.2). Substrate effects were also evaluated showing that increasing the size of R decreases ketone yields, but the electronic nature of R has little impact on the rate. The mechanism which best fits their studies is conversion of Co(IV) species, **7**, to product through a concerted intramolecular hydride transfer (Scheme 5.3). Even this comprehensive study alludes to the mechanistic complexity associated with Mukaiyama systems and that achieving a more-informed approach in modifying these systems is clearly difficult.

5.2.2
Coupled Catalytic Systems

The reoxidation rate of the reduced metal is important in an efficient catalyst system. The metal-centered catalyst may have the proper redox potential for O_2 oxidation, but if the rate of catalyst decomposition is competitive, then loss of the active catalyst results. To overcome issues of low turnover number (TON) and turnover frequency (TOF), added reagents are often used to couple multiple catalytic cycles in a biomimetic fashion [18]. Transition metals known to activate O_2 based on biological precedence are commonly used as coupling reagents, including Cu(II), Co(II), Fe(II) and Mn(II) [2]. Many other organic molecules, such as quinones and catechols, can also act as redox couples.

5.2.2.1 Organic Cocatalysts

Many cocatalysts are responsible only for electron and proton transfer between the metal catalyst and O_2, but examples of cocatalyst interaction with the substrate also exist. In the example below, Marko and coworkers have shown that the catalytic use of copper in an aerobic alcohol oxidation is achieved through the catalytic addition of an azo compound [19]. The mechanistic studies support the pathways presented in Scheme 5.4 with a Cu(I)-azo oxidant, **8**, in which the azo compound abstracts the hydride from the alcohol, **9**.

5.2.2.2 Metal Cocatalysts

5.2.2.2.1 Copper

Copper has been commonly utilized as a redox couple primarily due to the ease of O_2 activation by copper in proteins and coordination complexes [20, 21]. A quintessential application of this process is the Wacker oxidation, which is a highly useful reaction both synthetically and industrially (Scheme 5.5) [22]. Reoxidation of Pd(0) to Pd(II) with $CuCl_2$ was accomplished under the industrial conditions discovered by Smidt using high concentrations of both HCl and $CuCl_2$ [23]. The concentrations of Cu and HCl presumably enhance the rate of Pd(0) reoxidation by Cu(II), thereby avoiding palladium decomposition. Although an O-atom is incorporated into the product, water is the source of the oxygen, not O_2, therefore Wacker oxidations are considered oxidase reactions [22]. In this process, CuCl is reoxidized by O_2 and can easily be recycled [2, 23]. Unfortunately for synthetic organic applications, the highly acidic aqueous conditions and the copious amounts of $CuCl_2$ lead to solubility problems for longer chain olefins (more than four carbons), functional group intolerance, competi-

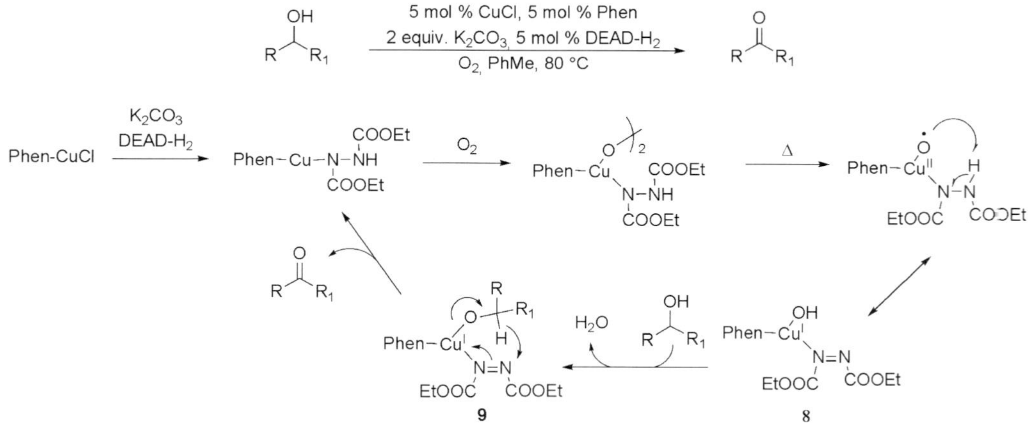

5 mol % CuCl, 5 mol % Phen
2 equiv. K_2CO_3, 5 mol % DEAD-H_2
O_2, PhMe, 80 °C

Phen-CuCl —(K_2CO_3, DEAD-H_2)→ Phen−Cu−N(COOEt)−NH(COOEt) —(O_2)→ ... —(Δ)→ ...

H₂O

8

9

Scheme 5.4 PhenCuCl-catalyzed aerobic oxidation of alcohols coupled with DEAD turnover catalyst.

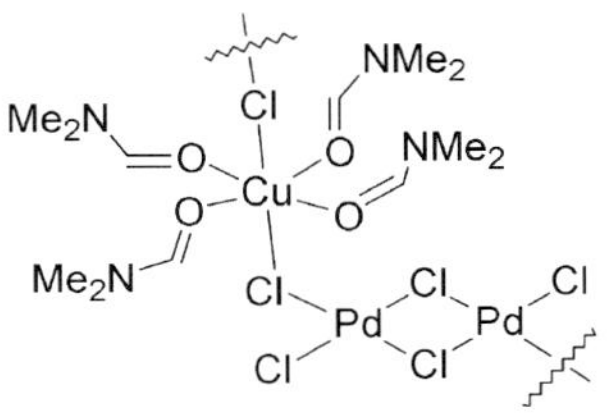

Fig. 5.3 Pd/Cu bimetallic catalytically active species **10** formed under anaerobic conditions with DMF.

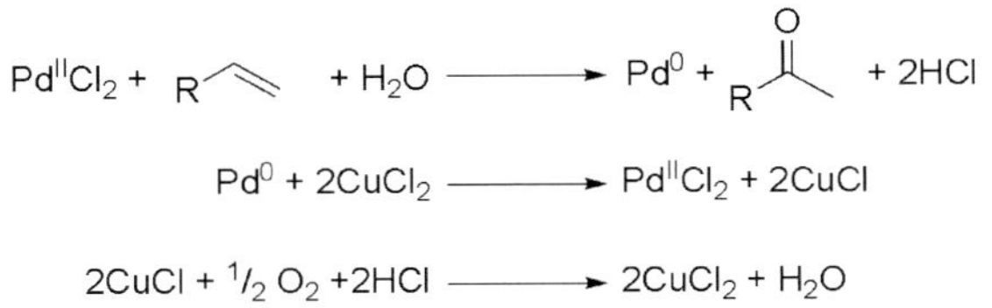

$$Pd^{II}Cl_2 + \text{R-CH=CH}_2 + H_2O \longrightarrow Pd^0 + \text{RC(O)CH}_3 + 2HCl$$

$$Pd^0 + 2CuCl_2 \longrightarrow Pd^{II}Cl_2 + 2CuCl$$

$$2CuCl + \tfrac{1}{2} O_2 + 2HCl \longrightarrow 2CuCl_2 + H_2O$$

Scheme 5.5 Wacker oxidation: a Pd/Cu-coupled oxidation of olefins via oxidase catalysis.

tive isomerization and the formation of chlorinated byproducts. To broaden the scope of the reaction, a mixed solvent system using 85% dimethylformamide (DMF)/H_2O was developed by Clement and Selwitz, and further explored by Tsuji and coworkers (Scheme 5.6) [2b, 22c, 24]. Wacker reactions performed in DMF showed high chemoselectivity for methyl ketone formation with low degrees of substrate isomerization compared to reactions performed in other water miscible solvents. Interestingly, the unique success of the DMF/H_2O system may not be based solely on solubility and Pd/Cu redox chemistry. Hosokawa and coworkers observed that $PdCl_2$, CuCl and DMF form two distinct complexes under an O_2 atmosphere [25]. The first is a bimetallic polymeric structure with the palladium and copper units linked by chlorine atoms (**10**, Fig. 5.3). Under aerobic conditions, complex **10** was found to catalyze the oxidation of decene, wherein water increases the rate and turnover number of the catalyst. The second structure, based on spectroscopic analysis, was formulated as $(PdCl_2)_x(CuO)_y(DMF)_z$ and was found to be competent under anaerobic conditions to oxidize decene in 90% yield with no H_2O present. These findings suggest copper may play a more complex role in Wacker oxidations than acting solely as an outer-sphere redox catalyst.

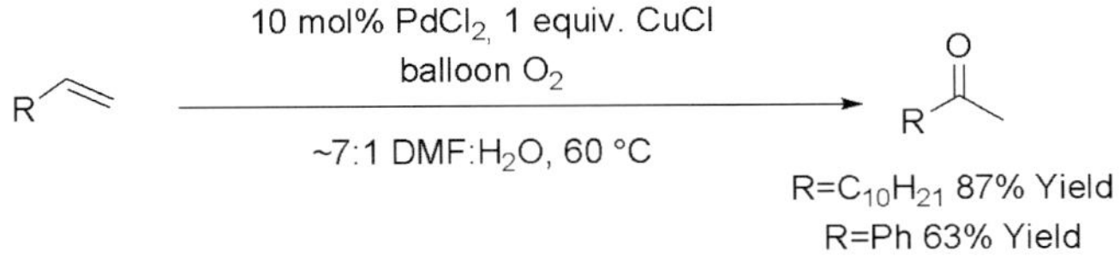

$$\text{R-CH=CH}_2 \xrightarrow[\sim 7:1\ DMF:H_2O,\ 60\ ^{\circ}C]{10\ mol\%\ PdCl_2,\ 1\ equiv.\ CuCl,\ balloon\ O_2} \text{RC(O)CH}_3$$

R=$C_{10}H_{21}$ 87% Yield
R=Ph 63% Yield

Scheme 5.6 Additional examples of Pd/Cu-coupled aerobic oxidation of sp^2 and sp^3 carbon centers.

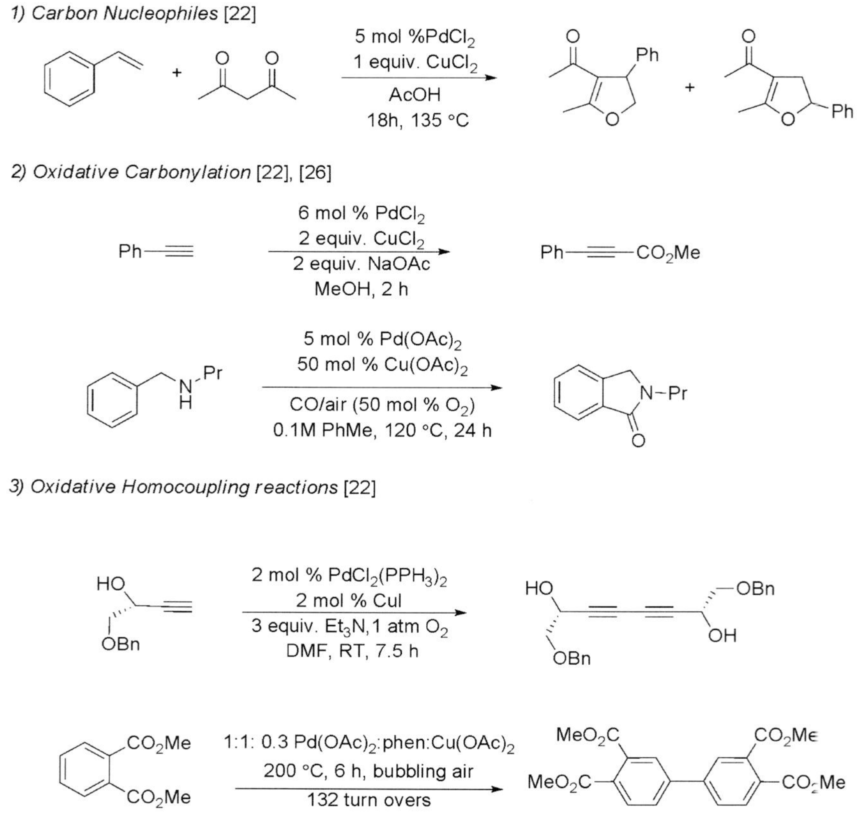

1) Carbon Nucleophiles [22]

5 mol % $PdCl_2$
1 equiv. $CuCl_2$
AcOH
18h, 135 °C

2) Oxidative Carbonylation [22], [26]

Ph—≡ → Ph—≡—CO_2Me

6 mol % $PdCl_2$
2 equiv. $CuCl_2$
2 equiv. NaOAc
MeOH, 2 h

5 mol % $Pd(OAc)_2$
50 mol % $Cu(OAc)_2$
CO/air (50 mol % O_2)
0.1M PhMe, 120 °C, 24 h

3) Oxidative Homocoupling reactions [22]

2 mol % $PdCl_2(PPH_3)_2$
2 mol % CuI
3 equiv. Et_3N, 1 atm O_2
DMF, RT, 7.5 h

1:1: 0.3 $Pd(OAc)_2$:phen:$Cu(OAc)_2$
200 °C, 6 h, bubbling air
132 turn overs

Scheme 5.7 DMF/H_2O mixed-solvent system in synthetic Wacker applications.

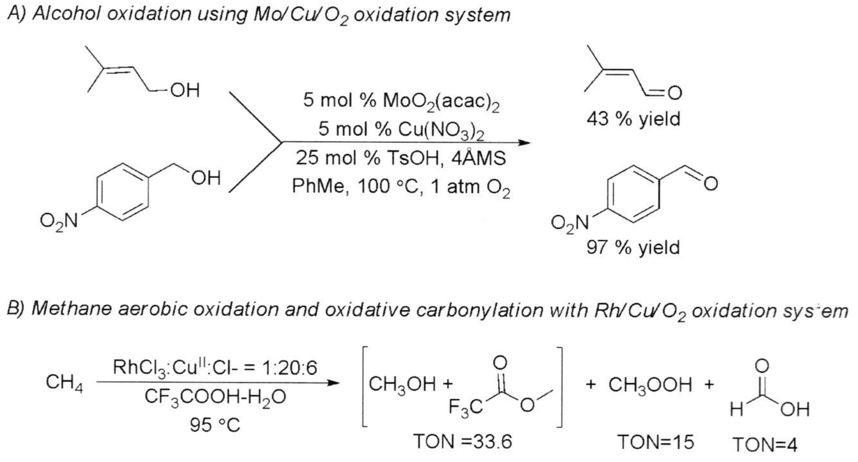

A) Alcohol oxidation using Mo/Cu/O_2 oxidation system

5 mol % $MoO_2(acac)_2$
5 mol % $Cu(NO_3)_2$
25 mol % TsOH, 4ÅMS
PhMe, 100 °C, 1 atm O_2

43 % yield

97 % yield

B) Methane aerobic oxidation and oxidative carbonylation with Rh/Cu/O_2 oxidation system

CH_4 → [CH_3OH + CF_3COOCH_3] + CH_3OOH + HCOOH

$RhCl_3$:Cu^{II}:Cl- = 1:20:6
CF_3COOH-H_2O
95 °C

TON =33.6 TON=15 TON=4

Scheme 5.8 Copper cocatalysis in molybdenum and ruthenium aerobic oxidation systems.

Pd/Cu-coupled catalysis has been used in many Wacker-type olefin oxidations other than those that involve Markovnikov methyl ketone formation from terminal olefins [1a,b, 21]. Pd/Cu-coupled aerobic oxidation systems have also been widely applied to other sp^2 and sp^3 carbon oxidations. Selected examples of these oxidations, including those involving carbon nucleophiles, oxidative carbonylations and oxidative coupling reaction, are pictured in Scheme 5.7 [22, 26]. The vast success found with Pd(0) and Cu(II) reoxidation led to Cu(II) being tested as an additive in other metal-catalyzed aerobic oxidations, including Rh(I)-catalyzed methane oxidation [27] and Mo(VI)-catalyzed alcohol oxidation [28] (Scheme 5.8).

5.2.2.2.2 **Multicomponent Coupled Catalytic Cycles**

Biomimetic electron transport mediation allows for the splitting of large redox potentials into smaller coupled systems with lower energy barriers [29]. Utilizing electron transport mediators in oxidase-type catalysis allows for indirect coupling catalyst reoxidation using O_2, which has a relatively low redox potential, with substrate oxidation, requiring a high redox potential. As an example, benzoquinone can be substituted for Cu(II) in Wacker-type oxidations [22, 30]. A greater number of coupled cycles can also be utilized. An elegant illustration of this concept has been executed for both alcohol and amine oxidation reactions using a palladium or ruthenium catalyst as the substrate oxidant and a Co/benzoquinone (BQ)/O_2-coupled catalyst oxidant (Scheme 5.9) [31]. Specifically, Bäckvall and coworkers have shown that the ruthenium precatalyst **11** is effective in oxidizing amines to imines (Scheme 5.10). Initial studies on the system utilized one equivalent of quinone as the oxidant. However, addition of **13**, a Co(salen)-type complex, allows the use of catalytic amounts of quinone. The catalytic system using 2 mol% **11**, 2 mol% **13** and 20 mol% quinone gave very little background reaction (less than 10% at 1 h versus more than 30% with only quinone). This methodology allows for the formation of both ketimines and aldimines under mild conditions.

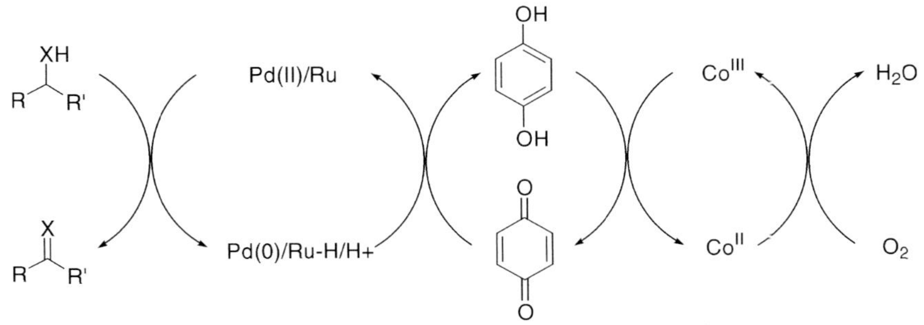

Scheme 5.9 Biomimetic coupled amine and alcohol oxidation using a Co/BQ/O_2 system.

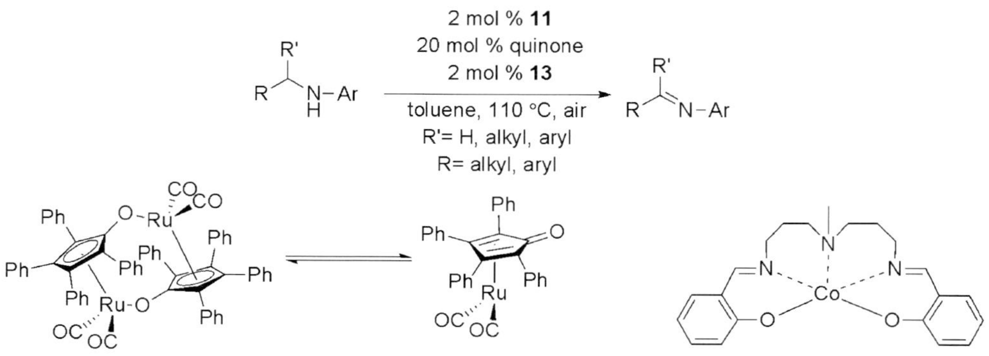

Scheme 5.10 Ruthenium-catalyzed aerobic oxidation of amines using a coupled catalytic system.

Coupled redox systems have also been successfully employed in asymmetric catalysis. Krief and coworkers were the first to use air as a terminal oxidant with the Sharpless dihydroxylation catalyst by using benzyl phenyl selenide as a photosensitizer [32–34]. The dihydroxylation of α-methylstyrene using the selenide sensitizer decreases the oxidant waste by over 50-fold compared to the standard oxidation conditions using AD-Mix ($K_3Fe[CN]_6$ is the terminal oxidant in AD-Mix) with similar enantiomeric excess and yield [34]. This example of a coupled reaction showcases the potential waste reduction provided by this approach.

5.3 Ligand-modified Catalysis

We have already discussed additives of catalytic aerobic processes that use stoichiometric coreductants as with the Mukaiyama systems and coupled oxidation cycles. Unfortunately, these modifications often defeat the goal of aerobic catalysis by adding undesirable byproducts and waste to the oxidation system. Reducing the need for a coreductant requires a metal catalyst that chemoselectively performs the oxidation and is readily reoxidized by O_2 – a challenging goal [35]. Nature is the inspiration as it can perform amazingly selective tasks with highly tuned enzyme catalysts [36]. Nature's triumph necessitates complex protein structures and a cascade of events which a synthetic chemist has a difficult time mimicking in small-molecule systems. However, as in biological systems, a ligand on the metal can play a vital role in the ability of a metal complex to achieve a direct aerobic oxidation of an organic substrate. The following sections highlight specific advancements in direct O_2 catalysis using small-molecule ligands.

5.3.1 Porphyrin Catalysis

Macrocyclic tetrapyroles, including porphyrins, are used throughout nature for binding metals in enzymatic systems and the resulting complexes perform a wide range of tasks (Fig. 5.4) [37]. The versatility of porphyrin catalysis relies on the electronic character of the metal center, the geometry of the porphyrin ring, as well as the presence and nature of axial ligands. Porphyrin complexes with iron, cobalt, manganese, copper and nickel are known to activate O_2, and their reactivity in organometallic chemistry has been extensively probed [38]. Cited below are examples of modified porphyrin systems used in direct O_2-coupled catalytic oxidations.

Cytochrome P450, an iron porphyrin found throughout biological systems, is one of the most widely studied oxidation enzymes and has been a model for biomimetic oxidations [39, 40]. Model studies with iron porphyrin systems identified a pitfall of the porphyrin ligand under standard oxidative conditions: susceptibility to oxidative degradation [41]. If both faces of the Fe(II) porphyrin are not protected, irreversible oxidation occurs forming a catalytically incompetent μ-oxo Fe(III) dimer which severely limits the TON. To enhance TON, Nocera and coworkers recently synthesized the "pacman" bis-iron porphyrin catalyst **14**, increasing the overall TON to near 600 in the aerobic photooxidation of olefins (Scheme 5.11) [42]. This synthetic modification to the porphyrin increased the durability by adding significant steric bulk to the porphyrin ligand, inhibiting

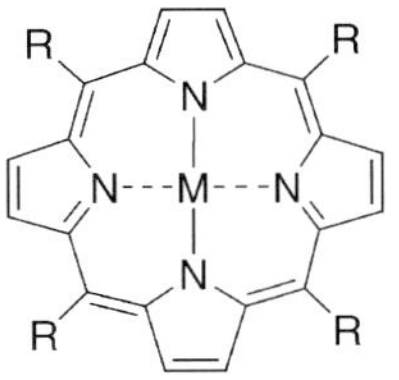

R R N N M N N R R

Fig. 5.4 General metal porphyrin structure.

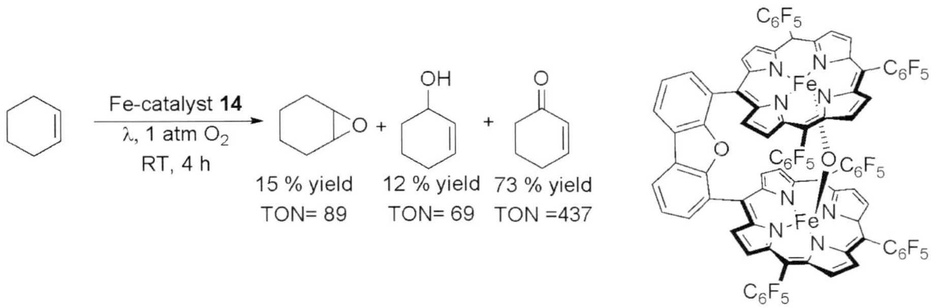

Fe-catalyst **14**
λ, 1 atm O_2
RT, 4 h

O + OH + O

15 % yield TON= 89
12 % yield TON= 69
73 % yield TON =437

C_6F_5 C_6F_5 Fe O C_6F_5 C_6F_5 Fe C_6F_5 C_6F_5 **14**

Scheme 5.11 "Pacman" iron porphyrin catalyst for aerobic olefin oxidation.

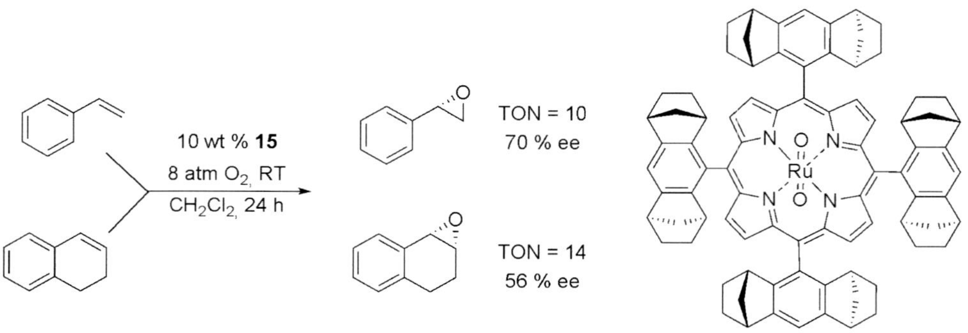

10 wt % **15**
8 atm O_2, RT
CH_2Cl_2, 24 h

TON = 10
70 % ee

TON = 14
56 % ee

15

Scheme 5.12 Asymmetric aerobic epoxidation of olefins with chiral ruthenium complex **15**.

oxidative dimer formation. This increases the applicability of iron porphyrin systems for use in aerobic catalysis.

Chiral porphyrin ligands have been used in asymmetric direct O_2-coupled catalytic oxidations. The first catalytic asymmetric homogeneous aerobic epoxidation, not requiring an aldehyde coreductant, was achieved with a Ru(VI)dioxo-(D_4-porphyrinato) complex, **15** (Scheme 5.12) [43]. Although these chiral ruthenium complexes have promoted good asymmetric induction in the product, the TON is still quite low [38]. Currently, a synthetically viable homogeneous aerobic epoxidation catalyst without coreductants has not been developed.

5.3.2
Schiff Bases

Due to their similarity to porphyrins and their ease of synthesis, Schiff bases have been extensively explored as ligands in oxidation catalysis [38, 44, 45]. Salen-type ligands are the most commonly used class of Schiff bases, and are synthesized from a diamine and two equivalents of a salicylaldehyde in which chiral derivatives are readily accessed (Fig. 5.5) [46]. Additionally, Schiff base ligands coordinate and stabilize a diverse range of metals in various oxidation states making them an excellent template to explore direct O_2-coupled catalytic oxidations.

In an effort to model the unique reaction mechanism of galactose oxidase (GAO), which entails a ligand-centered radical mechanism, rather than a direct

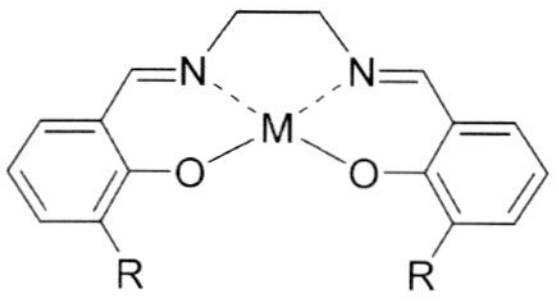

Fig. 5.5 General structure of M(salen) complexes.

metal-centered oxidation [47], Stack and coworkers designed and synthesized several Cu(salen)-type catalysts (**16–18**) to afford an aerobic alcohol oxidation (Scheme 5.13) [48, 49]. Oxidation of activated benzylic and allylic alcohols was accomplished using low catalyst loadings with a TON of up to 1300. X-ray crystal structural and EPR spectral analysis of **16** confirmed a nonsquare-planar geometry about copper caused by the binaphthyl backbone. Single-electron oxidation of species **B** is proposed to form a d^9 Cu(II) center and a ligand-based radical, similar to that of the active form of GAO. Further spectroscopic studies identified potential intermediates of the catalytic system (Scheme 5.13). A competitive kinetic isotope effect of 5.3 for $PhCD_2O^-$ and $PhCH_2O^-$ supports cleavage of the α-C–H bond during the rate-determining step. The mechanistic similarities between the synthetic system and the natural system showcase the possible success of using small-molecule systems to mimic biological aerobic oxidative transformations.

Also using a Schiff base ligand, Sain and coworkers recently reported two direct O_2-coupled catalytic oxidations: (i) the oxidation of secondary alcohols to ketones at room temperature under an O_2 atmosphere [50] and (ii) oxidative coupling of 2-naphthols in refluxing chlorobenzene (Scheme 5.14) [51]. Even though the TONs are not quite competitive with Stack's results, the simplicity of the system bodes well for further enhancement and applications.

Often the advancement of catalysis occurs through serendipitous observation as is the case with the oxidase system developed by Katsuki for photoactivated kinetic resolution of a secondary alcohol using a (NO)Ru(salen)-type catalyst, **20** or **21** [52]. Upon irradiation with visible light, the NO ligand dissociates, producing a coordinatively unsaturated Ru(III) complex. Initial exploration of this system using tetramethylpyrazine-*N,N'*-dioxide as cocatalyst for allylic alcohol oxidation resulted in a higher yield of oxidized substrate than consumption of oxygen atoms from the dioxide. This observation supports O_2 from air acting as

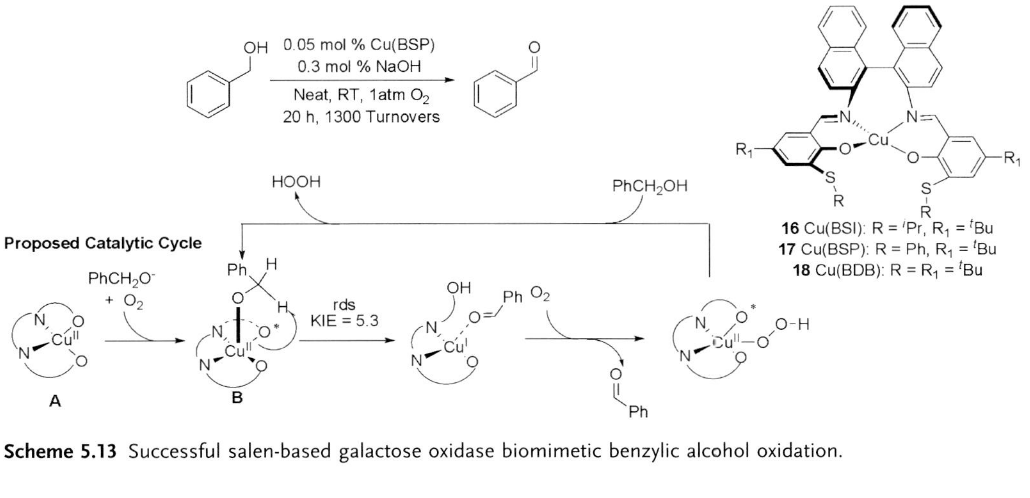

0.05 mol % Cu(BSP), 0.3 mol % NaOH, Neat, RT, 1atm O_2, 20 h, 1300 Turnovers

Proposed Catalytic Cycle

16 Cu(BSI): R = ^{i}Pr, R_1 = ^{t}Bu
17 Cu(BSP): R = Ph, R_1 = ^{t}Bu
18 Cu(BDB): R = R_1 = ^{t}Bu

Scheme 5.13 Successful salen-based galactose oxidase biomimetic benzylic alcohol oxidation.

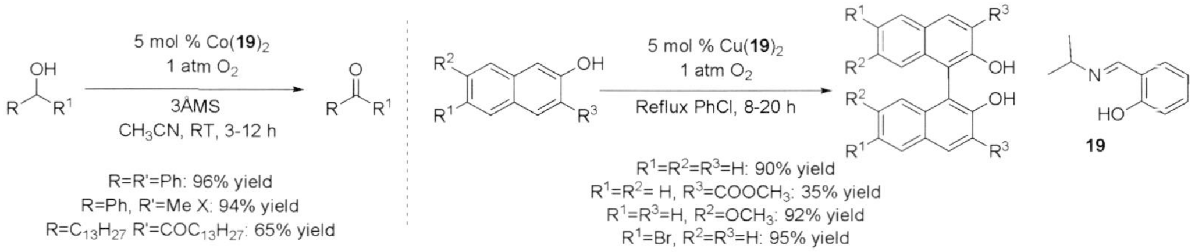

5 mol % Co(**19**)$_2$, 1 atm O_2, 3ÅMS, CH_3CN, RT, 3-12 h

R=R'=Ph: 96% yield
R=Ph, R'=Me X: 94% yield
R=$C_{13}H_{27}$ R'=$COC_{13}H_{27}$: 65% yield

5 mol % Cu(**19**)$_2$, 1 atm O_2, Reflux PhCl, 8-20 h

R^1=R^2=R^3=H: 90% yield
R^1=R^2= H, R^3=$COOCH_3$: 35% yield
R^1=R^3=H, R^2=OCH_3: 92% yield
R^1=Br, R^2=R^3=H: 95% yield

19

Scheme 5.14 Direct O_2-coupled oxidase catalysis using Schiff base ligands.

the terminal oxidant. Further evaluation of this system showed oxidation is not only chemoselective (alcohol oxidation over olefin oxidation), but can also occur in an enantioselective fashion (k_{rel} of 8–20 with **20**, Scheme 5.15) [53]. Preliminary hypotheses for the mechanism were based on substrate reactivity. For example, the lack of epoxide formation supports a reactive species other than a Ru-oxo. Also, α-cyclopropylbenzylalcohol yielded only alcohol oxidation with no observed cyclopropyl ring opening, ruling out a radical α-hydrogen atom abstraction. Ruthenium–alcohol binding and a metal-centered mechanism are proposed based on inhibition of oxidation by bulky α-substitution on the alcohol. The system's efficiency and efficacy in direct O_2 activation is exhibited through the need of only an air atmosphere and the observed wide range of oxidase-type catalysis, including kinetic resolution of secondary alcohols, enantioselective oxidative phenol coupling, oxygen radical cyclizations and oxidative desymmetrization of *meso* diols [52].

Chiral tridentate Schiff base ligands (O,N,O binding) were recently applied to the aerobic oxidative kinetic resolution of α-hydroxyesters using V(V) catalysts [54]. Tetradentate (O,N,N,O) salen-type ligands were incompetent in achieving asymmetric induction in these reactions, but use of tridentate Schiff base **22** lead to high k_{rel} values for a variety of α-hydroxy esters (Table 5.3). The catalyst was also chemoselective for alcohol oxidation with no observed epoxidation of olefin substrates (entries 4 and 5).

The diverse range of transformations catalyzed by M-salen systems with direct O_2 coupling is not limited to oxidase reactivity. Oxovanadium(IV) complexes

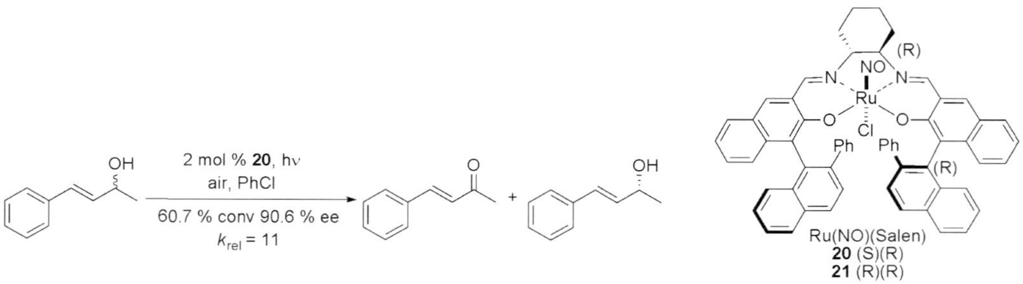

2 mol % **20**, hν, air, PhCl

60.7 % conv 90.6 % ee, k_{rel} = 11

Ru(NO)(Salen)
20 (S)(R)
21 (R)(R)

Scheme 5.15 Allylic alcohol aerobic kinetic resolution.

Table 5.3 Direct O_2 kinetic resolution of α-hydroxyesters catalyzed by V(V) and a tridentate Schiff base ligand

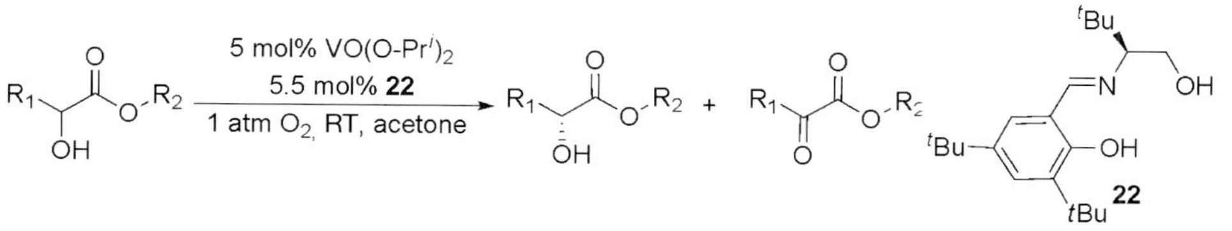

5 mol% $VO(O\text{-}Pr^i)_2$, 5.5 mol% **22**, 1 atm O_2, RT, acetone

Entry	R_1	R_2	Time	Conversion (%)	Yield (%)	*ee* (%)	k_{rel}
1	Ph	Et	10	51	49	99	>50
2	pMeOC_6H_4	Me	5.5	52	38	95	13
3	pCF$_3C_6H_4$	Me	4.0	57	35	98	29
4	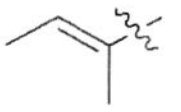	Bn	16	57	45	92	18
5	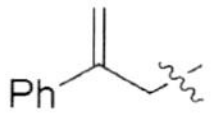 Ph	Et	72	48	48	90	34
6	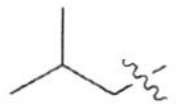	iPr	90	55	37	98	30
7	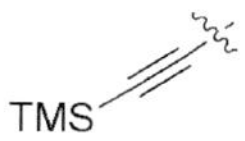 TMS	Et	16	47	53	50	6

O_2, V-Cat, 3:2 CH_3CN:DMF

Cat.	Epoxide	Cyclohexenol	Cyclohexenone
23	35	32	33
24	61	30	9
25	40	30	30
26	54	27	19

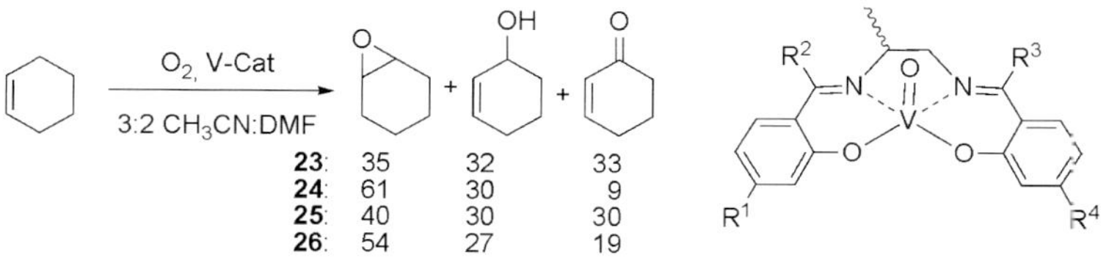

23 $R^1=R^2=R^3=R^4=H$
24 $R^1=R^2=R^4=H$ $R^3=CH_3$
25 $R^1=R^2=H$ $R^3=CH_3$ $R^4=OH$
26 $R^1=H$ $R^2=R^3=CH_3$ $R^4=OH$

Scheme 5.16 Oxovanadium aerobic epoxide catalysis.

with unsymmetrical salen ligands (**23–26**) were used in the aerobic oxidation of cyclohexene (Scheme 5.16) [55]. The highest reactivity was observed with catalysts **23** and **26**. The use of M-salen catalysts for direct O_2 activation in oxygenase systems remains poorly defined, but the accessibility of these catalysts make them excellent targets for further study.

5.3.2.1 **Industrial Considerations**

A key issue for industrial applications of direct O_2 aerobic oxidation catalyst systems is the generally decreased TOF compared to classic oxidations. An example of an aerobic oxidation that would have applications in industry is the oxidation

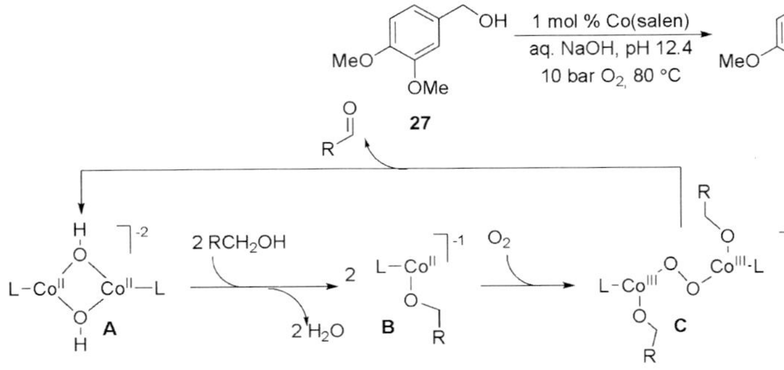

Intermediate	ATR-IR cm^{-1}	UV/Vis nm
A	984	316, 375, 490
B	993	339, 392
C	--	324, 373

Scheme 5.17 Proposed mechanism for veratryl alcohol oxidation with Co(salen) under alkaline conditions with spectroscopic data.

of veratryl alcohol **27** to aldehyde, a model reaction for lignin oxidation, which is used in pulp bleaching. In order to enhance the oxidation TOF, Weckhuysen and coworkers studied the mechanism [56]. The spectroscopic analysis of observed species compared to known species supports the mechanism presented in Scheme 5.17. The compilation of their data suggests that a dinuclear peroxo-$[Co(salen)]_2$ complex, **C**, rather than a superoxo-Co(salen), as previously hypothesized, is responsible for the catalytic activity. Their system was unable to oxidize benzyl alcohol, suggesting the electronic nature of veratryl alcohol enhances the overall reactivity.

5.3.3
Nitrogen-based Ligands

Nitrogen-based ligands are another class which has seen widespread utilization in aerobic oxidation catalysis. As with salen ligands, amines are generally resistant to the mild oxidation conditions of aerobic catalysis, coordinate a wide range of transition metals, and are readily accessible. Of particular note, nonrigid polyamine ligands for metal catalysis have been evaluated for both oxidase and oxygenase activity. As an example, benzylic C–H oxidation of ethylbenzene has been explored using the ligands in Fig. 5.6 with Fe(II) (Table 5.4) [57]. While low selectivity and reactivity is observed, tetradentate ligands were more selective for ketone formation (around 2:1) than pentadentate ligands (around 1.3:1).

A recent example where amine ligands have made a significant impact is direct O_2 palladium-catalyzed oxidations [2c, 58, 59]. A key contribution which inspired much of the growth in the field was reported in 1998 by Uemura and coworkers where they discovered the combination of $Pd(OAc)_2$ and pyridine as a system for aerobic alcohol oxidation (Scheme 5.18 A) [60]. A greater than 2:1 ratio of pyridine to $Pd(OAc)_2$ is required to avoid decomposition of the catalyst. Mechanistic studies performed indicate pyridine not only acts to support Pd(0) during reoxidation with dioxygen, but also inhibits the rate of alcohol oxidation

Table 5.4 Multidentate amine ligands in direct O_2-coupled iron-catalyzed benzylic oxidation

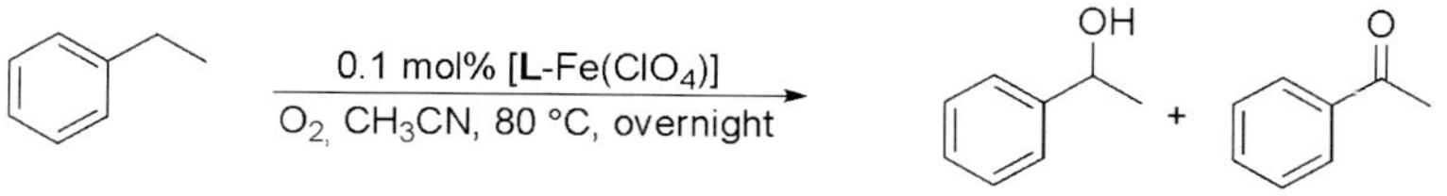

0.1 mol% [**L**-Fe(ClO_4)], O_2, CH_3CN, 80 °C, overnight

Entry	L	Alcohol (% yield)	Ketone (% yield)
1	**27**	3.7	4.8
2	**28**	3.1	3.0
3	**29**	2.3	1.5
4	**30**	3.1	6.1
5	**31**	2.5	5.2
6	**32**	2.3	2.3
7	**33**	<0.1	0.1

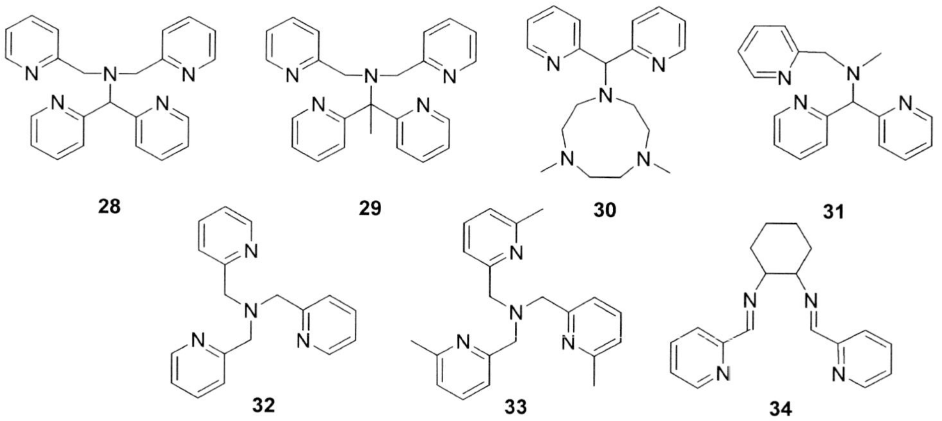

28 **29** **30** **31** **32** **33** **34**

Fig. 5.6 Sample of amine ligands screened for benzylic C–H bond oxidation.

[61]. A related system has been reported by Sigman and coworkers where NEt_3 is substituted for pyridine. This allows for a significant reduction in reaction temperature (25 versus 80 °C) (Scheme 5.18 B) [62]. The ease of ligand dissociation of Et_3N versus pyridine is thought to enhance the rate of β-hydride elimination [63]. Interestingly, the rate-limiting step using pyridine is β-hydride elimination, whereas the use of Et_3N leads to rate-limiting deprotonation of the bound alcohol. Stahl and coworkers also discovered that Et_3N has an impact on catalysis in a direct O_2-coupled oxidative amination of styrene (Scheme 5.18 C) [64]. Finally, Sheldon and coworkers have developed a unique aerobic alcohol oxidation using a water-soluble phenanthroline palladium catalyst, **35** (Scheme 5.18 D) [65]. This catalyst displays a high TOF (up to 100 h^{-1}) and TON (up to 1000) for aerobic alcohol oxidation.

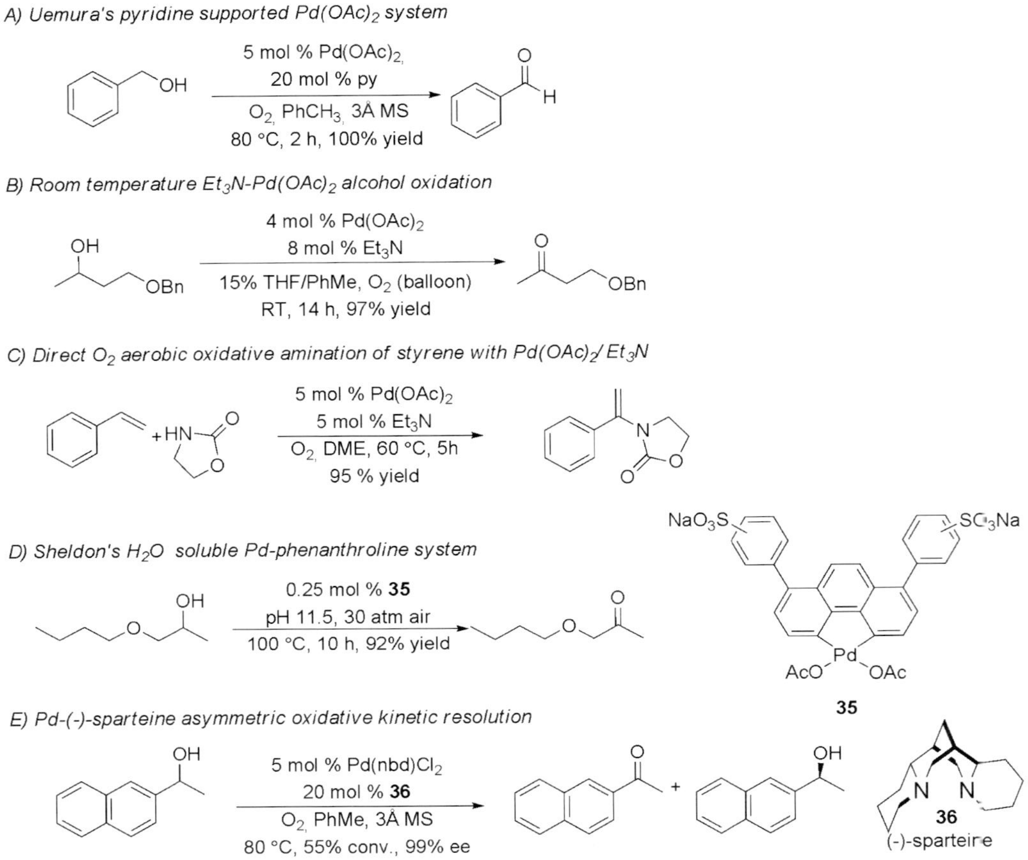
A) Uemura's pyridine supported Pd(OAc)2 system
OH
5 mol % Pd(OAc)2,
20 mol % py
O2, PhCH3, 3Å MS
80 °C, 2 h, 100% yield
O
H
B) Room temperature Et3N-Pd(OAc)2 alcohol oxidation
OH
OBn
4 mol % Pd(OAc)2
8 mol % Et3N
15% THF/PhMe, O2 (balloon)
RT, 14 h, 97% yield
O
OBn
C) Direct O2 aerobic oxidative amination of styrene with Pd(OAc)2/Et3N
+HN
O
O
5 mol % Pd(OAc)2
5 mol % Et3N
O2, DME, 60 °C, 5h
95 % yield
N
O
O
D) Sheldon's H2O soluble Pd-phenanthroline system
OH
O
0.25 mol % 35
pH 11.5, 30 atm air
100 °C, 10 h, 92% yield
O
O
NaO3S
SO3Na
Pd
AcO
OAc
35
E) Pd-(-)-sparteine asymmetric oxidative kinetic resolution
OH
5 mol % Pd(nbd)Cl2
20 mol % 36
O2, PhMe, 3Å MS
80 °C, 55% conv., 99% ee
O
+
OH
N
N
36
(-)-sparteine

Scheme 5.18 Direct O_2-coupled oxidations utilizing Pd(amine) systems.

Uemura's discovery also opened the door to direct O_2-coupled asymmetric catalytic oxidative kinetic resolution (OKR) of alcohols with catalytic amounts of chiral ligands. Simultaneously both Stoltz and Sigman reported the aerobic OKR of secondary alcohols using catalytic Pd[(–)-sparteine]Cl_2 with the addition of catalytic (–)sparteine, **36** (Scheme 5.18E) [66]. Mechanistic studies on this system performed by Sigman and coworkers have found the chiral amine to have a dual role as a ligand and an exogenous base [66e, h]. This system has been further applied to asymmetric Wacker-type cyclizations [2c, 22a, b].

Palladium-catalyzed oxidations are not the only systems to benefit from the use of amine ligands, as copper systems have also utilized diamine ligands for direct O_2-coupled aerobic oxidations. As an example, (–)-Phbox, and (+)-PMP ligands have been successfully used in an asymmetric copper coupling of 1 1′-bi-2-naphthol units (Scheme 5.19) [67]. Very recently Porco and coworkers reported a Cu(I)/(–)-sparteine-mediated system for the enantioselective oxidative dearo-

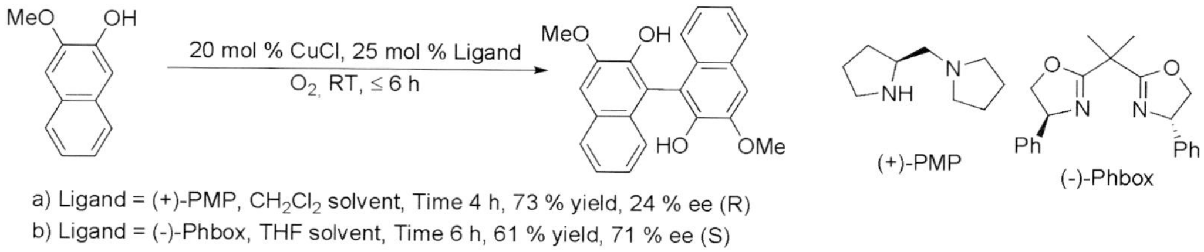

20 mol % CuCl, 25 mol % Ligand
O_2, RT, ≤ 6 h

(+)-PMP
(-)-Phbox

a) Ligand = (+)-PMP, CH_2Cl_2 solvent, Time 4 h, 73 % yield, 24 % ee (R)
b) Ligand = (-)-Phbox, THF solvent, Time 6 h, 61 % yield, 71 % ee (S)

Scheme 5.19 Asymmetric oxidase aerobic coupling reactions utilizing copper–amine complexes. (a) Ligand = (+)-PMP, CH_2Cl_2 solvent, time 4 h, 73% yield, 24% *ee* (*R*). (b) Ligand = (–)-Phbox, THF solvent, time 6 h, 61% yield, 71% *ee* (*S*).

2.2 equiv. $Cu(CH_3CN)_4PF_6$
2.4 equiv. (-)-sparteine
1.6 equiv. Hunig's Base
2.4 equiv. DMAP
CH_2Cl_2, O_2, -10 °C

KH_2PO_4/K_2HPO_4 aq. buffer
CH_3CN, RT

64% yield
95% ee

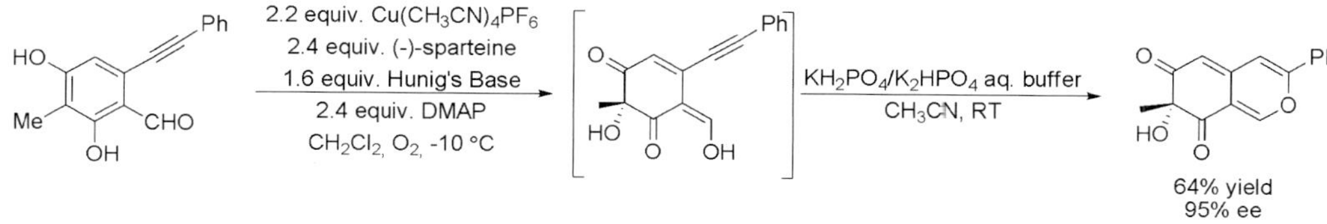

Scheme 5.20 Copper-mediated enantioselective oxygenase dearomatization using Cu(sparteine).

matization of catechols with up to 97% *ee* (Scheme 5.20) [68]. While a stoichiometric amount of the copper complex is necessary, the success of this oxidation reaction on a complex substrate leads to potentially exciting applications in synthesis.

The dihydroxylation of olefins with an osmium catalyst also utilizes chiral tertiary amine ligands to achieve high yields and enantioselectivity. Soon after Krief and coworkers reported on the coupled $O_2/PhSeCH_2Ph$ oxidation [34], Beller and coworkers discovered a direct O_2-coupled catalytic aerobic oxidation of olefins was possible using a phosphate-buffered pH 10.4 solution (Scheme 5.21) [69]. Under increased pressure with air rather than O_2, the catalyst remains active and a TOF of 40 h^{-1} is possible. This system does not quite achieve as high an enantioselectivity as the AD-Mix methods [70].

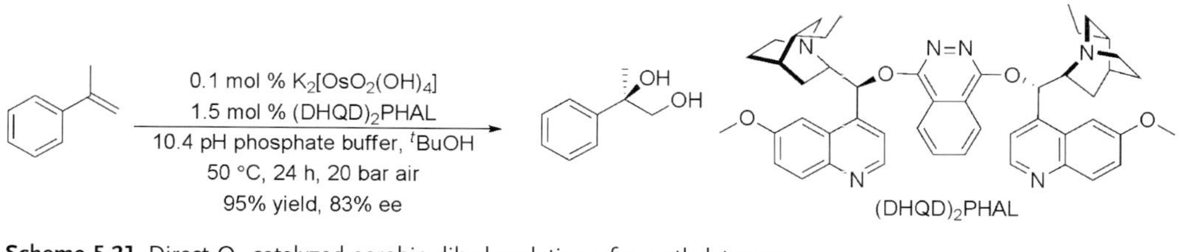

0.1 mol % $K_2[OsO_2(OH)_4]$
1.5 mol % $(DHQD)_2PHAL$
10.4 pH phosphate buffer, ^{t}BuOH
50 °C, 24 h, 20 bar air
95% yield, 83% ee

$(DHQD)_2PHAL$

Scheme 5.21 Direct O_2-catalyzed aerobic dihydroxylation of α-methylstyrene.

5.3.4
Other Ligand Systems

5.3.4.1 *N*-Heterocyclic Carbenes (NHCs)

Throughout this section, the focus has been on ligands inert to oxidation, but considering the vast use of phosphine ligands in other catalytic reactions their properties may enhance the scope of catalytic aerobic oxidation methods. Clearly though, phosphines undergo oxidation readily on or off the metal making them a poor ligand candidate. An emerging ligand class with features similar to phosphines (size and electronics) are NHC ligands that are relatively stable under oxidative conditions [71, 72]. Recent work has demonstrated the robust nature of metal/NHC complexes under oxidative conditions, as specifically seen in palladium-catalyzed Wacker cyclizations [22 b, 73], alcohol oxidation [74], kinetic resolution of alcohols [75], oxidative carbonylations and polymerizations [76, 77], and multicatalytic olefinations of alcohols (Fig. 5.7) [78]. Of specific note, the NHC-Pd catalyst for alcohol oxidation is capable of TONs of up to 1000 and can use air as the oxygen source. Additionally, the mechanism of catalyst turnover and palladium activation of O_2 has been debated as to whether a direct insertion of O_2 into a palladium hydride species occurs or whether Pd(0) is directly reoxidized to a Pd-peroxo [79]. NHCs have allowed for direct insight into this process by enabling the isolation of a Pd-peroxo from Pd(0) and subsequent protonation leading to a Pd-hydroperoxide (Scheme 5.22).

5.3.4.2 Polyoxometalates (POM)

An emerging area in oxidative aerobic catalysts is the use of POMs [80]. Due to the intricate nature of these complexes, it is difficult to classify them in terms of what we have outlined above (cocatalysts or ligands?). Use of these oxidatively stable complexes (the general structure of which is $LA[X_xM_mO_y]^{q-}$, where LA is a transition metal, X is an addenda, M is a metal, usually Mo or W, and O is oxygen) is becoming more prevalent [80 b]. POMs have been used with aldehyde coreductants for epoxidation with LA = Ru, Co, Mn [81].

These complexes have also been used in direct aerobic oxidation catalysis. A recent interesting addition in this area is the chemoselective oxidation of nitrobenzene to *p*-nitrophenol under aerobic conditions (Scheme 5.23) [82]. Based on

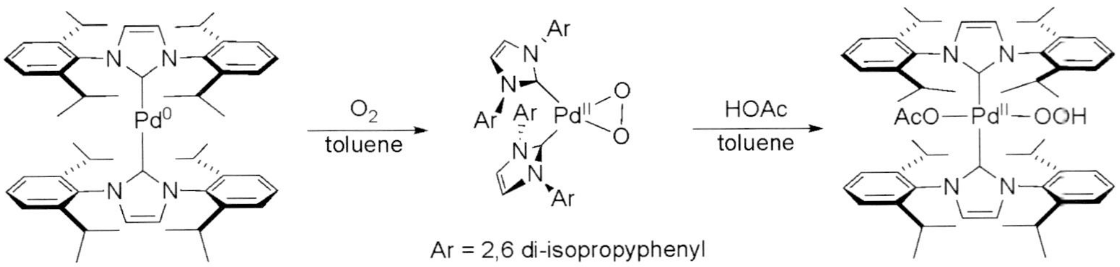

Pd^0 —(O_2, toluene)→ Pd^{II} peroxo —(HOAc, toluene)→ AcO—Pd^{II}—OOH

Ar = 2,6 di-isopropyphenyl

Scheme 5.22 Oxidative addition of O_2 to |Pd(0)| species followed Pd-peroxo formation.

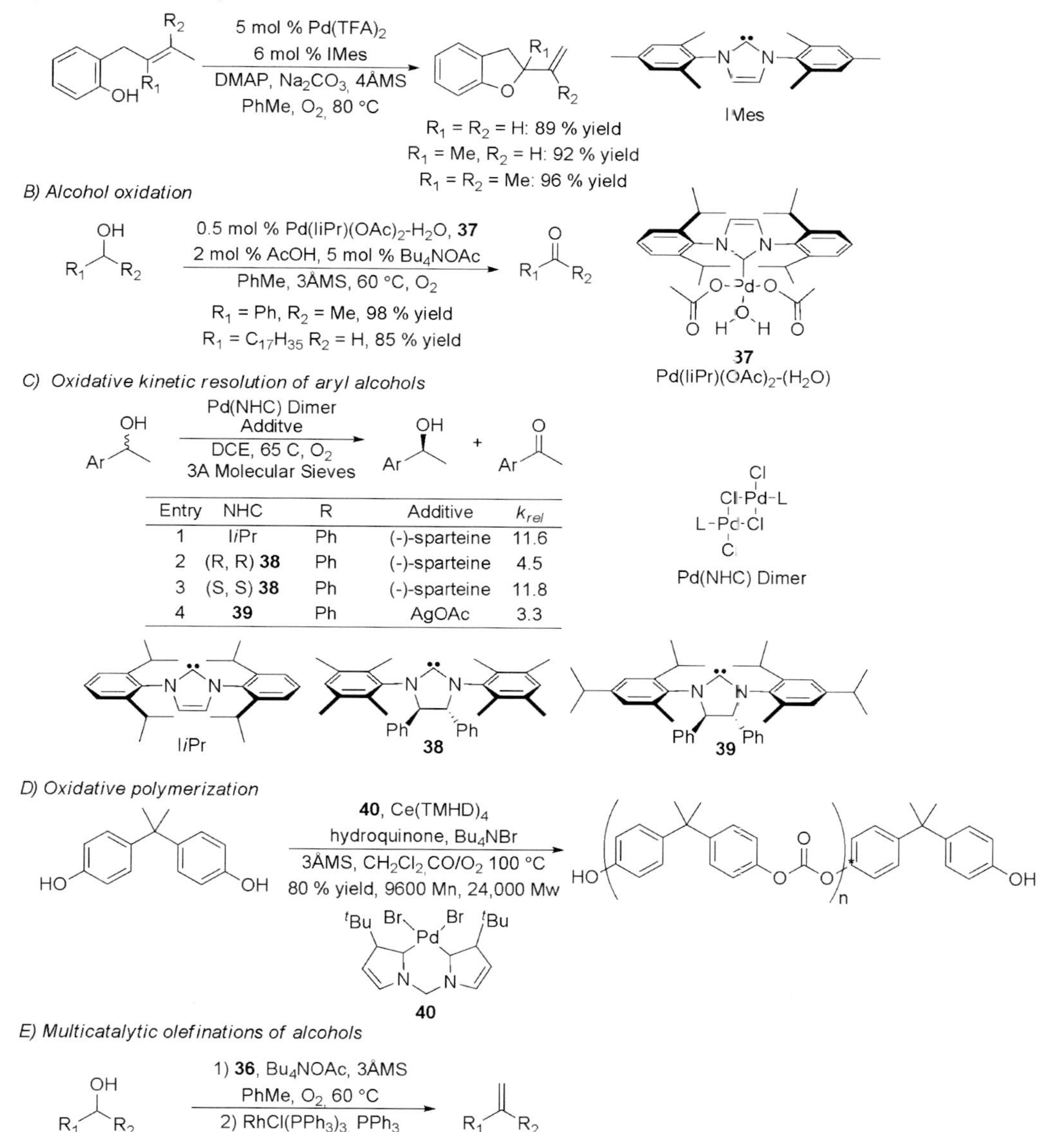

Fig. 5.7 Aerobic Pd(NHC)-catalyzed oxidation reactions.

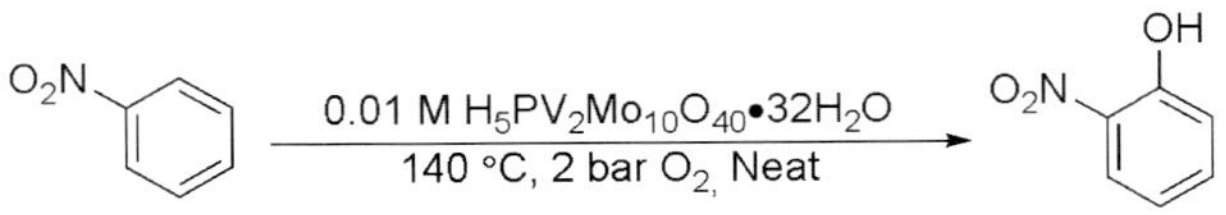

O_2N–C_6H_5 $\xrightarrow[140\ °C,\ 2\ bar\ O_2,\ Neat]{0.01\ M\ H_5PV_2Mo_{10}O_{40}•32H_2O}$ 2-nitrophenol (OH, O_2N)

Scheme 5.23 $H_5PV_2Mo_{10}O_{40}$-catalyzed aerobic regioselective *ortho* hydroxylation of nitrobenzene.

kinetic and mechanistic studies, Neumann and coworkers propose the selectivity of this process stems from an intramolecular regioselective oxidation process and not an autoxidation. An oxygenase system is implicated through $^{18}O_2$-labeling experiments with 86% incorporation of ^{18}O into the phenol oxygen.

5.4 Conclusions and Outlook

The activation of O_2 for aerobic oxidation catalysis continues to be a prominent goal in catalysis [1, 35]. The two general strategies discussed within this chapter include: (i) coupling oxygen activation and substrate oxidation using multiple reagents, and (ii) O_2 activation and substrate oxidation by a single catalyst. As demonstrated, additives and coupled redox cycles allow for oxidation of diverse substrates in different oxidation manifolds, i.e. oxidase and oxygenase reactions. However, this approach is often encumbered by the production of excessive byproducts associated with the coupled chemistry.

Initial forays into direct aerobic oxidation catalysis were inspired mainly by enzymes in which porphyrin mimics were utilized. These studies brought forth many important mechanistic models, but were unable to reproduce the level of activity/practicality necessary for use in synthesis. This can be attributed to difficulties in preparing the ligands, oxidative degradation of the complexes and lack of reaction chemoselectivity in the presence of a complex organic molecule.

More recent work has focused on utilizing catalyst systems based on nonbiological ligand constructs including amines, NHCs and POMs. Additionally, use of second- and third-row transition metal catalysts, not commonly found in biological systems, has increased TONs and TOFs of many aerobic catalytic oxidation systems. A question that remains is how can we compete with biological processes in a nonbiological system wherein protection of the metal and substrate specificity is not inherent? This question and many others will occupy us for the foreseeable future while we hope to develop new aerobic oxidation catalysts with broad applications both synthetically and industrially.

References

1 Vision 2020 Catalysis Report (www.ccrhq.org/vision/index/roadmaps/catrep.html).

2 For recent reviews on catalysis involving metal-mediated activation of dioxygen, see: (a) Ezhova, M. B., James, B. R. In *Advances in Catalytic Activation of Dioxygen by Metal Complexes*, Simandi, L. I. (Ed.), Kluwer, Dordrecht, **2003**, pp. 1–77; (b) Meunier, B., de Visser, S. P., Shaik, S. *Chem Rev* **2004**, *104*, 3947–3980; (c) Stahl, S. S. *Angew Chem Int Ed* **2004**, *43*, 3400–3420; (d) Bäckvall, J.-E. (Ed.) *Modern Oxidation Methods*, Wiley-VCH, Weinheim, **2004**; (e) Punniyamurthy, T., Velusamy, S., Iqbal, J. *Chem Rev* **2005**, *105*, 2329–2364; (f) Sigman, M. S., Jensen, D. R., Rajaram, S. *Curr Opin Drug Disc Dev* **2002**, *5*, 860–869; (g) Marko, I. E., Giles, P. R., Tsukazaki, M., Gautier, A., Dumeunier, R., Dodo, K., Philippart, F., Chelle-Regnault, I., Mutonkole, J.-L., Brown, S. M., Urch, C. J. In *Transition Metals for Organic Synthesis*, 2nd edn, Beller, M., Bolm, C. (Eds.), Wiley-VCH, Weinheim, **2004**, pp. 437–478.

3 Carlton, L., Read, F., Urgelles, M. *Chem Commun* **1983**, 586–588.

4 Yamada, T., Takahashi, K., Kato, K., Takai, T., Inoki, S., Mukaiyama, T. *Chem Lett* **1991**, 641–644.

5 Reiter, M., Ropp, S., Gouverneur, V. *Org Lett* **2004**, *6*, 91–94.

6 Nakajima, M., Miyoshi, I., Kanayama, K., Hashimoto, S.-I. *J Org Chem* **1999**, *64*, 2264–2271.

7 Jorgensen, K. A. *Chem Rev* **1989**, *89*, 3, 431–456.

8 Riley, D. P., Getman, D. P., Beck, G. R., Heintz, R. M. *J Org Chem* **1987**, *52*, 287–290.

9 (a) Takai, T., Yamada, T., Mukaiyama, T. *Chem Lett* **1990**, 1657–1659; (b) Mukaiyama, T., Takai, T., Yamada, T., Rhode, O. *Chem Lett* **1990**, 1660–1662.

10 Yamada, T., Imagawa, K., Mukaiyama, T. *Chem Lett* **1992**, 2109–2112.

11 Jacobsen, E. N., Zhang, W., Muci, A. R., Ecker, J. R., Deng, L. *J Am Chem Soc* **1991**, *113*, 7063–7034.

12 (a) Mukaiyama, T., Yamada, T., Nagata, T. Imagawa, K. *Chem Lett* **1993**, 327–330; (b) Imagawa, K., Nagata, T., Yamada, T., Mukaiyama, T. *Chem Lett* **1994**, 527–530; (c) Nagata, T., Imagawa, K., Yamada, T., Mukaiyama, T. *Inorg Chim Acta* **1994**, 283–287; (d) Nagata, T., Imagawa, K., Yamada, T., Mukaiyama, T. *Chem Lett* **1994**, 1259–1262.

13 (a) Bennett, S., Brown, S. M., Conole, G., Kessler, M., Rowling, S., Sinn, E., Woodward, S. *Dalton Trans* **1995**, 367–376; (b) Nam, W., Kim, H. J., Kim, S. H., Ho, R. Y. N., Valentine, J. S. *Inorg Chem* **1996**, *35*, 1045–1049; (c) Mastrorilli, P., Nobile, C. F., Suranna, G. P., Lopez, L. *Tetrahedron* **1995**, *51*, 7943–7950; (d) Wentzel, B. B., Gosling, P. A., Geiters, M. C., Nolte, R. J. M. *Dalton Trans* **1998**, 2241–2246.

14 Bryliakov, K. P., Kholdeeva, O. A., Vanina, M. P., Talsi, E. P. *J Mol Catal A* **2002**, *178*, 47–53.

15 (a) Mastrorilli, P., Nobile, C. F. *Tetrahedron Lett* **1994**, *35*, 4193–4196; (b) Giannandrea, R., Matrorilli, P., Nobile, C. F., Suranna, G. P. *J Mol Catal* **1994**, 27–36; (c) Mastrorilli, P. Nobile, C. F., Suranna, G. P. Lopez, L. *Tetrahedron* **1995**, *51* (29), 7943–7950; (d) Dell'Anna, M. M., Mastrorilli, P., Nobile, C. F. *J Mol Catal A* **1996**, *108*, 57–62; (e) Dell'Anna, M. M., Mastrorilli, P., Nobile, C. F., Taurino, M. R., Calo, V., Nacci, A. *J Mol Catal A* **2000**, *151*, 61–69; (f) Mastrorilli, P., Muscio, F., Nobile, C. F., Suranna, G. P. *J Mol Catal A* **1999**, *148*, 17–21.

16 (a) Tanaka, H., Kikuchi, R., Torii, S. *Tetrahedron* **1996**, *52*, 2343–2348; (b) Mandal, A. K., Iqbal, J. *Tetrahedron* **1997**, *53*, 7641–7648.

17 (a) Fernandez, I., Pedro, J. R., Rosello, A. L., Ruiz, R., Castro, I., Ottenwaelder, X., Journaux, Y. *Eur J Org Chem* **2001**, 1235–1247; (b) Blay, G., Fernandez, I., Marco-Aleixandre, A., Monje, B., Pedro, J. R., Ruiz, R. *Tetrahedron* **2002**, *58*, 8565–8571.

18 Deubel, D. V. *J Am Chem Soc* **2004**, *126*, 996–997.

19 (a) Marko, I. E., Giles, P. R., Tsukazaki, M., Brown, S. M., Urch, C. J. *Science,* **1996**, *274*, 2044–2045; (b) Marko, I. E., Giles, P. R., Tsukazaki, M., Chelle-Regnaut, I. Gautier, A., Brown, S. M., Urch, C. J. *J Org Chem* **1999**, *64*, 2433–2439; (c) Marko, I. E., Gautier, A., Dumeunier, R., Doda, K., Philippart, F., Brown, S. M., Urch, C. J. *Angew Chem Int Ed* **2004**, *43*, 1588–1591.

20 Karlin, K. D., Tyeklar, Z. (Eds.). *Bioinorganic Chemistry of Copper,* Chapman & Hall, New York, **1993**.

21 (a) Martell, A. E., Motekaitis, R. J., Menif, F., Rockcliffe, D. A., Llobet, A. *J Mol Catal A Chemical* **1997**, *117*, 205–213; (b) Karlin, K. D., Tolman, W. B., Kaderli, S., Zubervühler, A. D. *J Mol Catal A* **1997**, *117*, 214–222.

22 (a) Tsuji, J. (Ed.). *Palladium Reagents and Catalysts: New Perspectives for the 21st Century,* Wiley, New York, **2004**, pp. 22–103; (b) Takacs, J. M., Jiang, X. *Curr Org Chem* **2003**, *7*, 369–396; (c) Tsuji, J. *New J Chem* **2000**, *24*, 127–135; (d) Tsuji, J. *Synthesis* **1984**, *5*, 369–384; (e) Lyons, J. E. In *Oxygen Complexes and Oxygen Activation by Transition Metals,* Martell, A. E., Sawyer, D. T. (Eds.) Plenum Press, New York, **1988**, pp. 233–251.

23 (a) Smidt, J. *Chem Ind* **1962**, 54–62; (b) Smidt, J., Hafner, W., Jira, R., Sieber, R., Seldmeier, J., Sabel, A. *Angew Chem Int Ed* **1962**, *1*, 80–88.

24 Clement, W. H., Selwitz, C. M. *J Org Chem* **1964**, *29*, 241–243.

25 Hosokawa, T., Nomura, T., Murahashi, S. I. *J Organomet Chem* **1998**, 387–389.

26 (a) Batsanov, A. S., Collings, J. C., Fairlamb, I. J. S., Holland, J. P., Marder, T. B., Parsons, A. C., Ward, R. M., Zhu, J. *J Org Chem* **2005**, *70*, 703–706; (b) Orito, K., Horibata, A., Nakamura, T., Ushito, H., Nagasaki, H., Yuguchi, M., Yamashita, S., Tokuda, M. *J Am Chem Soc* **2004**, *126*, 14342–14343.

27 Chepaikin, E. G., Bezruchenko, A. P., Leshcheva, A. A. *Kinet Catal* **2002**, *43*, 507–513.

28 Lorber, C. Y., Smidt, S. P., Osborn, J. A. *Eur J Inorg Chem* **2000**, 655–658.

29 Samec, J. S. M., Éll, A. H., Bäckvall, J.-E. *Chem Eur J* **2005**, *11*, 2327–2334.

30 (a) Yokota, T., Fujibayashi, S., Nishiyama, Y., Sakaguchi, S., Ishii, Y. *J Mol Catal A Chem* **1996**, *114*, 113–122; (b) Miller, D. G., Wayner, D. D. M. *J Org Chem* **1990**, *55*, 2924–2927; (c) Takehira, K., Hayakawa, T., Orita, H. *Chem Lett* **1985**, 1835–1838; (d) Bäckvall, J.-E., Hopkins, R. B *Tetrahedron Lett* **1988**, *29*, 2885–2888; (e) Backvall, J.-E., Hopkins, R. B., Greenberg, H., Mader, M. M., Awasthi, A. K. *J Am Chem Soc* **1990**, *112*, 5160–5166.

31 (a) Wöltinger, J., Bäckvall, J.-E., Zsigmond, A. *Chem Eur J* **1999**, *5*, 1460–1467; (b) Csjernyik, G., Éll, A., Fadini, L., Pugin, B., Bäckvall, J.-E. *J Org Chem* **2002**, *67*, 1657–1662.

32 Kolb, H. C., Van Nieuwenhze, M. S., Sharpless, K. B. *Chem Rev* **1994**, *94*, 2483.

33 Sundermeier, U., Döbler, C., Beller, M. In *Modern Oxidation Methods,* Bäckvall, J.-E. (Ed.), Wiley-VCH, Weinheim, **2004**, pp. 1–20.

34 Krief, A., Colaux-Costillo, C. *Tetrahedron Lett,* **1999**, *40*, 4189–4192.

35 Stahl, S. S. *Science* **2005**, *309*, 1824–1826.

36 (a) Patel, R. N. (Ed.) *Stereoselective Biocatalysis,* Dekker, New York, **2000**; (b) Faber, K. (Ed.) *Biotransformations in Organic Chemistry,* Springer, Heidelberg, **2004**.

37 Shelnutt, J., Song, X., Ma, J., Jia, S., Jentzen, W., Medforth, C. J. *Chem Soc Rev* **1998**, *27*, 31–41.

38 Xia, Q.-H., Ge, H.-Q., Ye, C.-P., Lie, Z.-M., Su, K.-X. *Chem Rev* **2005**, *105*, 1603–1662.

39 (a) Ortiz de Montellano, P. R. (Ed.) *Cytochrome P450: Structure, Mechanism, and Biochemistry,* 3rd edn, Kluwer/Plenum, New York, **2005**; (b) Denisov, I. G., Makris, T. M., Sligar, S. G., Schlichting, I. *Chem Rev* **2005**, *105*, 2253–2277; (c) Meunier, B., de Visser, S. P., Shaik, S. *Chem Rev* **2004**, *104*, 3947–3980.

40 Rose, E., Andrioletti, B., Zrig, S., Quelquejeu-Etheve, M. *Chem Soc Rev* **2005**, *34*, 573–583.

41 Kikuchi, G., Yoshida, T., Noguchi, M. *Biochem Biophys Res Commun* **2005**, *338*, 558–567.

42 Rosenthal, J., Pistorio, B. J., Chang, L. L., Nocera, D. G. *J Org Chem* **2005**, *70*, 1885–1888.

43 Lai, T.-S., Zhang, R., Cheung, K.-K., Kwong, H.-L., Che, C.-M. *Chem Commun* **1998**, 1583–1584.

44 Cozzi, P. G. *Chem Soc Rev* **2004**, *33*, 410–421.

45 Venkataramanan, N. S., Kuppuraj, G., Rajagopal, S. *Coord Chem Rev* **2005**, 1249–1268.

46 Gennari, C., Piarulli, U. *Chem Rev* **2003**, *103*, 3071–3100.

47 Jazdzewski, B. A., Tolman, W. B. *Coord Chem Rev* **2000**, *200–202*, 633–685.

48 (a) Wang, Y., DuBois, J. L., Hedman, B., Hodgson, K. O., Stack, T. D. P. *Science* **1998**, *279*, 537–540; (b) Wang, Y., Stack, T. D. P. *J Am Chem Soc* **1996**, *118*, 13907–13908; (c) Mahadevan V., Klein-Gebbink, R. J. M., Stack, T. D. P. *Curr Opin Chem Biol* **2000**, *4*, 228.

49 Whittaker, M. M., Wittaker, J. W. *Biochemistry* **2001**, *40*, 7140–7148.

50 Sharma, V. B., Jain, S. L., Sain, B. *J Mol Catal A* **2004**, *212*, 55–59.

51 Sharma, V. B., Jain, S. L., Sain, B. *J Mol Catal A* **2004**, *219*, 61–64.

52 (a) Irie, R., Katsuki, T. *Chem Rec* **2004**, *4*, 96–109; (b) Shimizu, H., Onitsuka, S., Egami, H., Katsuki, T. *J Am Chem Soc* **2005**, *127*, 5396–5413.

53 Masutain, K., Uichida, T., Irie, R., Katsuki, T. *Tetrahedron Lett* **2000**, *41*, 5119–5123.

54 Radosevich, A. T., Musich, C., Toste, F. D. *J Am Chem Soc* **2005**, *127*, 1090–1091.

55 Mohebbi, S., Boghaei, D. M., Sarvestani, A. H., Salimi, A. *Appl Catal A* **2005**, 263–267.

56 Kervinen, K., Korpi, H., Mesu, J. G., Soulimani, F., Repo, T., Reiger, B., Leskelä, M., Weckhuysen, B. M. *Eur J Inorg Chem* **2005**, 2591–2599.

57 Klopstra, M., Hage, R., Kellogg, R. M., Feringa, B. L. *Tetrahedron Lett* **2003**, *44*, 4581–4584.

58 For recent reviews with excellent references, see: (a) Sheldon, R. A., Arends, I. W. C. E., Dijksman, A. *Catal Today* **2000**, *57*, 157–166; (b) Sheldon, R. A., Arends, I. W. C. E., ten Brink, G.-J., Dijksman, A. *Acc Chem Res* **2002**, *35*, 774–781; (c) Muzart, J. *Tetrahedron* **2003**, *59*, 5789–5816; (d) Nishimura, T., Ohe, K., Uemura, S. *Synlett* **2004**, 201–216.

59 Sigman, M. S., Schultz, M. J. *Org Biomol Chem* **2004**, *2*, 2551–2554.

60 (a) Nishimura, T., Onoue, T., Ohe, K., Uemura, S. *Tetrahedron Lett* **1998**, *39*, 6011–6014; (b) Nishimura, T., Onoue, T., Ohe, K., Uemura, S. *J Org Chem* **1999**, *64*, 6750–6755; (c) Nishimura, T., Maeda, Y., Kakiuchi, N., Uemura, S. *Perkin Trans 1* **2000**, 4301–4305.

61 Steinhoff, B. A., Stahl, S. S. *Org Lett* **2002**, *4*, 4179–4181.

62 (a) Schultz, M. J., Park, C. C., Sigman, M. S. *Chem Commun* **2002**, 3034–3035; (b) Schultz, M. J., Hamilton, S. S., Jensen, D. R., Sigman, M. S. *J Org Chem* **2005**, *70*, 3343–3352.

63 Schultz, M. J., Adler, R. S., Zierkiewicz, W., Privalov, T., Sigman, M. S. *J Am Chem Soc* **2005**, *127*, 8499–8507.

64 Timokhin, V. I., Anastasi, N. R., Stahl, S. S. *J Am Chem Soc* **2003**, *125*, 12996–12997.

65 (a) ten Brink, G.-J., Arends, I. W. C. E., Sheldon, R. A., *Science* **2000**, *287*, 1636–1639; (b) Sheldon, R. A., Arends, I. W. C. E., ten Brink, G.-J., Dijksman, A. *Acc Chem Res* **2002**, *35*, 774–781; (c) ten Brink, G.-J., Arends, I. W. C. E., Sheldon, R. A. *Adv Synth Catal* **2002**, *344*, 355–369; (d) ten Brink, G.-J., Arends, I. W. C. E., Hoogenraad, M , Verspui, G., Sheldon, R. A. *Adv Synth Catal* **2003**, *345*, 497–505.

66 (a) Ferreira, E. M., Stoltz, B. M. *J Am Chem Soc* **2001**, *123*, 7725–7726; (b) Bagdanoff, J. T., Ferreira, E. M., Stoltz, B. M. *Org Lett* **2003**, *5*, 835–837; (c) Trend, R. M., Ramtohul, Y. K., Ferreira, E. M., Stoltz, B. M. *Angew Chem Int Ed* **2003**, *42*, 2892–2895; (d) Jensen, D. R., Pugsley, J. S., Sigman, M. S. *J Am Chem Soc* **2001**, *123*, 7475–7476; (e) Mueller, J. A., Jensen, D. R., Sigman, M. S. *J Am Chem Soc* **2002**, *124*, 8202–8203; (f) Jensen, D. R., Sigman, M. S. *Org Lett* **2003**, *5*, 63–65; (g) Mandal, S. K., Jensen, D. R., Pugsley, J. S., Sigman, M. S. *J Org Chem* **2003**, *68*, 4600–4603; (h) Mueller, J. A., Sigman, M. S. *J Am Chem Soc* **2003**, *125*, 7005–7013; (i) Mandal, S. K., Sigman, M. S. *J Org Chem* **2003**, *68*, 7535–7537.

67 Habaue, S., Ajiro, J., Yoshii, Y., Hirasa, T. *J Polymer Sci A Chem* **2004**, *42*, 4528–4534.

68 Zhu, J., Grigoriadis, N. P., Lee, J. P., Porco, J. A. *J Am Chem Soc* **2005**, *127*, 9342–9343.

69 (a) Döbler, C., Mehltretter, G., Beller, M. *Angew Chem Int Ed* **1999**, *38*, 3026–3028; (b) Döbler, C., Mehltretter, G., Sundermeier, U., Beller, M. *J Organomet Chem* **2001**, *621*, 70–76.

70 Kolb, H. C., Van Nieuwenhze, M. S., Sharpless, K. B. *Chem Rev* **1994**, *94*, 2483–2547.

71 For studies on the stability of carbenes, see: Bourissou, D., Guerret, O., François, P. G., Bertrand, G. *Chem Rev* **2000**, *100*, 39–91 and references therein.

72 For studies on the stability of metal/NHC complexes, see: Crudden, C. M., Allen, D. P. *Coord Chem Rev* **2004**, *248*, 2247–2273.

73 Muñiz, K. *Adv Synth Catal* **2004**, *346*, 1425–1428.

74 (a) Jensen, D. R., Schultz, M. J., Mueller, J. A., Sigman, M. S. *Angew Chem Int Ed* **2003**, *42*, 3810–3813; (b) Schultz, M. J., Hamilton, S. S., Jensen, D. R., Sigman, M. S. *J Org Chem* **2005**, *70*, 3343–3352; (c) Mueller, J. A., Goller, C. P., Sigman, M. S. *J Am Chem Soc* **2004**, *126*, 9724–9734.

75 Jensen, D. R., Sigman, M. S. *Org Lett* **2003**, *5*, 63–65.

76 Okuyama, K., Sugiyama, J., Nagahata, R., Asai, M., Ueda, M., Takeuchi, K. *J Mol Catal A* **2003**, *203*, 21–27.

77 (a) Okuyama, K., Sugiyama, J., Nagahata, R., Asai, M., Ueda, M., Takeuchi, K. *Macromolecules* **2003**, *36*, 6953–6955; (b) Okuyama, K., Sugiyama, J., Nagahata, R., Asai, M., Ueda, M., Takeuchi, K. *Green Chem* **2003**, *5*, 563–566.

78 Lebel, H., Paquet, V. *J Am Chem Soc* **2004**, *126*, 11152–11153.

79 (a) Konnick, M. M., Guzei, I. A., Stahl, S. S. *J Am Chem Soc* **2004**, *126*, 10212–10213; (b) Yamashita, M., Goto, K., Kawashima, T. *J Am Chem Soc* **2005**, *127*, 7294–7295.

80 (a) Kozhevnikov, I. V. *Catalysis by Polyoxometalates*, Wiley, Chichester, **2002**; (b) Neumann, R. In *Modern Oxidation Methods*, Bäckvall, J.-E. (Ed.), Wiley-VCH, Weinheim, **2004**, pp. 223–251.

81 (a) Neumann, R., Dahan, M. *Chem Commun* **1995**, 171–172; (b) Neumann, R., Dahan, M. *J Am Chem Soc* **1998**, *120*, 11969–11976.

82 Khenkin, A. M., Winer, L., Neumann, R. *J Am Chem Soc* **2005**, *127*, 9988–9989.

6
Dioxygen Binding and Activation: Reactive Intermediates

Andrew S. Borovik, Paul J. Zinn, and Matthew K. Zart

6.1
Introduction

Oxidation chemistry has attracted the attention, and in many cases the passion, of chemists for well over a century. This interest has been fueled by several competing areas of science, with organic and biochemistry being the most prominent today. A link into these research areas is transition metal ions, which are intimately involved in several oxidative transformations. Understanding the relationships between metal ions and oxidants is necessary in the development of more efficient and clean oxidative processes.

The current state of oxidation chemistry, especially related to industrial processes, is at a crossroad. Commodity oxidations produce multimegatons of products worth tens of billions of dollars. However, it has been estimated that stoichiometric reagents account for nearly 70% of bulk chemical oxidations [1]. Reagents such as permanganate and chromate have become undesirable because of high costs and environmental concerns. It would be far better to use catalytic systems, especially if they can be coupled with dioxygen as the terminal oxidant. The advantages for using O_2 are that it is relatively plentiful, inexpensive and clean – it is commonly referred to as the only true "green oxidant".

Dioxygen is potentially reactive from a thermodynamic standpoint; however, under ambient conditions it coexists with combustible substances such as wood, fossil fuels and carbohydrates instead of converting them to the thermodynamically stable products of H_2O and CO_2. The basis of this lack of reactivity is kinetic, resulting from the triplet ground state of O_2. This kinetic barrier of reactivity can be overcome when O_2 is partnered with species that can undergo single-electron transfer reactions or that contain unpaired electrons. Such partners include radicals ($S=1/2$), photochemically produced triplet excited states ($S=1$) or paramagnetic transition metal centers ($S \geq 1/2$). All of the above-mentioned reactivity patterns are known in biology; however, in particular, the use of metal complexes is extensively exploited in biological systems to bind and activate dioxygen.

Activation of Small Molecules. Edited by William B. Tolman

ISBN: 3-527-31312-5

From a catalytic perspective, coupling a transition metal complex with O_2 is fraught with problems. Dioxygen, being a noninnocent ligand, undergoes reduction upon coordinating to a transition metal ion. This often leads to O–O bond cleavage and uncontrolled formation of metal oxides, in which the oxo groups are bridged between multiple metal centers. These species tend to be unreactive and thus useless for oxygen-atom (O-atom) transfer to substrates.

Many chemists have turned to biology to find ways to design complexes that bind and activate dioxygen. Advances in spectroscopy and X-ray diffraction (XRD) over the last 30 years have enabled scientists to perform detailed structure–function analyses on metalloproteins. This has been particularly helpful in probing metallooxygenases, whose redox-active metal centers can be readily monitored during turnover. The picture that has emerged from these studies is one of effective control for structural parameters that govern the metal–dioxygen center(s), leading to highly efficient and selective catalysts. While synthetic systems have not yet achieved this level of control, much progress has been made. It is the aim of this chapter to discuss some of the recent progress in metal–dioxygen chemistry for synthetic systems, with emphasis on structure–function relationships of species produced upon the initial binding of O_2. Along the way, important design concepts will be highlighted.

6.1.1 An Example: Cytochromes P450

6.1.1.1 Mechanism

Cytochromes P450 (P450) are monooxygenases found in nearly all living systems [2]. They catalyze a diverse set of oxidations, including hydroxylation of C–H bonds in hydrocarbons. P450s are the best characterized oxygenases and therefore serve as an instructive example of how dioxygen is activated by a metal complex. Four electrons are necessary to cleave dioxygen to formally produce two equivalents of oxo anions (i.e. $[O]^{2-}$). One of these anions combines with two protons to form water, while the other oxo anion is inserted into a C–H bond of the substrate, yielding product.

The active site of a P450 contains an iron protoporphyrin IX complex, in which one axial site on the iron has a coordinated sulfur atom from cysteine. The most thoroughly studied enzyme in this class is $P450_{cam}$, which catalyzes the regio- and stereospecific hydroxylation of camphor to 5-*exo*-hydroxycamphor. The $P450_{cam}$ system consists of three protein components: a hydroxylase protein that contains the iron center and is the site of catalysis, a NADH/FAD-dependent reductase and putidaredoxin (a Fe_2S_2 protein). An accepted mechanism involving $P450_{cam}$ is outlined in Fig. 6.1.

In the resting state, the iron center is six-coordinate with an axially bonded water occupying the final coordination site (**A**). The catalytic cycle starts with binding of the camphor substrate within the active site, which causes dissociation of the water from the Fe(III) complex. Camphor is not directly bonded to the iron, but docks proximal to the metal center through hydrogen bonds.

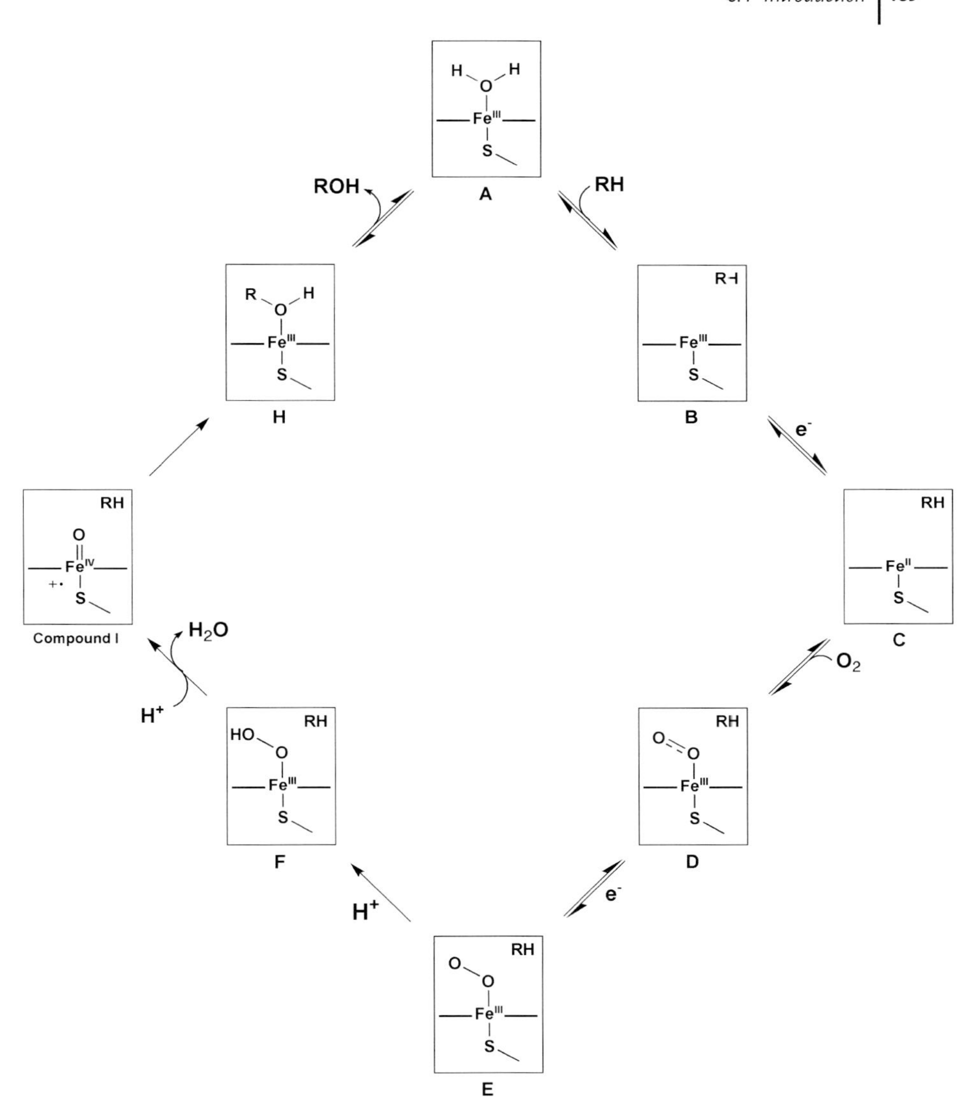

Fig. 6.1 Proposed mechanism for hydroxylation by P450.

The dissociation of water produces a five-coordinate Fe(III) complex (**B**) that can be reduced to an Fe(II) analog. Ferrous $P450_{cam}$ (**C**) readily binds dioxygen to form the oxy form of the enzyme. Dioxygen binds in a bent end-on mode (η^1-O_2) with the iron center formally oxidized to Fe(III) and the O_2 reduced by one electron to superoxide (**D**). Addition of a second electron from putidaredoxin follows to further reduce the complex to an Fe(III)-peroxo species (**E**). Proton transfer is required before the cleavage of the O–O bond: an Fe(III)-hydroperoxo complex (**F**) is proposed, in which a proton is shuttled to the distal O-atom of the peroxo ligand. A second proton is transferred into the active site with concomitant heterolytic cleavage of the O–O bond, forming water and a high-valent iron center with a terminal oxo ligand. This complex, referred to as compound **I**, is formulated as an Fe(IV)=O cation radical (from the porphyrin or the thiolate) and is the putative oxidant. The exact step(s) for O-atom transfer to the C_5-position on camphor are still controversial and will be discussed later in this chapter; however, it is clear that the proper alignment of substrate and timely injection of dioxygen, electrons and protons into the active site are crucial for activity.

6.1.1.2 The Role of the Secondary Coordination Sphere in Catalysis

Figure 6.1 does not detail the primary and secondary coordination spheres of the iron center in $P450_{cam}$; however, regulation of both spheres is essential for the observed reactivity. The role of the primary sphere has received much attention, specifically the effect(s) of thiolate ligation on activity [2]. Numerous spectroscopic, structural and enzymological studies illustrate the importance of the secondary coordination sphere [3]. For instance, substrate binding in $P450_{cam}$, which initiates the catalytic cycles, utilizes hydrogen bonds (H-bonds) formed between camphor and the amino acids within the active site. The proton shuttle proposed in Fig. 6.1 is also mediated by the microenvironment surrounding the iron complex. An H-bond network, created with amino acid residues and structural water molecules near the iron center, provides a pathway for protons to traverse through the active site. Furthermore, this H-bond network is essential for O–O bond scission that leads to the formation of compound I (Fig. 6.2) [4]. These results illustrate the crucial functional effects of noncovalent interactions within the active sites of metalloproteins.

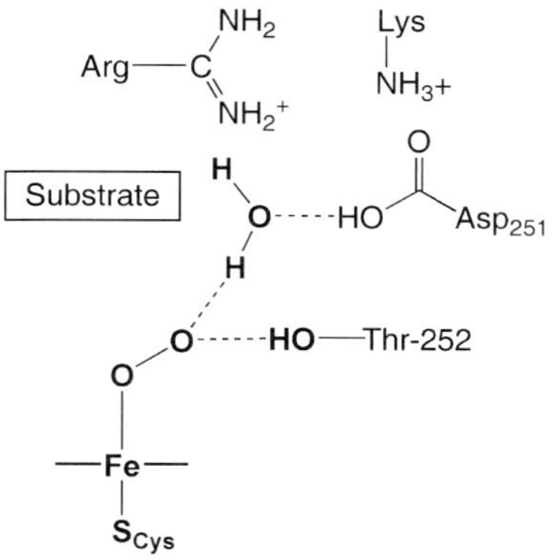

Fig. 6.2 Proposed H-bond network within the active site in P450.

6.1.2
Effective O_2 Binders and Activators in Biology

Synthetic chemistry has advanced significantly in the last 50 years to the point where placement of nearly any donor group around metal ions is possible. This has led to better control of the primary coordination sphere around metal ions, which enables the regulation of various functional properties, such as the stereochemistry, electronic structure, redox processes and Lewis acidity. However, as illustrated in the above example for $P450_{cam}$, sustainable function depends on several other factors, some of which are outlined below.

6.1.2.1 Accessibility

The active sites of oxygenases are often buried within the interior of the protein, distant from the molecular surface. The isolation of the metal sites prevents unwanted, and often deleterious, reactions with other species. Access to the metal center is controlled by several factors, including channels that connect the active site to other sites within the enzymes. Some oxygenases only function when assembled with other proteins. These additional proteins can serve a variety of roles, including regulating entry into the active site. This type of influence has been proposed in methane monooxygenase (MMO), a nonheme oxygenase that catalyzes the oxidation of methane to methanol [5]. The MMO system consists of three proteins, the oxygenase (MMOH) containing the nonheme active site, a reductase (MMOR) with a FAD and Fe_2S_2 cluster, and a relatively small effector protein MMOB that does not contain a cofactor. The proposed role of MMOB is to gate dioxygen and substrate into the active site of the hydroxylase component MMOH [6]. By binding to MMOH near the active site, MMOB can restrict assess, thereby allowing only the relatively small substrate methane to enter and be oxidized.

6.1.2.2 Secondary Coordination Sphere

Interaction within the secondary coordination sphere (microenvironment) between the protein architecture and active-site metal complexes is an integral feature of metalloenzyme function. As illustrated in Fig. 6.2 for $P450_{cam}$, noncovalent interactions such as H-bonds are acknowledged as being important forces in controlling reactivity at the metal centers in metalloenzymes [7, 8]. H-bonds are significantly weaker interactions than covalent bonds, yet appear to affect many aspects within active sites, particularly in proteins that bind and activate dioxygen. Control of the microenvironment in synthetic systems is not as defined as in proteins; however, including these weaker interactions could facilitate development of new reagents with improved function or, at the very least, allow for studies on the subtle but fundamental aspects that these weaker interactions have on metal-mediated processes.

6.1.2.3 Flow of Electrons and Protons

In the $P450_{cam}$ system, putidaredoxin is an effector protein because it supplies electrons into the active site. Although the first electron to form the ferrous center (**B**) can come from various sources, it is critical for prolonging catalysis that putidaredoxin deliver the second electron that leads to cleavage of the O–O bond (Fig. 6.1). The electron flow into the active site is coupled with proton shuttling that is mediated by the H-bond network surrounding the $Fe–O_2$ unit; failure in either of these processes leads to dysfunction [4]. It thus appears that metalloproteins have evolved to synchronize electron and proton transfer to ensure that oxidation of substrates occurs.

6.1.2.4 Lessons from Nature

The points outlined above for metalloproteins offer the basis for preparing synthetic systems that bind and activate dioxygen. The development of organic compounds that serve as ancillary ligands has aided in unraveling key features of metal-mediated dioxygen activation. These ligands allow for control of the primary coordination sphere(s) of the metal ion(s). Additionally, there must be some means to regulate assembly of multinuclear metal species – whereas oligomers can be advantageous in some cases, their prevention is often needed for oxidative function. Oligomerization typically can be stopped through appending groups onto the periphery of the ligand. Further functionalization of the ligand framework has proven successful in regulating the secondary coordination sphere.

6.2 Dioxygen Binders

6.2.1 Respiratory Proteins

There are three classes of proteins that reversibly bind dioxygen. Hemoglobins occur in a wide variety of organisms, having active sites with a single iron protoprophyrin IX cofactor. Hemerythrins are nonheme diiron proteins present in certain species of seaworms. The hemocyanins are dinuclear copper proteins found in the hemolymph of mollusks and arthropods. There have been extensive studies on the physical and structural properties of these proteins, and only a brief mention of their active site structures will be discussed here [9].

6.2.1.1 Hemoglobins

Dioxygen binding in hemoglobins occurs at the monomeric Fe(II) center in an η^1 manner – a similar binding mode as described for P450s. There is overwhelming biochemical and physical data to support that in most hemoglobins

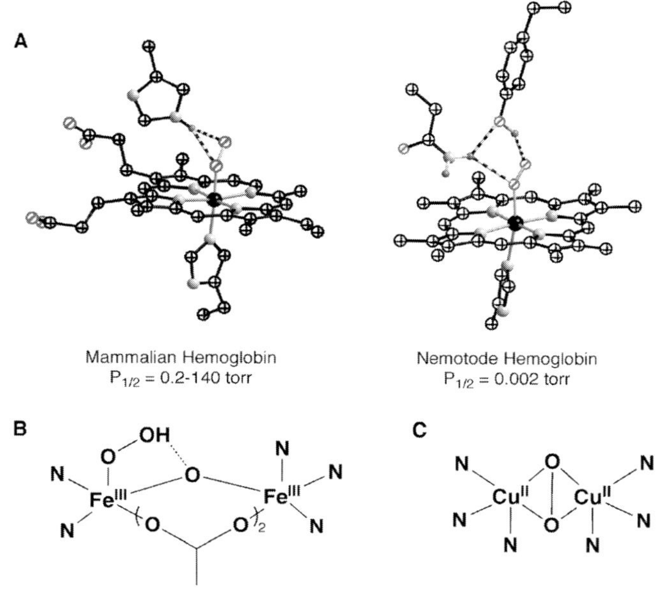

Fig. 6.3 Molecular structures of iron centers in the oxyhemoglobins from mammalians and the nematode (A), and descriptions of the active sites for oxyhemerythrin (B) and oxyhemocyanin (C).

there are additional intramolecular H-bond(s) between the coordinated dioxygen and amino acid residues within the active site. In most cases, the distal oxygen atom of O_2 is involved in H-bond formation, as shown schematically in Fig. 6.3 for the active site for mammalian hemoglobin [8a, b]. Changes in the H-bond network can influence function, as illustrated by the structure for the parasitic nematode *Ascaris suum* [8c]. A more extensive H-bond network surrounding the coordinated dioxygen is found, which partially contributes to the nearly 100-fold increase in O_2 affinity for *Ascaris* hemoglobin over mammalian hemoglobins.

6.2.1.2 **Hemerythrin**

The deoxy state of hemerythrin has an unsymmetrical dinuclear Fe(II) complex in its active site, in which one ferrous center is five-coordinate and the other is six-coordinate. The iron centers are connected through two 1,3-μ-carboxylato groups and one hydroxo bridging ligand. At the time of its discovery, this bridging motif was unique for iron and has since been duplicated in several synthetic systems [10]. Dioxygen binds only to the five-coordinate ferrous site in an end-on mode, causing the formal two-electron transfer from the diiron core to O_2. In addition, a proton transfer is proposed from the bridging hydroxo ligand to the distal oxygen of the coordinated dioxygen. The resulting diferric oxy complex thus has a bridging oxo ligand and a coordinated hydroperoxo ligand, which has an additional interaction with the oxo ligand via an H-bond (Fig. 6.3).

6.2.1.3 Hemocyanins

The dinuclear Cu(I) centers in the deoxy state of hemocyanin are three-coordinate, with each metal center having three imidazolyl ligands. Unlike in hemerythrin, there are no endogenous bridging ligands between the metal centers in the deoxy form of hemocyanin. Binding of dioxygen occurs between the two copper centers, resulting in a dicopper(II) complex with a μ-η^2:η^2 peroxo ligand [11]. This type of bridging mode for the peroxo ligand accounts for the strong antiferromagnetic coupling between the Cu(II) ions and the unusually low O–O vibration (Fig. 6.3).

6.2.2 Synthetic Analogs

Duplicating the structural and functional properties of respiratory proteins in synthetic systems has proven to be challenging. As already stated, difficulties arise because all the necessary primary and secondary sphere effects are often not included within a single synthetic system. For instance, simple iron porphyrin complexes are ineffective in modeling the reversible dioxygen binding as found in hemoglobins; they tend to form μ-oxo-bridged Fe(III) dimers, which are common end products of these reactions (Fig. 6.4). The formation of the oxo bridge species has been exhaustively studied and results at low temperature (below –80 °C) suggest dimer formation proceeds through an iron–dioxygen adduct. However, the first detectable species is not the desired monomer, but a μ-1,2-peroxide dimer. Cleavage of the O–O bond is promoted by the addition of

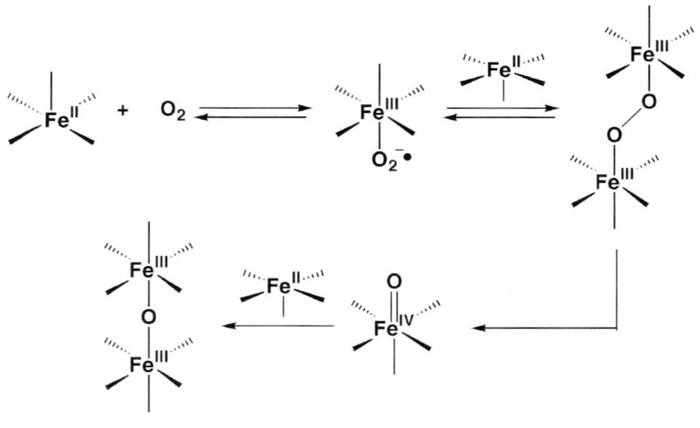

Fig. 6.4 Schematic of the proposed formation of Fe(III)-oxo-Fe(III) dimers from Fe(II) precursors and O_2.

a base to form two equivalents of an Fe(IV)porphyrin(base)(O) species, which was characterized by ^{1}H nuclear magnetic resonance (NMR) and optical spectroscopies [12]. The unstable Fe(IV)=O complexes produce the μ-oxo dimer upon warming to room temperature.

Numerous complexes have been reported that attempt to emulate the reversible dioxygen binding of these proteins, yet only a few accurately modeled the natural systems. The two general strategies most often used to probe reversible dioxygen binding are placement of bulky groups on the ancillary ligands to prevent unwanted bimolecular reactions and performing the binding studies at unnaturally low temperatures, such as temperatures below −20 °C.

6.2.2.1 **Hemoglobin Models**

An excellent example of a synthetic iron complex that reversibly binds dioxygen is the ferrous "picket-fence" porphyrin complex developed by Collman (Fig. 6.5) [13]. The two planar faces of the porphyrin ligand are differentiated because the four groups appended from the *meso* positions are oriented on either side of the ring. These groups provide enough steric bulk to regulate binding to the iron center and, more importantly, prevent any metal–metal interactions. Therefore

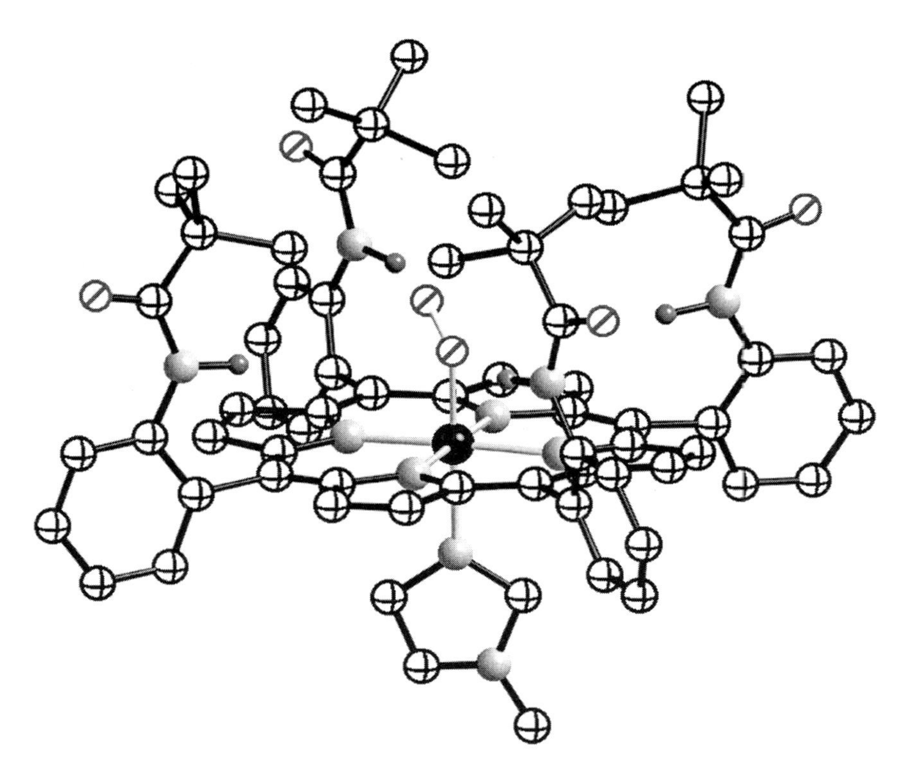

Fig. 6.5 Molecular structure of oxy picket-fence porphyrin.

the iron center is sufficiently isolated, as in hemoglobin, to eliminate peroxo or oxo bridge formation. Moreover, the amide groups, although not directly involved in H-bonding to the coordinated dioxygen, provide a dipolar contribution that undoubtedly aids in stabilizing O_2 adducts. The unhindered face of the porphyrin can bind an additional ligand; Collman's group found that imidazoles selectively bind to the coordination site on the unhindered side of the complex. A second equivalent of imidazole cannot bind because it is too big to fit into the binding site created by the picket fence. However, dioxygen does fit into this site and was shown to reversibly bind at room temperature. Collman's group and several others developed more elaborate systems based on this design concept and were able to evaluate several aspects of dioxygen binding to iron porphyrins. Although first published over 30 years ago, this pioneering work is still one of the most elegant examples of a synthetic complex designed to have functional properties similar to that of a metalloprotein.

6.2.2.2 **Hemerythrin Models**

There are no examples of a structurally characterized synthetic system that duplicates the binding of dioxygen in hemerythrin. The dearth of suitable synthetic models that match the unsymmetrical (μ-hydroxo)bis(μ-carboxylato) diiron(II) core in deoxyhemerythrin has been one reason for the lack of suitable models. A more important reason is the difficulty chemists have in limiting reduction of dioxygen to only the peroxo state. More common is the $4e^-$ reduction of O_2, producing two oxo ligands which often results in stable oxo-bridged diiron species (*vide supra*).

Lippard and Mizoguchi developed a diiron system that reproduces several of the spectroscopic properties of hemerythrin [14]. Their system relies on the formation of an unsymmetrical diiron(II) complex, with the necessary (μ-hydroxo)-bis(μ-carboxylato) diiron core. Assembly of this complex is promoted by a dicarboxylate ligand, in which the two carboxylated groups are appended from the 4- and 6-position of a dibenzofuran. Modeling studies by the authors suggested that the spacing of the carboxylates on the furan template was sufficient to form the desired triply bridged diiron core. This proved correct, as they were able to isolate a triply bridged diiron(II) complex with TMEDA that contains one five-coordinate and six-coordinate Fe(II) atom (the sixth ligand was a bound acetonitrile molecule). This complex reproduced many of the structural and spectroscopic properties of deoxyhemerthyrin (Fig. 6.6).

The open coordination site in the $[Fe(II)_2(\mu\text{-OH})(\mu\text{-OOR})_2]$ complex should be available to binding an exogenous ligand, such as dioxygen. This was realized when it was treated with O_2 at $T = -78\,^\circ C$ in the presence of a slight excess of *N*-methylimidazole. Two new optical features appeared at λ (ε_M) = 336 (7300) and 470 (2600) nm in the electronic absorbance spectrum, which are similar to those found in oxyhemerthyrin [λ (ε_M) = 330 (6800) and 500 (2200) nm]. Direct evidence for dioxygen binding came from resonance Raman studies that showed an O–O stretching frequency at 843 cm^{-1} that shifted to 797 cm^{-1} when

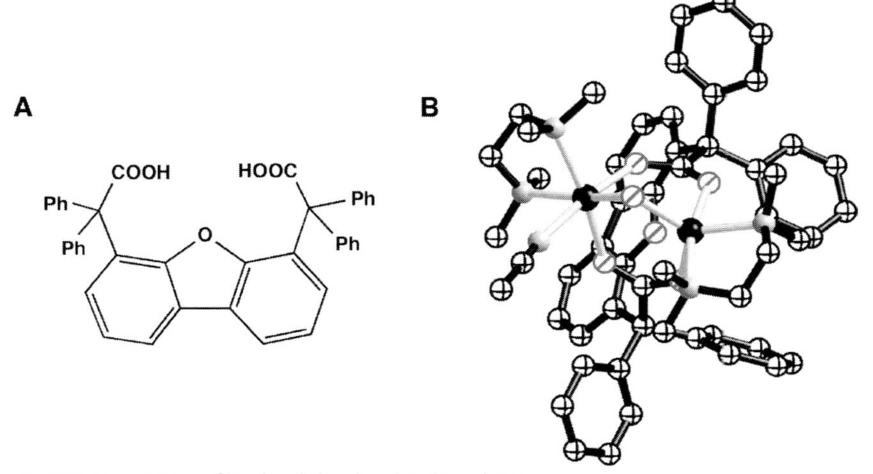

Fig. 6.6. Description of hindered dicarboxylate ligand (A) and molecular structure of the unsymmetrical $[Fe(II)_2(\mu\text{-}OH)(OOR)_2]$ complex (B).

prepared with $^{18}O_2$. Moreover, the Mössbauer parameters for this dioxygen-bonded complex are comparable to those found for oxyhemerthyrin. It has not yet been established that the dioxygen is bound as a hydroperoxide as in oxyhemerthyrin; this Fe_2–O_2 adduct is relatively unstable, thus preventing its isolation and subsequent release of dioxygen. Nonetheless, this is an important achievement in the understanding of the structural attributes needed to sequester dioxygen with diiron complexes.

6.2.2.3 **Synthetic μ-Peroxo Diiron Complexes**

There have been advances in the structural and physical properties of diiron complexes with dioxygen ligands. All of the well-defined structures have a μ-1,2-peroxo coordinated between two Fe(III) centers. These systems are not directly related to hemerythrin because of the mode of O_2 bonding; they are more relevant to the proposed structures in the active sites of ribonucleotide reductase (RNR) [5, 15] and MMOH [5] – two nonheme diiron-containing enzymes that activate dioxygen (*vide infra*).

Systems developed independently by Que and Kitajima provided the first evidence for the formation of nonheme diiron-peroxo complexes. Que and Kitajima used different synthetic approaches: Que utilized a diiron(II) complex containing a dinucleating ligand composed of terminal benzimidazole ligands and a bridging alkoxide (Fig. 6.7) [16], while Kitajima used a monomeric Fe(II) complex composed of a bidentate carboxylate and a hindered hydrotris(3,5-^{i}Pr-pyrazolyl)borate (Tp^{iPr}) ligand that assembled into an iron dimer (Fig. 6.8) [17]. In studies with both systems, the dioxygen binding to the Fe(II) complexes was monitored at temperatures less than –20 °C, which was necessary to assist in stabilizing the peroxo adducts and prevent iron-oxo species that are formed at room temperature. Addition of O_2 to the Fe(II) precursors at around $T = -50\,°C$ produced a solution with a strong absorbance band in the 600-nm region of the spectrum that is assigned to peroxo-to-Fe(III) charge transfer transitions. Resonance Raman studies support the presence of a μ-1,2-peroxo core: Que's complex has two features at 476 and 900 cm^{-1} that are assigned to ν(Fe–O) and ν(O–O), which were confirmed with isotopically labeled dioxygen ($^{18}O_2$). Similar ν(Fe–O) and ν(O–O) vibrations were observed in Kitajima's complex. Note that the 600-nm absorbance found in the μ-1,2-peroxo complexes is strongly red-shifted compared to the related charge transfer band in hemerythrin, suggesting the possibility for an optical method to determine the mode of peroxo coordination in nonheme iron dimers.

While Que and Kitajima's systems have similar spectroscopic properties, they differ in their ability to reversibly bind dioxygen. Que's complex irreversibly binds dioxygen – warming samples from $T = -80\,°C$ to room temperature leads to decomposition of the complex. In contrast, the dioxygen reaction with Kitajima's system is temperature dependent. At $T = -50\,°C$ the formation of the oxygenated complex is irreversible, yet at $T = -30\,°C$ reversible dioxygen binding is observed. Dioxygen is lost from the μ-1,2-peroxo diiron(III) complex upon bubbling argon which regenerates the Fe(II) starting compound. Repetitive experiments in toluene showed that more than 10 cycles of the reversible dioxygen binding are possible.

6.2.2.4 Structurally Characterized μ-Peroxo Diiron Complexes

During the long history of iron–dioxygen chemistry, there was no definitive structural proof of a bridging peroxo ligand within a diiron complex. This changed in amazing fashion during 1996, when three independent X-ray structural studies on μ-1,2-peroxo diiron(III) complexes were reported. Two of the

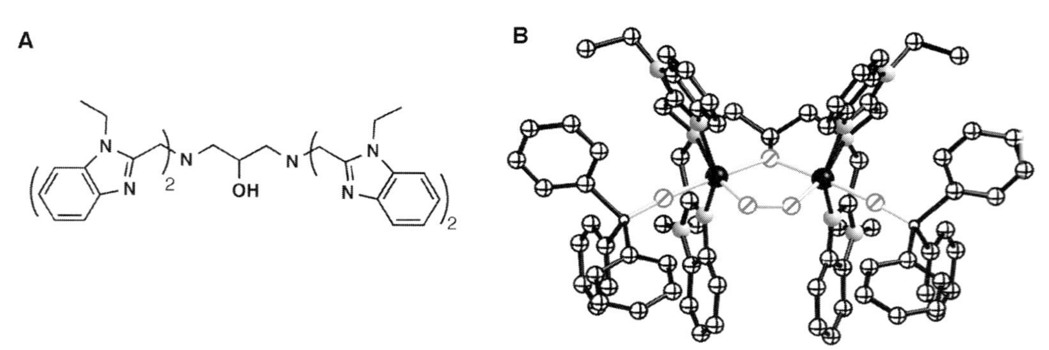

Fig. 6.7 Description of bidentate benzimidazole ligand (A) and the molecular structure of a $Fe_2(1,2\text{-}\mu\text{-}O_2)$ complex (B).

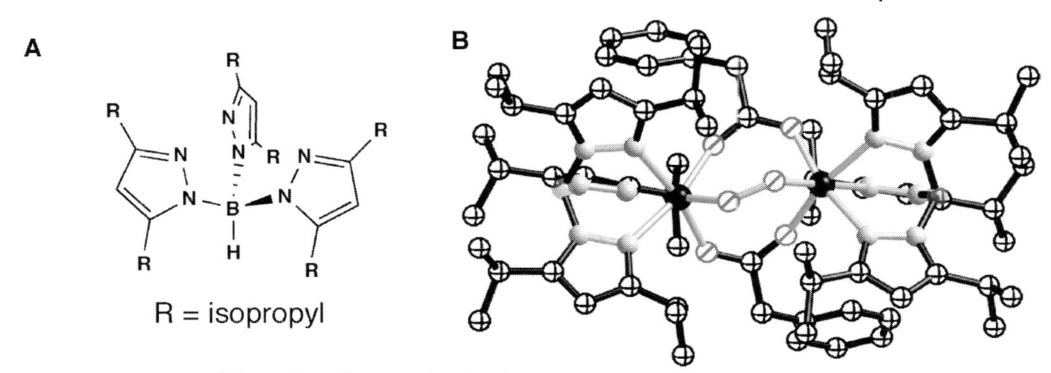

Fig. 6.8 Description of the Tp ligands (A) and molecular structure of a $Fe_2(1,2\text{-}\mu\text{-}O_2)$ complex (B).

systems published are closely related to the ones just described in the previous section. Que found that his μ-1,2-peroxo diiron(III) complex could be stabilized at $T = -40\,^{\circ}C$ by adding triphenylphosphine oxide (Ph_3PO), allowing for the growth of single crystals [18]. The molecular structure of the complex (Fig. 6.7) shows that the μ-alkoxodiiron(III) core is supported by an additional μ-1,2-peroxo bridge. The O–O bond distance of 1.416(7) Å is consistent with a coordinated peroxo ligand, an assignment in agreement with resonance Raman studies (*vide supra*). The peroxo binds in a *cis* manner and the Fe···Fe distance is 3.462(2) Å. Noticeably missing in the μ-1,2-peroxo diiron(III) species is the carboxylate ligand that was present in the diiron(II) starting complex – this ligand is replaced by the Ph_3PO groups that bind as monodentate ligands to each iron center. The role of the Ph_3PO ligands in stabilizing peroxo bridge ligation is not yet known.

The molecular structure of Kitajima's diiron–dioxygen adduct was determined and reported by Lippard, and it too matches the structural prediction from spectroscopy [19]. The diiron(III) core is triply bridged with two μ-1,3-carboxylate ligands and one *cis*-μ-1,2-peroxo, resulting in a long Fe···Fe separation of 4.004(2) Å (Fig. 6.8). The O–O length of 1.407(9) Å is nearly identical to that found in Que's system. Each iron center is six-coordinate, with the remaining coordination sites being filled by the pyrazole donors of the Tp ligands.

Perhaps the most remarkable of the three μ-1,2-peroxo diiron(III) systems is the one reported by Suzuki (Fig. 6.9), which uses a phenolate-bridged binucleating ligand [20]. This complex has similar structural properties as found in the other two systems: it has a *cis*-μ-1,2-peroxo ligand, a 1,3-μ-carboxylate group and a μ-phenolate bridge. The O–O distance of 1.426(6) Å is expected for a peroxo moiety and the Mössbauer parameters support the presence of two Fe(III) ions. In addition, all three complexes have ancillary ligands with nitrogen donors, similar to what is found in hemerythrin: Que's system has benzimidazoles, Kitajima uses pyrazoles and Suzuki's complex contains imidazoles. However, only Suzuki's complexes can reversibly bind dioxygen at room temperature. In

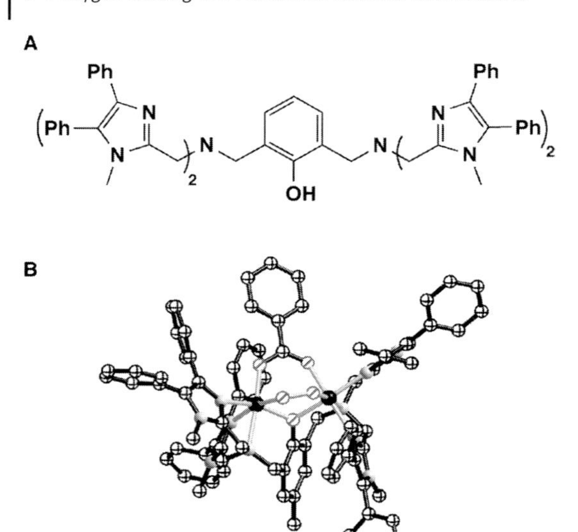

Fig. 6.9 Hindered bidentate ligand used to prepare a stable $Fe_2(1,2\text{-}\mu\text{-}O_2)$ complex (A) and molecular structure of the $Fe_2(1,2\text{-}\mu\text{-}O_2)$ complex (B).

fact, liberation of dioxygen is achieved by boiling the oxygenated complex in acetonitrile! Suzuki attributes this unusual stability to the binucleating ligand that invokes relatively weak Fe–N interactions *and* provides a hydrophobic cavity around the dioxygen-binding site. Each imidazole group has phenyl rings at the 4- and 5-positions that are proposed to cause an elongation of the Fe–N(imidazolyl) bonds through steric repulsion. This elongation weakens the electron donor ability of the imidazole groups, thus preventing irreversible oxidation of the metal centers and reduction of the peroxo ligand. Furthermore, these sterically bulky substituents form a hydrophobic cavity that aids in stabilizing the μ-1,2-peroxo diiron(III) unit. This constrained microenvironment prevents intermolecular interactions with other metal complexes, yet is sufficient to allow dioxygen access to the iron centers. These factors contribute to the prevention of the normal irreversible oxidation that occurs at room temperature with most iron–dioxygen complexes. Suzuki's system again highlights the need to control both spheres around the metal center(s) to regulate metal–dioxygen interactions.

6.2.2.5 Monomeric Nonheme Iron–Dioxygen Adducts

Numerous groups over the years have attempted to make stable iron–dioxygen adducts from O_2. Nearly all systems at room temperature undergo autoxidation to yield Fe(III) species, usually with bridging oxo ligands. Busch developed one of the best examples of nonheme iron–dioxygen species that is stable for around

6 h at $T=20\,^{\circ}C$ in 1:1:1 acetone/pyridine/water [21]. The ligand system is a "strapped" macrocycle, composed of a cyclidene with a xylylene bridge that spans one face of the ligand. Metal ion binding produced a five-coordinate Fe(II) complex with the fifth ligand being either a solvent molecule or a chloride ion. Binding of the dioxygen is believed to occur on the strapped face of the complex, which serves to protect the Fe–O_2 unit. In an elegant structure–function study, Busch showed how variations in the strap length and composition caused a change in the stability of the iron–dioxygen adduct. Autoxidation is the primary cause of the instability of these systems. The products of these processes are not well characterized – electron paramagnetic resonance (EPR) spectra have signals for both high and low spin Fe(III) species. Ferric-peroxo complexes are suggested but the necessary spectroscopic characterizations are lacking.

More recently several groups have prepared monomeric ferric-peroxo complexes from Fe(II) precursors and H_2O_2 [22]. These complexes often have limited stability at room temperature, which has prevented structural characterization via XRD methods. Nevertheless, information obtained from detailed spectroscopic studies has produced a consensus picture for two general species: an Fe(III)-η^1-hydroperoxo adduct and an Fe(III)-η^2-peroxo species (Fig. 6.10). The complexes have distinct magnetic properties that make their evaluation amenable to EPR spectroscopy. The Fe(III)-η^1-hydroperoxo complexes are low-spin ($S=1/2$) with rhombic EPR spectra having g values centered around 2. The Fe(III)-η^2-peroxo complexes are high-spin with EPR spectra that also have rhombic character. The optical, vibrational and magnetic properties of these complexes support the different modes of dioxygen binding. Interconversion between the two forms was accomplished by either treating Fe(III)-η^1-hydroperoxo with bases (such as triethylamine) or reacting the Fe(III)-η^2-peroxo complexes with acids.

Although derived from H_2O_2, these complexes have relevance to iron species that bind and activate dioxygen. For instance, Fe(III)-hydroperoxo units are present in the active site of hemerythrin. In addition, the Fe(III)-hydroperoxo complexes are related to an active form of bleomycin, a glycoprotein antibiotic that is used in anticancer treatment [23]. Its function is to cleave double-stranded DNA and exhibits the highest activity when an Fe(II) ion is present. In the presence of reducing agents, Fe(II) bleomycin [Fe(II)BLM] reacts with dioxygen to produce the activated form (ABLM), which is a low-spin Fe(III)-η^1-OOH species. Reactions of ABLM with DNA involve H-atom abstraction, most likely from a sugar moiety. A high kinetic isotope effect (k_H/k_D) of around 2–7 makes it unlikely that hydroxyl radicals, generated from homolytic cleavage of an Fe(III)-η^1-

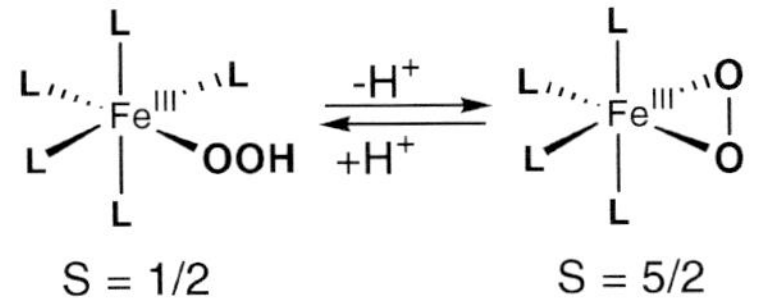

Fig. 6.10 Description of known monomeric ferric-peroxo complexes.

OOH unit, are involved. One suggestion that is based on detailed spectroscopic and theoretical studies is that H-atom abstraction occurs directly from the DNA sugar to the ABLM [24]. Interestingly, the direct pathway and one involving O–O homolysis are closely related, having the same overall reaction energy. Their differing pathways lead to distinct potential energy surfaces, resulting in different kinetics. Many of the synthetic Fe(III)-η^1-OOH complexes also cleave DNA and other C–H bonds. However, the mechanisms for these processes appear to involve the homolytic cleavage of the O–O bond.

6.2.2.6 **Models for Hemocyanin**

Efforts to model the active site of hemocyanin began before the details of the structure were determined via XRD. Information obtained from spectroscopic methods produced proposals for the structures of the active site, which served as initial targets to model. The majority opinion was that deoxyhemocyanin had three-coordinate Cu(I) ions with nitrogen donors from the imidazolyl residues of histidine and oxyhemocyanin contained a μ-1,2-peroxo dicopper(II) core. Another bridging ligand, most probably an endogenous μ-phenolate from tyrosine, was predicted because the μ-1,2-peroxo ligand could not mediate the observed strong antiferromagnetic coupling between the Cu(II) ions. These predictions led to the development of synthetic dicopper systems with a bridging phenolate ligand; however, while some elegant systems were prepared, none were able to reproduce the physical and functional properties of hemocyanin [25].

Experiments later eliminated phenolate as an endogenous bridge to the copper centers. The XRD studies of deoxyhemocyanin revealed that a tyrosine was not near the active site [26]. In addition, Karlin prepared a synthetic dicopper(II)-peroxo complex without a phenolate that accurately reproduced the electronic

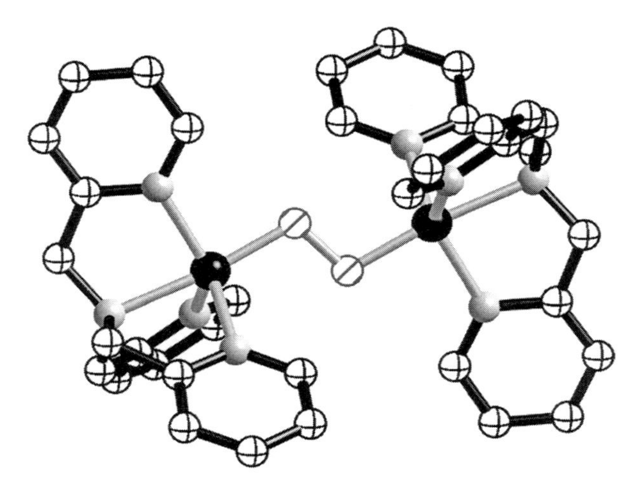

Fig. 6.11 Molecular structure of a [Cu(II)(1,2-μ-O_2)Cu(II)].

absorbance features of oxyhemocyanin [λ_{max} (ε_M) = 345 (20 000) and 570 (1000) nm] [27]. Karlin also reported the first structure of a dicopper–dioxygen adduct that contained a trans-μ-1,2-peroxo dicopper(II) core and Cu···Cu separation of 4.359(1) Å (Fig. 6.11) [28]. The ancillary ligands were from the tripodal ligand, tris(pyridylmethyl)amine and the complex was stable at temperatures less than –50 °C. However, this synthetic μ-1,2-peroxo dicopper(II) complex had unique optical properties with features at λ_{max} (ε_M) = 525 (11 500) and 590 (shoulder) nm that do not match those observed for oxyhemocyanin.

The binding mode of the peroxo ligand in oxyhemocyanin was resolved through the work of Kitajima, who prepared a synthetic complex that reproduces many of the protein's physical properties [29]. His strategy was to use [Tp^{iPr}Cu(I)], a monomeric Cu(I) complex containing hydrotris(3,5-^{i}Pr-pyrazolyl)-borate, that had isopropyl groups appended from the 3- and 5-position of the pyrazole rings. Attempts to make stable copper–dioxygen complexes with Tp ligands and O_2 had been reported with some encouraging results, but none of these previous systems had properties related to oxyhemocyanin [30]. Kitajima's system differed from others by using appended isopropyl groups – they adequately control the secondary coordination sphere at low temperatures by providing enough room for the assembly of a peroxo bridge copper(II) dimer, while preventing further intermolecular oxidative processes.

Kitajima's key finding was that addition of dioxygen to [Tp^{iPr}Cu(I)] at T = –78 °C produced a deep purple-colored complex that was isolable. This complex has nearly identical properties as oxyhemocyanin, including λ_{max} (ε_M) = 349 (21 000) and 551 (800) nm and a ν(O–O) at 741 cm^{-1} that shifts to 698 cm^{-1} in the ^{18}O-labeled isotopomer [oxyhemocyanin: $\nu(^{16}O_2)$ = 749 cm^{-1}]. The structure of this complex was surprising, as the peroxo coordinates in a μ-η^2:η^2 manner,

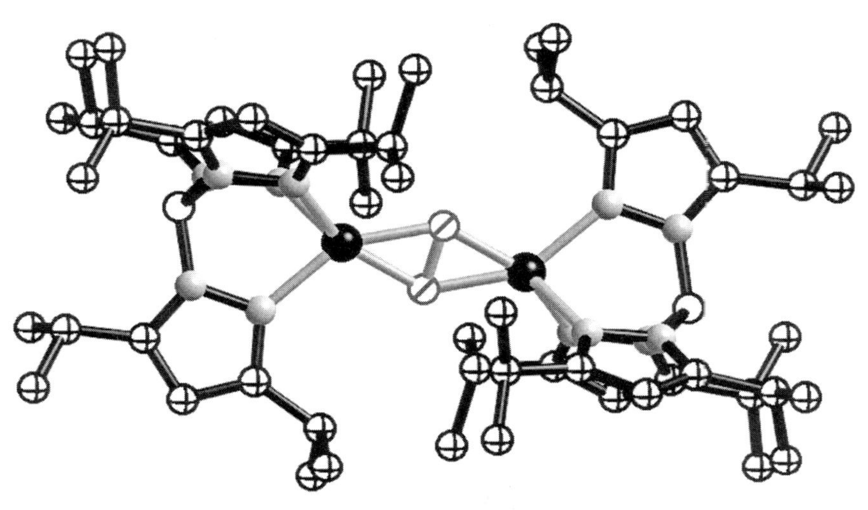

Fig. 6.12 Molecular structure of [(Cu(II)Tp^{iPr})(μ-η^2:η^2-O_2)Cu(II)Tp^{iPr}].

with O–O and Cu···Cu distances of 1.412(12) and 3.560(3) Å, respectively (Fig. 6.12) [29, 31]. At the time of its publication, this type of peroxo coordination was unknown in the chemistry of copper and was used to predict the active site structure in oxyhemocyanin. The μ-η^2:η^2-peroxo dicopper(II) core explains the physical and structural properties observed in oxyhemocyanin without evoking another bridging ligand. The structure of oxyhemocyanin was subsequently determined and its active site contained the predicted μ-η^2:η^2-peroxo dicopper unit [32]. There are now several examples of dicopper complexes with "side-on" bonded peroxo ligands and, as will be discussed in subsequent sections, this complex set the stage for others to explore the activation of dioxygen with dicopper complexes.

6.2.2.7 Monomeric Copper–Dioxygen Adducts

The tendency for copper–dioxygen species to dimerize has made it difficult to prepare mononuclear copper–dioxygen complexes. Thompson reported analytical and spectroscopic evidence that [Cu(I)Tp^{Me}(C_2H_4)] [Tp^{Me}=hydrotris(3,5-methylpyrazolyl)-borate] reversibly bound dioxygen to form a Cu(II)-superoxo complex at $T=-14\,°C$ [33]. His most compelling datum was a noticeable color change for a dichloromethane solution of the Cu(I) complex in the presence of dioxygen. Isotopic labeling experiments were also performed and the infrared (IR) spectrum contained a relatively weak band at 1015 cm^{-1} for the putative [Cu(I)Tp^{Me}($^{18}O_2$)] isotopomer, which was assigned to the ν(^{18}O–^{18}O). However, the related band predicted for the ^{16}O-labeled isotopomer was not observed because strong Tp^{Me} vibrations occur at the same frequency (1060 cm^{-1}). Without additional vibration information it was difficult to make an accurate assessment of these results; nevertheless, these findings point the way toward stabilizing a monomeric copper–dioxygen adduct.

Ten years after Thompson's work appeared, Kitajima reported definitive proof for a mononuclear copper–dioxygen species. Kitajima had great success isolating μ-η^2:η^2-peroxo dicopper cores with complexes containing Tp^{iPr}. He further reasoned that increasing the steric bulk on each pyrazole ring would prevent dimer formation and produce Cu(II)-superoxo complexes. This was accomplished using Cu(I) complexes of Tp^{tBu,iPr}, in which *tert*-butyl groups are in the 3-positions of the rings [34]. Treating a methylene chloride solution of [Cu(I)Tp^{tBu,iPr}(DMF)] with 1 atm of O_2 at $T=-50\,°C$ afforded [Cu(II)Tp^{tBu,iPr}(O_2)], which has absorption bands at λ_{max} (ε_M) = 352 (2330) and 510 (230) nm. The IR spectrum of a solid sample of [Cu(II)Tp^{tBu,iPr}($^{16}O_2$)] has a band at 1112 cm^{-1}, which is in the region expected for coordinated superoxide. Unfortunately, the [Cu(II)Tp^{tBu,iPr}($^{18}O_2$)] isotopomer should have a ν(^{18}O–^{18}O) at around 1060 cm^{-1}, but this region of the spectrum is obscured by vibrations from the Tp ligand, a similar problem encountered by Thompson (*vide supra*). However, resonance Raman experiments support the assignment of coordinated superoxo in [Cu(II)Tp^{tBu,iPr}(O_2)] by the appearance of bands at 1112 and 1060 cm^{-1} in the ^{16}O- and ^{18}O-labeled isotopomers [35].

The molecular structure of [Cu(II)Tp^{tBu,iPr}(O_2)], determined using XRD methods, supports a monomeric Cu(II)–O_2 complex (Fig. 6.13) [34]. The reported struc-

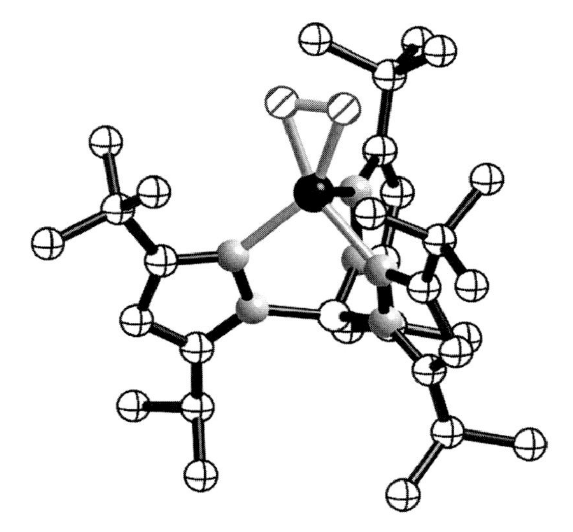

Fig. 6.13 Molecular structure of $[Cu(II)Tp^{tBu,iPr}(O_2)]$.

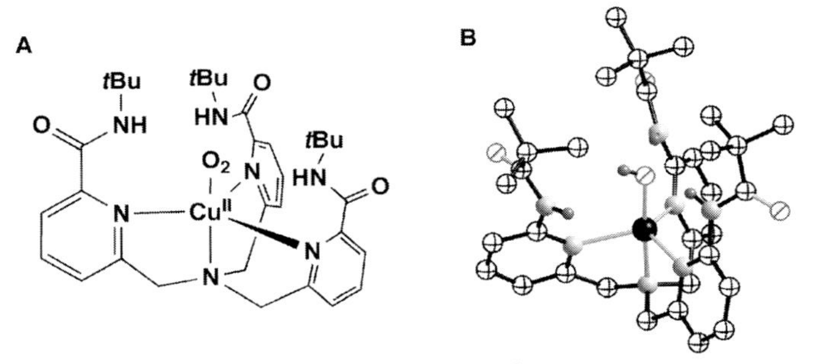

Fig. 6.14 Description of the proposed $[Cu(II)(\eta^1\text{-}O_2)]$ (A) and molecular structure of the [Cu(II)(OH)] complex (B).

ture is not high quality, but some important conclusions can be made. The coordination geometry about the copper ion is best described as a square pyramid, with one long Cu–N_{py} bond (above 2.20 Å) and the superoxo ligand coordinated side-on within the basal plane. The superoxo ligand has an O–O distance of 1.22(3) Å; however, recent theoretical studies on $[Cu(II)Tp^{tBu,iPr}(O_2)]$ and related systems suggested that this distance may be short by around 0.1 Å because the X-ray data was collected at room temperature [36]. Regardless of these new issues, the existing evidence still supports the assignment of $[Cu(II)Tp^{tBu,iPr}(O_2)]$ as being a monomeric Cu(II)-η^2-superoxo complex.

Masuda reported the isolation of another mononuclear Cu(II)-superoxo complex generated from dioxygen that appeared soon after $[Cu(II)Tp^{tBu,iPr}(O_2)]$ was published [37]. In an effort to simulate the secondary coordination found in re-

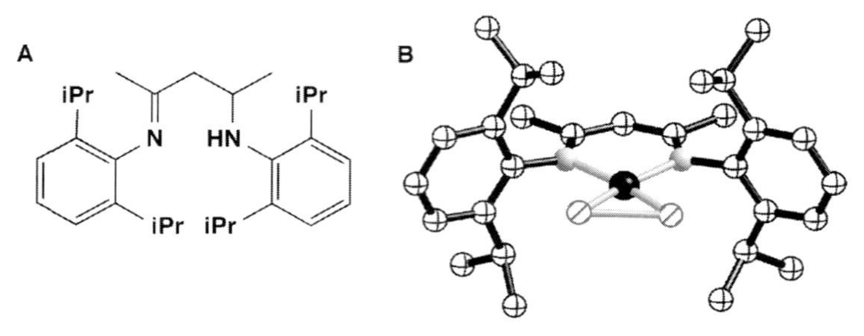

Fig. 6.15 Description of a β-diketiminate ligand (A) and the molecular structure of a $[Cu(II)(\eta^2\text{-}O_2)]$ complex (B).

spiratory proteins, this second complex utilized a new tripodal ligand containing H-bond donating groups. The ligand used the common tripodal ligand, tris(pyridylmethyl)amine as a foundation and appended pivaloylamido groups from the 6-position on the pyridine rings (tppa) (Fig. 6.14). Bubbling dioxygen into a methanolic solution of $[Cu(I)tppa]^+$ at $T = -80\,^\circ C$ produced a color change and a diamagnetic system. The authors proposed that this complex was $[Cu(II)tppa(O_2)]^+$. XRD studies showed a monomeric copper complex with an end-on (η^1) bonded superoxo ligand. Noteworthy are the proposed intramolecular H-bonds between the amide group and the coordinated dioxygen. While this is indeed an intriguing report, key supporting evidence is lacking, such as vibration data on the $^{18}O/^{16}O$-labeled isotopomers and manometric O_2 uptake measurements.

Subsequent investigations by Tolman strongly indicated that this complex does not contain a coordinated superoxo ligand, but rather a bound hydroxo ligand (i.e. $[Cu(II)tppa(OH)]^+$) [38]. Compelling evidence was provided, including a new solution to the X-ray data of Masuda that gave a better solution for $[Cu(II)tppa(OH)]^+$ and a reproduction of the reported synthesis of Masuda that produced the $[Cu(II)tppa(OH)]^+$ complex (Fig. 6.14), which has the same optical properties as those reported for the alleged $[Cu(II)tppa(O_2)]^+$ complex. This controversy emphasizes the difficulty in studying metal–dioxygen systems and the need to do multiple experiments before definitive assignments can be made.

More recently, Tolman has reported the formation of mononuclear $Cu\text{–}O_2$ adducts using β-diketiminate ligands, which are dependent on the steric properties of these ancillary ligands (Fig. 6.15) [39]. Relatively unhindered β-diketiminate ligands afford complexes with $Cu(III)(\mu\text{-}O)_2Cu(III)$ cores, whereas with sufficiently bulky ligands, 1:1 adducts are formed with discreet $Cu(\eta^2\text{-}O_2)$ units. Resonance Raman data support side-on coordination, with bands at 968 and 917 cm^{-1} for the $^{18}O_2/^{16}O_2$-labeled isotopomers. In addition, the mixed isotope adduct has only a single peak at $\nu(^{16}O^{18}O) = 943\ cm^{-1}$. Disorder problems hamper the molecular structures determined by XRD on the initially reported complexes, yet more recent studies confirm the formation of the $Cu(\eta^2\text{-}O_2)$ unit with an O–O distance of 1.392(3) Å [40]. Moreover, these complexes can be used as synthons to prepare unsymmetrical $Cu(III)(\mu\text{-}O)_2Cu(III)$ complexes. While

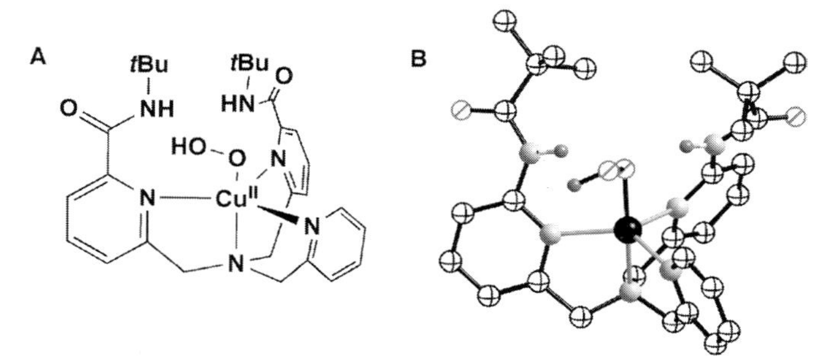

Fig. 6.16 Description of the Cu(II)(η^1-OOH) complex (A) and its molecular structure (B).

the η^2-coordination mode for the dioxygen in the monomeric adduct is apparent, the oxidation level of the dioxygen does not fit a routine description. In particular, the vibrational properties and density functional theory (DFT) studies suggest that this complex has intermediate features between a Cu(II)-superoxide and a Cu(III)-peroxide.

The final contribution on monomeric Cu–O_2 adducts involves a species not derived directly from dioxygen, but H_2O_2. Masuda has been able to isolate a Cu(II)-hydroperoxo complex at room temperature using Cu(II) precursors and a large excess of H_2O_2 [41]. These conditions are harsh and no reported yields of the isolated Cu(II)-hydroperoxo complex are given. Vibration data support peroxide coordination: a strong resonance-enhanced Raman band was observed at 856 cm^{-1}, which shifted to 810 cm^{-1} when $H_2^{18}O_2$ was used in the preparation. In addition, the complex gives an axial EPR spectrum that is consistent with a trigonal bipyramidal Cu(II) species. The molecular structure determined by XRD methods also supports this assignment (Fig. 6.16). The hydroperoxide is coordinated in an end-on fashion with an O–O distance of 1.460(6) Å and a Cu–O–O angle of 114.5°. The ancillary ligand in this complex is similar to tppa used previously by Masuda, in that it contains a tris(pyridylmethyl)amine tripod with pivaloylamido groups at the 2-positions on two of the pyridines. The structural results also reveal that the proximal oxygen of the hydroperoxo ligand is H-bonded to the two appended NH groups of the tripodal ligand: the O···N distances are less than 2.8 Å and the N–H vectors are positioned toward the coordinated oxygen atom (N–H···O angle around 160°). The authors correctly assert that these intramolecular H-bonds add stability that is essential in the isolation of this unique Cu(II)–OOH complex.

6.3
Reactive Intermediates: Iron and Copper Species

Most of the metal–dioxygen adducts previously discussed have limited stability and therefore serve as the starting point in our discussion of reactive intermediates. Although many of the complexes are difficult to work with under ambient conditions, there has been much progress in the area of dioxygen-derived reactive intermediates over the last 15 years. Two general technological innovations have aided in the search for these species. Low-temperature spectroscopic methods, while still fraught with difficulty, are no longer limited to the specialists, allowing nearly all chemists the means for detecting the initial species formed after dioxygen binding and activation. Advances in XRD methods have also led to accurate structural determinations of complexes that have limited stability, especially at room temperature. Many of the examples of metal–dioxygen adducts illustrated how these methods are helpful in elucidating structure–function relationships.

6.3.1 Reactive Species with Fe-oxo Motifs

6.3.1.1 Reactive Species from Monomeric Heme Iron–Dioxygen Complexes

Studies on P450 have dominated this field for the last 25 years and, while much has been discovered about the intermediates formed during turnover, controversies still exist. One of the central issues involves the hydroxylation step in the mechanism (Fig. 6.1). Mechanistic studies have led researchers to propose that the [Fe(IV)=O(ligand radical)] complex (compound **I**) homolytically cleaves the C–H bond on the substrate to give an Fe(IV)–OH species and an alkyl radical intermediate (Fig. 6.17 A). The next step, termed "oxygen rebound", has the alkyl radical attacking the hydroxo ligand to form the iron-bound product, which is subsequently displaced by water [42].

Evidence for the formation of a protein-derived Fe(IV)–OH complex was reported by Green who examined the active-site properties of chloroperoxidase (CPO), an enzyme having the same primary coordination sphere around its iron center as that in P450 (i.e. protoporphyrin IX and an axial thiolate ligand) [43]. CPO also forms a compound **I** intermediate which can convert to compound **II** and is believed to be a different Fe(IV) species. Examination of X-ray absorption spectroscopy (XAS) data for compound **II** suggests it is an Fe(IV)–OH complex: the X-ray absorption near-edge structure (XANES) region supports an Fe(IV) center and analysis of the X-ray absorption fine-structure (EXAFS) spectrum found an Fe–O bond length of 1.82 Å. This distance is too long for an Fe(IV)–O bond length, which is typically less than 1.7 Å. Additional support for this assignment comes from DFT results, which found a similar Fe–O bond length of 1.81 Å. The assignment of an Fe(IV)–OH center in compound **II** is intriguing but counter-intuitive because one would expect this unit to be strongly acidic and difficult to observe. An argument given to account for this unexpected result involved the axial thiolate ligand, which is a good π donor, and the overall anionic character of compound **II**. When taken together, these would effectively reduce the acidity of the coordinated hydroxide.

The important functional role of the axial thiolate ligand in P450 has been acknowledged for several years. Green recently proposed that the thiolate is directly involved in the formation of compound **I** [44]. A series of DFT studies indicated that the thiolate acquires radical character during compound **I** formation.

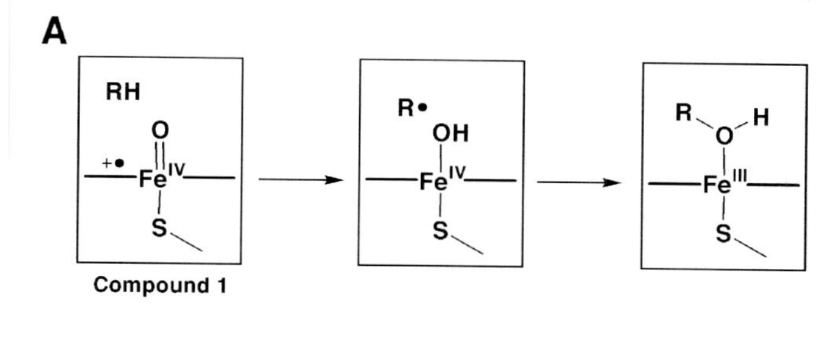

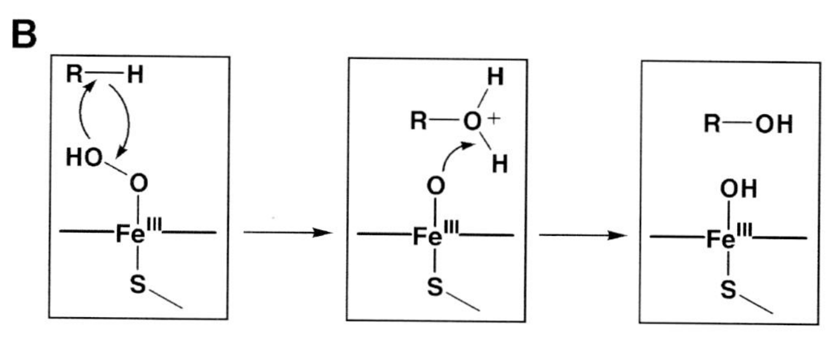

Fig. 6.17 Possible hydroxylation steps in P450: the H-atom abstraction/oxygen rebound pathway (A) and cationic pathway (B).

This opposes the normal depiction of this species, which has the ligand-radical centered on the porphyrin. Compound **I** has not been cleanly isolated and thus it is difficult to discern these possibilities. Green argues that the [Fe(IV)=O(thiolate radical)] provides a reasonable explanation for the unusual electronic coupling observed in CPO–compound **I**, yet this issue will not be reconciled until better experimental data are obtained.

Recently mechanistic studies have questioned the role of the [Fe(IV)=O(ligand radical)] intermediate and viability of the oxygen-rebound step in the hydroxylation process [42, 45]. The major issue is whether the Fe(III)–OOH species generated prior to compound **I** formation is directly involved in hydroxylation. The two possible scenarios are that (i) the Fe(III)–OOH complex acts as an electrophilic oxidant to insert into the C–H bond of the substrate in a similar manner as proposed for the [Fe(IV)=O(ligand radical)] intermediate and (ii) the Fe(III)–OOH species inserts an OH^+ ion into the substrate, producing protonated alcohols by a mechanism called cationic rearrangement (Fig. 6.17) [42, 46]. Mechanistic evidence does not clearly differentiate between the various possibilities, leaving the question open as to what is the competent oxidant in P450 [47].

6.3.1.2 **Reactive Species from Monomeric Nonheme Iron–Dioxygen Complexes**

The last 20 years has seen tremendous growth in our understanding of biological and synthetic nonheme iron systems that activate dioxygen. Central to many systems is that the cleavage of O_2 occurs after coordination to the metal center.

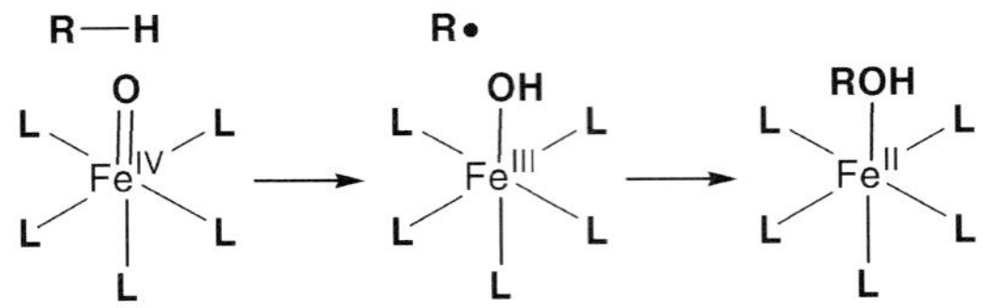

Fig. 6.18 Proposed steps for O-atom transfer in nonheme iron enzymes.

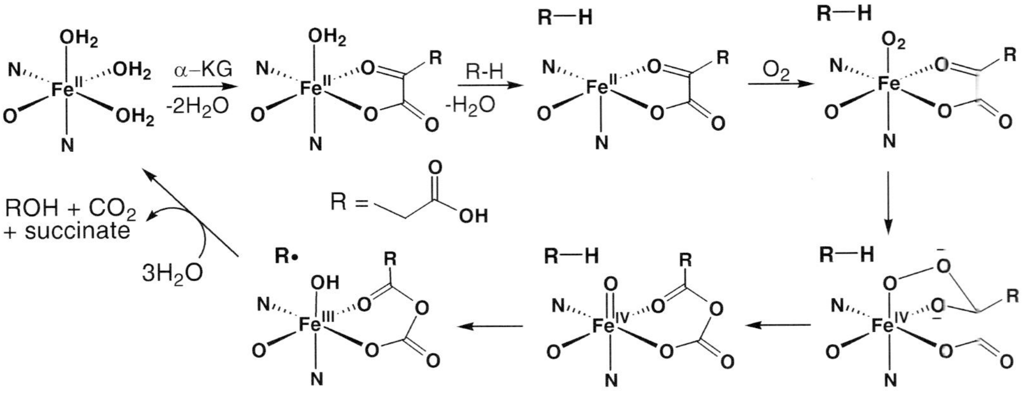

Fig. 6.19 A mechanism proposed for hydroxylation by α-KG-dependent nonheme iron enzymes.

The fate of the iron species is believed to mimic that found in iron porphyrin systems, in that high-valent iron-oxo complexes are postulated. However, there are important mechanistic differences between heme and nonheme iron systems. An Fe(IV)=O(ligand radical) species is formed after dioxygen activation in heme systems, which reacts with substrate producing an Fe(IV)–OH complex and an organic radical after homolytic C–H bond cleavage. Formation of the C–O bond of the product follows with the iron center returning to the ferric state. In nonheme systems the iron center is formally not as oxidized – the competent oxidant is an Fe(IV)=O species, which in some cases yields the corresponding Fe(III)–OH complex after C–H bond scission. A rebound step is also proposed, yet in nonheme systems this involves an Fe(III)–OH intermediate (Fig. 6.18). The factors leading to O_2 binding, activation and O-atom transfer are only beginning to become known; some of these new findings are highlighted here.

Only since 2003 has there been spectroscopic evidence to support that monomeric, nonheme iron oxygenases cleave O_2 and form Fe(IV)=O species. These findings came from studies on taurine dioxygenase (TauD), a member of the Fe(II)/α-ketoglutarate (α-KG)-dependent hydroxylase superfamily of enzymes [48]. A generic mechanism proposed for hydroxylation is shown in Fig. 6.19 and includes an Fe(IV)=O intermediate as the competent H-atom abstractor.

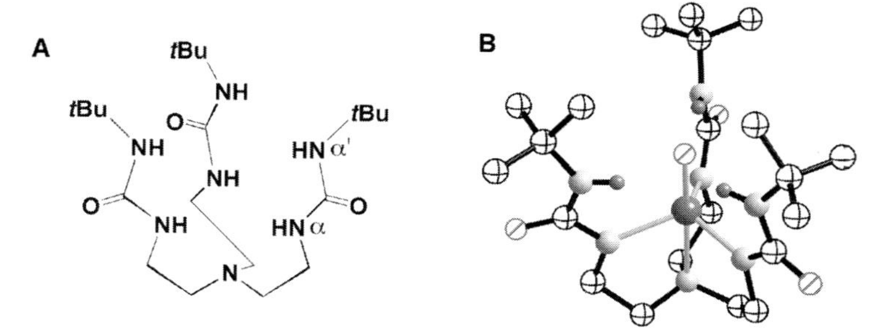

Fig. 6.20 Description of H_6buea (A) and molecular structure of $[Fe(III)H_3buea(O)]^{2-}$ (B).

Stopped-flow spectroscopy revealed the presence of an intermediate species with a $\lambda_{max}=318$ nm that forms around 20 ms after mixing dioxygen with taurine/α-KG/Fe(II)TauD at $T=5\,^{\circ}$C [49a]. Mössbauer and EPR studies suggest that this intermediate contains a Fe(IV) center with an $S=2$ ground state. Continuous-flow resonance Raman measurements at $T=-38\,^{\circ}$C further show that this species has an isotope-sensitive vibration at 821 cm^{-1} (^{16}O isotopomer), which shifts to 787 cm^{-1} in the ^{18}O isotopomer [50]. Taken together, the optical, magnetic and vibration data suggest the presence of an Fe(IV)=O species along the reaction pathway for TauD. Moreover, the decay rate for this Fe(IV)=O intermediate in the presence of taurine shows a substantial kinetic isotope effect of 37, indicating it is directly involved in C–H bond cleavage from the substrate [49b]. These important findings are the first set of consistent data to support the notion that nonheme iron enzymes can form an Fe(IV)=O species directly from O_2 activation.

As discussed earlier, efforts to synthetically model this process have been difficult, mostly because Fe(II) complexes and dioxygen tend to form oxo bridge dimers. Although nonheme iron complexes with terminal oxo ligands have been invoked in numerous processes, few have been prepared and characterized. Our group was the first to prepare and structurally characterize an iron complex with a terminal oxo ligand derived directly from O_2 activation [51]. To accomplish this task, we developed a urea-based tripodal compound, tris[(*N'*-*tert*-butylureayl)-*N*-ethylene]amine (H_6buea) that binds metal ions through the deprotonated α-nitrogen atoms to form trigonal pyramidal complexes (Fig. 6.20). Metal ion coordination also establishes a rigid H-bond donating cavity that supports up to three intramolecular H-bonds to an additional ligand (i.e. an oxo group) residing within the cavity. The inspiration for using H-bonds to stabilize a Fe–O unit came from metalloproteins, whose active sites are replete with H-bond donors/acceptors that are essential for oxidative function.

The Fe(III)–O complex, $[Fe(III)H_3buea(O)]^{2-}$, was obtained in 60% crystalline yield by treating $[Fe(II)H_2buea]^{2-}$ with dioxygen. Isotopic labeling studies with $^{18/16}O_2$ confirm that the oxo ligand arises directly from O_2 activation: $[Fe(III)H_3buea(^{16}O)]^{2-}$ has a ν(Fe–O) band at 671 cm^{-1} that shifts to 645 cm^{-1} in

$[Fe(III)H_3buea(^{18}O)]^{2-}$. The X-band EPR spectrum of $[Fe(III)H_3buea(O)]^{2-}$ is axial ($E/D=0$) with g values consistent with an $S=5/2$ ground state. Structural and vibrational data indicate the oxo ligand forms three intramolecular H-bonds with the α'-NH moiety of $[H_3buea]^{3-}$ (Fig. 6.21). We have also isolated $[Fe(III)H_3buea(OH)]^-$ directly from activation of dioxygen. This Fe(III)–OH complex is also high spin and has intramolecular H-bonds involving the coordinated hydroxo ligand.

We proposed that cleavage of dioxygen is accomplished through a 1,2-μ-peroxo Fe(III) dimer, producing an Fe(IV)=O intermediate, which has the capability to abstract an hydrogen atom. Bond energy evaluations found that the O–H bond dissociation energy (BDE) for $[Fe(III)H_3buea(OH)]^-$ is 115 kcal mol^{-1} [52], supporting the notion of an Fe(IV)=O complex having a strong thermodynamic driving force to abstract hydrogen atoms to form an Fe(III)–OH complex. We have also found that $[Fe(III)H_3buea(O)]^{2-}$ can cleave C–H bonds with relatively weak BDE (less than 80 kcal mol^{-1}), such as those in 9,10-dihydroanthracene and 1,4-cyclohexadiene.

The Fe(III)–O unit in $[Fe(III)H_3buea(O)]^{2-}$ has unusual spectroscopic, structural and theoretical properties compared to other metal-oxo complexes. For instance, $[Fe(III)H_3buea(O)]^{2-}$ is a high-spin complex, which is rare among metal-oxo systems; interestingly, the Fe(IV)=O complex in TauD is the only other high-spin metal-oxo system. In addition, DFT studies suggest that a single bond is the best description for the Fe(III)–O interaction. These differences from the norm are undoubtedly caused by the placement of the Fe(III)–O unit within a rigid H-bond donating cavity. The strong intramolecular H-bonds would affect the properties and bonding within the Fe(III)–O unit. The π-basicity of the oxo ligand will be weakened, leading to the observed longer bond distances and lower iron-oxo bond order. The potential regulatory effects of H-bonds on metal oxo complexes could have important ramifications in both synthetic and biological systems. This is particularly relevant for metalloproteins, which commonly have H-bond networks within their active sites. These structure–function relationships are just beginning to emerge and should lead to a more detailed understanding of oxidative processes.

6.3.1.3 **Reactive Intermediates: Nonheme Fe(IV)-oxo Species**

Authentic synthetic nonheme Fe(IV)=O complexes have eluded isolation until very recently. Wieghardt reported the formation of an Fe(IV)=O species, postulated to be [Fe(IV)(O)cyclam-acetate] [cyclam-acetate=1-(carboxymethyl-1,4,8,11-tetraazacyclo-tetradecane)] at $T=-80\,^{\circ}C$ prepared from O_3 [53]. This species only comprised around 20% of the iron present, and its characterization was limited to its optical and Mössbauer spectroscopies. Nam and Que used a derivative of this complex to isolate the first Fe(IV)=O species, which is prepared from O-atom transfer reagents rather than dioxygen [54]. [Fe(II)(tmc)(OTf)](OTf) (tmc=1,4,8,11-tetramethyl-1,4,8,11-tetraazadodecane; $OTf=CF_3SO_3^-$) reacts with H_2O_2 or iodosylbenzene (PhIO) at $T=-40\,^{\circ}C$ to produce a temperature-sensitive

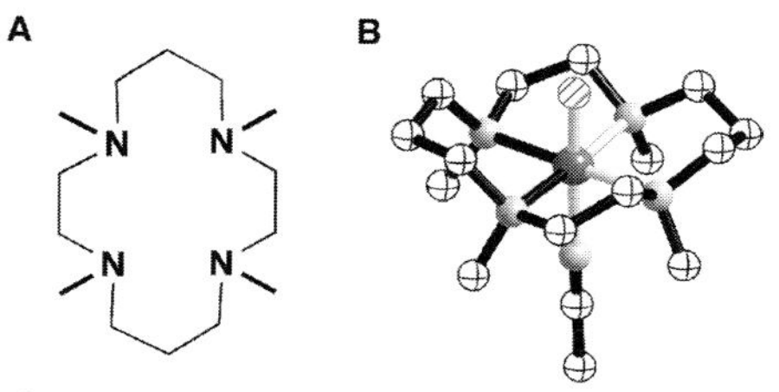

Fig. 6.21 Description of tmc ligand (A) and molecular structure of $[Fe(IV)(TMC)(O)(NCCH_3)]^{2+}$ (B).

green complex that is $[Fe(IV)(TMC)(O)(NCCH_3)]^{2+}$. The Fe(IV)=O vibration is at 834 cm^{-1}, shifting the predictable amount to 800 cm^{-1} in the ^{18}O isotopomer. Unlike the iron centers in TauD and $[Fe(III)H_3buea(O)]^{2-}$, $[Fe(IV)(TMC)(O)(NCCH_3)]^{2+}$ has an $S=1$ ground state that was determined using Mössbauer spectroscopy. The molecular structure was determined by XRD methods and confirmed the presence of a discreet Fe(IV)=O unit (Fig. 6.21). The Fe–O bond distance of 1.646(3) is significantly shorter than in $[Fe(III)H_3buea(O)]^{2-}$, suggesting appreciable multiple bond character. Theoretical studies support multiple bonding within the Fe(IV)=O unit with strong covalent oxo-to-iron π-donor and σ-donor bonds [55].

Que has since reported several derivatives of $[Fe(IV)(TMC)(O)(NCCH_3)]^{2+}$, which show different spectroscopic properties and reactivity [56]. For instance, changing the axial ligand from acetonitrile to a trifluoroacetate produces the less stable complex $[Fe(IV)(TMC)(O)(O_2CCF_3)]^{+}$, whose half-life is an order of magnitude lower than that of the parent complex at room temperature [57]. Moreover, $[Fe(IV)(TMC)(O)(O_2CCF_3)]^{+}$ reacts faster with external substrates, such as 9,10-dihydroanthracene; in contrast, with phosphines, the two Fe(IV)=O complexes react with similar rates. Que proposes that anionic ligands destabilize the Fe(IV)=O complexes by allowing better access to the more reactive $S=2$ spin state. This is an intriguing proposal, yet further supporting studies are needed.

The properties of Fe(IV)=O complexes have also been altered by changing the multidentate ligands [58]. Iron(IV)-oxo complexes of neutral pentadentate ligands, such as *N,N*-bis(2-pyridylmethyl)-bis(2-pyridyl)methylamine (N4Py) and *N*-benzyl-*N,N′,N′*-tris(2-pyridylmethyl)-1,2-daminoethane (Bn-tpen), show different stability and reactivity compared to those with the TMC ligand (Fig. 6.22). Both Fe(IV) complexes have enhanced stability at room temperature with half-lives of 60 and 6 h for $[Fe(IV)(N4Py)(O)]^{2+}$ and $[Fe(IV)(Bn\text{-}tpen)(O)]^{2+}$. Despite their stability at room temperature, these complexes react with hydrocarbons at room temperature. For instance, $[Fe(IV)(N4Py)(O)]^{2+}$ reacts with triphenylmethane to produce the Fe(II) precursor, $[Fe(II)(N4Py)(NCCH_3)]^{2+}$ and triphenylmethanol in 90% yield. Reactivity is also observed with other hydrocarbons, including cyclohexane, albeit at much slower rates and with lower yields (11%). A large kinetic isotope effect (KIE) of 30 was found for the reaction of $[Fe(IV)(N4Py)(O)]^{2+}$ with ethyl benzene and its d_{10} isotopomer; this value is similar to those found in MMOH and TauD, and suggests a hydrogen tunneling

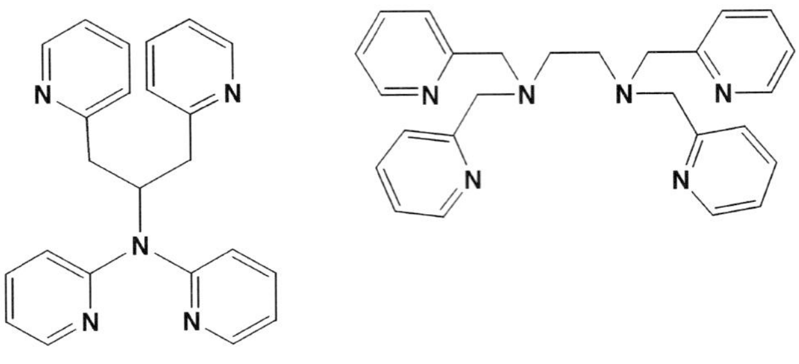

N4Py Bn-tpen

Fig. 6.22 Polypyridine ligands used in generating nonheme Fe(IV)=O complexes.

mechanism. In addition, a linear correlation was found between the BDE_{C-H} of the substrates and the second order rate constants for the reactions. Curiously, these reactions were done in acetonitrile, which has a BDE_{C-H} of 94 kcal mol^{-1} that is *less* than some of the substrates examined. Furthermore, acetonitrile would not fit on the above mentioned log k versus BDE_{C-H} plot, in fact it would be "a dramatically outlying point." Wolczanski has attributed this discrepancy to the difficulty in oxidizing acetonitrile and in a replot of Que's data as log k versus ionization energy (eV) found a linear correlation for all substrates, including acetonitrile [59]. This implies a mechanism involving proton-coupled electron transfer is operative, rather than simple H-atom transfer process. While further studies are needed to completely understand these mechanistic questions, they again illustrate the exciting avenue of research that has transpired as a result of these new iron-oxo complexes.

Oxidants other than dioxygen were used to generate the Fe(IV)=O complexes just described. A more recent report by Nam suggests that in certain solvent mixtures, dioxygen can convert an Fe(II) complex to an Fe(IV)=O intermediate [60]. An acetonitrile solution of [Fe(II)(tmc)(OTf)](OTf) does not react with dioxygen, whereas solutions prepared with 1 : 1 mixtures of acetonitrile and either ethanol, butyl ether or tetrahydrofuran (THF), react to form an intermediate that has properties similar to $[Fe(IV)(tmc)(O)(OTf)]^+$, and reacts catalytically with PPh_3, thioanisole and benzyl alcohol. While the performance of the catalyst is limited to less than 10 total turnovers, these results show the possibility of obtaining functional iron-based oxidation catalysts using dioxygen as the oxidant.

Many of the above examples utilize Fe(II) precursors to react with dioxygen – this is the usual method because higher valent iron complexes are normally unreactive with O_2. However, a new report by Collins demonstrated that certain Fe(III) complexes can bind and react with dioxygen. Collins has developed a family of highly anionic tetradentate ligands (denoted TAML) that contain four coplanar deprotonated amide moieties within a macrocyclic framework [61]. These ligands were developed in order to stabilize highly oxidized metal ions and through iterative engineering of the ligand design, Fe(III) systems have

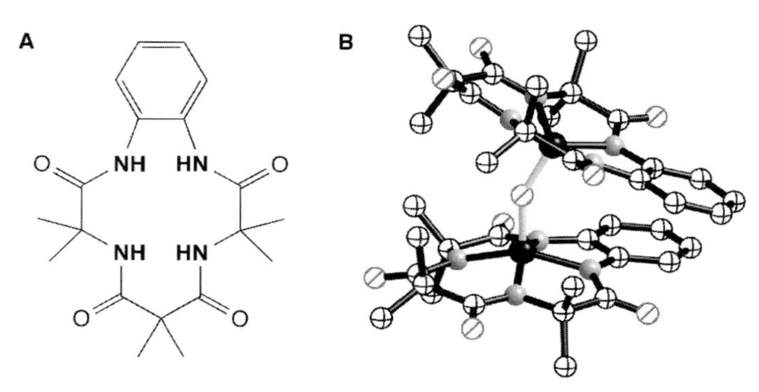

Fig. 6.23 Example of a TAML ligand (A) and the molecular structure of $\{[(TAML)Fe(IV)]_2\text{-}\mu\text{-}O\}^{2-}$ (B).

been prepared that are robust in the oxidation of various substrates [62]. In particular, the Fe(III) precursor, $[Fe(III)TAML(OH_2)]^-$, reacts with dioxygen under mild conditions to produce $\{[(TAML)Fe(IV)]_2\text{-}\mu\text{-}O\}^{2-}$, an unprecedented complex having an Fe(IV)–O–Fe(IV) core (Fig. 6.23). $\{[(TAML)Fe(IV)]_2\text{-}\mu\text{-}O\}^{2-}$ has Mössbauer parameters consistent with two $S=1$ Fe(IV) centers that are exchange coupled with a J value of greater than $100\ cm^{-1}$. Moreover, $\{[(TAML)Fe(IV)]_2\text{-}\mu\text{-}O\}^{2-}$ catalytically converts benzylic alcohols to their corresponding aldehydes at $T=100\,^\circ C$. Modest turnovers were observed under these conditions: 60 equiv. of benzaldehydes from benzyl alcohol in 1 h was the highest conversion obtained. In addition, $\{[(TAML)Fe(IV)]_2\text{-}\mu\text{-}O\}^{2-}$ was converted to its Fe(III) precursor as the reaction proceeded.

6.3.2 Reactive Iron and Copper Intermediates with $M(\mu\text{-}O)_2M$ Motifs

Copper and iron complexes with $M(\mu\text{-}O)_2M$ cores were not known before 1994, which is somewhat surprising considering its prevalence with other metal ions, such as manganese. These cores were discovered in copper and iron complexes at nearly the same time and now are touted for their roles in oxidative processes. Their discovery grew out of biological chemistry, but quickly attracted the attention of synthetic chemists. Through the combined efforts of researchers in both disciplines, there are now several examples of complexes with $M(\mu\text{-}O)_2M$ cores having functional oxidative properties.

6.3.2.1 Reactive Intermediates with $Cu(III)(\mu\text{-}O)_2Cu(III)$ Motifs

The isolation of complexes with $\mu\text{-}\eta^2\text{:}\eta^2$-peroxo dicopper(II) core sparked a renaissance in copper–dioxygen chemistry [63, 64]. Numerous researchers began investigating the formation of these complexes under various experimental conditions, using different ancillary ligands, solvents and temperatures. For instance, changing the ancillary ligands to hindered tacn^{iPr} (tacn^{iPr}=1,4,7-triisopropyl-1,4,7-triazacyclononane) [65] or triimidazolylphosphines [66] produces dimeric copper systems that reversibly bind dioxygen at low temperature. Spectroscopic studies indicate that μ-η^2:η^2-peroxo dicopper(II) complexes are formed at T=–80 °C; reversion back to the starting Cu(I) complexes occurs by rapidly warming the solutions under vacuum or N_2 purge. During the course of one of these investigations, Tolman uncovered an unprecedented finding related to the μ-η^2:η^2-peroxo dicopper(II) core [67]. Initial studies on [(Cu(II)tacn^{iPr})$_2\mu$-η^2:η^2-peroxo]$^{2+}$ indicated that solution of the complex at T= –80 °C had some unusual properties, leading to the idea that more than one species may be present. The identity of another dioxygen-derived species became clearer when the chemistry of [Cu(I)tacn^{Bn}]$^+$ was examined – the tacn ligand in this complex has benzyl groups instead of isopropyl substituents. The oxidized complex of [Cu(I)tacn^{Bn}]$^+$ had properties that were indeed different from those found for complexes with known μ-η^2:η^2-peroxo dicopper(II) cores. This new complex was found to be [(Cu(III)tacn^{Bn})$_2$(μ-oxo)$_2$]$^{2+}$, a species with a Cu(III)(μ-O)$_2$Cu(III) core.

The optical spectra of Cu(III)(μ-O)$_2$Cu(III) complexes are unique with intense charge transfer bands at around 300 and 400 nm. Resonance Raman studies on [(Cu(III)tacn^{Bn})$_2$(μ-oxo)$_2$]$^{2+}$ found isotopically sensitive bands at 602 and 612 cm^{-1} that shifted to a single peak at 583 cm^{-1} when the complex was prepared with $^{18}O_2$. This isotopic shift is too small to be from a peroxo ligand but is consistent for a Cu–O vibration. The Cu–O vibration in the ^{16}O isotopomer is split into a Fermi doublet, giving rise to two observable peaks. Conclusive structural evidence was obtained from low-temperature XRD measurements that revealed Cu···Cu and Cu–O distances of 2.794(2) and 1.81 Å (Fig. 6.24) – these distances are significantly shorter than those found in complexes with Cu(II)(μ-OH)$_2$Cu(II) or Cu(II)(μ-η^2:η^2-OO)Cu(II) cores.

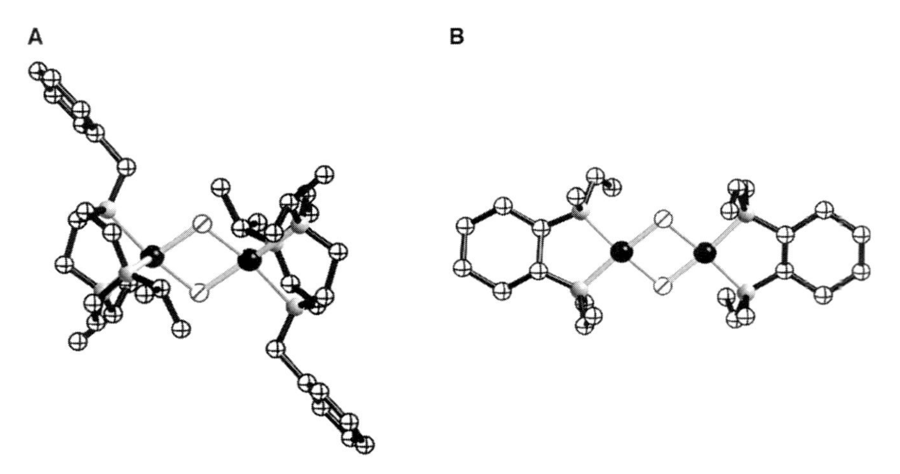

Fig. 6.24 Molecular structures of [(Cu(III)tacn^{iPr,Bn})$_2$(μ-O)$_2$]$^{2+}$ (A) and [(Cu(III)^{Me,Me}cd)$_2$(μ-O)$_2$]$^{2+}$ (B).

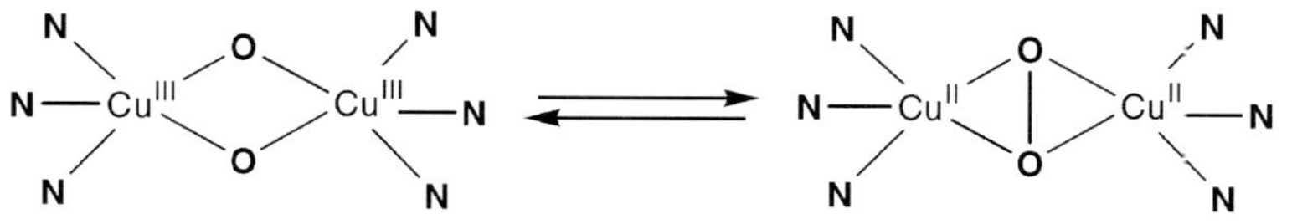

Fig. 6.25 Equilibrium between complexes with Cu(III)(μ-O)$_2$Cu(III) and Cu(II)(μ-O$_2$)Cu(II) cores.

Further exploration into the properties of these Cu(II)(μ-η^2:η^2-OO)Cu(II) and Cu(III)(μ-O)$_2$Cu(III) units led to the remarkable discovery that these cores interconvert (Fig. 6.25) [67, 68]. The factors that govern the interconversion are numerous and their relationships are complicated. For example, [Cu(I)tacn^{iPr}-(NCCH$_3$)]X (X=PF_6^-, ClO_4^-) in CH_2Cl_2 gave the peroxo bridge product upon treatment with O_2; however, the same reactions performed in THF produced [(Cu(III)tacn^{iPr})$_2$(μ-oxo)$_2$]$^{2+}$. Interconversion between the two species is accomplished by interchanging the solvents. For instance, a greater than 10-fold dilution of CH_2Cl_2 solution of peroxo complex with THF at T=–80 °C caused conversion to the Cu(III)(μ-O)$_2$Cu(III) species; similar experiments, but diluting the THF solution of the Cu(III)(μ-O)$_2$Cu(III) complex with CH_2Cl_2, afforded the peroxo complex.

The importance of the ancillary multidentate ligand became apparent with the different reactivities of the Cu(I) complexes with tacn^{iPr} and tacn^{Bn}. The degree of substitution at the α-C-atom to the N-donors has an important effect on the formation of the oxidized metal species: the peroxo bridge complex is favored when the tacn ligand has substituents whose α-C-atoms are tertiary, while the Cu(III)(μ-O)$_2$Cu(III) complex is produced if one of the α-C-atoms is less substituted. Steric interactions are the obvious explanation for this effect – bulkier ligands prevent the close approach of the two copper centers that is needed to form the Cu(III)(μ-O)$_2$Cu(III) core. A similar trend was observed in copper complexes with bidentate ligands, whereby the more sterically constrained systems favored formation of complexes with Cu(II)(μ-η^2:η^2-OO)Cu(II) cores.

These findings are the first demonstration of reversible formation and scission of an O–O bond within a dinuclear complex. Furthermore, the interconversion of the core structure again highlights the importance of ligand design in regulating metal-based reactivity. The ability to tune the secondary coordination sphere allowed researchers to obtain definitive evidence for Cu(III)(μ-O)$_2$Cu(III) and Cu(II)(μ-η^2:η^2-OO)Cu(II) cores. It should be noted that other ligands have now been reported that support the inter-conversion between these cores structures [64].

6.3.2.2 Reactive Intermediates with Cu$_3$(μ-O)$_2$ Motifs

Complexes with this motif are rare and were first reported for a synthetic system by Stack in 1996 [69, 70]. The Cu(I) complex [Cu(I)(^{Me,Me}cd)(NCCH$_3$)]$^+$ [^{Me,Me}cd, *N*-permethylated (1*R*,2*R*)-cyclohexanediamine] was treated with dioxygen at $T = -80\,^\circ C$ in CH_2Cl_2 to produce the thermally sensitive complex, $[(^{Me,Me}cd)_3Cu_3(\mu_3\text{-}O)_2]^{3+}$. Low-temperature XRD (Fig. 6.26) and gas uptake measurements established the formation of the $[Cu_3(\mu_3\text{-}O)_2]$ core from the cleavage of dioxygen. The copper ions have square planar coordination geometries with N_2O_2 donors. One copper center is unique by having significantly shorter Cu–O bonds of 1.83 Å, consistent with a Cu(III) oxidation state. The other Cu–O distances of 1.98 Å are comparable to those found for Cu(II) centers. The presence of shorter Cu–O bonds were also observed in the Cu K-edge EXAFS for solid and solution samples. A logical mechanism involves the assembly of the core via O_2 cleavage promoted by the $2e^-$ oxidation of one copper complex and the $1e^-$ oxidation of two other copper species. $[(^{Me,Me}cd)_3Cu_3(\mu_3\text{-}O)_2]^{3+}$ is paramagnetic with an $S = 1$ ground state, and has strong absorbance bands between 290 and 620 nm.

The formation of this $[Cu_3(\mu_3\text{-}O)_2]$ has been explored at low temperatures and is dependent on the ancillary ligands. In general, the least sterically encumbering ligands produce Cu(I) complexes that assemble $[Cu_3(\mu_3\text{-}O)_2]$ cores – a trend that was also observed for formation of Cu(III)(μ-O)$_2$Cu(III) complexes. The formation of $[Cu_3(\mu_3\text{-}O)_2]$ cores appears to be more sensitive to steric constraints: changing to an ethyl group on each nitrogen donor of the bidentate ligand produces a complex with a Cu(II)(μ-η^2:η^2-OO)Cu(II) unit. Other factors also affect core formation, such as the initial concentrations of the Cu(I) complexes and O_2. For instance, assembly of the $[Cu_3(\mu_3\text{-}O)_2]$ core is favored when the concentration of $[Cu(I)(^{Me,Me}cd)(NCCH_3)]^+$ is much greater than that of O_2. However, the reverse situation promotes the formation of a species with a Cu(II)(μ-η^2:η^2-OO)Cu(II) unit.

6.3.2.3 Reactive Intermediates with Fe(μ-O)$_2$Fe Motifs

Complexes with Fe(III)(μ-O)Fe(III) cores are well known in coordination chemistry, but there are few systems with Fe(μ-O)$_2$Fe units. The first suggestions that these motifs exist were made in conjunction with the metalloenzymes MMOH and RNR, both of which contain diiron active sites and activate dioxygen in the

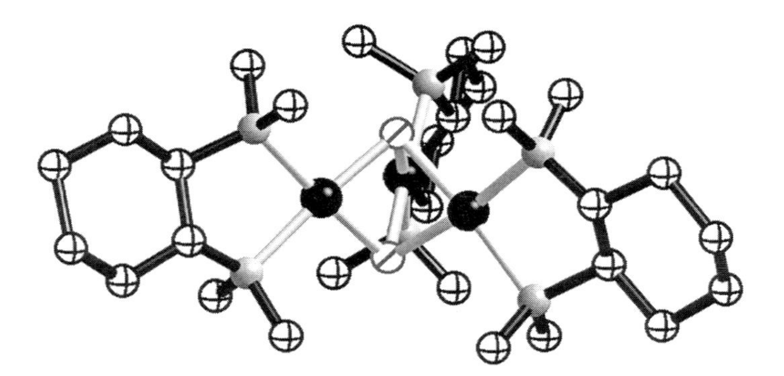

Fig. 6.26 Molecular structure of the $[Cu_3(\mu_3\text{-}O)_2]$ complex.

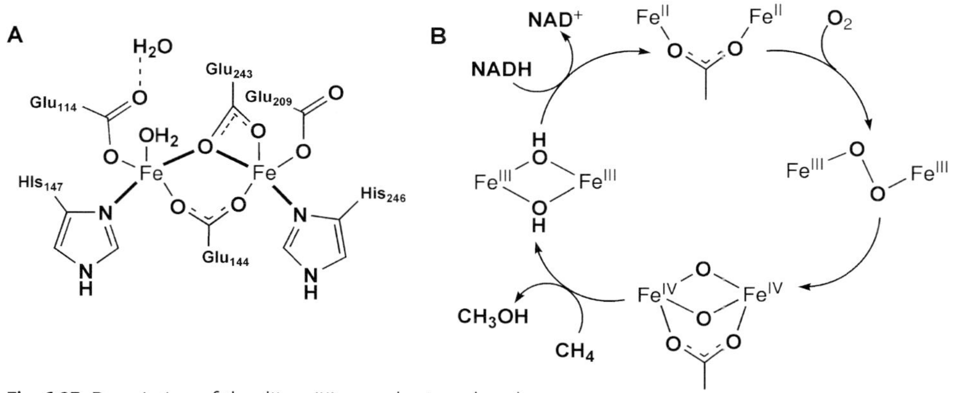

Fig. 6.27 Description of the diiron(II) complex in reduced MMOH (A) and a proposed mechanism for MMOH (B).

course of their catalytic function. In particular, MMOH has received a large amount of attention because of the industrial importance of transforming methane to methanol using dioxygen as the terminal oxidant. Figure 6.27 shows the primary coordination environment around the iron centers in the reduced form of MMOH determined from XRD measurements [5 b]. The primary coordination sphere of the iron centers in MMOH is appreciably more electron-rich than that found in hemerythrin – these structural differences correlate to their differing functions: MMOH binds and activates O_2, whereas hemerythrin reversibly binds dioxygen. The generally accepted catalytic cycle for MMOH is also shown in Fig. 6.27 [5]. Dioxygen binds to the diferrous form producing the peroxo intermediate MMOH-P, which converts to a species referred to as compound MMOH-Q, which is believed to have a Fe(IV)(μ-O)$_2$Fe(IV) core. Evidence for this structure comes from detailed spectroscopic EXAFS measurements that found short Fe–O bonds of less than 1.8 Å and Fe···Fe distances of less than 3 Å. MMOH-Q is proposed to be the competent oxidant, reacting with methane to yield methanol and the reduced core, Fe(III)(μ-OH)$_2$Fe(III). Further reduction of the core with NADH completes the catalytic cycle. A similar mechanism can be rationalized for RNR, but with a mixed valent Fe(III)(μ-O)$_2$Fe(IV) core as the bis-oxo-bridged intermediate.

The above proposals for MMOH and RNR have stimulated synthetic efforts to prepare complexes with Fe(μ-O)$_2$Fe cores. The current structurally characterized examples are derived from oxidants other than dioxygen, yet they serve as important species in understanding the structural requirements needed to support such motifs. The first systems with Fe(μ-O)$_2$Fe cores were prepared from Fe(III)(μ-OH)(μ-O)Fe(III) complexes to produce Fe(III)(μ-O)$_2$Fe(III) compounds [71]. Fig. 6.28 illustrates the molecular structure of [(6-Me$_3$tpa)$_2$Fe(III)$_2$(μ-O)$_2$]$^{2+}$ (6-Me$_3$tpa = tris[(6-methyl)pyridylmethyl]amine), which has the expected short Fe···Fe distance of 2.716 Å; the core is unsymmetrical because of dissimilar Fe–O bond lengths [1.844(3) and 1.916(4) Å].

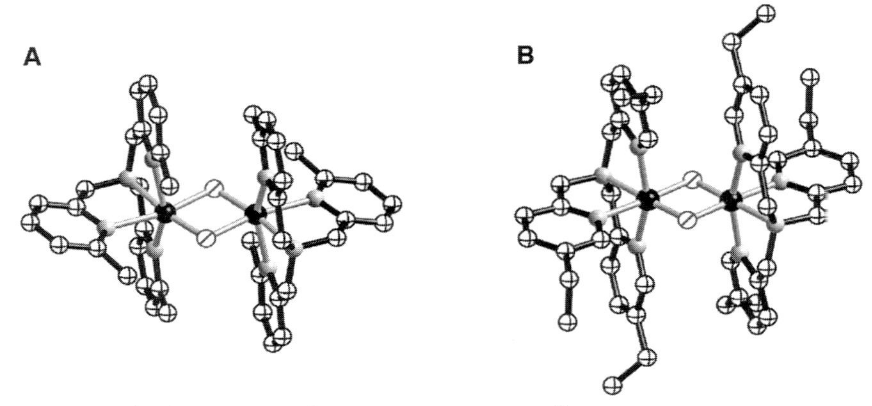

Fig. 6.28 Molecular structures of $[(6\text{-Me}_3\text{tpa})_2\text{Fe(III)}_2(\mu\text{-O})_2]^{2+}$ (A) and $[(5\text{-Et}_3\text{tpa})_2\text{Fe(III)Fe(IV)}(\mu\text{-O})_2]^{3+}$ (B).

Related mixed valent $\text{Fe(III)}(\mu\text{-O})_2\text{Fe(IV)}$ complexes can be synthesized by treating $[\text{L}_2\text{Fe(III)}_2(\mu\text{-O})(\text{OH})(\text{H}_2\text{O})]^{3+}$ complexes with H_2O_2 at $T = -40\,^\circ\text{C}$. XRD studies on $[(5\text{-Et}_3\text{tpa})_2\text{Fe(III)Fe(IV)}(\mu\text{-O})_2]^{3+}$ (5-Et_3tpa = tris[(5-ethyl)-pyridylmethyl]-amine) established the $\text{Fe}(\mu\text{-O})_2\text{Fe}$ core, which has an Fe···Fe distance of 2.681(1) Å, and Fe–O bond lengths of 1.805(3) and 1.860(3) Å (Fig. 6.29) [72]. Note that the differences in the Fe–O distances also cause an unsymmetrical $\text{Fe}(\mu\text{-O})_2\text{Fe}$ core structure similar to what is found in $[(6\text{-Me}_3\text{tpa})_2\text{Fe(III)}_2(\mu\text{-O})_2]^{2+}$.

Antiferromagnetic coupling is observed between the two high-spin ferric centers in the $\text{Fe(III)}(\mu\text{-O})_2\text{Fe(III)}$ complexes. For example, $[(6\text{-Me}_3\text{tpa})_2\text{Fe(III)}_2(\mu\text{-O})_2]^{2+}$ has a $J = +54(8)\ \text{cm}^{-1}$ ($\mathbf{H} = J\mathbf{S}_1\mathbf{S}_2$), which is considerably lower than the 180–250 cm^{-1} values reported for high-spin diiron(III) complexes with a single oxo bridge [73]. This difference has been attributed to the $\text{Fe(III)}(\mu\text{-O})_2\text{Fe(III)}$ complexes having smaller Fe–O–Fe angles and longer Fe–O bonds.

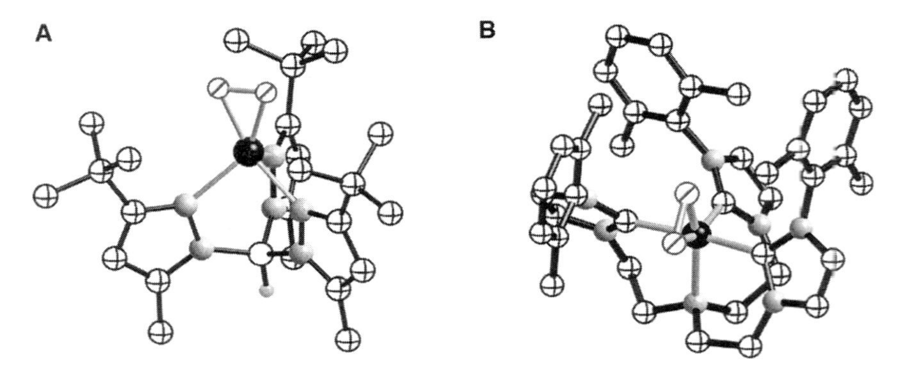

Fig. 6.29 Molecular structures of $[\text{CoTp}^{t\text{Bu,Me}}(\eta^2\text{-O}_2)]$ (A) and Meyer's $[\text{Co}(\eta^2\text{-O}_2)]$ (B).

The electronic properties of the Fe(III)(μ-O)$_2$Fe(IV) complexes are more varied because of different spin states of the iron centers. One group of complexes has high-spin iron centers that are antiferromagnetically coupled to produce an $S=1/2$ ground state. The X-band EPR spectrum has an isotopic signal at $g=2$, which is consistent with one unpaired electron. In addition, the Mössbauer properties are suggestive of a valence-trapped system because two quadrupole doublets are observed – one of which is characteristic of a high-spin Fe(III) species and the other for a high-spin Fe(IV) center. A second group of Fe(III)(μ-O)$_2$Fe(IV) complexes have X-band EPR spectra with features at $g=4.3$, 3.7 and 2; these values are typical for an $S=3/2$ ground state. This spin state can arise from double exchange coupling of a low-spin ($S=1/2$) Fe(III) center with an $S=1$ Fe(IV) site. The Mössbauer spectra for these contain a single quadrupole doublet, indicating that these Fe(III)(μ-O)$_2$Fe(IV) cores are valence delocalized.

6.4 Cobalt–Dioxygen Complexes

Dioxygen binding to cobalt complexes is well known and has been reviewed extensively [74]. The most widely studied systems are cobalt porphyrin and salen complexes [salen, *N,N'*-bis(salicylidene)-1,2-ethylene-diaminato(2–)], with the latter dating back to the time of Werner [75]. The reactions of these systems most often involve Co(II) complexes and the formation of stable cobalt-(η^1-superoxo) complexes or species with Co(III)(μ-1,2-peroxo)Co(III) cores. Recent work has focused on more reactive cobalt–dioxygen species and will be highlighted here.

6.4.1 Cobalt-η^2-Dioxygen Complexes

Theopold reported the first 1:1 η^2-O_2 adduct of cobalt [76]. This complex was prepared from dioxygen using Co(I)Tp^{tBu,Me}(N_2), which has *tert*-butyl groups at the 3-position of the pyrazole rings. While vibrational spectroscopic studies confirm the side-on coordination, the ν(O–O) was observed at 961 cm^{-1}, an energy that is between those normally reported for superoxo-containing (1200–1070 cm^{-1}) and peroxo-containing (930–740 cm^{-1}) complexes. An XRD study performed at room temperature revealed a monomeric [CoTp^{tBu,Me}(η^2-O_2)] complex with an O–O bond length of 1.262(8) Å (Fig. 6.29). This distance suggested a superoxo moiety with the cobalt center formally in the 2+ oxidation state. Variable temperature magnetic susceptibility measurements showed a temperature-independent magnetic moment of 3.88 μ_{BM} that was interpreted to arise from the antiferromagnetic coupling between the Co(II) center ($S=3/2$) and the superoxo ligand ($S=1/2$).

Although all the necessary experiments were performed correctly, the assignment of the η^2-superoxo ligand in [CoTp^{tBu,Me}(η^2-O_2)] was nonetheless controversial because many thought side-on dioxygen adducts of cobalt should yield peroxo complexes [36]. Moreover, inconsistencies became apparent as more data accumulated on other monomeric metal–dioxygen species. For instance, the C–O bond lengths in [CoTp^{tBu,Me}(η^2-O_2)] and [Cu(II)Tp^{tBu,iPr}(η^2-O_2)] are nearly identical, but their ν(O–O) differ by around 150 cm^{-1}. In addition, a DFT study predicted that the O–O bond length in [CoTp^{tBu,Me}(η^2-O_2)] should be longer by almost 0.1 Å. This same study also found a linear correlation between the ν(O–O) and O–O bond lengths for M(η^2-O_2) adducts – [CoTp^{tBu,Me}(η^2-O_2)] was one of the few compounds that did not fit this correlation.

These issues prompted a redetermination of the X-ray structure for [CoTp^{tBu,Me}(η^2-O_2)] at $T = -123\,°C$; this was 100 °C cooler than the temperature used to collect the data for the initial structure [36]. The high- and low-temperature structures were identical except for the O–O bond length, which increased significantly to 1.355(3) Å when measured at $T = -123\,°C$. The authors plausibly attribute the differences to librational motion of the coordinated O_2 at high temperature that resulted in an underestimation of the O–O bond length. Librational motion is caused by a η^2-O_2 unit rotating/oscillating around an intramolecular axis defined by the cobalt center and the midpoint of the O–O bond. This motion often has a negligible effect on molecular structures; however it appears to have a significant effect in [CoTp^{tBu,Me}(η^2-O_2)] because the coordinated dioxygen is relatively unconstrained. Therefore only the low-temperature data is credible, giving an O–O distance that closely agrees with that predicted by theory (1.380 Å).

Meyer recently reported a Co(η^2-O_2) system using a carbene-based tripodal ligand (TIMEN = tris[2-(3-arylimidazol-2-ylidene)ethyl]amine) developed in his laboratory [77]. He also prepared a Co(I) complex and found that it cleanly reacted with dioxygen to produce the 1 : 1 cobalt dioxygen adduct. The dioxygen is coordinated as a peroxo ligand based on the O–O stretching frequency that is found at 890 cm^{-1} and shifts 50 cm^{-1} in the ^{18}O-labeled compound. Further support for peroxo coordination comes from XRD studies that show a six-coordinate cobalt complex with a side-on O_2 ligand having an O–O distance of 1.429(3) Å (Fig. 6.29). The O–O bond length lies within the range typically found for other cobalt-peroxo complexes and correlates with the O–O stretching frequency. Similar to the Co(III)-peroxo complex of Tp ligands, this complex couples the electronic-donating ability of the tripod with a relatively rigid cavity structure surrounding the dioxygen-binding site that prevents unwanted bimolecular reaction pathways.

6.4.2
Dinuclear Cobalt-μ-superoxo Complexes

Theopold also examined the effects of changing the secondary coordination sphere on the dioxygen reactivity of the Co(I)Tp complexes [78]. The straightforward change to Tp^{iPr,Me}, which substitutes isopropyl groups for the *tert*-butyl group on the pyrazole rings, led to Co(I) complexes that were sterically less constrained. [CoTp^{iPr,Me}(O_2)] could be generated and is stable for hours at temperatures less

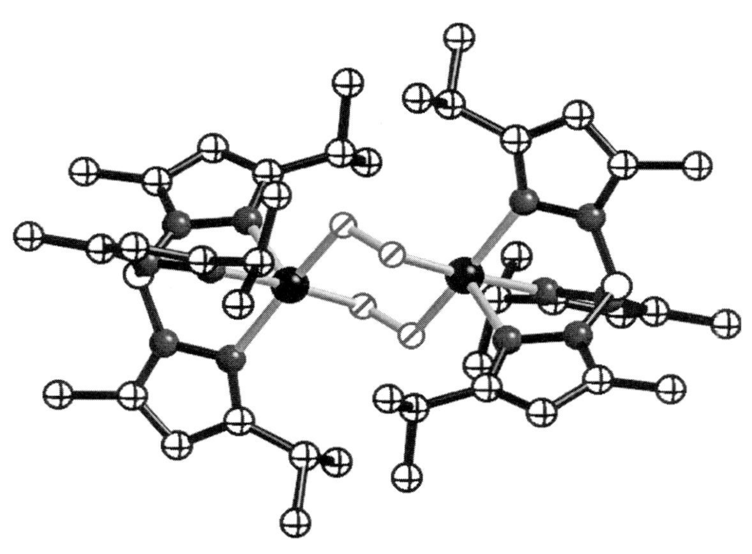

Fig. 6.30 Molecular structure of $[CoTp^{iPr,Me}(\mu\text{-}1,2\text{-}O_2)_2CoTp^{iPr,Me}]$.

than $T = -10\,^\circ C$. However, at room temperature $[CoTp^{iPr,Me}(O_2)]$ stability was greatly reduced compared to $[CoTp^{tBu,Me}(\eta^2\text{-}O_2)]$ and it converts to a transient intermediate proposed as $[CoTp^{iPr,Me}(\mu\text{-}O_2)CoTp^{iPr,Me}]$. While the structure of this complex has not been determined definitively, it is reasonable to assume that it resembles the analogous copper complexes. Cooling $[CoTp^{iPr,Me}(\mu\text{-}C_2)\text{-}CoTp^{iPr,Me}]$ below $T = -30\,^\circ C$ produces $[CoTp^{iPr,Me}(\mu\text{-}1,2\text{-}O_2)_2CoTp^{iPr,Me}]$, a unique species containing two bridging dioxygen ligands (Fig. 6.30). Based on the O–O distances of 1.354(5) Å, the authors assigned the bridging groups as superoxo ligands, yet were unable to obtain vibrational data needed to corroborate this assignment. Support for this assignment was provided by Suzuki and coworkers who reported structural and vibrational data on $[Ni_2(\mu\text{-}1,2\text{-}O_2)_2]$ – a related complex [79].

At room temperature, crystals of $[CoTp^{iPr,Me}(\mu\text{-}1,2\text{-}O_2)_2CoTp^{iPr,Me}]$ rapidly produce a gas, possibly dioxygen, and a pink solid that has similar properties to those of $[CoTp^{iPr,Me}(\mu\text{-}O_2)CoTp^{iPr,Me}]$, the mono-bridge dioxygen system. An interesting reaction pathway was proposed, whereby $[CoTp^{iPr,Me}(O_2)]$ dimerizes to $[CoTp^{iPr,Me}(\mu\text{-}1,2\text{-}O_2)_2CoTp^{iPr,Me}]$ which readily loses one dioxygen ligand to produce $[CoTp^{iPr,Me}(O_2)CoTp^{iPr,Me}]$ [80]. The mono-bridge complex is prone to C–H bond cleavage involving the methine centers of the appended isopropyl groups. Detailed kinetic and isotopic studies suggested that $[CoTp^{iPr,Me}(O_2)CoTp^{iPr,Me}]$ is capable of abstracting hydrogen atoms, producing Co(OH) units and alkyl radicals that disproportionate to isopropyl and isopropenyl substituents. The measured kinetic isotope effect (k_H/k_D) was 22, an unusually high value, suggestive of a tunneling contribution to the cleavage event. It is worth mentioning that other metal-based systems having similarly large k_H/k_D have been reported, including metalloenzymes involved in hydrogen-transfer processes.

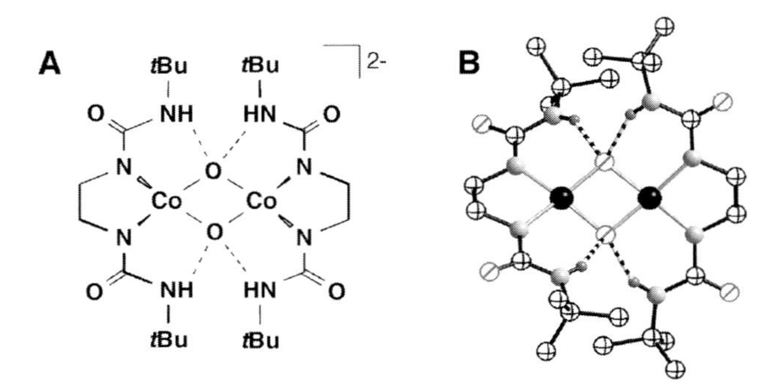

Fig. 6.31 Description of $[Co(III)H_2bade(O)]_2^{2-}$ (A) and its molecular structure (B).

The results with $[CoTp^{tBu,Me}(O_2)]$ and $[CoTp^{iPr,Me}(O_2)]$ further illustrate important relationships between regulating the primary and secondary coordination spheres in the binding and activation of dioxygen. $[CoTp^{tBu,Me}(O_2)]$ has the correct electronic properties and coordination geometry to bind dioxygen; its constrained secondary sphere prevents interactions with other metal complexes, contributing to its observed stability at room temperature. In contrast, $[CoTp^{iPr,Me}(O_2)]$ is relatively unstable at room temperature, even though it has nearly the same electronic properties as $[CoTp^{tBu,Me}(O_2)]$. The instability of $[CoTp^{iPr,Me}(O_2)]$ is a result of its ability to dimerize to μ-1,2-peroxo species that lead to C–H bond cleavage. The less congested secondary coordination sphere allows for dimer formation, opening reactivity paths for ligand modification. Thus, accessibility to $Co(O_2)$ is a key parameter: too little access leads to stable cobalt-peroxo complexes, while too much causes dimer formation and ligand modification. The next challenge in this system is to further manage the secondary sphere to produce a synthetic system that reacts intermolecularly on organic substrates rather than with a coordinated ligand.

Increasing the access to a metal center, while still controlling its secondary coordination sphere, prompted us to develop the bidentate ligand with H-bond donors ($[H_2bade]^{2-}$, Fig. 6.31) [81]. We found that high-spin $[Co(II)(H_2bade)_2]^{2-}$ reacts with excess dioxygen to initially produce a species that is formulated as a monomeric cobalt–dioxygen species based on its spectroscopic properties. This intermediate species is not stable at room temperature and changes to the $[Co(III)(\mu\text{-}O)_2Co(III)]$ complex, $[Co(III)H_2bade(O)]_2^{2-}$. Isotopic labeling studies with $^{18}O_2$ confirm the oxo ligands are derived from dioxygen. The molecular structure corroborates the $[Co(III)(\mu\text{-}O)_2Co(III)]$ core structure, with each Co(III) being four-coordinate with distorted square planar geometry. The oxo ligands form intramolecular H-bonds to the NH moieties of the urea groups of $[H_2bade]^{2-}$. This H-bond network, in conjunction with the strong anionic ligand field, appears to be sufficient to stabilize the $[Co(III)(\mu\text{-}O)_2Co(III)]$ at room temperature. The observed thermal stability of $[Co(III)H_2bade(O)]_2^{2-}$ contrasts with that of many other $[M(\mu\text{-}O)_2M]$ that are only stable at temperatures below

$T = 0\,^{\circ}C$ [63, 82]. In fact, $[Co(III)H_2bade(O)]_2^{2-}$ seems to be *too stable* because there is no observed cleavage of C–H bonds nor O-atom transfer. The $[H_2bade]^{2-}$ ligand, with its H-bonds, may ultimately reduce accessibility to the $[Co(III)(\mu\text{-}O)_2Co(III)]$ core, shutting down any potential reactivity.

6.5 Manganese–Dioxygen Complexes

Manganese complexes are well known as stoichiometric reagents and catalysts in the oxidation of organic compounds. There are, however, surprisingly few examples of structurally characterized manganese–dioxygen adducts. Wieghardt has reported the only system isolated directly from dioxygen: a dinuclear Mn(IV) complex with a μ-1,2-peroxo bridge [83]. The molecular structure of the complex revealed a triple bridge core (Fig. 6.32), with the two additional bridging oxo groups, rendering a short Mn···Mn distance of 2.531(7) Å. The complex is stable at ambient temperature for hours but rapidly releases O_2 under anaerobic conditions. The proposed mechanism for O_2 release involves $2e^-$ intramolecular electron transfer to afford dioxygen and a Mn(III) dimer that disproportionates to a stable observable $[Mn(IV)_2(\mu\text{-}O)_3]$ species and an unobserved Mn(II) dimer. This latter complex could dissociate to produce a Mn(II)-aquo complex that can be rapidly oxidized by O_2 to a Mn(III)-hydroxo complex.

Mn(II) tetraphenylporphyrin [Mn(II)TPP] binds dioxygen at low temperatures, forming the 1:1 adduct [84]. EPR spectra have signals indicative of a $S = 3/2$ ground state, which has been interpreted as arising from a Mn(IV)-peroxo complex. Moreover, the presence of six-line hyperfine features supports the assignment of a monomeric complex. Matrix isolation IR experiments at 15 K found vibrational bands at 983, 958 and 933 cm^{-1} when Mn(II)TPP was co-condensed with a 1:2:1 mixture of $^{16}O_2$, $^{16}O^{18}O$ and $^{18}O_2$. The bands at 983 and 933 cm^{-1} correspond to the ν(O–O) of the $^{16}O_2$ and $^{18}O_2$ isotopomers, while the middle band is from the mixed isotopomer. The energies of these vibrations are diagnostic for peroxo ligation and the existence of three bands supports side-on binding.

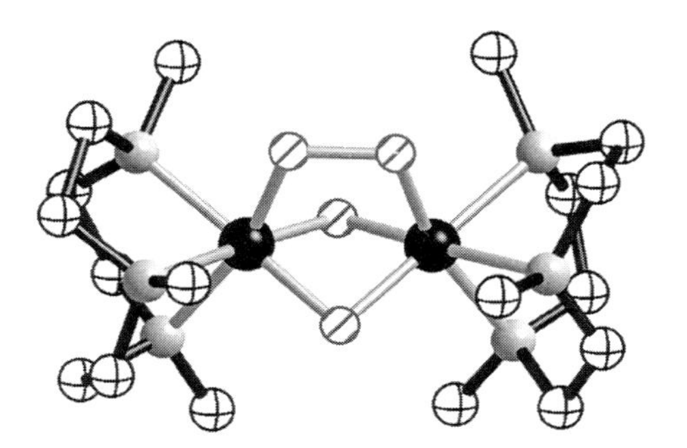

Fig. 6.32 Molecular structure of $[(Mn(IV)Me_3\text{-tacn})_2(1,2\text{-}\mu\text{-}O_2)(O)_2]^{2+}$.

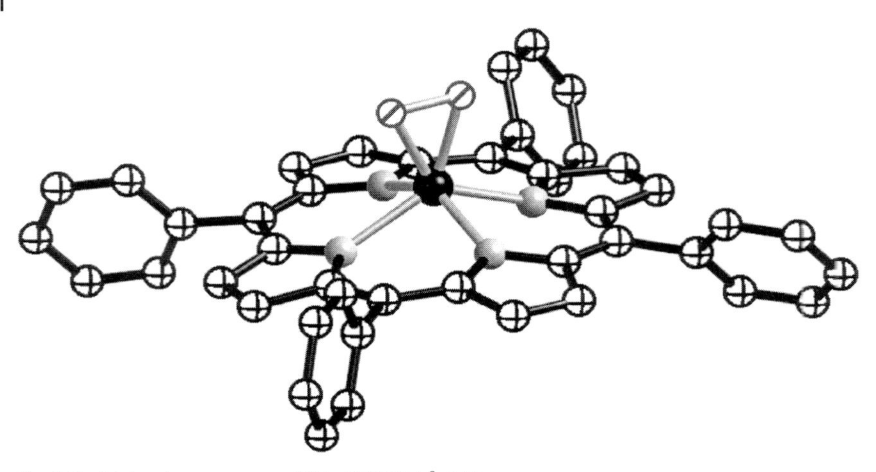

Fig. 6.33 Molecular structure of [Mn(III)TPP(η^2-O_2)].

Valentine prepared and structurally characterized a related system, [Mn(III)TPP(O_2)]$^-$, by treating [Mn(III)TPP(Cl)] with 2 equiv. of superoxide ions [85]. The first equivalent of superoxide reduces the [Mn(III)TPP(Cl)] to [Mn(II)TPP], which reacts with the second equivalent to afford [Mn(III)TPP(η^2-O_2)]$^-$. The structure of [Mn(III)TPP(η^2-O_2)]$^-$ contains a side-on O_2 whose O–O distance is 1.421(5) Å (Fig. 6.33). The peroxo binds in such a way that it eclipses two of the pyrrole nitrogen atoms of the TPP ligand. This complex is noteworthy because it was the first structural example of a complex consisting of a discrete manganese-peroxo unit.

Nonheme Mn(III)-peroxo complexes have been prepared at low temperatures by Kitajima, using Tp^{iPr} and an 3,5-diisopropylpyrazole (3,5-^{i}PrpzH) as the ancillary ligands [86]. Two isomeric forms were observed: a brown species was found at around $T = -20\,°C$, while a blue complex was present at a temperature approaching $-78\,°C$. The complexes were evaluated via XRD methods and both are six-coordinate with a side-on coordinated peroxo ligand (Fig. 6.34), similar to that found for the Mn(III)TPP(O_2) complex. The brown and blue isomers differ by the H-bonding network surrounding the Mn(III)(η^2-O_2) unit. The molecular structure of the blue complex appears to contain two intramolecular H-bonds between the coordinate peroxo and the NH group of 3,5-^{i}PrpzH; N···O distances of 2.99(2) and 2.82(2) Å are observed, which are indicative of H-bond formation. In contrast, the brown complex has N···O distances that are greater than 3.0 Å, too long to be considered H-bonds. Additional support for these assignments was provided by IR experiments that showed the blue species having a broadened $\nu(N\text{–}H_{pz})$ vibration, which is shifted to lower energy compared to that found for the brown species. The isomers have identical ν(O–O) vibrations at 841 cm^{-1}, which is unexpected and still unexplained. Finally, warming the complex does cause O-atom transfer to occur to phosphines and sulfoxides, albeit in yields less than 20%.

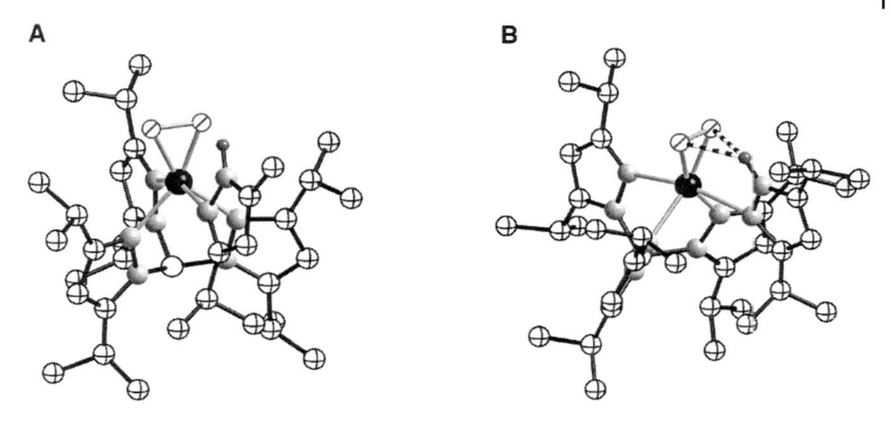

Fig. 6.34 Molecular structures of the two isomeric forms of $[MnTp^{iPr}(\eta^2\text{-}O_2)]^+$: brown isomer (A) and blue isomer (B). Intramolecular H-bonds are indicated in structure B.

The H-bond network surrounding the $Mn(III)(\eta^2\text{-}O_2)$ in the blue complex is unique for a synthetic system, and is reminiscent of the active sites in hemoglobins and myoglobins. These results show that it is possible to prepare H-bonded metal-peroxo systems. Still unclear is whether these noncovalent interactions aid in the stabilization of the metal-peroxo unit. In this regard, it is surprising that the lower temperature species contains H-bonds, which may suggest that the overall rigidity of the complex is not sufficient to promote H-bonds at elevated temperatures. How these structural parameters affect H-bond formation and factor into the stability of metal–dioxygen adducts remains to be elucidated.

6.6 Nickel–Dioxygen Complexes and Their Reactive Intermediates

The interaction of dioxygen with nickel complexes has been investigated for well over 30 years, with most reports focused on the reactivity of subsequent $Ni–O_2$ adducts. For instance, Ni(II) azamacrocycles activated dioxygen to afford products that are consistent with C–H bond cleavage [87]. However, little is known about the possible intermediates that can accomplish this chemistry. Recent work from Riordan has greatly increased our understanding in this area, through his investigations on the dioxygen reactivity of Ni(I) complexes with phenyltris[(alkylthio)methyl]borate ($PhTt^R$) ligands that contain three sulfide donors [88]. Control of dioxygen reactivity was achieved through adjustments in the size of the R groups on the $PhTt^R$ ligands. A 1:1 $\eta^2\text{-}O_2$ complex was obtained when $[PhTt^{ad}Ni(I)CO]$ (R = adamantyl) was treated with dioxygen at $T = -70\,^\circ C$ (Fig. 6.35) [89]. This complex has a rhombic EPR spectrum at 4 K that is reminiscent of a species with a Ni(III) center, yet XAS and DFT results favor a

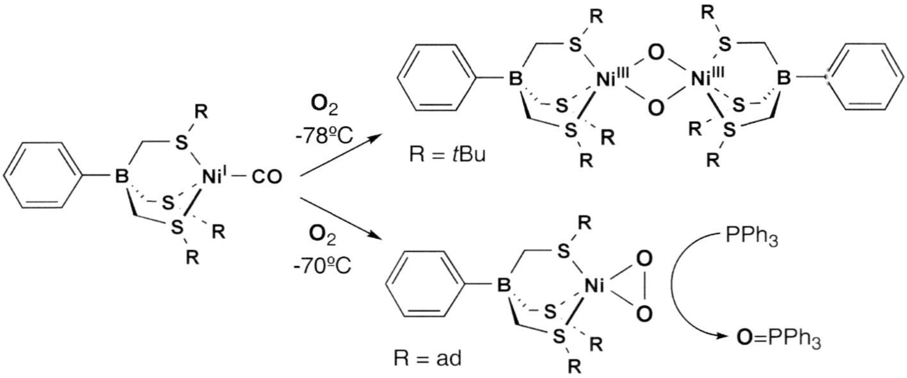

Fig. 6.35 Summary of the dioxygen chemistry for [PhTt^{R}Ni(I)R].

Ni(II)-superoxo description, with antiferromagnetic coupling between the five-coordinate Ni(II) center and the side-on superoxo radical. [PhTt^{ad}Ni(II)η^2-O_2] also reacts with some external substrates including PPh_3 and NO to form O=PPh_3 and NO_3^-, respectively.

In contrast, the reaction of [PhTt^{tBu}Ni(I)(CO)] with dioxygen at $T = -78\,°C$ produced [PhTt^{tBu}Ni(III)]$_2$(μ-O)$_2$, a complex with a Ni(III)(μ-O)$_2$Ni(III) rhomb (Fig. 6.35) [90]. The difference in reactivity to [PhTt^{ad}Ni(I)CO] is attributed to the smaller *tert*-butyl groups on the PhTt^{tBu} ligand allowing sufficient access to the nickel sites for dimer formation. The close approach of the nickel centers causes the four-electron reduction of dioxygen producing the bridging oxo ligands. Isotopic labeling studies confirmed that dioxygen was the source of the oxo ligands in [PhTt^{tBu}Ni(III)]$_2$(μ-O)$_2$. The importance of the ligands' steric properties was further illustrated in the step-wise formation of a Ni(III)(μ-O)$_2$Ni(III) complex in which each nickel center has a different borate ligand. This was accomplished by treating the Ni(II)-superoxo complex, [PhTt^{ad}Ni(II)η^2-O_2], that was generated at low temperature, with [PhTt^{tBu}Ni(I)(CO)]. The resulting Ni(III)(μ-O)$_2$Ni(III) complex was obtained in 85% yield and contained a different PhTt^R ligand on each nickel center. Note that a similar strategy was used to prepare unsymmetrical Cu(III)(μ-O)$_2$Cu(III) complexes [39b] and systems with Cu(III)(μ-O)$_2$Ni(III) rhombs [91].

Riordan has also observed a complex containing a Ni(II)(1,2-μ-peroxo)Ni(II) core that was generated from dioxygen and $[Ni(I)(tmc)]^+$ in CH_3CN at $T = -45\,°C$ [92]. Although it is relatively unstable at this temperature, with a half-life of 4 min, magnetic and resonance Raman studies support the assignment of a bridge peroxo ligand to two Ni(II) centers. The complex is EPR silent at 77 and 4 K, precluding a 1Ni:1O_2 adduct which has either an $S = 1/2$ or 3/2 ground state. The resonance Raman spectrum contains two oxygen-sensitive vibrations at 479 and 778 cm^{-1} that are assigned to the ν(Ni–O) and ν(O–O), respectively. Comparison of the vibrational data to those for copper and cobalt systems suggested a *trans*-1,2-μ-peroxo conformation. Warming of the peroxo complex afforded $[Ni(II)tmc(OH)]^+$, which has a terminal hydroxo ligand. The oxygen atom of the coordinated hydroxide is derived directly from O_2 and solvent labeling studies point to CH_3CN being the source of the hydrogen atom.

Some of the above species have also been observed using oxidants other than dioxygen; for instance, complexes with Ni(III)(μ-O)$_2$Ni(III) cores have been prepared with H_2O_2 [93]. Nevertheless, the observations of species with Ni(II)-superoxo, Ni(III)-1,2-μ-peroxo-Ni(III) and Ni(III)(μ-O)$_2$Ni(III) motifs derived *directly from dioxygen* are important in establishing these species as viable intermediates in O_2 activation. These findings extended the scope of metal-mediated cleavage of O_2 to include defined nickel complexes whose properties can be tuned to support a variety of dioxygen adducts.

6.7 Summary

This chapter has outlined some of the recent advances in metal-mediated dioxygen binding and activation. The intention was to examine the metal complexes generated upon the binding of dioxygen and then explore the resultant species produced after O–O bond cleavage. Included in the discussions were important systems that utilized oxidants other than dioxygen. Particular attention was given to structural factors that can control reactivity and these were highlighted throughout the chapter. The emphasis was on 3d metal chemistry with examples from biology introduced when appropriate.

Our knowledge of the factors that contribute to the binding and activation of O_2 has grown immensely within the last 15 years. We understand the molecular basis for the reversible binding of dioxygen to the three major respiratory proteins found in biology. Furthermore, many of the features have been duplicated in synthetic systems, which can mimic the functional aspects of proteins. Advances in structural, spectroscopic and theoretical measurements produce important new findings into how protein structure affects the activation of dioxygen. The structure–function relationship emerging from these studies can act as guidelines for the development of synthetic oxidation catalysts.

For all of the exciting advances, we are still far away from routinely producing systems that fully utilize dioxygen in oxidative transformations under a variety of experimental conditions. There are several metal-based systems that transfer O-atoms to substrates, yet few are able to do it directly from dioxygen. Moreover, the efficiencies and selectivities of these systems pale in comparison to metalloproteins. Many of the systems presented in this chapter are only observable and functional at low temperature, which makes them impractical for use. It is now recognized that factors enumerated in Section 6.1.2 need to be incorporated into synthetic complexes in order to produce systems that function in a comparable manner to enzymes. The challenges are numerous, yet given the great successes recently achieved, the future appears to be quite promising.

Acknowledgments

Acknowledgment is made to the NIH for financial support.

References

1 Davis, D.D., Kemp, D.R. In *Kirk-Othmer Encyclopedia of Chemical Technology*, 4th edn, vol. 1, Kroschwitz, J.I. (Ed.), Wiley, New York, **1991**, pp. 466.

2 (a) Ortiz de Montellano, P.R. *Acc Chem Res* **1987**, *20*, 289; (b) Sono, M., Roach, M.P., Coulter, E.D., Dawson, J.H. *Chem Rev* **1996**, *96*, 2841; (c) Ortiz de Montellano, P.R. (Ed.), *Cytochrome P450: Structure, Mechanism, and Biochemistry*, 3rd edn, Plenum Press, New York, **2004**.

3 Dawson, J.H. *Science* **1988**, *240*, 433.

4 (a) Martinis, S.A., Atkins, W.M., Stayton, P.S., Sligar, S.G. *J Am Chem Soc* **1989**, *111*, 9252; (b) Gerber, N.C., Sligar, S.G. *J Am Chem Soc* **1992**, *114*, 8742; (c) Schlichting, I., Berendzen, J., Chu, K., Stock, A.M., Maves, S.A., Benson, D.E., Sweet, R.M. Ringe, D., Pestko, G.A., Sligar, S.G. *Science* **2000**, *287*, 1615.

5 (a) Wallar, B.J., Lipscomb, J.D. *Chem Rev* **1996**, *96*, 2625; (b) Merkx, M., Kopp, D.A., Sazinsky, M.H., Blazyk, J.L., Müller, J., Lippard, S.J. *Angew Chem Int Ed* **2001**, *40*, 2782 and references therein.

6 Brazeau, B.J., Wallar, B.J., Lipscomb, J.D. *Biochem Biophys Res Commun* **2003**, *312*, 143.

7 (a) Holmes, M.A., Stenkamp, R.E. *J Mol Biol* **1991**, *220*, 723; (b) Fülöp, V., Phizackerley, R.P., Soltis, S.M., Clifton, I.J., Wakatuski, S., Erman, J., Hajdu, J., Edwards, S.L. *Structure* **1994**, *2*, 201; (c) Mukai, M., Nagano, S., Tanaka, M., Ishimori, K., Morishima, I., Ogura, T., Watanabe, Y., Kitagawa, T. *J Am Chem Soc* **1997**, *119*, 1758 and references therein; (d) Dunietz, B.D., Beachy, M.D., Cao, Y., Whittington, D.A., Lippard, S.J., Friesner, R.A. *J Am Chem Soc* **2000**, *122*, 2828; (e) Gherman, B.F., Dunietz, B.D., Whittington, D.A., Lippard, S.J., Friesner, R.A. *J Am Chem Soc* **2001**, *123*, 3836; (f) Du Bois, J., Mizoguchi, T.J., Lippard, S.J. *Coord Chem Rev* **2000**, *200–202*, 443; (g) Berglund, G.I., Carlsson, G.H., Smith, A.T., Szöke, H., Henriksen, A., Hajdu, J. *Nature* **2002**, *417*, 463.

8 (a) Perutz, M.F., Fermi, G., Luisi, B., Shaanan, B., Liddington, R.C. *Acc Chem Res* **1987**, *20*, 309; (b) Springer, B.A., Sligar, S.G., Olsen, J.S., Philips, G.N., Jr. *Chem Rev* **1994**, *94*, 699; (c) Yang, J., Kloek, A.P., Goldberg, D.E., Mathews, F.S. *Proc Natl Acad Sci USA* **1995**, *92*, 4224.

9 Kurtz, D. In *Comprehensive Coordination Chemistry II*, vol. 8, McCleverty, J.A., Meyers, T.J. (Eds.), Elsevier, Oxford, **2003**, p. 229.

10 (a) Shiemke, A.K., Loehr, T.M., Sanders-Loehr, J.S. *J Am Chem Soc* **1986**, *108*, 2437; (b) Stenkamp, R.E. *Chem Rev* **1994**, *94*, 715.

11 Magnus, K.A., Ton-That, H., Carpenter, J.E. *Chem Rev* **1994**, *94*, 727.

12 (a) Chin, D.H., La Mar, G.N., Balch, A.L. *J Am Chem Soc* **1980**, *102*, 5945; (b) Chin, D.H., Balch, A.L., La Mar, G.N. *J Am Chem Soc* **1980**, *102*, 1446; (c) Balch, A.L., Chan, Y.W., Cheng, R.J., La Mar, G.N., Latos-Grazynski, L., Renner, M.W. *J Am Chem Soc* **1984**, *106*, 7779.

13 (a) Jameson, G.B., Molinaro, F.S., Ibers, J.A., Collman, J.P., Brauman, J.I., Rose, E., Suslick, K.S. *J Am Chem Soc* **1978**, *100*, 6769; (b) Collman, J.P., Brauman, J.I., Doxsee, K.M., Halbert, T.R., Bunnenberg, E., Linder, R.E., LaMar, G.N., Del Gaudio, J., Lang, G., Spartalian, K. *J Am Chem Soc* **1980**, *102*, 4182; (c) Collman, J.P., Brauman, J.I., Iverson, B.L., Sessler, J.L., Morris, R.M., Gibson, Q.H. *J Am Chem Soc* **1983**, *105*, 3052; (d) Collman, J.P., Zhang, Z., Wong, E., Brauman, J.I. *J Am Chem Soc* **1994**, *116*,

6245; (d) Momenteau, M., Reed, C.A. *Chem Rev* **1994**, *94*, 659.

14 Mizoguchi, T., Lippard, S.J. *J Am Chem Soc* **1998**, *120*, 11022.

15 (a) Que, L., Jr., Dong, Y. *Acc Chem Res* **1996**, *29*, 190; (b) Solomon, E.I., Bunold, T.C., Davis, M.I., Kemsley, J.N., Lee, S.-K., Lehnert, N., Neese, F., Skulan, A. J. Yang, Y.-S. Zhou, J. *Chem Rev* **2000**, *100*, 235.

16 (a) Menage, S., Brennan, B.A., Juarez-Garcia, C., Münck, E., Que, L., Jr. *J Am Chem Soc* **1990**, *112*, 6423; (b) Dong, Y., Menage, S., Brenna, B.A., Elgren, T.E., Jan, H.G., Pearce, L.L., Que, L., Jr. *J Am Chem Soc* **1993**, *115*, 6423.

17 Kitajima, N., Tamura, N., Amagai, H., Fukui, H., Moro-oka, Y., Mizutani, Y., Kitagawa, T., Mathur, R., Heerwegh, K., Reed, C.A., Randall, C.R., Que, L., Jr., Tatsumi, K. *J Am Chem Soc* **1994**, *116*, 6423.

18 Dong, Y., Yan, S., Young, V.G., Jr., Que, L., Jr. *Angew Chem Int Ed* **1996**, *108*, 618.

19 Kim, K., Lippard, S.J. *J Am Chem Soc* **1996**, *118*, 4914.

20 Ookkubo, T., Sugimoto, H., Nagayama, T., Masuda, H., Sato, T., Tanaka, Y., Maeda, Y., Okawa, H., Hayashi, Y., Uehara, A., Suzuki, M. *J Am Chem Soc* **1996**, *118*, 701.

21 (a) Herron, N., Cameron, J.H., Neer, G.L., Busch, D.H. *J Am Chem Soc* **1983**, *105*, 298; (b) Busch, D.H., Alcock, N.W. *Chem Rev* **1994**, *94*, 585.

22 (a) Lubben, M., Meetsma, A., Wilkinson, E.C., Feringa, B., Que, L., Jr. *Angew Chem Int Ed* **1995**, *34*, 1512; (b) Ho, R.Y.N., Roefles, G., Feringa, B.L., Que, L., Jr. *J Am Chem Soc* **1999**, *121*, 264; (c) Simaan, A.J., Banse, F., Girerd, J.-J., Wieghardt, K., Bill, E. *Inorg Chem* **2001**, *40*, 6538; (d) Mialane, P., Novorojkine, A., Pratviel, G., Azema, L., Slany, M., Godde, F., Simann, A., Banse, F., Kargar-Grisel, T., Bouchoux, G., Sainton, J., Horner, O., Guilhem, J., Tchertanov, L., Meunier, B., Girerd, J.-J. *Inorg Chem* **1999**, *38*, 1085; (e) Roelfes, G., Vrajmasu, V., Chen, K., Ho, R.Y.N., Rohed, J.-U., Zondervan, C., Crois, R.M., Schudde, E.P., Lutz, M., Spek, A.L., Hage, R., Feringa, B.L., Münck, E., Que, L., Jr. *Inorg Chem* **2003**, *42*, 2639.

23 (a) Stubbe, J., Kozarich, J.W., Wu, W., Vanderwall, D.E. *Acc Chem Res* **1996**, *29*, 322; (b) Burger, R.M. *Chem Rev* **1998**, 1153.

24 Neese, F., Zaleski, J.M., Loeb-Zaleski, K., Solomon, E.I. *J Am Chem Soc* **2000**, *122*, 11703.

25 (a) Karlin, K.D., Gultneh, Y. *Adv Inorg Chem* **1987**, *35*, 219; (b) Sorrell, T.N. *Tetrahedron* **1989**, *40*, 3.

26 Gaykema, W.P., Hol, W.G.J., Vereijken, J.M., Soeter, N.M., Bak, H.J., Beintema, J.J. *Nature* **1984**, *309*, 23.

27 Karlin, K.D., Haka, M.S., Cruse, R.W., Gultneh, Y. *J Am Chem Soc* **1985**, *107*, 5828.

28 Jacobson, R.R., Tyeklar, Z., Farooq, A., Karlin, K.D., Liu, S., Zubieta, J. *J Am Chem Soc* **1988**, *110*, 3690.

29 Kitajima, N., Fujisawa, K., Moro-oka, Y. *J Am Chem Soc* **1989**, *111*, 8975.

30 A copper Tp^{Me} complex with an absorbance spectrum similar to that of oxyhemocyanin has been reported. It was prepared with H_2O_2 and proposed to contain a peroxo bridge between two copper centers: Kitajima, N., Koda, T., Hashimoto, S., Kitagawa, T., Moro-oko, Y. *Chem Commun* **1988**, 151.

31 Kitajima, N., Fujisawa, K., Fujimoto, C., Moro-oka, Y., Hashimoto, S., Kitagawa, T., Toriumi, K., Tatsumi, K. Nakamura, A. *J Am Chem Soc* **1992**, *114*, 1277.

32 Magnus, K.A., Ton-That, H., Carpenter, J.E. *Chem Rev* **1994**, *94*, 727.

33 Thompson, J.S. *J Am Chem Soc* **1984**, *106*, 4057.

34 Fujisawa, K., Tanaka, M., Moro-oka, Y., Kitajima, N. *J Am Chem Soc* **1994**, *116*, 12079.

35 Chen, P., Root, D.E., Campochiaro, C., Fujisawa, K., Solomon, E.I. *J Am Chem Soc* **2003**, *125*, 466.

36 Cramer, C.J., Tolman, W.B., Theopold, K.H., Rheingold, A.L. *Proc Natl Acad Sci USA* **2003**, *100*, 3635.

37 Harata, M., Jitsukawa, J., Masuda, H., Einage, H. *J Am Chem Soc* **1994**, *116*, 10817.

38 Berreau, L.M., Mahapatra, S., Halfen, J.A., Young, V.G., Jr., Tolman, W.B. *Inorg Chem* **1996**, *35*, 6339.

39 (a) Spencer, D.J.E., Aboelella, N.W., Reynolds, A.M., Holland, P.L., Tolman,

W. B. *J Am Chem Soc* **2002**, *124*, 2108; (b) Aboelella, N. W., Lewis, E. A. Reynolds, A. M., Brennessel, W. W., Cramer, C. J., Tolman, W. B. *J Am Chem Soc* **2002**, *124*, 10660; (c) Aboelella, N. W., Kryatov, S. V., Gherman, B. F., Brennessel, W. W., Young, V. G., Jr., Sarangi, R., Rybak-Akimova, E. V., Hodgson, K. O., Hedman, B., Solomon, E. I., Cramer, C. J., Tolman, W. B. *J Am Chem Soc* **2004**, *126*, 16896.

40 Reyonds, A. M., Gherman, B. F., Cramer, C. J., Tolman, W. B. *Inorg Chem* **2005**, *44*, 6989.

41 Wada, A., Harata, M., Hasegawa, K., Jitsukawa, K., Masuda, H., Mukai, M., Kitagawa, T., Einaga, H. *Angew Chem Int Ed* **1998**, *37*, 798.

42 Limberg, C. *Angew Chem Int Ed* **2003**, *42*, 5932 and references therein.

43 Green, M. T., Dawson, J. H., Gray, H. B. *Science* **2004**, *304*, 1653.

44 Green, M. T. *J Am Chem Soc* **1999**, *121*, 7939.

45 (a) Schröder, D., Shaik, S., Schwarz, H. *Acc Chem Res* **2000**, *33*, 139; (b) Shaik, S., de Visser, S. P., Ogliaro, F., Schwarz, H., Schröder, D. *Curr Opin Chem Biol* **2002**, *6*, 556.

46 Newcomb, M., Shen, R., Choi, S.-Y., Toy, P. H., Hollenberg, P. F., Vaz, A. D. N., Coon, M. J., *J Am Chem Soc* **2000**, *122*, 2677.

47 (a) Newcomb, M., Shen, R., Lu, Y., Coon, M. J., Hollenberg, P. F., Kopp, D. A., Lippard, S. J. *J Am Chem Soc* **2002**, *124*, 6879; (b) Auclair, K., Hu, Z., Little, D. M., Ortiz de Montellano, P. R., Groves, J. T. *J Am Chem Soc* **2002**, *124*, 6020.

48 Hausinger, R. P. *Crit Rev Biochem Mol Biol* **2004**, *39*, 21.

49 (a) Price, J. C., Tirupati, B., Bollinger, J. M., Krebs, C. *Biochemistry* **2003**, *42*, 7497; (b) Price, J. C., Barr, E. W., Glass, T. E., Krebs, C., Bollinger, J. M. *J Am Chem Soc* **2003**, *125*, 13008; (c) Riggs-Gelasco, P. J., Price, J. C., Guer, R. B., Brehm, J. H., Barr, E. W., Bollinger, J. M. Krebs, C. *J Am Chem Soc* **2004**, *126*, 8108.

50 Proshlyakov, D. A., Henshaw, T. F. Monterosso, G. R., Ryle, M. J., Hausinger, R. P. *J Am Chem Soc* **2005**, *126*, 1022.

51 (a) MacBeth, C. E., Golombek, A. P., Young, V. G., Jr., Yang, C., Kuczera, K., Hendrich, M. P., Borovik, A. S. *Science* **2000**, *289*, 938; (b) Gupta, R., Borovik, A. S. *J Am Chem Soc* **2003**, *125*, 13234; (c) MacBeth, C. E., Gupta, R., Mitchell-Koch, K. R., Young, V. G., Jr., Lushington, G. H., Thompson, W. H., Hendrich, M. P., Borovik, A. S. *J Am Chem Soc* **2004**, *126*, 2556; (d) Borovik, A. S. *Acc Chem Res* **2005**, *38*, 54.

52 Gupta, R., Borovik, A. S. *J Am Chem Soc* **2003**, *125*, 13234.

53 Grapperhaus, C. A., Mienert, B., Bill, E., Weyhermüller, T., Wieghardt, K. *Inorg Chem* **2000**, *39*, 5306.

54 Rohde, J.-U., In, J.-H., Lim, M. H., Brennessel, W. W., Bukowski, M. R., Stubna, A., Münck, E., Wam, W., Que, L., Jr. *Science* **2003**, *299*, 1037.

55 (a) Decker, A., Rohde, J.-U., Que, L., Jr., Solomon, E. I. *J Am Chem Soc* **2004**, *126*, 5379; (b) Decker, A., Solomon, E. I. *Angew Chem Int Ed* **2005**, *44*, 2252.

56 (a) Lim, M. H., Rohde, J.-U., Stubna, A., Bukowski, M. R., Costas, M., Ho, R. Y. N., Münck, E., Nam, W., Que, L., Jr. *Proc Natl Acad Sci USA* **2003**, *100*, 3665; (b) Sastri, C. V., Park, M. J., Ohta, T., Jackson, T. A., Stubna, A., Seo, M. S., Lee, J., Kim, J., Kitagawa, T. Münck, E., Que, L., Jr., Nam, W. *J Am Chem Soc* **2005**, *127*, 12494; (c) Bukowski, M. R., Koehntop, K. D., Stubna, A., Bominaar, E. L., Halfen, J. A., Münck, E., Nam, W., Que, L., Jr. *Science* **2005**, *310*, 1000.

57 Rohde, J.-U., Que, L., Jr. *Angew Chem Int Ed* **2005**, *44*, 2.

58 Kaizer, J., Klinker, E. J., Oh, N. Y., Rohde, J.-W., Song, W. J., Stubna, A., Kim, J., Münck, E., Nam, W., Que, L., Jr. *J Am Chem Soc* **2004**, *126*, 472.

59 Sydora, O. L., Goldsmith, J. I., Vaid, T. P., Miller, A. E., Wolczanski, P. T., Abruna, H. D. *Polyhedran* **2004**, *23*, 2841.

60 Kim, S. O., Sastri, C. V., Seo, M. S., Kim, J., Nam, W. *J Am Chem Soc* **2005**, *127*, 4178.

61 (a) Collins, T. J., Nichols, T. R., Uffelman, E. S. *J Am Chem Soc* **1991**, *113*, 4708; (b) Collins, T. J., Powell, R. D., Slebodnick, C., Uffelman, E. S. *J Am Chem Soc* **1991**, *113*, 8419; (c) Collins, T. J. *Acc Chem Res* **2002**, *35*, 782–790; (d) Horwitz, C. P., Fooksman, D. R., Vuocolo, L. D., Gordon-Wylie, S. W., Cox, N. J.,

Collins, T. J. *J Am Chem Soc* **1998**, *120*, 4867.
62 Ghosh, A., de Oliveira, F. T., Yano, T., Nishioka, T., Beach, E. S., Kinoshita, I., Münck, E., Ryabov, A. D., Horwitz, C. P., Collins, T. J. *J Am Chem Soc* **2005**, *127*, 2505.
63 Que, L., Jr., Tolman, W. B. *Angew Chem Int Ed* **2002**, *41*, 1114.
64 (a) Mirica, L. M., Ottenwaelder, X., Stack, T. D. P. *Chem Rev* **2004**, *104*, 1013; (b) Lewis, E. A., Tolman, W. B. *Chem Rev* **2004**, *104*, 1047.
65 Mahapatra, S., Halfen, J. A., Wilkinson, E. C., Que, L., Jr., Tolman, W. B. *J Am Chem Soc* **1994**, *116*, 9785.
66 (a) Lynch, W. E. Kurtz, E. M., Jr., Wang, S., Scott, R. A. *J Am Chem Soc* **1994**, *116*, 11030; (b) Sorrell, T. N., Allen, W. E., White, P. S. *Inorg Chem* **1995**, *34*, 952.
67 (a) Halfen, J. A., Mahapatra, S., Wilkinson, E. C., Kaderli, S., Young, V. G., Jr., Que, L., Jr., Zuberbühler, A. D., Tolman, W. B. *Science* **1996**, *271*, 1397; (b) Tolman, W. B. *Acc Chem Res* **1997**, *30*, 227.
68 Mahapatra, S., Halfen, J. A., Wilkinson, E. C., Pan, G., Wang, X., Young, V. G., Jr., Cramer, C. J., Que, L., Jr., Tolman, W. B. *J Am Chem Soc* **1996**, *118*, 11555.
69 Cole, A. P., Root, D. E., Mukherjee, P., Solomon, E. I., Stack, T. D. P. *Science* **1996**, *273*, 1848.
70 For another example, see: Taki, M., Teramae, S., Nagatomo, S., Tachi, Y., Kitagawa, T., Itoh, S., Fukusumi, S. *J Am Chem Soc* **2002**, *124*, 6367.
71 Zheng, H., Zang, Y., Dong, Y., Young, V. G., Jr., Que, L. *J Am Chem Soc* **1999**, *121*, 2226.
72 Hsu, H.-F., Dong, Y., Shu, L., Young, V. G., Jr., Que, L., Jr. *J Am Chem Soc* **1999**, *121*, 5230.
73 Kurtz, D. M., Jr. *Chem Rev* **1990**, *90*, 585.
74 (a) Jones, R. D., Summerville, D. A., Basolo, F. *Chem Rev* **1979**, *79*, 139 and references therein; (b) Niederhoffer, E. C., Timmons, J. H., Martell, A. E. *Chem Rev* **1984**, *84*, 137; (c) Norman, J. A., Pez, G. P., Roberts, D. A. In *Oxygen Complexes and Oxygen Activation by Transition Metals*, Martell, A. E., Sawyer, D. T. (Eds.), Plenum Press, New York, **1988**, p. 107.
75 Werner, A., Mylius, A. Z. *Anorg Allg Chem* **1898**, *16*, 245.
76 Egan, J. W., Jr., Haggerty, B. S., Rheingold, A. L., Sendlinger, S. C., Theopold, K. H. *J Am Chem Soc* **1990**, *112*, 2445.
77 Hu, X., Castro-Rodriguez, I. Meyer, K. *J Am Chem Soc* **2004**, *126*, 13464.
78 Reinaud, O. M., Yap, G. P. A., Rheingold, A. L., Theopold, K. H. *Angew Chem Int Ed* **1995**, *34*, 2051.
79 Shiren, K., Ogo, S., Fujinami, S., Hayashi, H., Suzuki, M., Uehara A., Watanabe, Y., Moro-oka, Y. *J Am Chem Soc* **1999**, *122*, 254.
80 Reinaud, O. M., Theopold, K. H. *J Am Chem Soc* **1994**, *116*, 6979.
81 Larsen, P. L., Parolin, T. J., Powell, D. R., Hendrich, M. H., Borovik, A. S. *Angew Chem Int Ed* **2003**, *42*, 85.
82 Hikichi, S., Yoshizawa, M., Sasakura, Y., Komatsuzaki, H., Moro-oka, Y., Akita, M., *Chem Eur J* **2001**, *7*, 5012.
83 Bossek, U., Weyhermüller, T., Wieghardt, K., Nuber, B., Weiss, J. *J Am Chem Soc* **1990**, *112*, 6387.
84 (a) Weschler, C. J., Hoffman, B. M., Basolo, F. *J Am Chem Soc* **1975**, *97*, 5278; (b) Hoffman, B. M., Weschler, C. J., Basolo, F. *J Am Chem Soc* **1976**, *98*, 5473.
85 VanAtta, R. B., Strouse, C. E., Hanson, L. K., Valentine, J. S. *J Am Chem Soc* **1987**, *109*, 1425.
86 Kitajima, N., Komatsuzaki, H., Hikichi, S., Osawa, M., Moro-oka, Y. *J Am Chem Soc* **1994**, *116*, 11596.
87 (a) Kimura, E., Sakonaka, A., Machida, R. *J Am Chem Soc* **1982**, *104*, 4255; (b) Kimura, E., Machida, R. *J Chem Commun* **1984**, 499; (c) Kushi, Y., Machida, R., Kimura, E. *Chem Commun* **1985**, 216; (d) Chen, D., Martell, A. E. *J Am Chem Soc* **1990**, *112*, 9411; (e) Cheng, C. C., Gulia, J., Rokita, S. E., Burrows, C. J. *J Mol Catal A* **1996**, *11*, 379.
88 (a) Schelber, P. J., Mandimutsira, B. S., Riordan, C. G., Liable-Sands, L. M., Incarvito, C. D., Rheingold, A. L *J Am Chem Soc* **2001**, *123*, 331; (b) Fujita, K., Rheingold, A. L., Riordan, C. G. *Dalton Trans* **2003**, 2004.
89 Fujita, K., Schenker, R., Gu, W., Brunold, W., Cramer, S. P., Riordan, C. G. *Inorg Chem* **2004**, *43*, 3324.

90 Mandimutsira, B. S., Yamarik, J. L., Brunold, T. C., Gu, W., Cramer, S. P., Riordan, C. G. *J Am Chem Soc* **2001**, *123*, 9194; (b) Schenker, R., Mandimutsira, B. S., Riordan, C. G., Brunold, T. C. *J Am Chem Soc* **2002**, *124*, 13842.

91 Aboelella, N. W., York, G. T., Reynolds, A. M., Fujita, K., Kinsinger, C. R., Cramer, C. J., Riordan, C. G., Tolman, W. B. *Chem Commun* **2004**, 1716.

92 (a) Kieber-Emmons, M. T., Schenker, R., Yap, G. P. A., Brunold, T. C., Riordan, C. G. *Angew Chem Int Ed* **2004**, *43*, 6716; (b) Schenker, R., Kieber-Emmons, M. T., Riordan, C. G., Brunold, T. C. *Inorg Chem* **2005**, *44*, 1752.

93 Hikichi, S., Yoshizawa, M., Sasakura, Y., Akita, M., Moro-oka, Y. *J Am Chem Soc* **1998**, *120*, 10567.

7
Methane Functionalization

Brian L. Conley, William J. Tenn, III, Kenneth J. H. Young, Somesh Ganesh, Steve Meier, Vadim Ziatdinov, Oleg Mironov, Jonas Oxgaard, Jason Gonzales, William A. Goddard, III, and Roy A. Periana

7.1
Methane as a Replacement for Petroleum

Today and into the foreseeable future, our world will continue to run on the conversion of fossilized hydrocarbons to energy and materials (Fig. 7.1) Energy production (fuels for propulsion, electrical power generation, heating, etc.) is the major use of hydrocarbon feedstocks but materials (chemical, petrochemical, plastics, rubber industries) are also essential for modern life. Given the potential issues with CO_2 emissions and global warming, there is a need to develop non-carbon based energy sources. However, any alternatives will take many

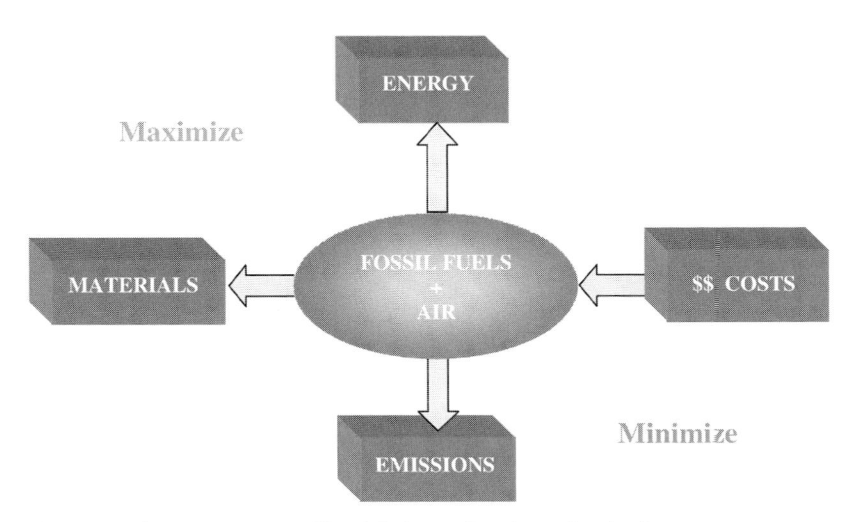

Fig. 7.1 Oxidative conversion of fossil fuels is a foundational technology.

Activation of Small Molecules. Edited by William B. Tolman

ISBN: 3-527-31312-5

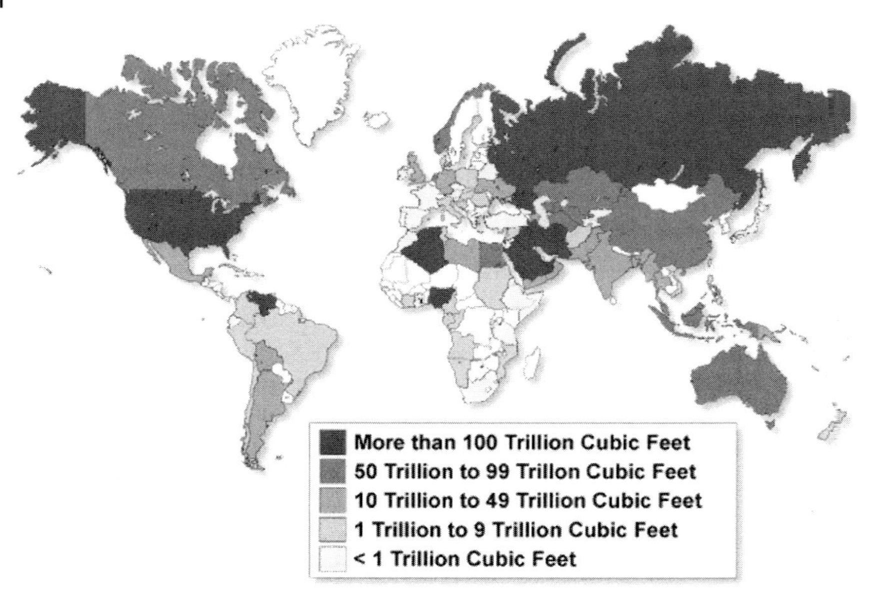

Fig. 7.2 World natural gas reserves, 2004 (from Radler, M. *Oil & Gas Journal* **2003**, *101*, 43–47).

decades to develop and implement and with rising demand for hydrocarbons, dwindling reserves, and geopolitical considerations it is imperative that we develop improved technologies for the conversion of hydrocarbons to energy as well as materials. As emphasized in Figure 7.1, the key will be to reduce emissions as well as costs while increasing energy and materials output.

Today, petroleum is our primary hydrocarbon feedstock. It is estimated [1] that at the beginning of 2004, worldwide reserves were 1.27 trillion barrels of oil. At current usage this supply is projected to last ~50 years. The other important hydrocarbon feedstocks are coal and natural gas. Methane, CH_4, is the major component of natural gas (75–90%) and as shown in Fig. 7.2, is one of the most abundant hydrocarbons on the planet and is widely distributed. Current estimates are that natural gas reserves (6,100 trillion cubic feet) are comparable to oil on an energy content basis. If the estimates of natural gas in methane hydrate deposits in the deep oceans are considered, the amount of methane on the planet could far exceed the known oil reserves! Given the abundance, wide distribution, higher hydrogen to carbon ratio and low sulfur content compared to oil and coal, methane could become the hydrocarbon feedstock of choice that could lead to a "Methane Economy" for the next several decades.

Electricity is one of the largest scale products from hydrocarbons and the primary source of CO_2 emissions that is a basis for serious ecological concerns. Today the most efficient engines for the production of energy from methane (natural gas) are turbines. Turbines currently operate at ~30% efficiency with respect to converting the chemical energy in natural gas (largely methane) to electrical energy and are very capital intensive and complex to operate. Consequently, there is substantial room for improvements with respect to reducing emissions and costs in the production of this basic commodity. Thus, if efficiencies for power generators can be increased from ~30% to ~85%, both usage of hydrocarbon feedstocks *and* emissions could be reduced by more than 300% while maintaining the present quality of life!

In spite of the large reserves and desirable characteristics of natural gas, petroleum remains the primary feedstock for chemicals and fuels. One reason for this is that natural gas cannot be economically transported due the low boiling point of methane and the low energy density in the gaseous state at standard conditions. Additionally, in spite of the high carbon yields (>50%) of commercially available technologies for conversion of natural gas to liquids that can be efficiently transported, such as the methane to methanol or methane to liquid paraffins (the Fischer-Tropsch process), these commercial processes are too expensive to compete with materials made from petroleum. Consequently, in order to have natural gas augment or replace petroleum as the primary feedstock, new technologies that are *both* substantially less costly and more energy and atom-efficient need to be developed to lessen the World's dependence on oil [2].

7.2 Low Temperature is Key to Economical Methane Functionalization

7.2.1 Lower Temperature Leads to Lower Costs

Unlike lower volume products such as drugs, the bulk of the costs of producing large scale products such as electricity, fuels and commodity chemicals result from the capital costs of the plant. Thus, in the case of converting methane to methanol or paraffins, >65% of the costs are related to the capital. This is because the current processes for conversion of methane to these liquid products are based on high-temperature (~900 °C) chemistry for the conversion of methane to syngas, followed by the conversion of the syngas to methanol or paraffins. As a result of the high temperatures utilized in the generation of the syngas, the specialized reactor and process controls required are capital intensive. In the case of electricity production, the methane is combusted at ~1000 °C in a gas turbine (a heat engine). Here also the high temperatures lead to capital intensive, complicated plants. The ~30% efficiency of the gas turbine results from the practical upper limit of the temperature in the combustion zone. The key to moving to a methane economy will be the development of lower temperature methane conversion processes that can be practiced in less capital intensive plants and with high energy- and atom-efficiencies.

7.2.2 Methane Functionalization by CH Hydroxylation

The chemistry with the greatest technical challenges and the broadest potential for developing new technologies for both the production of power as well as materials from methane is the selective, hydroxylation of the alkane C-H bond with O_2 (oxy-functionalization of methane). Thus, as shown in Fig. 7.3, chemistry for the conversion of the methane C-H bond to a C-OH bond (hydroxylation) with air (O_2) could enable next generation technologies for the direct, selective, low temperature conversion of methane to methanol and other materials. Significantly, as shown in Fig. 7.3, the low temperature complete oxidation of methane to CO_2 can be viewed as the complete, anodic, hydroxylation of the CH bonds on methane in liquid water, coupled with the reduction of oxygen. Viewed this way, the connection to CH hydroxylation is clear as a necessary goal to developing the next generation of maximally efficiently, direct, low temperature, methane fuel cells. The key to implementing such CH hydroxylation chemistry with oxygen for both methanol and electricity production will be to develop the next generation of catalysts that will facilitate CH hydroxylation reactions with low activation barriers (<25 kcal/mol) and with high selectivity.

7.2.3 Methane as the Least Expensive Reductant on the Planet

In addition to new technologies involving the direct conversion of methane to methanol and electricity as shown in Fig. 7.3, if new, low temperature, selective, direct CH conversion chemistry could be developed for methane, less obvious advances (such as the ones shown in Fig. 7.3) could be developed that reduce the number of process steps and overall costs for the production of some of the other large scale commodities on the planet such as ammonia, acetic acid and

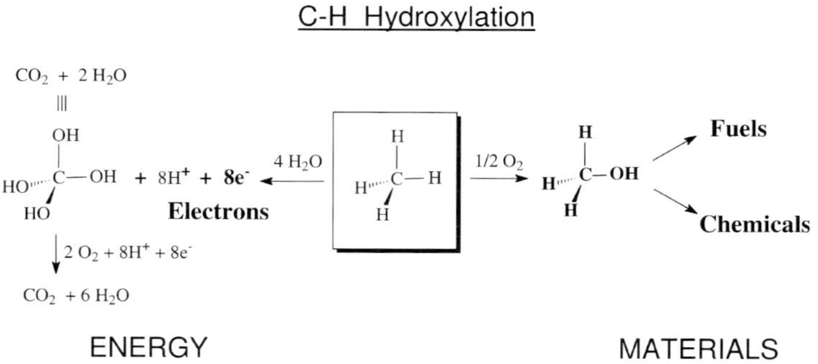

Fig. 7.3 Hydroxylation of the methane CH bond with O_2 can enable next generation, low cost, efficient energy and material production.

hydrogen peroxide. The direct reactions of methane shown in Figure 7.4 serve to illustrate that, in addition to being one of the most abundant, cheapest, carbon-based raw materials on the planet, it can also be considered the least expensive reductant, replacing hydrogen. The key to unleashing the largely underutilized potential of this molecule is increasing the rate of reaction; that is, making methane a kinetically facile reductant.

The challenges involved in developing an inexpensive methane to methanol processes that can compete with petroleum exemplify the challenges to utilizing methane and will be the focus of the chapter with emphasis on our work on CH hydroxylation of methane. The primary reason that new technologies for direct conversion of methane to methanol remain a challenge is that the current commercial catalysts for alkane oxidation (typically solid metal oxides) are not sufficiently active for the oxy-functionalization of alkane CH bonds and high temperatures or expensive reactive reagents must be utilized that lead to low reaction selectivity and inefficient chemistry. The development of next generation catalysts that would allow the selective functionalization of methane to methanol at low temperatures (~200–250 °C) in inexpensive reactors, with fewer steps and in high yields could provide a basis for this paradigm change in the petrochemical industry. Nature has evolved enzymes that are capable of selectively converting methane to methanol by using highly potent oxidizing metal centers; these have been reviewed elsewhere [3], but will not be the focus of this article, which focuses instead on commodity technology development.

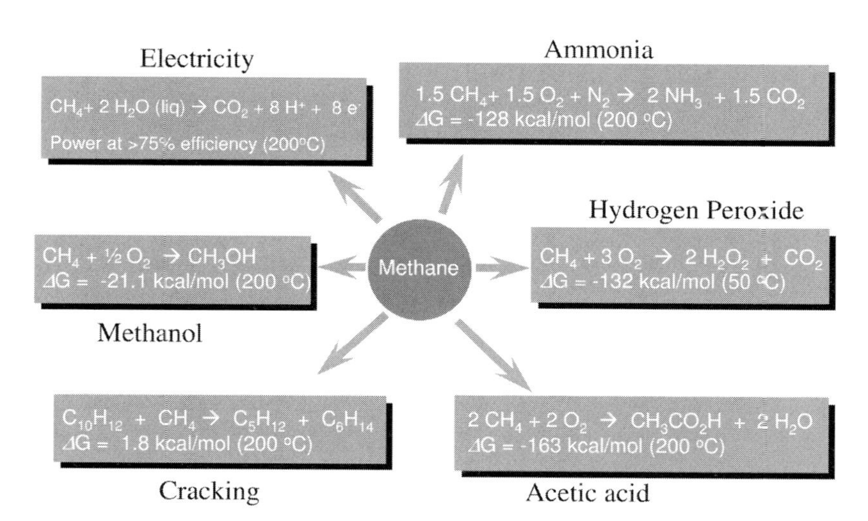

Fig. 7.4 Thermodynamically feasible pathways involving direct use of methane.

7.2.4
Selectivity is the Key to Methane Functionalization by CH Hydroxylation

The selective hydroxylation of methane to methanol with oxygen in high yield is challenging as a result of the high homolytic CH bond strength of methane (~105 kcal/mol) relative to that of methanol (93 kcal/mol). Additionally, while oxygen is considered to be a reactive molecule, it typically reacts with the generation of free-radicals and is relatively unreactive in non-free radical reactions. Typical oxidation catalysts today operate by free-radical reactions, and as a result of the correlation between free-radical reaction rates (and activation barriers) and homolytic bond strengths, methanol is substantially more reactive than methane with known free-radical based oxidation catalysts. This can be seen in Fig. 7.5, where the rate for hydrogen atom abstraction from methane (2.4×10^8 s^{-1}) by a hydroxyl radical is compared to that for methanol (8×10^8 s^{-1}). However, it should be possible to reverse this relatively reactivity by use of non-free radical species and/or by the use of functional groups that modify or "protect" the methanol. As shown in Fig. 7.5, the concept of "protection" is well established in organic chemistry. Thus, replacing a hydrogen atom on benzene with a nitro group "protects" the resulting nitrobenzene from further oxidation to poly-nitro benzenes. Similarly, replacing the hydroxyl group of methanol with an electron withdrawing cyanide group reduces the rate of hydrogen atom abstraction in methyl cyanide by over two orders of magnitude. Thus, it could be anticipated that reversible modification of the hydroxyl group in methanol by e.g. protonation or esterification could similarly reduce the rate of reaction with the CH bonds of the "protected" methanol.

As shown in Fig. 7.6, to develop commercially viable technology for the hydroxylation of methane, new generations of catalysts will need to be designed that will reduce the activation barrier for the oxidation of methane below that for methanol.

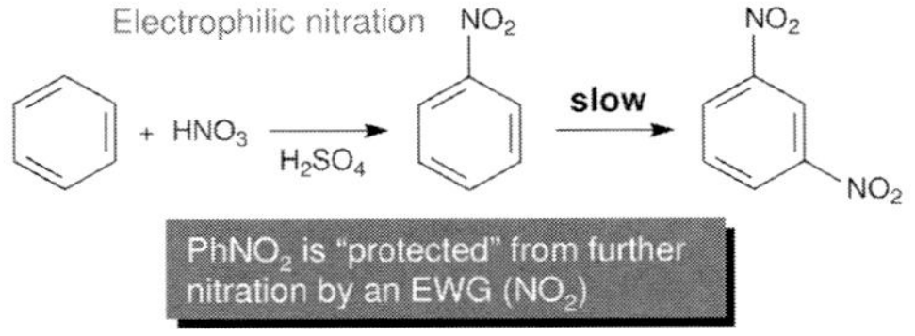

Electrophilic nitration

$$C_6H_6 + HNO_3 \xrightarrow{H_2SO_4} C_6H_5NO_2 \xrightarrow{\text{slow}} C_6H_4(NO_2)_2$$

PhNO$_2$ is "protected" from further nitration by an EWG (NO$_2$)

$$CH_4 + {\bullet}OH \longrightarrow {\bullet}CH_3 + H_2O \qquad k = 2.4\cdot10^8$$

$$CH_3OH + {\bullet}OH \longrightarrow {\bullet}CH_2OH + H_2O \qquad k = 8\cdot10^8$$

$$CH_3CN + {\bullet}OH \longrightarrow {\bullet}CH_2CN + H_2O \qquad k = 3.5\cdot10^6$$

Fig. 7.5 Modification of products by "protecting" groups can reduce reaction in both free-radical and coordination reactions.

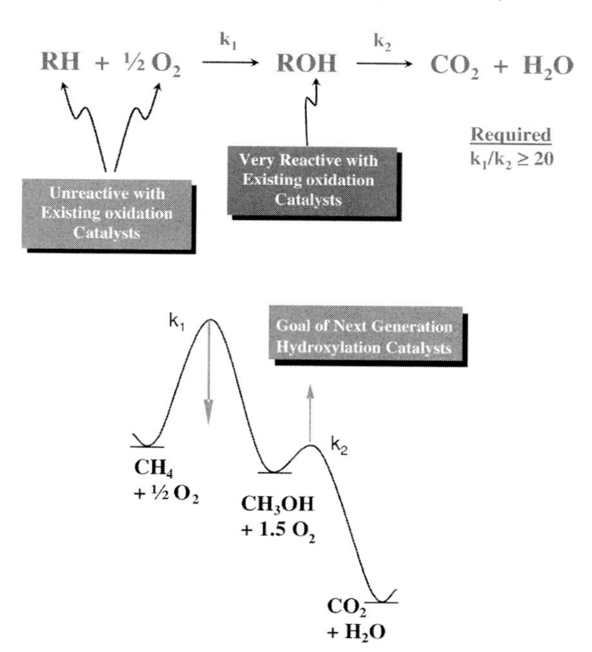

Fig. 7.6 Direct conversion of alkanes to alcohols requires new hydroxylation catalysts.

The issues of selectivity in methane hydroxylation are heightened by the low solubility of methane (1 mM at 1 atm at 25 °C) relative to the complete miscibility of methanol in suitable reaction solvents such as water. Thus, as reaction proceeds and methanol is generated the intrinsic reactivity of methane must be much higher than that of methanol to maintain high reaction selectivity. Assuming a reactor with a 1:1 gas to liquid ration and 500 psig methane pressure, kinetic models show that the relative rate constants for reaction of methane to methanol and methanol to CO_2, k_1 and k_2 respectively in Fig. 7.6, must be at least 20:1 for high reaction selectivity to methanol to be maintained at conversions of methane over 15%. Given this requirement to reverse the typical reactivities of methane and methanol it is clear that the next generation of catalysts must operate by non-free radical reactions.

7.2.5
Requirements of Methane Functionalization Chemistry Influenced by Plant Design

In developing this next-generation direct methane to methanol process, several key considerations are important. As a general consideration it should be expected that any new process not only substantially reduces the capital costs for the process (ideally by >50%) but also meets (and desirably exceeds) the overall yield and atom-efficiency of the existing process. Significantly, giving the economic challenges, the only other starting material that can be utilized in any large scale process to functionalize methane is air. Indeed, even pure or enriched oxygen (obtained from air) could be too expensive. Thus, other oxidants such as hydrogen peroxide, nitrous oxide, ozone, peroxides, or persulfates would not be useful since these materials cannot be regenerated from air.

Since capital cost of the plant is critical to developing the next generation methane to methanol processes, it is instructive to consider the type of process design, and the impact on the required chemistry, that could lead to sufficiently low capital costs to compete with petroleum. Many schemes can be considered for reacting air and methane (assuming that suitable catalysts are available) to generate methanol at low costs. Reactions involving direct combination of the alkane and air would seem to be theoretically ideal. However, practical considerations, such as avoiding explosive gaseous mixtures and minimizing likely free radical reactions due to the triplet ground state of oxygen, suggest that a Wacker type scheme, that employs inexpensive, liquid-phase bubble column reactors and a stoichiometric, air recyclable oxidant in direct reaction with the methane could be preferred. A simplified process diagram for such a scheme is shown in Fig. 7.7. In the original Wacker system for the oxidation of ethylene to

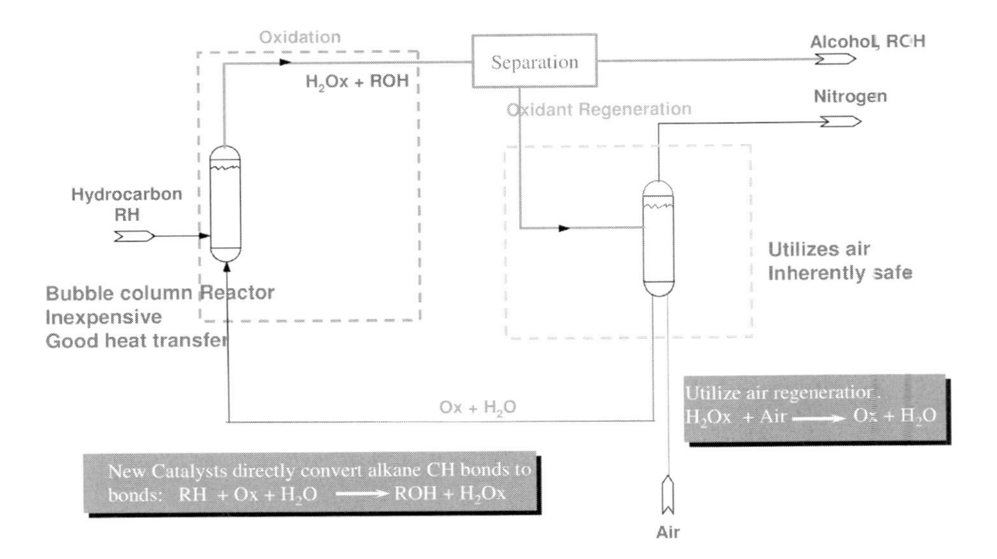

Two Step Reaction:

$$RH + Ox + H_2O \rightarrow ROH + H_2Ox \quad (1)$$

$$H_2Ox + \tfrac{1}{2} O_2 \rightarrow Ox + H_2O \quad (2)$$

$$RH + \tfrac{1}{2} O_2 \rightarrow ROH \quad (Net)$$

Fig. 7.7 Wacker type system for conversion of alkanes to alcohols.

acetaldehyde catalyzed by Pd(II), the role of this stoichiometric, air recyclable oxidant was met with Cu(II).

As can be seen, two key advantages of utilizing this process design are: A) air, instead of pure oxygen, can be utilized for the reoxidation and B) inexpensive gas-liquid bubble column reactors can be employed. Since these reactors can be designed to operate at comparable pressures and temperatures, there is no disadvantage to separating the alkane reaction from reactions with air.

Process studies show that the single most important requirement in developing a direct, air-based process for the conversion of methane to methanol is that the reaction be highly selective (>90%) at a minimum level of conversion (≥20%) per pass through the reactor with respect to both methane and the oxidant. This emphasis is critical because selectivity sets the maximum yield. A less obvious but important consideration is that loss of selectivity leads to the generation of CO_2 as the major side product which can lead to increased separations and heat management that can substantially increase the process and capital costs. Another key consideration to reducing the capital costs of a new process is to utilize lower reaction temperatures in the process because this can lead to simpler and less expensive reactors. However, an unobvious consideration with respect to the desired temperature is that while lower temperatures are desirable, the temperature range from 200–250 °C is the preferred range for an efficient, practical process. The basis for the upper temperature limit is that the reactor system can be maintained in the liquid phase at moderate pressures which allow the use of simple, inexpensive bubble-column reactors (essentially large pipes). The basis for the lower limit of ~200 °C (rather than lower) is that the direct oxidation of methane to methanol is an exothermic process (–26 kcal/mol at 220 °C). Consequently, heat will be released when the net reaction is carried out and this heat must be removed to maintain the reaction temperature. Carrying out the reaction between 200–250 °C allows the heat to be economically removed with room temperature water. In contrast, if the reaction was to be carried out at room temperature expensive refrigeration would be required to remove the heat of reaction. Moreover, the superheated steam generated in maintaining the 200–250 °C reaction temperature can also be utilized beneficially, thereby decreasing the cost and increasing the efficiency of the overall process. These considerations emphasize the need for thermally stable catalysts but also somewhat mitigate the challenge of designing fast catalysts.

For the catalyst to be useful it must meet minimum cost requirements. Most efficient reactor systems operate at volumetric productivities or Space Time Yields (STY) of $\sim 10^{-6}$ mol/cc · s (this is the amount of product that must be generated per unit reactor size per unit time). The reactor STY is related to catalyst activity and concentration by the relationship STY = TOF × [Cat], where TOF is the catalyst turnover frequency and [Cat] is the catalyst concentration. Typically, for nobel metal catalysts, given the enormous daily productivity of world-scale plants, the desired catalyst concentration to be economical is ~ 1 mM. This leads to a required catalyst rate (TOF) of ~ 1 s^{-1} in order to be cost effective (Table 7.1). To put these challenges in perspective, validation that these guidelines

Table 7.1 System guidelines for a commercializable methane to methanol catalyst

Engineering Guidelines
- $>20\%$ Methane conversion *per pass*
- $>90\%$ Product selectivity (overall carbon yield $>70\%$)
- $>20\%$ Oxidant conversion *per pass*
- Ultimate oxidant should be air
- Few steps
- Inexpensive reactors
- Facile product isolation
- Pressure <500 psig
- Temperature $>200\,^{\circ}C$ but $<300\,^{\circ}C$
- Reactor volumetric productivity (STY) $\sim 10^{-6}$ *mol* s^{-3}

Key Catalyst Guidelines
- TOF $\sim 1\ s^{-1}$
- TON $>10^3$
- Catalyst concentration of 1 mM at TOF $=1\ s^{-1}$ to be cost effective
- At 1:1 gas:liquid should generate 2 M MeOH in ~ 1.5 h

had been met would require a reactor with a 1 mM catalyst concentration and a total volume ratio of gas CH_4:liquid reactant of 1:1 to generate a ~2 M solution of methanol after ~1.5 h of reaction time.

7.2.6
Strategy for Methane Hydroxylation Catalyst Design

In considering the *de novo* design of any new catalyst it is important to consider that *all* catalysts *must simultaneously* meet minimum performance requirements related to the *three* key catalyst characteristics of *stability, rate, and selectivity* (Fig. 7.8; meeting any one or two would *not* lead to *useful* catalysts. This repre-

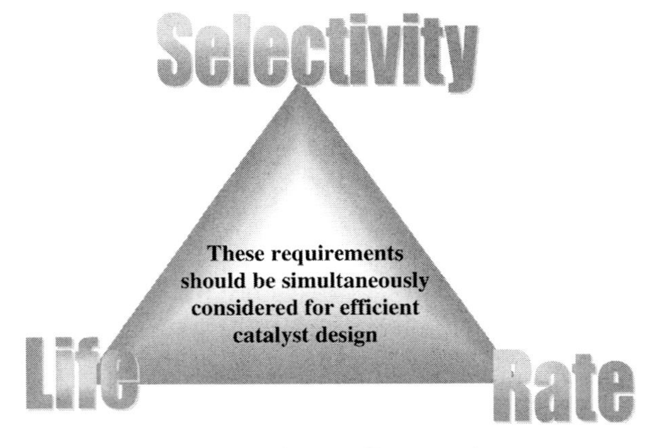

Fig. 7.8 Key requirements for any efficient catalyst.

sents a key challenge to the rational design of *any* efficient catalyst. This is because these three requirements are all interdependent on the molecular structure and composition of the catalyst and reaction system. This is an important consideration because a common approach to *de novo* catalyst design is to focus on meeting one requirement, e.g., the rate *or* selectivity by modifying the catalyst with the implicit assumption that after that target is met the catalyst can subsequently be further modified to meet the other two performance requirements.

Significantly, such a sequential approach to catalyst design may not be the most efficient because modifications to meet one requirement can lead to catalyst motifs that can not be efficiently (or indeed, at all) modified to meet the other performance requirements. This consideration would suggest that an effective strategy to developing useful catalysts would involve *simultaneous* consideration of all structure-function relationships that relate to stability, rate and selectivity. This can be challenging because this can severely restrict the range of catalyst structures and compositions to be examined. One approach to addressing the *de novo* design of catalysts could be to focus efforts on catalysts that are designed to operate by a specific reaction mechanism to that is known to meet one or more of these requirements of rate, selectivity and life. Since the CH activation reaction has been shown to exhibit extraordinary selectivity, focusing efforts on designing catalysts that operate by this mechanism could be expected to lead to selective catalysts. This could then allow focus on meeting the other requirements of stability and rate.

7.3 CH Activation as a Pathway to Economical Methane Functionalization via CH Hydroxylation

7.3.1 CH Activation is a Selective, Coordination Reaction

Many chemical approaches, summarized in Fig. 7.9, are being considered to develop technologies to convert methane. However, the approaches that have shown the greatest potential for developing low temperature, selective processes for the hydroxylation of the CH bond are based on the CH Activation reaction using homogeneous catalysts.

Homogeneous transition metal catalysis has had a substantial impact on chemistry. Thus, from polymerizations to hydrogenations there are few aspects of organic chemistry that have not been touched by this field of research. The majority of these catalytic reactions take place in the inner or first coordination sphere of the homogeneous metal catalyst and in many cases lead to the formation of organometallic, M-C, intermediates. The advantage of these inner sphere, coordination, organometallic reactions is that the reactants are bound to the catalyst center during conversion to products and as a result, the catalyst

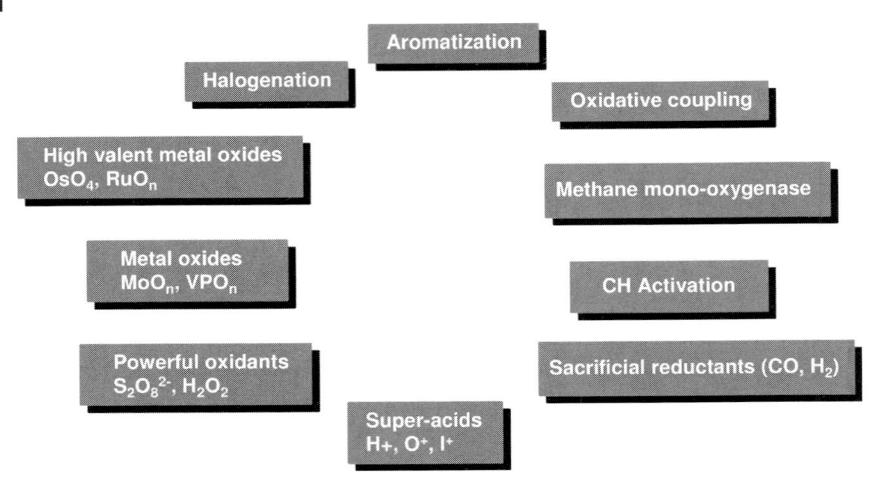

Fig. 7.9 Broad classification of the various approaches being examined for the direct conversion of methane.

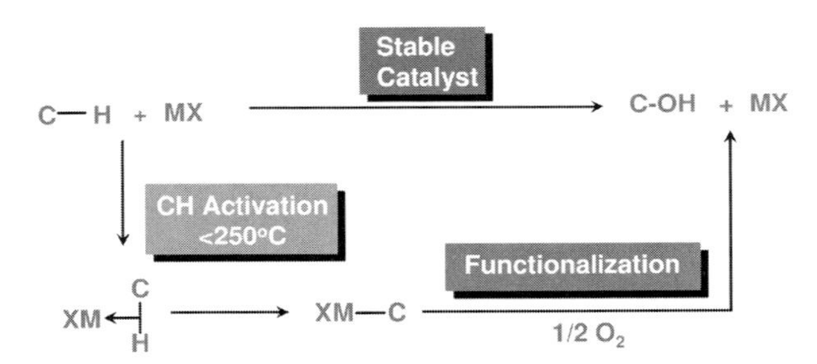

Fig. 7.10 General scheme for oxidation catalysis based on CH activation.

can effectively mediate both rate and selectivity in the conversion of the reactants to desired products.

There are many definitions of the "CH activation" reaction. As shown in Fig. 7.8, we define CH activation as a facile CH cleavage reaction with an "MX" species that proceeds by coordination of an alkane to the inner-sphere of "M" (either via an intermediate "alkane complex" or a transition state) leading to a M-C intermediate [4]. Emphasized in this definition is the requirement that during the CH cleavage, the hydrocarbyl species remains in the inner-sphere and under the influence of "M" [5].

The key characteristic of inner-sphere reactions that can lead to high rates and selectivities is the strong covalent forces between the CH bond and "M". The CH bond of methane is strong (homolytic bond strength of ~105 kcal/mol) and an important feature of the CH activation reaction is the formation of strong M-C bonds that compensate for breaking the CH bond. This is one of

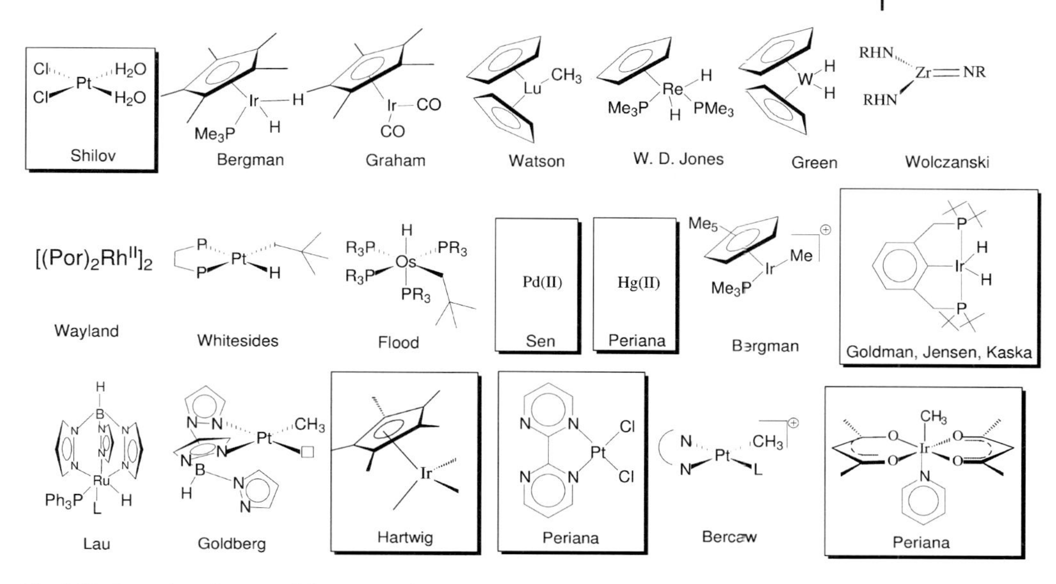

Fig. 7.11 Examples of some of the reported CH activation systems with catalytic systems that generate functionalized products highlighted.

the key bases for the involvement of second and third row transition metals in the many CH activation reactions that have been reported, Figure 7.11. Interestingly, while the involvement of these metals facilitate the CH cleavage, these metals are among the kinetically least labile and, as will be discussed, can lead to slow rates for the CH activation reaction in the presence of good ligands.

As might be expected from the wide variety of complexes that cleave CH bonds by the CH activation reaction (Fig. 7.11), there are several recognized mechanistic classifications as shown in Fig. 7.12. These are all related by the requirement that the alkane displace a ligand, L, and coordinate to the inner-sphere of the metal center either as an intermediate or in a transition state leading to the formation of organometallic M-C intermediates. The specifics of the actual mode of cleavage depend on the electronic configuration of the metal and the X group; several variations of these classifications have been distinguished. Of these, the most common modes are Electrophilic Substitution (ES), Oxidative Addition (OA) and σ-Bond Metathesis. In all cases high reaction selectivities are observed. To capitalize on the unique properties of the CH activation reaction the M-C species must be more easily functionalized than the CH bond to yield a useful C-X product with regeneration of the MX species. Ideally, to maintain high reaction selectivity and catalyst control it is likely that the functionalization also occur within the coordination sphere of "M".

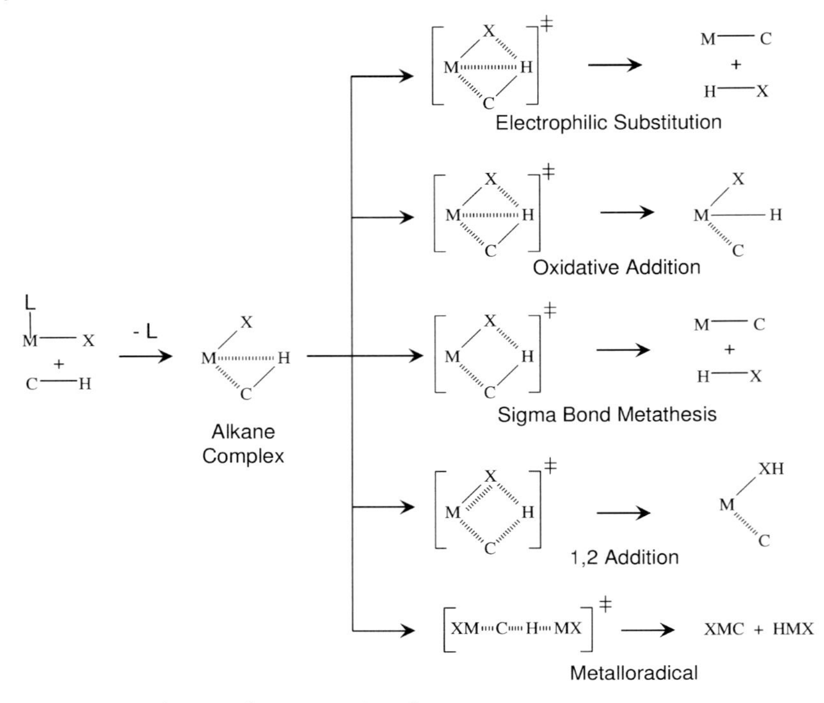

Fig. 7.12 Classification of various modes of CH activation.

7.3.2
Comparison of CH Activation to Other Alkane Coordination Reactions

The cleavage of the CH bond mediated via formation of an M-C species can be contrasted to other coordination reactions of alkane CH bonds such as those based on free radical, carbocation or carbanion intermediates. Since these reactions are based on highly energetic species these reactions require either very reactive conditions, such as extreme temperatures, or very energetic (and expensive) reagents such as superacids or peroxides that are not desirable [2a]. Additionally, in these "organic-type" CH bond-breaking reactions, unlike that for the CH activation reaction as defined above, the reactive alkyl fragments that are generated are not under the influence of the catalyst (because they are not strongly bound to the catalyst center) and consequently, can exhibit intrinsic reactivities that are generally undesirable. For example, free-radical or carbocations can eliminate hydrogen atoms or protons respectively, rather than react to generate alcohols. Of course, the primary issue with these systems are that alcohols are much more reactive than alkanes. This may be a reason that the commercially available oxidation catalysts based on metal oxides and utilized at high temperatures are not selective for the conversion of methane to methanol. Thus, the key advantage of the CH activation reaction over these more classical

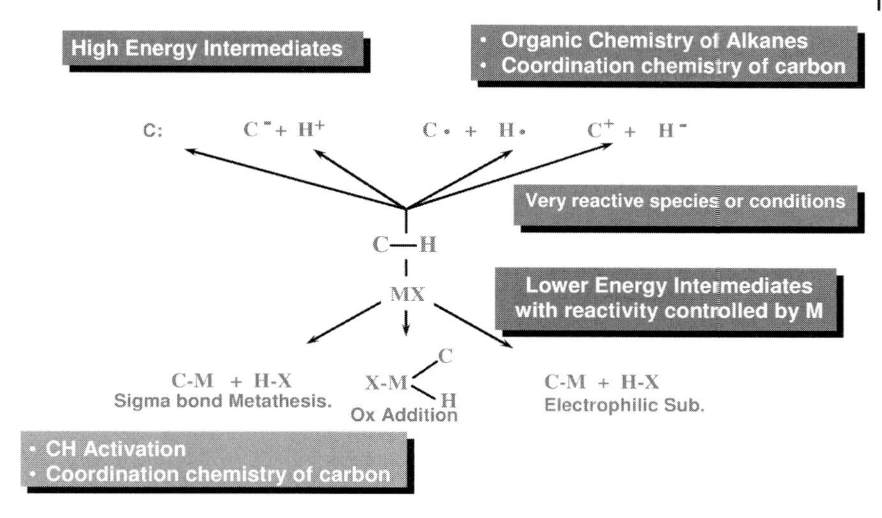

Fig. 7.13 A key advantage of the CH activation reaction is that, unlike classical reactions of the CH bond, the reaction can occur with moderate energetic materials and conditions.

coordination reactions of alkanes is that it is the only coordination reaction of the CH bond that leads to cleavage with the use of moderately energetic conditions and reagents and that generates an alkyl fragment whose reactivity is mediated by the catalyst center. This can lead to practical systems that can functionalize alkanes with high selectivities.

To illustrate the inner-sphere characteristics of the CH activation chemistry, an analogy can be made between CH activation by coordination of an alkane CH bond to a metal center and the known catalysis resulting from coordination of olefins via the CC double bond (note that the nature of the orbitals involved in bonding are quite different). It is well known that coordination of olefins to electrophilic metal centers can activate the olefin to nucleophilic attack and conversion to organometallic, M-C, intermediates. The M-C intermediates thus formed can then be more readily converted to functionalized products than the uncoordinated olefin. An important example of this in oxidation catalysis is the Wacker oxidation of ethylene to acetaldehyde. In this reaction, catalyzed by Pd(II) as shown in Fig. 7.14, ethylene is activated by coordination to the inner-sphere of an electrophilic Pd(II) center. This leads to attack by water and facile formation of an organometallic, palladium alkyl intermediate that is subsequently oxidized to acetaldehyde. The reduced catalyst is reoxidized by Cu(II) to complete the catalytic cycle. The Wacker reaction is very rapid and selective and it is possible to carry out the reaction is aqueous solvents. This is largely possible because of the favorable thermodynamics for coordination of olefins to transition metals that can be competitive with coordination to the water solvent. The reaction is very selective presumably because the bonds of the product (po-

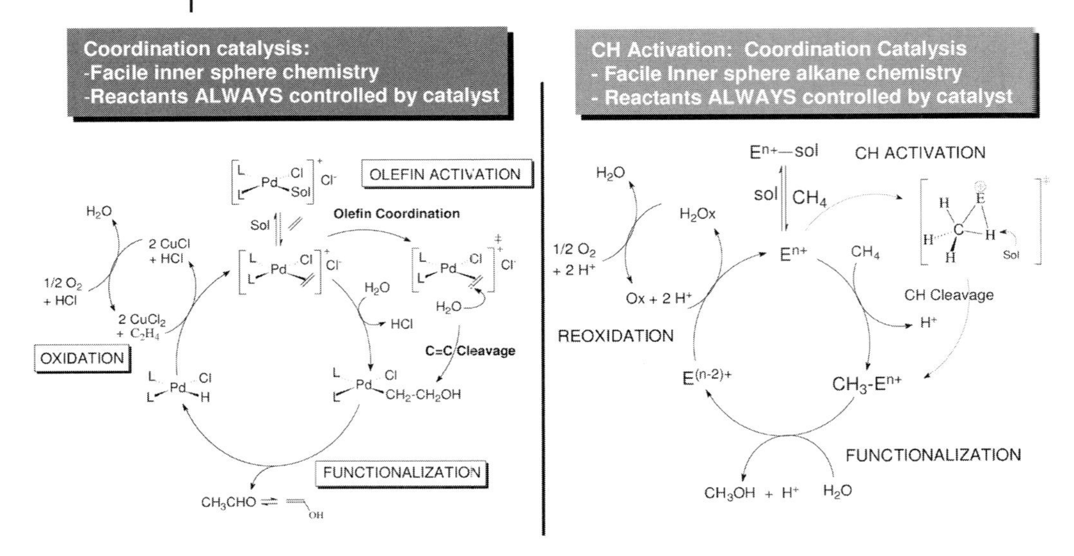

Fig. 7.14 Comparison showing that electrophilic CH activation-based oxidation and Wacker catalysts operate by coordination chemistry.

tentially reactive CC double bonds are still available from tautomerization of the acetaldehyde, Fig. 7.14) are not as readily activated. The most likely explanation is that the product forms much less stable olefin complexes to electrophilic Pd(II) due to the presence of the electron-withdrawing OH group. With the discovery of the Wacker reaction, the earlier processes based on classical chemistry such as the acid-catalyzed hydration of ethylene to ethanol, followed by high temperature, free-radical auto-oxidation are no longer practiced.

The useful comparison between the Wacker process and alkane CH activation is that all the coordination steps identified in the Wacker reaction (activation, functionalization and reoxidation) have parallels in catalytic, alkane CH activation and functionalization systems that operate with electrophilic catalysts. Thus, the coordination of the double bond of the olefin to electrophilic Pd(II) followed by cleavage by nucleophilic attack of water can be compared to CH activation of CH_4 by an electrophilic substitution (ES) pathway.

As shown in Fig. 7.14, in the CH Activation by an ES pathway the coordination of the CH bond of the alkane to an electrophilic center followed by loss of a proton can also be described as a nucleophilic attack of "sol" on the coordinated CH bond, leading to CH cleavage and generation of an intermediate E-CH_3 species. Of course, it would be expected that as the CC double bond is considerably more electron rich than an alkane CH bond, more reactive electrophilies would be required for similar coordination and cleavage of the CH bond. Given the low energy, σ-symmetry and low polarizability of the HOMO of the CH bond of methane, frontier orbital considerations of the interaction between

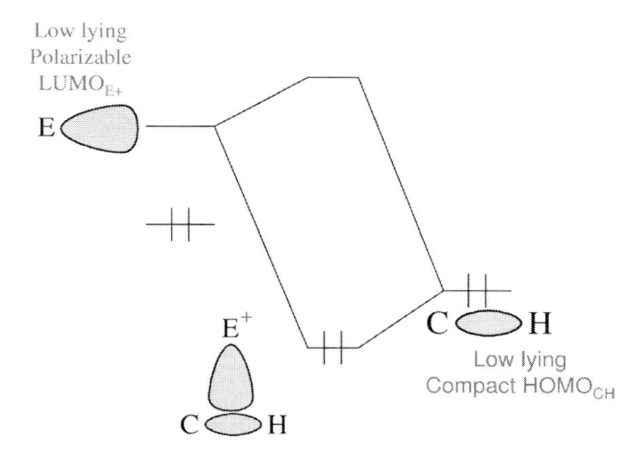

Fig. 7.15 Orbital interaction between the low lying compact HOMO of a CH bond and the low lying, polarizable LUMO of a "soft" electrophile.

the CH bond and electrophiles (Fig. 7.15) would indicate that 'soft", electrophiles characterized by low lying, polarizable LUMO's with σ-symmetry would be effective for this mode of CH Activation.

This has been found to be the case with the "soft", powerful electrophilic species such as $[XHg]^+$ that can be generated by dissolving HgX_2 salts in strongly acidic solvent such as sulfuric acid or triflic acid [6b]. As shown in Fig. 7.16, the reaction mechanism is proposed to occur by an electrophilic substitution (ES) pathway that involves coordination of methane to the inner-sphere of the poorly coordinated XHg^+ species (solvated by liquid sulfuric acid) either via a transition state or intermediate methane complex and subsequent loss of a proton to generate a $[CH_3\text{-}Hg]^+$ intermediate. The $[CH_3\text{-}Hg]^+$ intermediate is converted to methanol and the reduced catalyst is reoxidized. Interestingly, given the poor basicity of the sulfuric acid solvent, the coordination of the CH bond to $[XHg]^+$ followed by "nucleophilic attack" by sulfuric acid shows that the coordinated methane must be considered quite an acidic species as this is, in effect, the protonation of sulfuric acid (albeit it is likely that low concentration of HSO_4^- or water may be involved). Similar increases in acidity of hydrogen upon coordination to electrophilic metal centers have also been reported [7].

The $Hg(II)/H_2SO_4$ system is among the most effective catalyst for the conversion of methane to methanol in 96% sulfuric acid solvent. Thus, a 20 mM concentration of $Hg(HSO_4)_2$ in sulfuric acid reacts with methane at ~180 °C to generate ~1 M methanol with yields of over 40% based on added methane and selectivities >90% (Eq. 1). However, in spite of the high yields, selectivity and low cost of the catalyst, this system cannot be commercialized because the maximum attainable concentration of methanol, ~1 M, is too low for cost-effective product separation. Process studies show that the system can be practical if ~5 M concentrations of methanol can be obtained. However, since the catalyst is essentially completely inhibited by methanol or water above ~1 M concentrations, higher concentrations of methanol cannot be obtained by this catalyst system.

$$CH_4 + H_2SO_4 \xrightarrow{Hg(II)/H_2SO_4} CH_3OH + H_2O + SO_2 \quad (1)$$

In this system both the reaction rate and selectivity can be directly attributed to the relative rates of CH activation of the CH bonds present in the reaction system by the poorly solvated $[XHg]^+$ catalyst. Calculations show that the poorly solvated species, $[XHg]^+$, reacts with a ~29 kcal/mol barrier via a transition state in which methane is coordinated to a two-coordinate cationic mercury species, $[XHgCH_4]^+$, that loses a proton to the solvent to generate CH_3HgX. This correlates well with the experimental activation barrier of ~28 kcal/mol and the direct observation of $[HgCH_3]^+$ as an intermediate in the catalytic cycle. Both calculations and experimental investigations show that the $[HgCH_3]^+$ species readily reacts to generate methanol and the reduced catalyst, $Hg_2(II)$, is rapidly reoxidized by hot sulfuric acid.

Along with the relatively low barriers for breaking the CH bond of methane by an ES CH Activation mechanism, the electrophilic $[XHg]^+$ cation also exhibits extraordinary selectivity for methane relative to methanol in liquid sulfuric acid [5 b]. Thus, kinetic analysis of the relative rate constants (assuming a 1 mM CH_4 solubility at 1 atm and 25 °C in concentrated sulfuric acid) in a 1:1 gas to liquid volume system indicates that the active catalyst $[XHg]^+$ reacts with CH bonds of methane with a rate constant that is at least 1000 times greater than those for

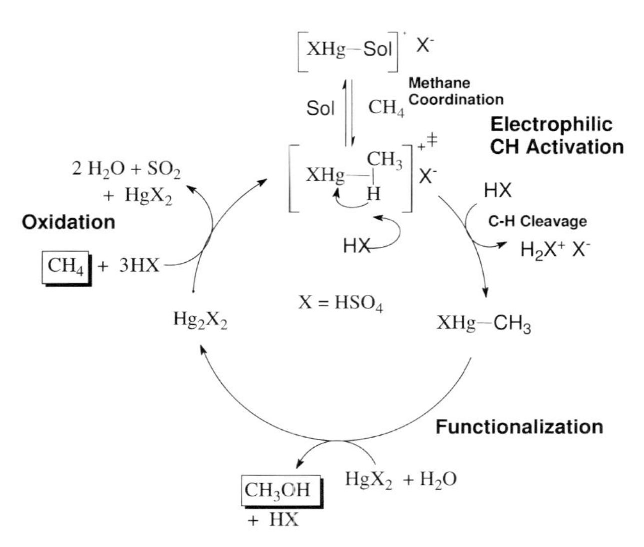

Fig. 7.16 Proposed electrophilic CH activation mechanism for the oxidation of methane to methanol by the $Hg(II)/H_2SO_4$ system.

CH_3OH (which exist primarily as the protonated or sulfated forms, $[CH_3OH_2]^+$ or CH_3OSO_3H, respectively, in sulfuric acid solvent). The greater reactivity of the CH bonds of methane compared to those of methanol can be traced to substantially lower reactivity of the electrophilic $[XHg]^+$ catalyst with the CH bonds of methanol which due to the electron-withdrawing effect of protonation or sulfation are substantially less electron rich than those of those of methane.

These considerations show that a key reason for focus on the CH activation reaction (as defined above) to develop the next generation of hydroxylation catalysts is that the reactions can allow the cleavage and functionalization of CH bonds with extraordinary selectivity and high rates. If it can be presumed that these high selectivities and rates can be retained for any CH activation system when coupled to efficient M-C functionalization reactions this provides a compelling rationalization for designing oxidation catalysts based on the CH Activation reaction. Strategically, as suggested earlier, this restriction of the reaction mechanism to a CH activation pathway could greatly simplify *de novo* catalyst design as this can effectively fix the catalyst structure-selectivity (and potentially rate) relationship and allows for focus on structure-stability and structure-rate modifications with a high probability that the resulting catalyst could be made stable, selective and rapid. Of course, this places restrictions on the catalyst structures and compositions that will be examined, but given the wide variety of metals and ligand motifs (Fig. 7.11) that show activity for the CH activation reaction, this should not be an insurmountable issue [8].

7.3.3
Some Key Challenges and Approaches to Designing Hydroxylation Catalysts Based on the CH Activation Reaction

Three general requirements for developing useful methane hydroxylation catalysts based on the CH activation reaction can be considered: A) A reactive species must be generated that can react rapidly with the CH bond via the CH activation reaction without the need for high energy conditions or reagents and, critically, that is not inhibited in the presence of required reactants (typically the oxidant), the solvent (desirably a species such as water), or the reaction products (methanol or water); B) the CH activation reaction must be coupled in a catalytic cycle to an oxidative functionalization reaction that generates methanol, regenerates the catalyst and utilizes only air, or an air-recyclable oxidant and C) the catalyst should be stable to the reaction system.

The *simultaneous* requirement of stability, rate and functionalization catalyst functions could provide an explanation to why, despite the large number of systems known to cleave CH bonds by the CH activation reaction, only relatively few systems have been found to function as catalysts to convert methane to methanol and *none of them* meet the target guidelines discussed above. While several reasons can be attributed to this lack of efficient catalytic systems, key ones are slow rates, lack of stability and/or no efficient functionalization reactions. For example, for systems that can be classified as operating by the metalloradical, 1,2 addition and sigma bond metathesis pathways, no efficient functionalization pathways have yet been identified that generate useful products while regenerating the catalyst. The other two mechanistic classifications, oxidative addition (OA) and electrophilic substitution (ES), have been successfully coupled to functionalization pathways that regenerate the catalyst. However, in general these catalytic systems still suffer primarily from either poor rates and/or stability issues and have not met the performance goals required to be useful.

Meeting the requirements of increasing rate and stability of CH activation-based functionalization catalysis represents a significant challenge given the state of knowledge in the field. However, a useful strategy for the rational design of new catalysts could be to utilize structurally and compositionally flexible motifs that can be expected to be thermally stable to the protic, oxidizing conditions required for functionalization and that can be expected to operate via the CH activation reaction. With such selected, stable motifs in hand, efforts could then be focused on modifications to increase rates and coupling of the CH activation and oxidative functionalization steps with a high probability that the catalyst will be sufficiently stable to show useful chemistry or catalysis.

7.3.3.1 Stable Catalyst Motifs for CH Activation

An important common feature of the known catalyst systems shown in Figure 7.11 that leads to their effectiveness is the unique stability of the systems rather than a particular ligand motif or CH cleavage mechanism. In the case of the Pd(II) and Hg(II) cationic systems that operate in strongly acidic media by ES mechanisms, the stability is imparted by the lack of fragile ligands and simplicity of these catalysts that minimizes the probability of degradation of complex catalytic structures. The primary mode of decomposition of these catalysts is irreversible formation of the reduced metallic state or other inactive states such as polymeric salts. In the case of Pd(II) this is a significant issue as Pd black is stable. However, in the case of Hg(II) salts in sulfuric acid media, this is not a critical issue as the reduced metal or oxidation states, Hg_2(II), is readily reoxidized to the active catalytic state, Hg(II).

The primary issue with the Shilov system [9], $PtCl_2(H_2O)_2$, that is proposed to operate via an oxidative addition (OA) pathway, is catalyst instability due to irreversible decomposition to Pt metal or insoluble, polymeric Pt salts such $(PtCl_2)_n$. This can be addressed by the use of ligands as has been done with the high yield, Pt(bpym)Cl_2/H_2SO_4 system (bpym = κ^2-{2,2′-bipyrimidyl}) [6]. This system is stable and active for the conversion of methane to methanol in concentrated sulfuric acid (Eq. 2). Yields of over 70% methanol (based on methane) with selectivities of >90% and turn-overs of ~1000 have been observed with turn-over-frequencies of ~10^{-3} s^{-1}.

$$\underset{500\,\text{psig}}{CH_4 + H_2SO_4} \xrightarrow[\substack{220\,^\circ\text{C},\ 2.5\ \text{hrs} \\ 150\,\text{ml}\ 98\%\ H_2SO_2}]{\substack{(\text{bpym})PtCl_2\ (20\ \text{mM}) \\ \text{TON}\geq 1000,\ \text{TOF}\sim 10^{-3}\ s^{-1}}} \underset{\sim 1\,\text{M},\ \sim 70\%\ \text{yield},\ >95\%\ \text{sel}}{CH_3OH + H_2O + SO_2} \quad (2)$$

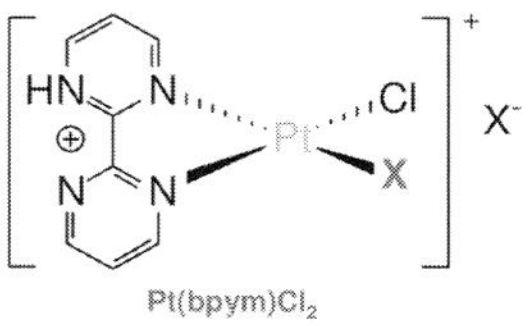

Fig. 7.17 The unique structure of the bpym ligand affords its thermal and protic stability.

In this system, experimental and theoretical studies show that this complex is thermally stable to strong acid solvents due to the unique structure and composition of the bpym (bipyrimidine) ligand (Fig. 7.17) [6, 5].

Theoretical and experimental studies [5a, 10] show that protonation of the bpym ligand in strongly acidic media (Fig. 7.19), imparts both thermal and acid stability by stabilizing the active Pt(II) oxidation state relative to irreversible formation of insoluble $(PtCl_2)_n$ or Pt black. This has been shown to be due to the presence of the two nitrogens in the same aromatic ring in the bipyrimidine ligand. Consistent with this, replacement of the bpym ligand with simpler NH_3 ligands or bipyridine ligands [6a] leads to systems that are active for the catalytic oxidation of methane to methanol but are unstable and lead to rapid catalyst decomposition. Theoretical studies [5] confirm the experimental observations that $Pt(NH_3)_2Cl_2$ is thermodynamically *less* stable than $(PtCl_2)_n$ whereas the protonated $Pt(bpym)Cl_2$ complex, $[(Hbpym)PtCl_2]^+$, is thermodynamically *more* stable.

Theoretical and experimental studies show that, as in the case of the electrophilic $[XHg]^+$ system, this cationic Pt(II) catalyst also operates via an electrophilic substitution reaction pathway as shown in Fig. 7.18 and Fig. 7.19. This switch

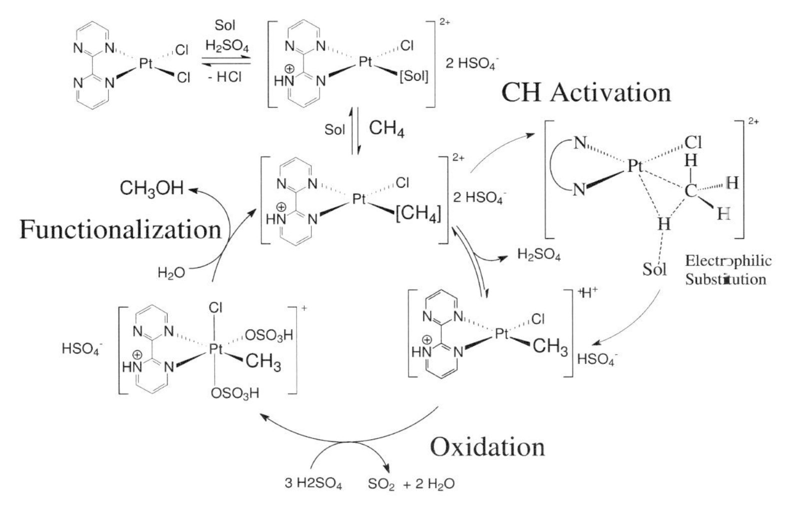

Fig. 7.18 Proposed electrophilic substitution CH activation mechanism for the $Pt(bpym)Cl_2/H_2SO_4$ system for methane oxidation to methanol.

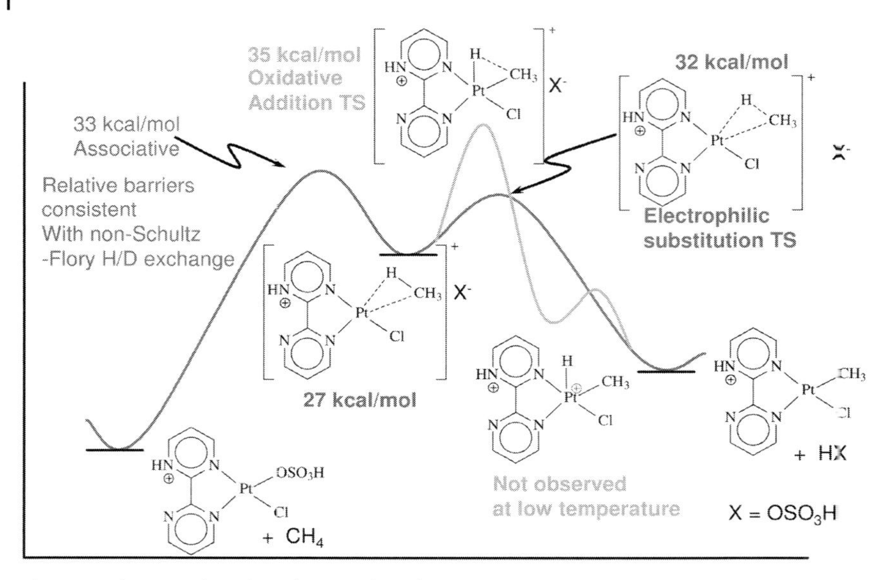

Fig. 7.19 Theoretical studies showing the relative energies of the ES and OA pathways for CH activation with $[(Hbpym)PtClX]^+$.

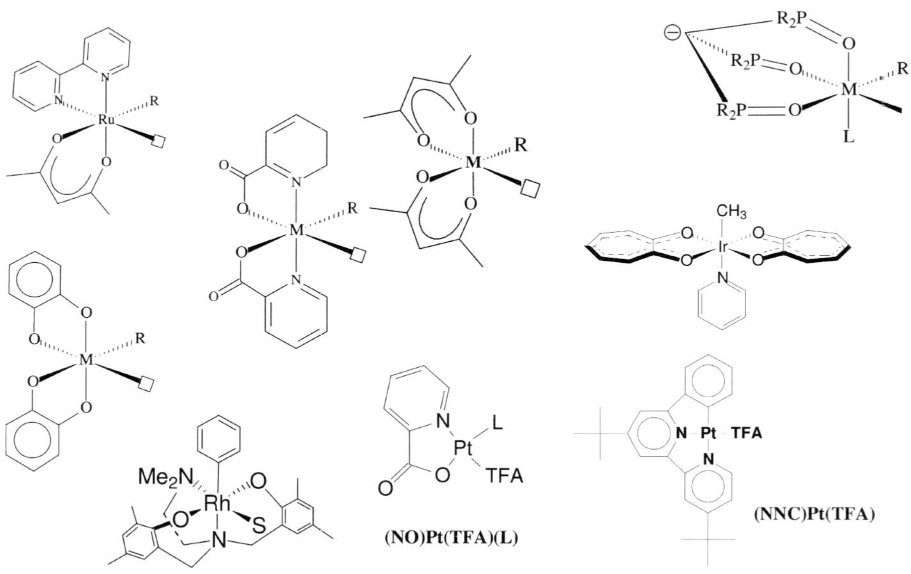

Fig. 7.20 Examples of ligand motifs that are being investigated for designing new CH activation complexes that could be expected to be stable to conditions required for functionalization.

to an ES pathway, whereas the Shilov Pt(II) system is proposed to operate via an OA pathway, results from the increased electrophilicity of the metal imparted by the protonation of the bpym ligand and the electron-withdrawing HSO_4 groups. In spite of the high efficiency, this system, as is the case for the $[XHg]^+$ system, is also strongly inhibited by the reaction products methanol and water. In this case, the inhibition limits the product concentration to ~1.5 M and remains too low for commercialization.

In an effort to design catalysts systems that may not be as susceptible to water inhibition but retain the remarkable stability of the bpym-ligated Pt(II) system, we have begun to explore O-donor and other C- and N-donor chelating ligand systems such as those shown in Fig. 7.20. The fundamental concept behind the use of O-donor ligands as possible stable ligand motifs is that O-donor ligands, unlike P and C ligands, could be expected to be less susceptible to oxidation.

7.3.3.2 **Slow Rates of CH Activation-based Catalysts**

7.3.3.2.1 **Catalyst Inhibition by Ground State Stabilization**

As shown above, Fig. 7.11, many systems that activate CH bonds are now known and it is possible that if these systems can be made stable that some could be used as the basis for development of catalysts for alkane oxidation. In many cases, the CH activation rates of these reported systems are quite rapid when rates are extrapolated to temperatures of ~200 °C. On the basis of these observations and assuming that stable motifs can be identified that would facilitate CH activation and functionalization, it might be assumed that acceptable catalysis rates can be readily obtained by simply basing catalyst designs on these CH activation systems. However, this is not likely to be the case and other considerations need to be taken into account.

As discussed for the Hg(II) and (bpym)Pt(II) systems, an issue with many of the reported CH activation systems is severe inhibition of the catalyst by water and/or methanol. Indeed, the CH activation is typically a rapid reaction only when the reaction system is carefully chosen such that the alkane is the most (or only) coordinating species present. However, under conditions where useful products such as alcohols or other coordinating species can be produced, many of the reported CH activation systems, even if they can be made stable, would exhibit very low rates. Significantly, this type of inhibition is likely to be a general issue with oxidation systems based on the CH activation reaction (as defined above).

One fundamental reason for this is that the alkane CH bond, unlike CC double bonds of olefins or other functional groups, are among the poorest known ligands and unlikely to compete well with other more coordinating species for the metal center. This may well be the "Achilles heel" of hydroxylation catalysts based on the CH activation reaction. Consistent with the poor ligating capability of alkanes, other than by spectroscopic methods and mechanistic studies, only

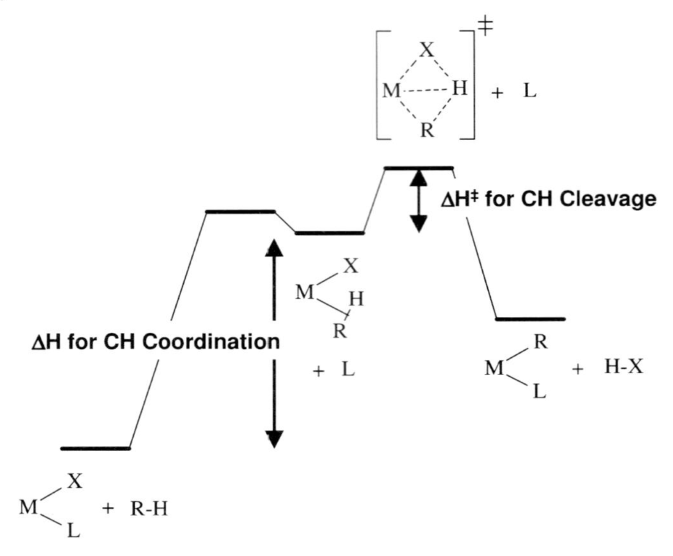

Fig. 7.21 Generalized energy diagram emphasizing the two key steps involved in the CH activation reaction: RH coordination and CH cleavage.

one experimental observation of a transient alkane complex, and that by low temperature NMR, has been reported [11, 12].

The CH activation reaction involves two distinctly different steps: A) coordination of the alkane and B) cleavage of the CH bond as shown in Fig. 7.21. The coordination of an alkane to the metal center (either leading to an intermediate alkane complex or a transition state that leads directly to CH cleavage) can be viewed as an inner sphere ligand displacement or interchange process where the alkane is one of the ligands and the other is the ligand initially occupying that coordination site, L. Typically, the ligand to be displaced by the alkane will be the most nucleophilic (ligating) species present in the reaction system and can be either reactants other than the alkane, solvent or the products. There is some consideration as to whether this ligand exchange involving the alkane is dissociative or associative [13]. This would be expected to depend on the binding constant of the ligand being displaced by the alkane but given the poor binding characteristics of alkanes, it is plausible that there will be substantial dissociative character to the displacement reaction in all cases except with extremely poorly coordinating ligands.

These considerations point to a fundamental issue that must be overcome in developing hydroxylation catalyst systems based on the CH activation reactions: coordination reactions with alkanes (whether associative or dissociative) leading to weakly bound, intermediate alkane complexes, or directly to a transition state leading to CH cleavage, can be expected to be subject to severe ground state inhibition by the desired products such as methanol or by most media or reactants that would be required for oxidation. This ground state stabilization fundamentally arises from strong binding of other possible ligands to the catalyst in the reaction system and the resulting stabilization of the catalyst resting state. As can be seen from Fig. 7.21, the more stable this state is, the higher the expected activation barrier for CH Activation.

It is challenging to imagine how methane coordination could occur to a sufficient extent to allow efficient catalysis in a solvent such as liquid water given the excellent coordinating properties of water, the poor solubility of methane in most useful media and the high concentration of the solvent. Of course, the reaction does not have to be carried out in solvents as ligating as water. However, since the objective is the hydroxylation of methane to methanol, then at a minimum, (if as desired, the catalyst is expected to operate at high turnover numbers before separation of product) methanol, which can be expected to bind more tightly to the catalyst than methane, will be present in the system and catalyst inhibition by the methanol could be expected.

This issue of ground state inhibition is observed in many catalytic alkane functionalization systems that operate by the CH activation reaction. Thus, this is the fundamental reason for the inhibition observed in the electrophilic CH activation-based Hg(II)/H_2SO_4 and (bpym)Pt(II)/H_2SO_4 systems that limit the maximum concentration of methanol to ~1 M. Theoretical and experimental studies show that the inhibition of the (bpym)Pt(II) system results from "ground state stabilization" from preferential coordination of water or methanol to the electrophilic Pt(II) catalyst (Fig. 7.22). This type of "ground state stabilization" is also the basis for the inhibition observed in the Hg(II)/H_2SO_4 system.

Indeed, other electrophilic systems, such as the Sen Pd(II)/CF_3CO_2H system [8i, j] (Fig. 7.22), exhibit ground state inhibitions that can be traced to water (or methanol) binding. In other CH activation/functionalization systems that operate by mechanisms other than ES, such as the Kaska/Goldman/Jensen system [14] for the dehydrogenation of alkanes to olefins that operates by an OA mechanism, inhibition is also observed. For example, in this case, the reaction cannot be carried in the presence of ethylene, an ideal, inexpensive, hydrogen acceptor, due to ground state inhibition from olefin binding. The slow rates of the Shilov system (TOF of ~10^{-7} s^{-1} at 60 °C) that is proposed to operate by an OA mechanism is also most likely due to strong ground state inhibition from solvent (water) binding.

We have been examining two approaches to this problem: A) use of strongly acidic solvents and B) altering the electronic properties of the metal center so as to minimize this ground state inhibition. The fundamental idea behind the use of an acidic solvent (this can be a Lewis or Bronsted acid) is that the strongest base that can exist in such an acidic solvent is the conjugate base of the acid. In the case of a strong acid, both the acid and the conjugate base will be weakly basic and expected to be poorly coordinating. This will minimize ground state stabilization by the solvent and it's conjugate base. Equally important, the use of a strong acid as a solvent rather than in stoichiometric amounts relative to the catalyst is central to preventing catalyst inhibition by products or reactants

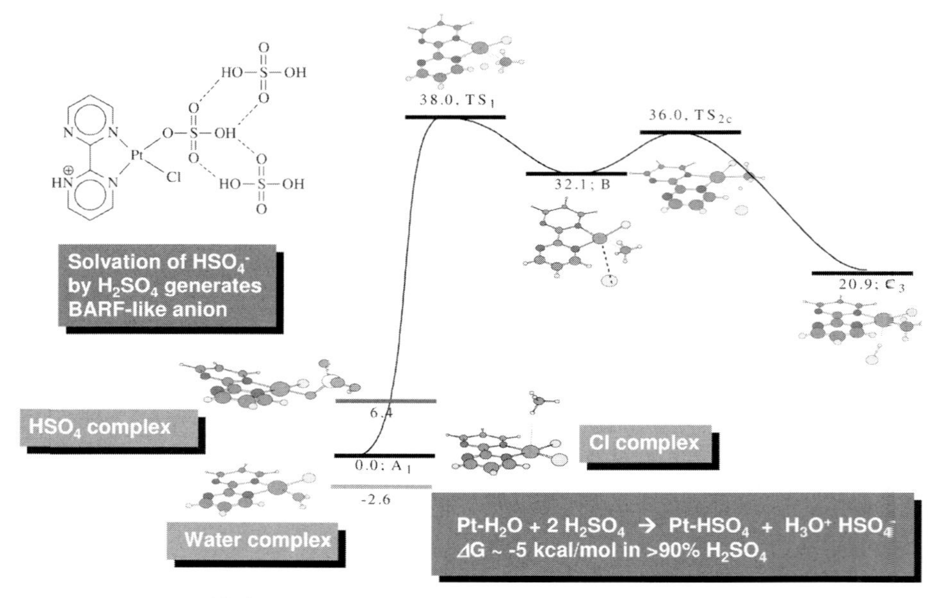

Fig. 7.22 Simplified energy diagram showing the inhibition of the Pt(bpym)X_2 catalyst by water.

as these materials will also be present in large excess over the catalyst in any useful catalyst system.

7.3.3.2.2 Use of Acidic Solvents to Minimize Catalyst Inhibition by Ground State Destabilization

The general use of Lewis or Bronsted acids to facilitate coordination of reactants is a well-known phenomenon in coordination chemistry [15]. Some of the most active systems reported for the stoichiometric CH activation can be seen as complexes that have been activated by addition of a Lewis acid. Thus, one of the most active complexes known for catalytic CH activation developed by Bergman [16], $[Cp^*Ir(PMe)_3Me(CH_2Cl_2)]^+$ $[MeB(C_6F_5)_3]^-$, is generated by reaction of the Lewis acid, $B(C_6F_5)_3$, with $Cp^*Ir(PMe)_3Me_2$ in a poorly coordinating solvent, CH_2Cl_2. One reason that this complex reacts rapidly with methane (at –10 °C) is that all the possible competing ligands in the reaction system, $[MeB(C_6F_5)_3]^-$ and CH_2Cl_2, are poorly coordinating species that minimize ground state stabilization and allow methane to effectively compete for coordination to the metal center. This strategy of stoichiometric use of a weakly coordinated complex can lead to very active catalysts in reactions where no strongly coordinating reactants, solvents, or products are present. However, such complexes could not be expected to remain active in the presence of stoichiometric or greater amounts of coordinating species such as water or methanol. In the presence of these materials, all weakly coordinated groups will be readily displaced leading to severe ground state inhibition of the catalyst. Thus, this approach of stoichiometric use of weakly coordinating groups would not be suitable for catalytic systems where the desired product is methanol and many catalyst turnovers are required.

One approach that could be considered is to carry out the reaction in liquid $B(C_6F_5)_3$ as solvent. Under these conditions essentially all the methanol formed should complex to the Lewis acid and free methanol would be unavailable for coordination to the metal, minimizing ground state stabilization and catalyst inhibition. The key issue with this strategy is that $B(C_6F_5)_3$ is expensive and unstable and most likely it would not be cost effective to attempt to separate the $MeOH:B(C_6F_5)_3$ complex and recycle the $B(C_6F_5)_3$. However, if the Lewis acid utilized is inexpensive, thermally robust and the methanol: Lewis acid adduct formation can be made inexpensively reversible this could be potentially useful strategy.

This is the key basis behind the use of inexpensive sulfuric acid solvent for facilitating the selective oxidation of methane with the (bpym)Pt(II) and Hg(II) systems already discussed [6]. Liquid sulfuric acid, at concentrations $> \sim 85\%$, is a polar, strongly Lewis acidic, poorly nucleophilic liquid in which the strongest nucleophile that can exist is HSO_4^- which is substantially less coordinating that water or methanol to Pt(II). Above this concentration of acid solvent, any water or methanol generated (or any other species more basic than HSO_4^-) is essentially fully protonated and not available for coordination and inhibition of the metal center by ground state inhibition. Below this acid concentration the solvent acidity drops rapidly [17] and water or methanol can become available for coordination to the metal center leading to inhibition of the CH activation reaction. This role of the acid solvent in preventing ground state stabilization and resulting inhibition of the catalyst can be visualized by the effect of added concentrated H_2SO_4 solvent to a system where the ground state is the water complex (Fig. 7.23). As shown, addition of H_2SO_4 (HX) to all points in the reaction coordinate where free H_2O is present would result in an ~ -12 kcal/mol drop in energy (due to the heat of hydration of concentrated H_2SO_4) with the net result that in concentrated sulfuric acid solvent the catalyst ground state is the labile Pt-HSO_4 complex, rather than the stable Pt-H_2O complex.

The key challenge to utilizing this strategy is the identification of catalysts, reactants and products that are thermally stable in such strongly acid media. Both methane and methanol are thermally stable in sulfuric acid at temperature below 250 °C and several catalyst systems have been identified that are stable in this media for the selective oxidation of methane to methanol. As noted above both the Hg(II) and $Pt(bpym)Cl_2$ systems have been found to be efficient catalysts for methane oxidation to methanol at ~ 200 °C in this 96% sulfuric acid solvent as both are very stable in this media. Consistent with the concept of basicity leveling, theoretical studies show that, at sulfuric acid concentrations $>85\%$, the ground state of these catalysts, $[(Hbpym)PtCl(HSO_4)]^+$ and $Hg(HSO_4)_2$, are coordinated to weakly binding HSO_4^- that is most likely extensively hydrogen bonded to solvent H_2SO_4 molecules (Fig. 7.24). As shown, the coordination of sulfuric

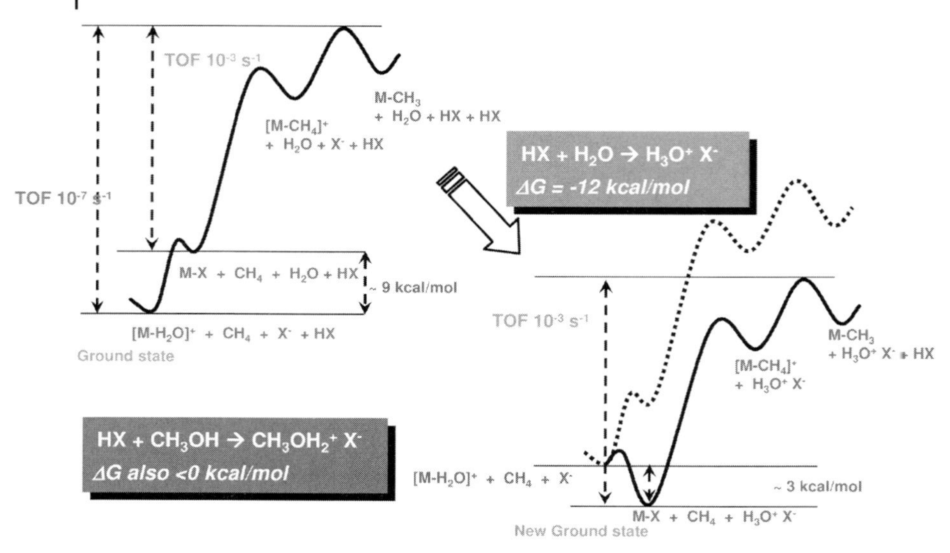

Fig. 7.23 Addition of an acidic species, HX, destabilizes the ground state for the CH activation reaction.

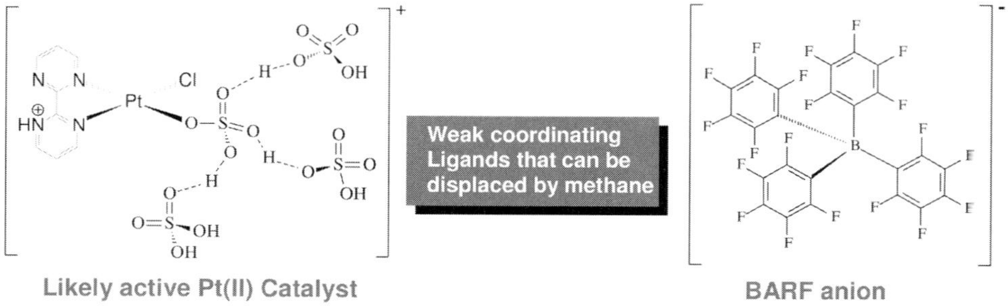

Likely active Pt(II) Catalyst

Weak coordinating Ligands that can be displaced by methane

BARF anion

Fig. 7.24 Comparison of the likely active catalyst in the $Pt(bpym)Cl_2/H_2SO_4$ system with the weakly coordinating BARF anion.

acid to HSO_4^- leads to a highly dispersed anion that is similar to the weakly coordinating anion, $B(C_6F_5)_3Me]^-$, and that can be expected to be displaced by methane more readily than water.

Calculations show [10] that replacement of the HSO_4^- ligand by methane in the $Pt(bpym)Cl_2/H_2SO_4$ system is uphill by 24 kcal/mol with a barrier of ~33 kcal/mol (Fig. 7.22). This is comparable to the ~28–30 kcal/mol barrier obtained experimentally for the CH activation step by carrying out the reaction in D_2SO_4. Theoretical and experimental studies show that the overall activation energy for CH activation with the $Pt(bpym)Cl_2/H_2SO_4$ system results from the two key

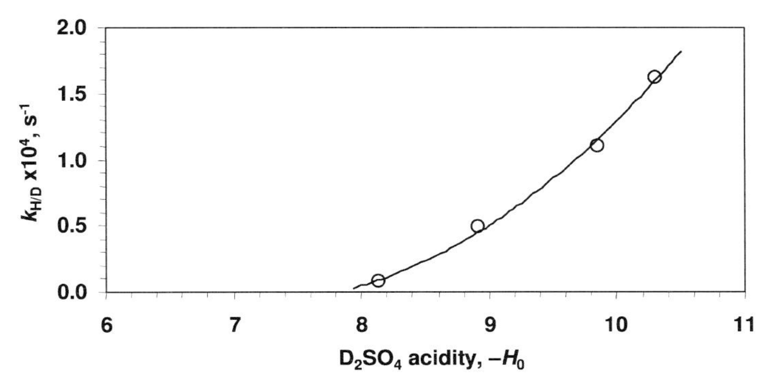

Fig. 7.25 Correlation of CH activation rate (measured as CH_4/D_2SO_4 H-D exchange turnover frequency at 150 °C) with solvent acidity.

steps shown in Fig. 7.21: A) CH coordination followed by B) CH cleavage. Of these two contributions, the ΔH for methane coordination (27 kcal mol^{-1}, L=X=HSO_4^-) far outweighs that for ΔH‡ for CH cleavage (5 kcal mol^{-1}). Further, calculations confirm that the inhibition by H_2O is a result of ground state stabilization (rather than transition state destabilization of the CH cleavage step) that increases the ΔH for methane coordination by ~9 kcal mol^{-1}.

Importantly, as a result of the large excess of sulfuric acid solvent (the catalyst concentration is typically 5–50 mM), substantially more than one equivalent of methanol (relative to the catalyst) can be generated in this system before catalyst inhibition due to water or methanol binding slows reaction to impractical rates. Thus, with a catalyst concentration of 50 mM, and starting with 100% sulfuric acid solvent, ~300 turn-overs have been demonstrated with the generation of >1.5 M methanol at ~80% conversion of methane with >90% selectivity to methanol [5a]. At these high levels of water and methanol concentrations the activity of water and methanol are considerably higher because the sulfuric acid concentration is reduced. Experimental studies show (Fig. 7.25), that at acid concentrations below 80% sulfuric acid the reaction rates are too low to be useful (~10^{-7} s^{-1}) at 200 °C. In this acid concentration range, the CH activation step is rate limiting and is at least 10000 times slower than at 96% sulfuric acid solvent concentration.

The basis for this large difference in rate can be explained by calculations [10] (Fig. 7.22), that show that the water complex, $[(Hbpym)PtCl(H_2O)]^{2+}$ is ~9 kcal/mol more stable than the $[(Hbpym)PtCl(HSO_4)]^+$ complex. Consistent with the expected dependence on solvent acidity, as can be seen from Fig. 7.25, the drop off in rate below 85% sulfuric acid solvent correlates well with the solvent acidity. As shown schematically in Fig. 7.26, this decrease in rate is consistent with a kinetic model that show that the added ground state stabilization in the presence of water (~9 kcal/mol) would be expected to decrease the TOF by a pre-equilibrium factor of ~10^{-4}.

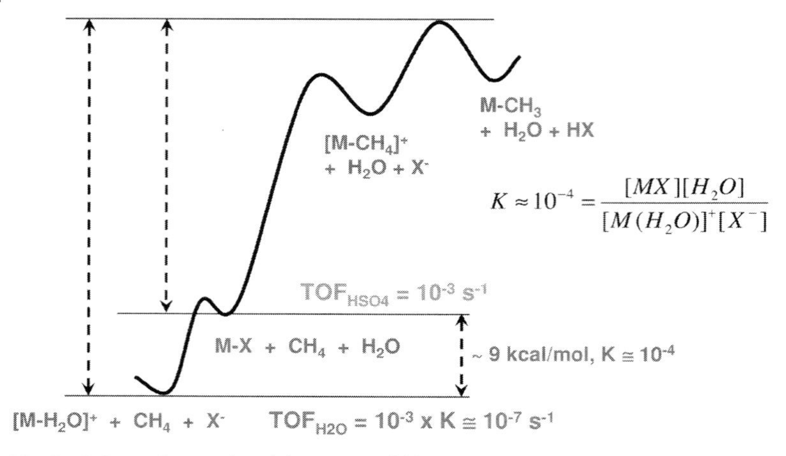

Fig. 7.26 Ground state destabilization could have dramatic influence on rate of reaction.

Critically, while the use of sulfuric acid allows the catalytic reaction to proceed efficiently, the rapid inhibition of the catalysts after ~1 M methanol is produced leads to uneconomical catalyst rates (for the Pt system) and high separation costs for the methanol (for both the Pt and Hg systems). Calculations show that if catalyst inhibition can be minimized to allow an ~5 M solution of methanol to be obtained, with an overall catalyst TOF of ~1 s^{-1} that a process based on the use of sulfuric acid could potentially be commercially viable.

7.3.3.2.3 Catalyst Modifications that Minimize Catalyst Inhibition by Ground State Stabilization

The other approach to overcoming the inhibition of the catalyst from ground state inhibition that we have been investigating is to tune the electronics of the metal complex to minimize coordination to available ligands. This could allow catalysts to exhibit high rates in strong acid solvents and mitigate the product separation issues inherent to the use of strongly acidic solvents. Increasing electron density at the Pt(II) center could be expected to reduce H_2O or CH_3OH binding to σ-acceptor orbitals on Pt(II) since H_2O and CH_3OH are σ- or π-donors but not π-acids, thereby minimizing inhibition by destabilizing the ground state. Significantly, Pt(II), with filled d-orbitals in the valence level can interact with both σ-acceptor and π-donor orbitals in the TS for CH cleavage (Fig. 7.27). Cleavage of the CH bond by electrophilic substitution involves CH donor interactions to σ-acceptor orbitals on Pt(II). This mode of cleavage would likely be destabilized by increasing electron density at the metal center. However, CH cleavage can also occur by Pt(II) insertion into the CH bond (oxidative addition) followed by rapid proton loss. Such insertion reactions would involve interactions between the CH σ^* anti-bonding orbitals and π-donor orbitals on Pt(II).

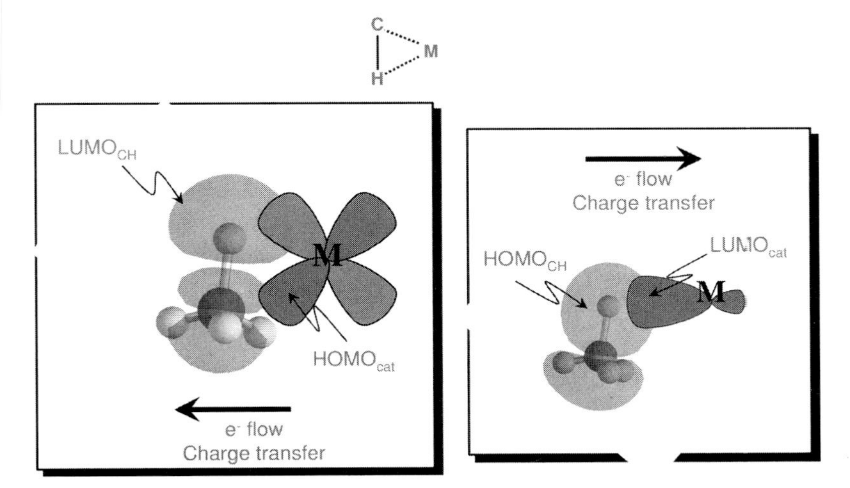

Fig. 7.27 Two modes of interaction between CH bonds and metals with occupied d-orbitals in the valence shell.

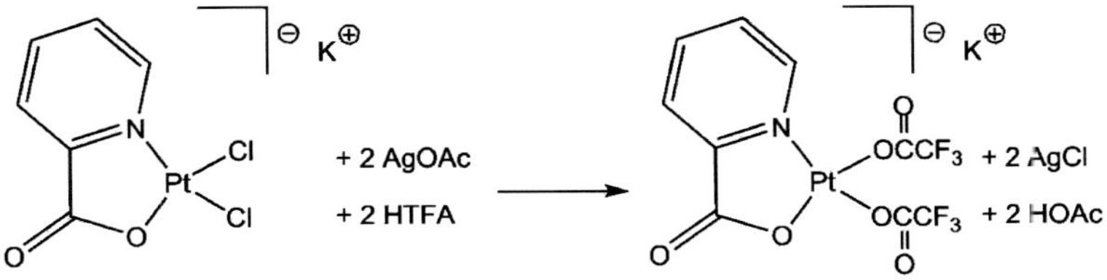

$[(pic)PtCl_2]^{\ominus} K^{\oplus}$ + 2 AgOAc + 2 HTFA ⟶ $[(pic)Pt(OC(O)CF_3)_2]^{\ominus} K^{\oplus}$ + 2 AgCl + 2 HOAc

Fig. 7.28 Synthesis of the (κ^2-N,O-picolinate) bis-TFA Pt(II) complex.

Thus, increasing electron density could potentially facilitate CH cleavage as well as *destabilize* water coordination to lower the overall barrier. To accomplish this we examined replacing the neutral N,N bipyrimidine ligand with a potentially more electron-donating, mono-anionic ligand, κ^2-N,O-picolinate (pic) [18].

In-situ NMR studies of the $[(pic)Pt(TFA)_2]K$ complex in TFA showed that the complex was thermally stable at 70 °C in air for several days. A slow reaction occurs at 100 °C with a $t_{1/2}$ of ~10 days to quantitatively generate $(pic)_2Pt$ and a black precipitate believed to be Pt black. The CH activation studies were conducted with benzene where reaction could be readily observed at 70 °C since reactions with methane were found to require temperatures above 100 °C. On the basis of GC/MS data analysis, the $[(pic)Pt(TFA)_2]K$ complex was shown to be 300 times *more* active for H/D exchange between benzene and CF_3CO_2D than $(bpym)Pt(TFA)_2$. Calculations of CH activation with these systems (Fig. 7.29) are consistent with the experimental results. According to calculations, $[(pic)Pt(TFA)_2]K$ activates benzene C-H bonds with a $\Delta H^{\ddagger}$ ~21 kcal/mol. This is significantly lower than the ~27 kcal/mol calculated for $Pt(bpym)TFA_2$

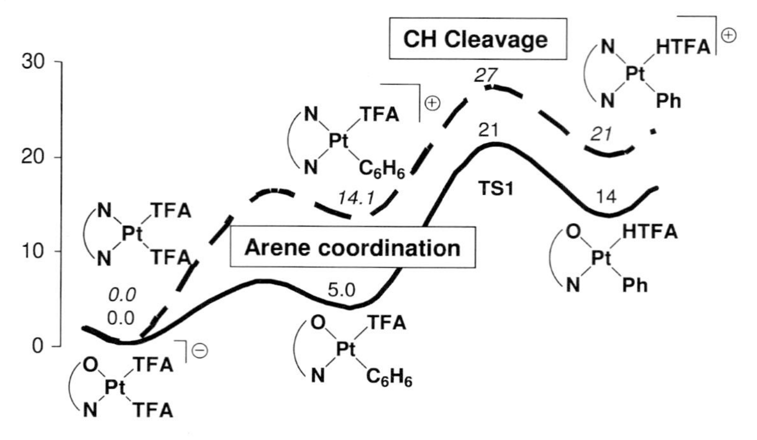

Fig. 7.29 Thermodynamics of calculated mechanism for the benzene C-H activation: (pic)Pt– solid line; (bpym)Pt – *dashed line.*

and comparable to the experimental activation barrier of ~23 kcal/mol obtained for [(pic)Pt(TFA)$_2$]K.

As can be seen in Fig. 7.29, the calculated resting state of the Pt(bpym) system is a bis-TFA complex with an overall neutral charge, Pt(bpym)TFA$_2$. The resting state of the picolinate system is a bis-TFA, anionic complex, [(pic)Pt(TFA)$_2$]$^-$. The calculated ΔH for loss of a TFA$^-$ and coordination of benzene is ~9 kcal/mol lower than that for the neutral Pt(bpym)(TFA)$_2$ complex and is consistent with our original expectation that increasing the electron density at Pt could facilitate substrate binding. However, as the ΔH‡ for CH cleavage shows an increase of only ~3 kcal/mol for the picolinate complex an overall decrease in the rate of CH activation for the picolinate complex would be expected.

The C-H cleavage occurs through a six-membered transition state, **TS1** (Fig. 7.30), where the covalent Pt-O and C-H bonds are broken while the Pt-C and O-H bonds are created. Both C-O bonds are 3/2 bond order, with the π-bond transforming into a σ-bond. Similar 6-membered, cyclic transition states

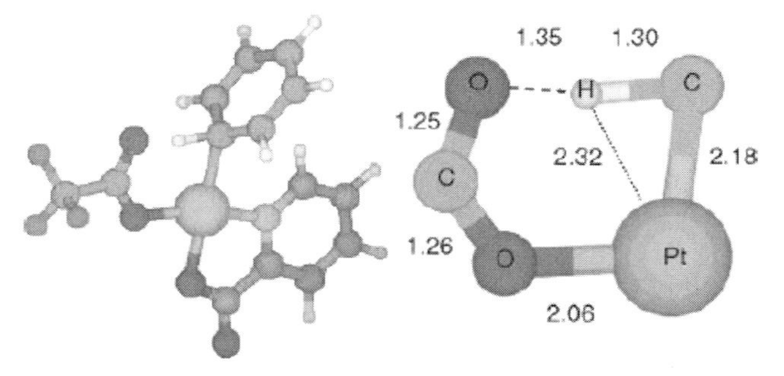

Fig. 7.30 C-H cleavage transition state TS1. Bond lengths in Å.

have previously been postulated in a recent computational study for cyclometalation of dimethylbenzylamine [19] and by Ryabov [20]. The reaction is best described as an electrophilic substitution that proceeds via addition of the Pt center to the arene ring. Similar transition states have been implicated in other theoretical works [19, 21]. We are continuing to study these systems to determine why the $\Delta H^{\ddagger}$ for CH cleavage for the Pt(pic) and the Pt(bpym) systems differ by only 3 kcal/mol given the 9 kcal/mol differences in the ΔH for benzene coordination (Fig. 7.30).

7.3.3.2.4 Heterolytic CH Activation with Electron-rich Metal Complexes

Heterolytic CH activation (Eq. 3), where X is a heteroatom, is well suited for incorporation into catalytic cycles for the oxy-functionalization of hydrocarbons and is relatively common with the electrophilic metals Pt(II), Pd(II), Hg(II), Au(III)/(I) [22].

$$\text{C-H} + \text{M-X} \rightarrow \text{M-C} + \text{H-X} \tag{3}$$

However, this type of CH activation is not common with more electron-rich metals [23] and we have been working to develop such systems as they may be less susceptible to the water inhibition that is characteristic of more electrophilic catalytic systems. The complex *trans*-(acac-O,O)$_2$Ir(III)(OH)(py), Ir-OH was synthesized from the methoxo complex (acac-O,O)$_2$Ir(III)(OCH$_3$)(CH$_3$OH) in quantitative yield by reaction with water at 70 °C, followed by treatment with pyridine [24].

Benzene CH activation with Ir-OH was carried out in neat C_6H_6 at 160 °C for 10 h and cleanly generates that *trans*-(acac)$_2$Ir(C_6H_5)(py), Ir-Ph in 71% yield. Typical of other well-defined CH activation reactions [4], reactions with toluene

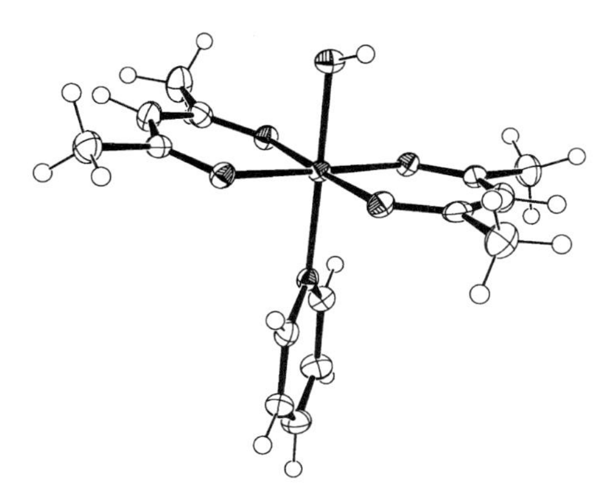

Fig. 7.31 ORTEP diagram of *trans*-(acac-O,O)$_2$Ir(OH)(py), showing ellipsoids at the 50% probability level. A molecule of cocrystallized $CHCl_3$ has been omitted for clarity. Selected bond distances (Å): Ir1-O5, 2.018(4); Ir1-N1, 2.044(5).

show that only the meta and para aromatic CH bonds are activated in a ~2:1 ratio. Both the CH activation rate and selectivity is insensitive to added oxygen suggesting that a free-radical mechanism is unlikely. Having established that Ir-OH can stoichiometrically activate the CH bonds of benzene, the complex was examined for catalytic CH activation of benzene since examples of metal-catalyzed H/D exchange with water are relatively rare and the $Pt(bpym)Cl_2$ is inactive in aqueous solvent systems [25].

The rates of H/D exchange of mixtures of C_6H_6 and D_2O catalyzed by Ir-OH (0.1 mol%) were measured at 190 °C. The system is stable over the time period studied (~86 h) and turn-over-numbers (TON) of 329 and turn-over-frequencies (TOF) of 1.1×10^{-3} s^{-1} were observed based on added Ir-OH. The CH activation is likely to proceed via a mechanism involving substrate coordination and CH cleavage by hydrogen transfer to the hydroxo group via electrophilic substitution (ES), insertion (oxidative insertion, OA), oxidative hydrogen migration (OHM) or sigma-bond metathesis (SBM). Ir-OH is a 6-coordinate, 18-electron complex and it is likely that the benzene coordination proceeds via a five-coordinate intermediate generated by dissociative loss of pyridine in a pre-equilibrium step. The observed linear dependence of the TOF for H/D exchange versus 1/[Py] is consistent with this proposal (Fig. 7.32). The aquo complex $(acac\text{-}O,O)_2Ir(OH)(\text{-}H_2O)$ could be expected to be a more effective catalyst and preliminary results on the synthesis and testing of this complex show that this is the case.

To further probe the mechanism of the CH activation reaction, the deuterium kinetic isotope effect (KIE) for reaction of Ir-OH with a mixture of C_6H_6/C_6D_6 with that for 1,3,5-trideuterobenzene was obtained. This comparison can allow a distinction between rate -determining benzene coordination and rate-determining CH cleavage to be made (assuming negligible secondary isotope effects) [26]. The KIE for 1,3,5-trideuterobenzene was found to be normal with $k_H/k_D = 2.65 \pm 0.56$. DFT calculations (B3LYP/LACVP** with ZPE and solvent corrections) [27] of this KIE ($k_H/k_D = 2.9$) compares well with this value. However, no KIE was observed ($k_H/k_D = 1.07 \pm 0.24$) for the C_6H_6/C_6D_6 mixture. These results

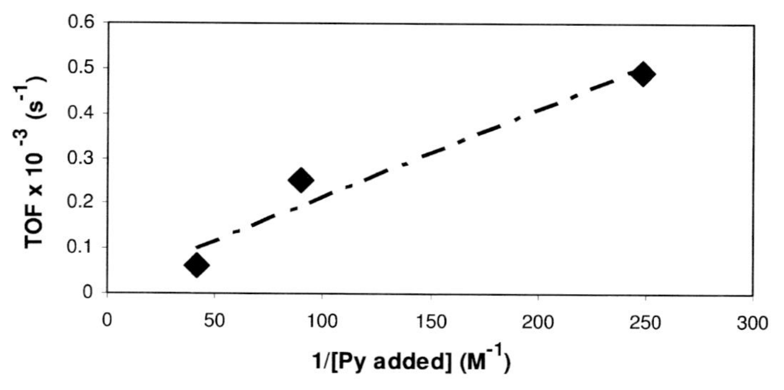

Fig. 7.32 Plot of TOF versus 1/[Py] for C_6H_6/D_2O H/D exchange with trans-5-coordinate intermediate (10 mM).

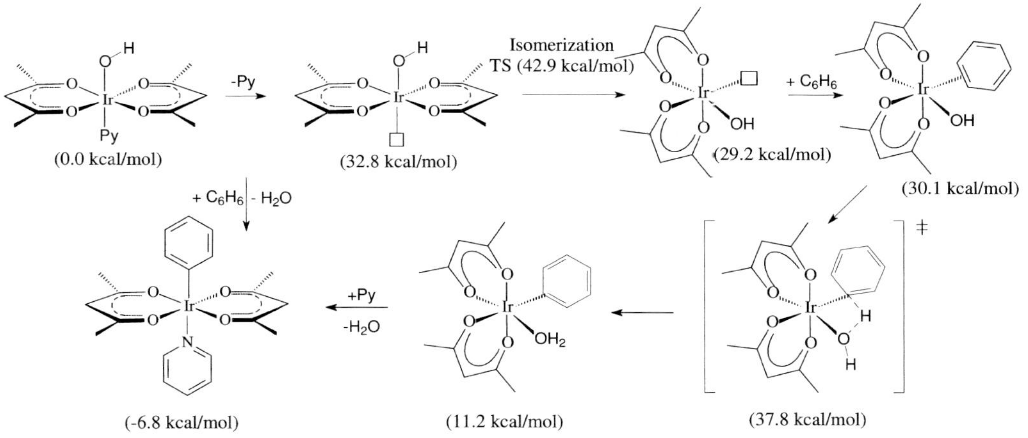

Fig. 7.33 Proposed mechanism for the reaction of trans-5-coordinate intermediate (values in parenthesis are calculated ΔG).

are most consistent with the CH activation of benzene proceeding via rate-determining benzene coordination followed by faster CH cleavage.

Consistent with these results, as shown in Fig. 7.33, the DFT calculations of the CH activation mechanism show that the lowest energy pathway involves pre-equilibrium, dissociative loss of pyridine to generate a trans-5-coordinate intermediate, followed by rate-determining trans-cis isomerization to generate the cis-benzene complex and fast CH bond cleavage by a sigma-bond metathesis transition state.

The calculations, unlike those reported recently for a Ru(II) hydroxo complex [28], show that the CH activation reaction is energetically favorable and is consistent with the formation of Ir-Ph from Ir-OH in good yield. The calculated overall barrier for reaction of Ir-OH with benzene of 42.9 kcal/mol and lower barrier for Py loss are consistent with the experimental results that show that the reaction proceeds at 190 °C with a $t_{1/2} \sim 4.8$ h and an inverse rate dependence on Py. The calculations also support the results of the KIE experiments that indicate that benzene coordination rather than CH cleavage is rate determining.

It is interesting that Ir-OH exhibits stable catalysis (TON of > 300 observed) and is not deactivated by the irreversible formation of bridging hydroxo species (assuming thermodynamic control) [29] that is common for metal hydroxo complexes. Our attempts at synthesis of these bridging hydroxo complexes have thus far been unsuccessful. Calculations show that the formation of a μ-hydroxo bridged complex from two molecules of Ir-OH is not very favorable ($\Delta G_{rxn} = -2$ kcal/mol) and may serve to explain the catalytic stability of Ir-OH.

7.3.3.3 Coupling CH Activation with Functionalization

As discussed above, on the basis of process considerations, the preferred mode of carrying out the hydroxylation of methane would be to utilize a "Wacker" type system using an air-recyclable oxidant. CH activation-based systems are quite amenable to such process considerations and both the Hg(II) and (bpym)Pt(II)/H_2SO_4 are examples of such systems, where S(VI)/S(IV) redox cycle plays the role of the "Wacker-type" oxidant. A generalized scheme of this type of system is shown in Fig. 7.34 which emphasizes the possibility of avoiding the generation of free-radicals in the methane oxidation reactor by the use of "singlet" air-recyclable oxidants, Ox.

7.3.3.3.1 Functionalization by Formal C-O Reductive Eliminations

If stable motifs can be identified that would allow CH activation to occur under conditions required for functionalization, i.e. at higher temperatures and in the presence of an oxidant, without being inhibited or destroyed by the reactants or products of the reaction, then efforts can be focused on oxidative conversion of the M-C intermediate to functionalized products. As indicated above, to maintain the influence of the catalyst on both the cleavage of the CH bond as well as the functionalization to the desired C-X product, the alkyl fragment of the M-C intermediate should also ideally be functionalized while remaining in the inner sphere of the metal. It is anticipated that in the functionalization step there could be a greater extent of electronic configuration change than in the CH activation reaction as the formal oxidation state of the alkyl fragment in the M-C intermediate will increase by two units due to the formal assignment of oxidation states to M-C and C-X species, where X is a heteroatom. This is, of course, a formalism, but it likely reflects some degree of reality. Depending on the actual extent of electron transfer, the driving force and structural changes around the M fragment, this reaction could be a facile or a challenging step. The conversion of M-C to C-X is well documented only for reactions involving formal reduction at M (typically electron-poor metals) and it is likely that other

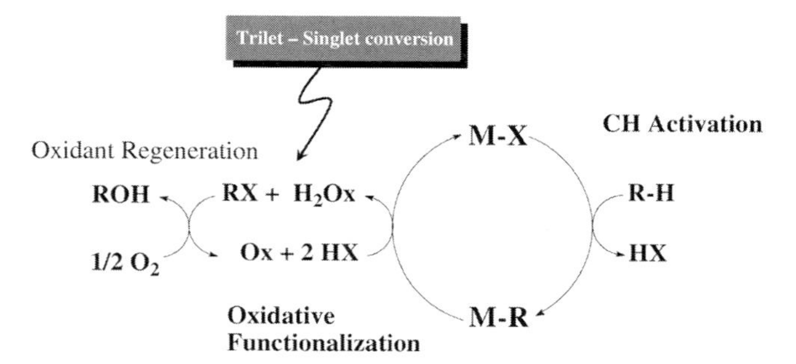

Fig. 7.34 Wacker type system can avoid issues with unselective free-radical reactions and likely explosive mixtures.

types of functionalization reactions will be required to develop new oxidation catalysts based on the CH activation reaction.

One of the simplest systems that illustrates the formal reductive elimination pathway is the $Hg(II)/H_2SO_4$ system [6b]. As noted above, this system operates by an ESCH Activation reaction mechanism. The MC intermediate resulting from CH activation, $[CH_3Hg]^+$ HSO_4^-, undergoes a formal reduction to generate CH_3OSO_3H and Hg(0). It is interesting to speculate on how the methanol is formed in this reaction. It is well known that Hg(II) with strong field ligands such as CH_3^- adopts a linear, two coordinate geometry. Consequently, a concerted reductive elimination is unlikely. On the basis of preliminary theoretical and experimental studies, we can speculate that the reaction occurs by solvent assisted heterolysis of the $[CH_3\text{-}Hg]^+$ species with simultaneous capture of the departing electrophilic CH_3 fragment by H_2SO_4 (or HSO_4^- and H_2O which can be expected to be present in concentrated sulfuric acid at 180 °C) to generate CH_3OSO_3H or CH_3OH and Hg(0). Notably, Hg(0) is not observed as it reacts rapidly with Hg(II) in a comproportionation reaction to generate Hg_2(II). The reaction is presumably aided by the polar sulfuric acid solvent but as this is not strong nucleophilic, the reaction is most likely largely dissociative in character. The driving force for this reaction must be the relatively high oxidation potential for Hg(II)/Hg(0) which in sulfuric acid is greater than the standard potential of 0.85 V. We initially speculated that this heterolysis may be aided by free Hg(II) in an electrophilic substitution to directly generate Hg_2(II) (which we observe in solution) and CH_3X as shown in Fig. 7.35. However, kinetic studies [5b] on $[CH_3Hg]^+$ in hot sulfuric acid show that the rate of formation of CH_3OH is independent of the concentration of added Hg(II). The Hg_2(II) species are rapidly reoxidized to Hg(II) by hot H_2SO_4.

This Hg(II) system is interesting in that the higher oxidation state seems to be the active catalyst (or species) for the CH activation. Thus, studies show that Hg_2(II) is not active for CH activation. Thus the properties of Hg(II) that lead to efficient oxidation with methane in strongly acidic media can be described as "soft", "redox active" and electrophilic. These properties are shared by the third row elements in the middle to right of the periodic table due to the high Z_{eff}, high principal quantum number and large size. Predictably, we have found that Tl^{III}, Au^{III}, $Pt^{(II)}$ and Pd^{II} are all active species for reaction with methane in concentrated H_2SO_4 and are inhibited below ~85% sulfuric acid presumably due to ground state stabilization as discussed above.

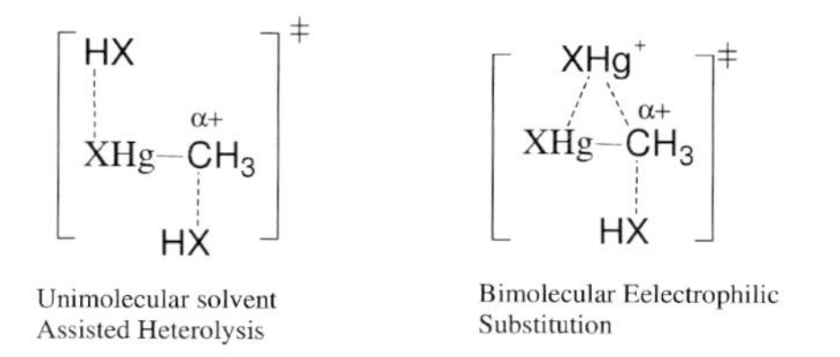

Fig. 7.35 Possible mechanisms for the functionalization of $[CH_3Hg]^+$.

The reactions with Tl^{III} are stoichiometric but cleanly convert methane to methanol by the general reaction shown in Eq. 4.

$$TlX_3 + CH_4 \rightarrow CH_3X + TlX + HX \quad \text{where} \quad X = HSO_4 \tag{4}$$

The reaction with Au(III) is interesting in that Au(0) is generated with clean formation of CH_3X by the stoichiometry shown in Eq 5.

$$2\,AuX_3 + 3\,CH_4 \rightarrow 3\,CH_3X + 2Au(0) + 3\,HX \tag{5}$$

Since Au(I) is isoelectronic with Hg(II), and Au(III) is isoelectronic with Pt(II), and both Hg(II) and Pt(II) have been found to be active, it is possible that both Au(I) and Au(III) could be capable of CH activation of methane in strong acids by ES mechanisms. As shown in Fig. 7.36, catalytic cycles can be written with either Au(I) or Au(III) involved in the CH activation reaction followed by oxidation and functionalization steps that lead to irreversible formation of Au(0). Preliminary theoretical studies suggest both mechanisms are energetically plausible. We have attempted to study the discrete reaction of CH_3AuX_2 by synthesizing this species in strongly acidic media. However, thus far, attempts to generate such species without soft ligands such as PR_3 have led to decomposition to Au(0).

As noted in Fig. 7.36, if the Au(0) could be reoxidized it should be possible to make the systems catalytic. In considering possible oxidants it was necessary to ensure that the reoxidation reaction would not destroy the methanol product and that the acidity of the solvent would not be reduced by the presence of the oxidant. A good choice is selenic acid, H_2SeO_4. This acid is almost as strong as sulfuric acid but is a much more powerful oxidant and importantly is known to dissolve gold metal. Using this reagent we have found that the Au(III) system

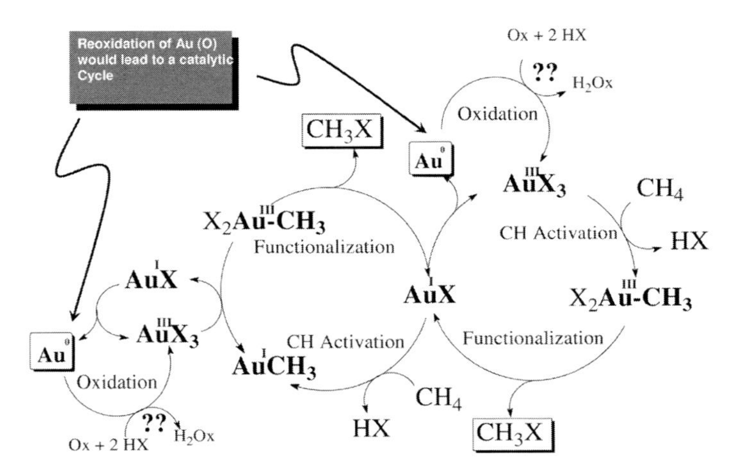

Fig. 7.36 Proposed mechanism for the oxidation of methane with Au(III) and Au(I).

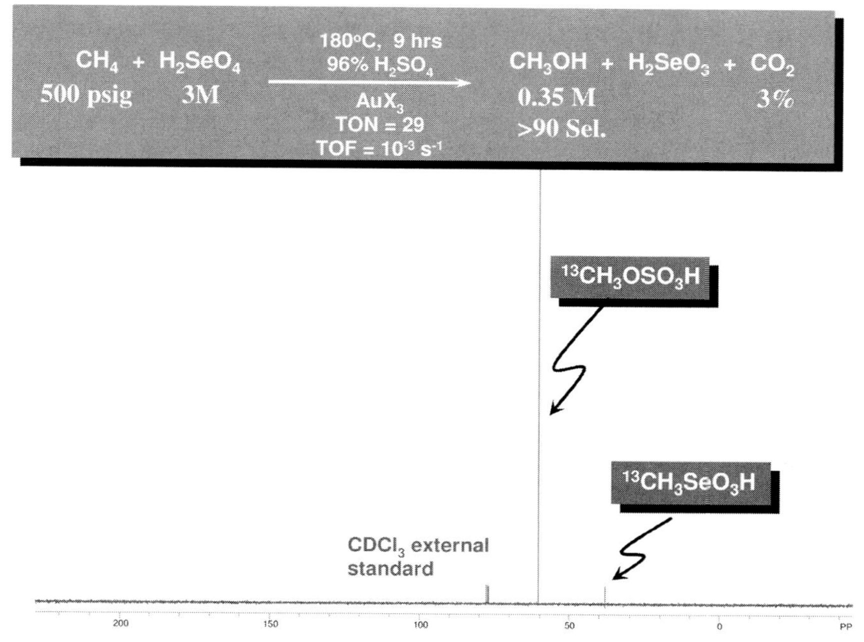

Fig. 7.37 ^{13}C NMR of the crude reaction mixture of methane with H_2SeO_4 dissolved in H_2SO_4.

can be made catalytic and we have observed ~30 turn-overs to generate a 0.35 M methanol solution with a TOF of ~10^{-3} s^{-1} at 500 psig of CH_4 at 180 °C as shown in Fig. 7.37.

The reaction is quite clean as shown by the ^{13}C NMR of crude reaction mixture from the reaction carried out with 100% enriched $^{13}CH_4$.

7.3.3.3.2 **Functionalization by Oxidative Insertion**

One of the key issues with using strongly acid solvents to generate products such as methanol is the high affinity between methanol and the solvent that can lead to costly separation unless high concentrations of the product can be obtained. An approach that we are exploring is to generate products that can potentially be more readily removed from strongly acid media, such as carboxylic acid. The simplest target could be acetic acid generated by oxidative carbonylation in sulfuric acid by the proposed mechanism shown in Fig. 7.38. In this reaction the M-CH_3 bond is functionalized by insertion of CO, followed by reduction at the metal center.

We have recently found that this reaction is feasible and can be carried out with both the (bpym)$PtCl_2$/H_2SO_4 system as well as with Pd(II) in sulfuric acid [30]. The reaction with Pd(II) is very interesting because, while this system can catalyze the additional oxidative insertion of CO into a CH bond of methane, it

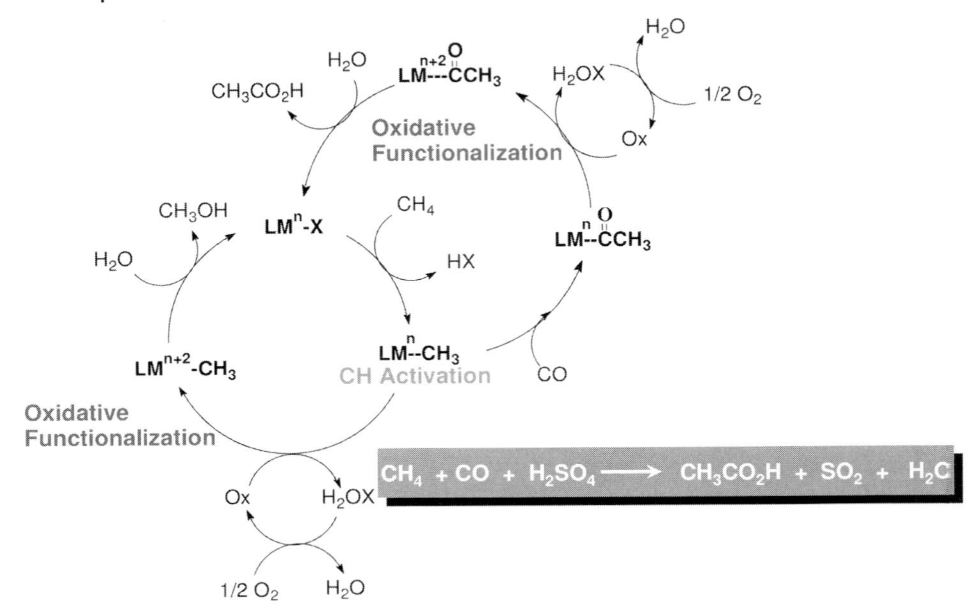

Fig. 7.38 Plausible mechanism for the oxidative carbonylation of methane via the CH activation reaction.

is also the first system that oxidatively couples two methane molecules to generate acetic acid, *without the need for added CO*. Thus, as shown in Fig. 7.39,, reaction of $^{13}CH_4$ at 180 °C with Pd(II) dissolved in H_2SO_4 cleanly leads to $^{13}CH_3{}^{13}CO_2H$ as the major product. The reaction is efficient and ~12% yield, based on added methane, can be obtained with a selectivity of ~90%. The reaction stops after ~20 turn-overs due to catalyst deactivation and formation of Pd metal.

We are currently studying this reaction mechanism and believe that the reaction proceeds via an ESCH activation reaction, followed by CO insertion, likely involving a $Pd(CO)_x^{2+}$ species that may be generated by overoxidation of the methanol. Evidence for this is that added ^{13}CO as well as added $^{13}CH_3OH$ reacts with $^{12}CH_4$ in separate experiments to both generate $^{12}CH_3{}^{13}CO_2H$ [30]. Consistently, addition of $^{12}CH_3OH$ to a $^{13}CH_4$ reaction mixture generated primarily $^{13}CH_3{}^{12}CO_2H$. Significantly, low levels of CO (added or produced by *in situ* methanol oxidation) is effective for the generation of acetic acid. However, higher concentrations of CO shuts down the reaction leading to rapid formation of Pd black and CO_2. These observations are consistent with the mechanism shown in Fig. 7.40. Thus, the observation that the reaction is effectively stopped by high concentrations of CO but facilitated by low levels of CO can be explained if the rate-limiting step is reoxidation of Pd(0) by sulfuric acid. From independent experiments we have found that at high CO concentrations the rate of formation of Pd(0) is much higher than the reoxidation and the overall rate

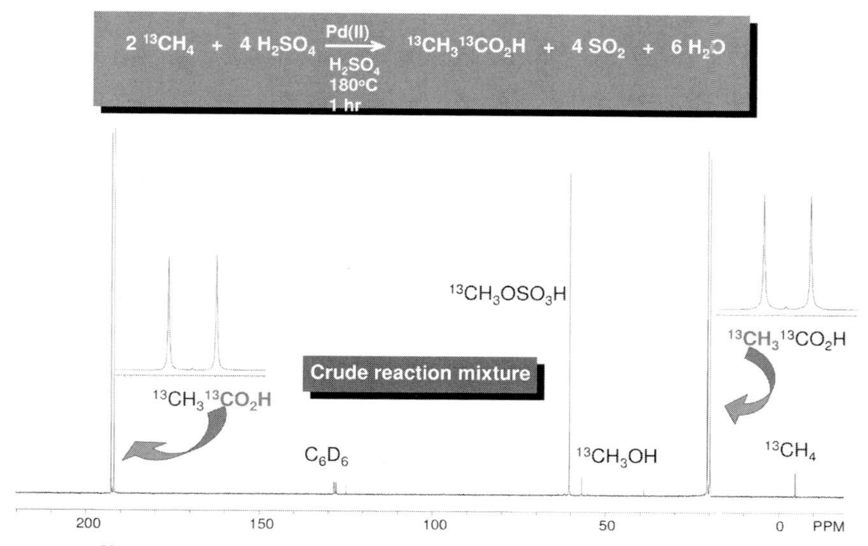

Fig. 7.39 ^{13}C NMR of the crude reaction mixture from the oxidation coupling of methane to acetic acid-catalyzed reaction by $Pd(II)/H_2SO_4$

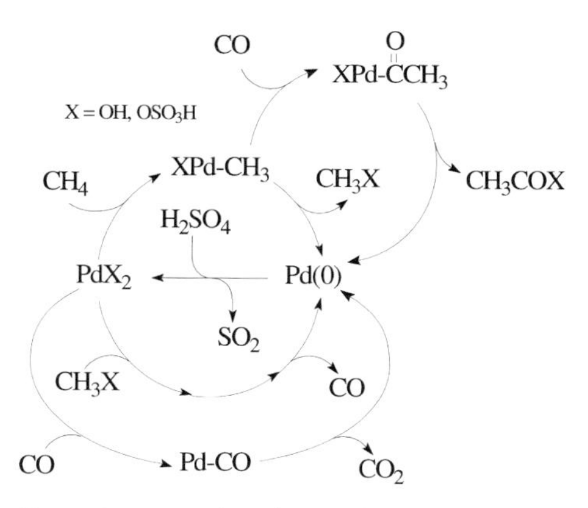

Fig. 7.40 Proposed tandem catalysis mechanism for the oxidative condensation of methane to acetic acid with $Pd(II)/H_2SO_4$.

of acetic acid formation falls off due to loss of the Pd(II) catalyst. Under low CO concentrations, the balance between the rates of Pd(0) reoxidation and Pd(II) reduction (from reactions with methane, methanol and CO) can be balanced and the reaction can be maintained.

7.3.3.3.3 Functionalization by O-Atom Insertion

7.3.3.3.3.1 Simultaneous CH Activation and Functionalization with M-OR Complex

The majority of homogeneous catalysts that have been shown to generate oxy-functionalized products via CH activation operate via reductive heterolysis or elimination reactions based on electronegative, redox-active cations (Hg(II)/(I), Au(III)/(I), Tl(III)/(I), Pd(II)/(0), Pt(IV)/(II) and I(I)/(0)) in poorly coordinating solvents Fig. 7.41 A]. Due to the high electronegativity and redox potentials of these cations, reductive-functionalization reactions are quite facile with this class of reagents. In contrast, these redox-type functionalization reactions are rarely observed with more electron rich, less oxidizing metals such as Ir, Ru, Zr, etc. in the low oxidation states typically required for CH activation. This is likely because reductive functionalization reactions are not thermodynamically favorable or that the systems are not stable to the very oxidizing conditions.

To begin to develop functionalization catalysts based on these more electron-rich systems we have been investigating the development of catalytic cycles based on the reaction of O-donor metal-alkoxo complexes with CH bonds as shown in Fig. 7.41 B). This reaction is intriguing because as shown, the reaction leads to *the simultaneous* CH activation of the hydrocarbon *as well as the formation of a desired oxy-functionalized product,* ROH, in one step. We recently reported [24] the first intermolecular example of such a transformation with *trans*-(acac-O,O)$_2$Ir(OCH$_3$)(CH$_3$OH), MeOH-Ir-OMe and the corresponding pyridine complex, Py-Ir-OMe. There is no precedent for this type of CH activation reaction with alkoxo complexes [31] and such complexes typically decomposition by facile β-hydride elimination reactions or formation of inert dinuclear complexes.

The arene CH activation with these complexes was carried out under an inert atmosphere in neat C_6H_6 (Fig. 7.42) and studies showed that the CH activation product, Ir-Ph [32], is produced. P$_7$-Ir-one provided Ir-Ph in comparable yields but required 4 h and 180 °C. Carrying out the reaction in benzene-d$_6$ led to the formation of labeled methanol, CH$_3$OD, in >95% yield (based on added MeOH-Ir-OMe) which was identified by gas chromatography-mass spectrometry (GC-MS) analyses. No other C$_1$ products were detected. The reaction is insensitive to

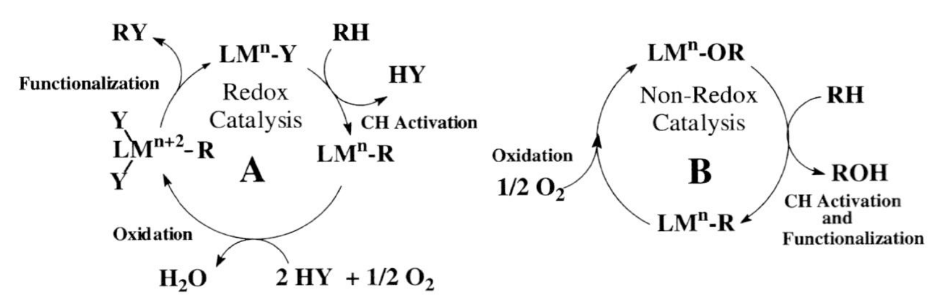

Fig. 7.41 Redox, A, and non-redox, B, catalytic sequences for functionalization of hydrocarbons via CH bond activation.

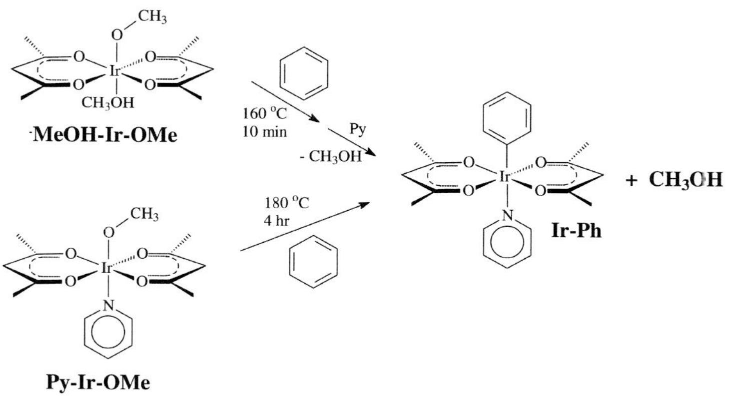

Fig. 7.42 CH activation of benzene by MeOH-Ir-OMe and Py-Ir-OMe.

added oxygen and only CH activation of the *m*- and *p*-CH bonds of toluene are observed. These results would suggest that non-free radical reactions are not involved. It is likely that these properties of the Ir center in an O-donor ligand field could minimize the expected irreversible side reactions of metal alkoxides such as: A) β-hydride elimination reactions by reduction of the electron density at the metal center or B) the formation of bridging alkoxo complexes by the cis-labilization effect of, albeit weakly, π-donor spectator O-ligands [33].

Calculations are consistent with the reaction proceeding via the coordination pathway shown in Fig. 7.43. The slower reaction of Py-Ir-OMe compared to MeOH-Ir-OMe is consistent with reversible loss of L since pyridine is a less la-

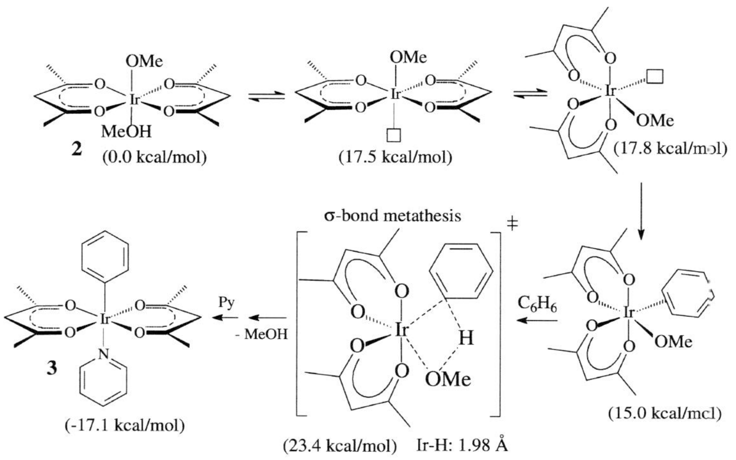

Fig. 7.43 Proposed mechanism for the reaction of 7 (values in parenthesis are calculated ΔH).

bile ligand than methanol. The favorable thermodynamics (–17.1 kcal/mol) and calculated barrier (23.4 kcal/mol) of the reaction with MeOH-Ir-OMe are consistent with the reaction proceeding in good yield and at 160 °C in ~10 min. The transition state for C-H cleavage is best described as involving a σ-bond metathesis as can be observed from the Ir-H distance of 1.98 Å (Scheme 2), which corresponds to classical σ-bond metathesis geometry. Alternative pathways involving oxidative addition, oxidative hydrogen migration [34], or ionization of the methoxide group were all found to be higher in energy.

Ir-H's are well documented to be highly active for CH activation reactions [35] and transition metal alkoxides are well known to decompose to metal hydrides via β-hydride elimination reactions [36]. However, for the CH activation reaction to proceed via a mechanism involving Ir-H's the formation of an Ir-H must necessarily be reversible to account for the stoichiometric formation of MeOH. This possibility was investigated by the reaction of (acac-O,O)$_2$Ir(O^{13}CH$_3$)(Py), with C_6D_6. Under these conditions, the Ir-H pathway would be expected to lead to generation of the D^{13}CH$_2$OD isotopomer whereas the proposed σ-bond metathesis would lead to ^{13}CH$_3$OH(D). The observation, by ^{13}C NMR spectroscopy of the crude reaction mixture showing that only ^{13}CH$_3$OH(D) was formed [37] supports the σ-bond metathesis mechanism shown in Fig. 7.43.

MeOH-Ir-OMe and Py-Ir-OMe both catalyze H/D exchange between D_2O and C_6H_6 at 160 °C. The reactions are stable over the time period studied (6 h) and turn-over-frequencies (TOF) of 2.7×10^{-3} s^{-1} were observed based on added MeOH-Ir-OMe.

7.3.3.3.3.2 **Conversion of M-R to M-OR by O-Atom Insertion**

The catalytic cycle shown in Fig. 7.41 B can be completed by O-atom insertion into a M-C bond to generate a M-OC complex. The conversion of M-R to M-OR is not well known and the few reported examples proceed with O_2 by free-radical pathways [38] or by slow redox reactions involving alkyl to metal oxo migration [39]. Pt(IV) or Hg(II) alkyls are M-C$^{\sigma+}$ polarized and readily undergo reductive functionalization with O$^{\sigma-}$ nucleophiles [6 b, 40], M-Rs of more electropositive metals such as Ir or Re are likely M-C$^{\sigma-}$ polarized in the lower oxidation states that are typically required for CH activation and do not undergo facile reductive functionalizations by reaction with O$^{\sigma-}$ nucleophiles. However, as observed in organic reactions such as the Baeyer-Villiger or alkyl boranes oxidations, Y-C$^{\sigma-}$-polarized intermediates readily undergo C-O functionalization reactions with electrophilic O-atom donors (YO) such as peroxides or iodosyl benzene (PhIO). The analogous organometallic reaction (Eq. 6), has not been reported but should be feasible and coupled with the reoxidation of Y to YO with O_2 could complete the overall catalytic cycle shown in Fig. 7.41 for the oxidation of methane to methanol.

$$\text{M-R} + \text{YO}_\text{M} \rightarrow \text{M-OR} + \text{Y} \qquad (6)$$

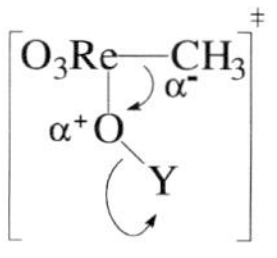

$$\left[\begin{array}{l} O_3Re - CH_3 \\ \quad \alpha^- \\ \alpha^+ O \\ \quad Y \end{array}\right]^{\ddagger}$$

Fig. 7.44 BV-type transition state for O-insertion into Re-CH_3 bond.

We have recently established experimental and theoretical evidence for a facile Re-R to Re-OR bond conversion with non-peroxo YOs that proceeds via a low energy, Baeyer-Villiger (BV) type, electrophilic O-atom insertion (Fig. 7.44).

BV and alkyl borane oxidation reactions to generate oxy-esters and alkoxy boranes, respectively, are well-known organic reactions involving electrophilic O-insertions with YOs. Significantly, both peroxo and non-peroxo YOs can be utilized and the reactions proceed without free-radicals or formal redox changes [41]. Methyltrioxorhenium, MTO, with peroxo YOs is well known to catalyze olefin epoxidation and other oxidation reactions likely via Re η^2-peroxo intermediates [42]. A reported observation that attracted our attention was that an undesirable side reaction is the decomposition of MTO to methanol at room temperature [43]. We were intrigued because, in spite of the high Re(VII) oxidation state, unlike Pt(IV) alkyls [4b, 44], treatment of MTO in basic or acidic water does not generate Re(V) and methanol. Consistent with the observations of the initial investigators [43], we find that the formation of methanol from MTO in water requires added H_2O_2 as the oxidant. The reaction is facile, selective, quantitative and, significantly, proceeds without a change in oxidation state of the Re to generate the ReO_4^- anion.

In the initial studies on decomposition of MTO to methanol, only H_2O_2 was investigated and two non-BV-type mechanisms proposed: reaction via a η^2-peroxo intermediate or by direct methyl migration to the hydroxo of Re-coordinated OOH^-. We considered that since Nature tends to conserve low energy pathways, the reaction may proceed via the BV-type pathway shown in Fig. 7.44, where the leaving group, Y, could be OH^- or H_2O. More significantly, given the ease of functionalization of the Re-CH_3 bond and the d^0 electron configuration, this system could be a useful model to determine if a BV-type pathway was viable without complication from metal-centered oxidations. Establishing that a BV-type pathway is feasible with M-Rs would be useful because, to our knowledge, this functionalization pathway has not been reported, it should be lower energy than η^2-peroxo pathways [45] and accessible with a broader range of potentially more practical, non-peroxo YOs.

To investigate this possibility we compared the reaction of MTO with H_2O_2 and three non-peroxo YOs: PhIO, PyO and IO_4^- in water. As can be seen in Table 7.2, PhIO and IO_4^- are as efficient as H_2O_2 for generation of methanol. Controls show that the selectivities and yields are independent of added O_2 and free-radicals are likely not involved. Facile methanol formation with the non-peroxo YOs is consistent with a low energy BV-type pathway and would rule against direct methyl migration to the β-atom of coordinated YO since the β-atom is not O. However, these observations alone cannot rule out a η^2-peroxo

Table 7.2 Reaction yields [a] and overall calculated barriers [b]

YO	% MeOH	BV TS	μ-Peroxo TS
H_2O_2	80	20	13
PyO	0	32	47
IO_4^-	100	17	25
PhIO	90	8	18

a) Yields based on added MTO (0.1 mM) with 2 eq YO at 25 °C for 1 h under air or argon.
b) B3LYP/LACVP**++ enthalpies in kcal/mol, implicitly solvated in water.

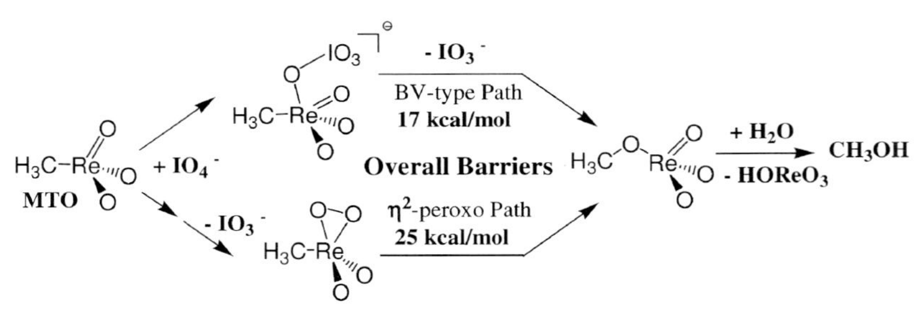

Fig. 7.45 B3LYP/LACVP** calculated low energy pathways for methanol formation from MTO and IO_4^- in H_2O.

pathway with non-peroxo YOs. Significantly, calculations [46] show that a BV-type pathway is both viable and, as shown in Table 7.2, lower in energy than the η^2-peroxo pathways for all the non-peroxo YOs.

For IO_4^- the BV-type and η^2-peroxo pathways, shown in Fig. 7.45, have calculated barriers of 17 and 25 kcal/mol, respectively. The products of the BV pathway are IO_3^- and the methoxo species, $MeORe(O)_3$, which readily hydrolyses to methanol and $Re(O)_3OH$.

The BV-type transition state involves concerted methyl migration and IO_3^- loss as observed by stretching of the C-Re bond from 2.168 Å to 2.516 Å and the I-O bond from 1.803 Å to 2.399 Å. Similar to BV or alkyl borane oxidation reactions in organic chemistry, this transition state can be described as a formal insertion of an electrophilic O into the Re-CH_3 bond. While it is possible that a more exhaustive investigation could lead to alternative low-energy pathways, these results emphasize that a BV-type pathway can be considered as a particularly facile route for M-R functionalizations.

The 17 kcal/mol activation energy calculated for IO_4^- is remarkably low for a M-C to M-O-C transformation given the significant change in electronic configurations. However, this value is consistent with the facile reaction observed at room temperature. As the BV-type transition state is calculated to be significantly favored over a η^2-peroxo pathway, the O in the MeOH product should be

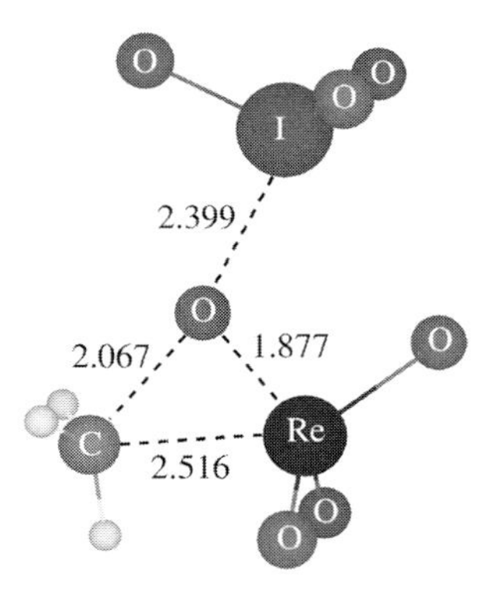

Fig. 7.46 B3LYP/LACVP/6-311G** BV-type transition state for MTO+IO_4^- (bond length values in angstroms, Å).

derived almost exclusively from YO and not from MTO. Consistently, the reaction of ^{16}O-MTO with $[I^{18}O_4]^-$ [47], followed by GC-MS analysis of the reaction mixture at low conversion of MTO showed that only $CH_3{}^{18}OH$ was formed. While this observation supports a BV-pathway, it does not rule out reaction proceeding via an unsymmetrical η^2-peroxo species.

The relatively high calculated BV barrier for PyO of 32 kcal/mol is consistent with the observation that methanol was not formed at room temperature. Realizing that the known $MeOReO_3$ complex [48] should be generated but not hydrolyzed in aprotic media at moderate temperature, we examined the reaction of MTO with one equivalent of PyO in THF-d_8 at 125 °C in the presence of excess pyridine-d_5 by ^1H NMR. It is known [43] and we observe that MTO is quantitatively converted to the MTO-Py-d_5 adduct [49] (s, 1.70 ppm) at room temperature. Upon heating, loss of this adduct is observed along with clean formation of free pyridine-h_5 and the $MeOReO_3$-Py-d_5 adduct (s, 4.48 ppm) based on comparison to the chemical shift of the known $MeOReO_3$-amine adduct [49]. While these results taken individually do not prove a specific mechanism, it is our belief that the convergence between the experimental and theoretical results strongly supports a BV-style mechanism for the functionalization of MTO by non-peroxo YOs.

Calculations of the reaction of MTO with H_2O_2 in water were found to be considerably more complicated than the reaction with non-peroxo YOs due to the multiple possible hydrogen and oxygen rearrangements. Nevertheless, the calculations show two low energy pathways: one via a η^2-peroxo and the other via a BV-type pathway Table 7.2. While the complete mechanism for MTO-H_2O_2 will be addressed in a more thorough study, it is clear that the BV mechanism is feasible even for peroxo YOs such as H_2O_2 [50].

These results are encouraging and may point to a facile pathway for heteroatom functionalization of M-R intermediates of more electron-rich metals via a BV-type pathway with electrophilic, O-atom donors, YO. However, there are some key considerations that must be addressed before we can determine if this pathway will be broadly applicable for M-R functionalizations. In MTO, rhenium is pseudo tetrahedral and formally a d^0 metal. Consequently, competitive oxidation of the metal center, versus O-atom insertion is not an issue in reactions of MTO with YO. Thus, a key question we are investigating is whether this type of concerted, low energy, BV-type transition state can be extended to a range of YOs and M-Rs with other geometries and electronic configurations and the feasibility of incorporation into catalytic cycles.

7.4
Conclusions and Perspective for Methane Functionalization

Today it is possible, for the first time, to quickly and efficiently convert methane to methanol in multi gram quantities by use of the Hg(II) or Pt(bpym)Cl_2/H_2SO_4 catalyst systems using typical equipment available in an organic laboratory. These processes are not economically competitive with commercial processes but serve to demonstrate that the chemistry of methane is advancing to the state where the Latin derivation of the word alkane meaning "inert" will no longer be applicable. The progression of hydrocarbon chemistry over time can be succinctly summarized as shown in Fig. 7.47 and it is highly likely that methane functionalization via CH activation will play a role in the next paradigm change in the power and petrochemical industry.

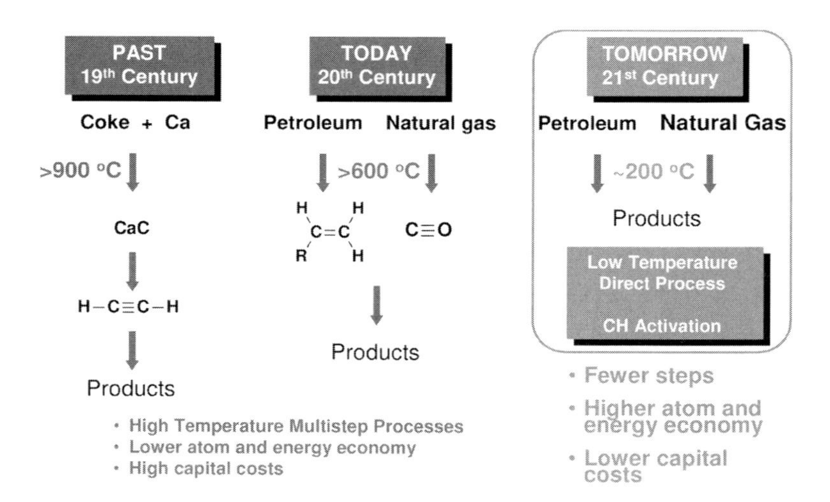

Fig. 7.47 Direct, low temperature methane functionalization is the next step.

References

1 Radler, M., *Oil and Gas Journal* **2003**, *101*, 43–47.

2 (a) *Methane Conversion by Oxidative Processes,* Wolf, E. E. Ed.; Van Nostrand Reinhold, New York, **1992**. (b) *Catalytic Activation and Functionalization of Light Alkanes. Advances and Challenges*, Derouane, E. G.; Haber, J.; Lemos, F.; Ribeiro, F. R.; Guisnet, M. Eds.; Nato ASI Series, Kluwer Academic Publishers, Dordrecht, The Netherlands, **1997**. (c) Lunsford, J. H. *Catalysis Today*, **2000**, *63*, 165. (d) Periana, R. A., *C&E News*, **2001**, *79*, 287. (e) *Natural Gas Conversion II*, Curry-Hyde, H. E.; Howe, R. F. Eds. Elsevier, New York, **1994**.

3 (a) Chen, P. P. Y.; Chan, S. I., *J. Inorg. Biochem.* **2006**, *100*, 801–809. (b) Lieberman, R.L.; Rosenzweig, A.C, *Nature* **2005**, *434*, 177–182. (c) Hanson, R. S.; Hanson, T. E. *Microbiol. Rev.* **1996**, *60*, 439–471. (d) Wallar, B. J.; Lipscomb, J. D., *Chem. Rev.* **1996**, *96*, 2625–2658. (e) Baik, M.-H.; Newcomb, M.; Friesner, R. A.; Lippard, S. J., *Chem. Rev.* **2003**, *103*, 2385–2419.

4 (a) Arndtsen, B. A.; Bergman, R. G.; Mobley, T. A.; Peterson, T. H., *Acc. Chem. Res.* **1995**, *28*, 154 and citations therein. (b) Periana, R. A.; Bhalla, G.; Tenn III, W. J.; Young, K. J.; H.; Liu, X. Y.; Mironov, O.; Jones, C.; Ziatdinov, V. R.; *J. Mol. Cat. A. Chem.* **2004**, *220*, 7 and citations therein.

5 (a) Xu, X.; Kua, J.; Periana, R. A.; Goddard, W. A. III, *Organometallics* **2003**, *22*, 2057. (b) "Stability and Thermodynamics of the PtCl2 Type Catalyst for Activating Methane to Methanol: A Computational Study." Kua, J.; Xu, X.; Periana, R. A.; Goddard, W. A. III, *Organometallics* **2002**, *21*, 511.

6 (a) Periana, R. A.; Taube, D. J.; Gamble, S.; Taube, H.; Satoh, T.; Fuji, H. *Science*, **1998**, *280*, 560. (b) Periana, R. A.; Taube, D. J.; Evitt, E. R.; Loffler, D. G.; Wentrcek, P. R.; Voss, G.; Masuda, T. *Science*, **1993**, *259*, 340. (c) Periana, R. A.; Taube, D. J.; Evitt, E. R.; Loffler, D. G.; Wentrcek, P. R.; Voss, G.; Masuda, T., *Stud. Surf. Sci. Catal.* **1994**, *81*, 533. (d) Periana, R. A. *Adv. Chem. Ser.* **1997**, *253*, 61.

7 (a) Jessop, P. G.; Morris, R. H. *Coord. Chem. Rev.* **1992**, *121*, 155. (b) Heinekey, D. M.; Oldham, W. J. Jr. *Chem. Rev.* **1993**, *93*, 913. (c) Kubas, G. J. *Acc. Chem. Res.* **1988**, *21*, 120. (d) Huhmann-Vincent, J.; Scott, B. L.; Kubas, G. J. *J. Am. Chem. Soc.* **1998**, *120*, 6808. (e) Kubas, G. J. *J. Organomet. Chem.* **2001** *635(1/2)*, 37.

8 (a) Waltz, K. M.; Hartwig, J. F. *Science*, **1997**, *277*, 211. (b) Chen. H. Y.; Hartwig, J. F. *Angew. Chem. Int. Ed. Engl.* **1999**, *38*, 3391. (c) Inverson, C. N.; Smith, M. R. *J. Am. Chem. Soc.* **1999**, *121*, 7696. (d) Chen. H. Y.; Schlecht, S.; Semple, T. C.; Hartwig, J. F. *Science*, **2000**, *287*, 1995. (e) Xu, W.; Rosini, G. P.; Krogh-Jespersen, K.; Goldman, A. S.; Gupta, M.; Jensen, C. M.; Kaska, W. C. *Chem. Commun.* **1997**, *23*, 2273 (f) Liu, F.; Pak, E. B.; Singh, B.; Jensen, C. M.; Goldman, A. S. *J. Am. Chem. Soc.* **1999**, *121*, 4086. (g) Kanzelberger, M.; Singh, B.; Czerw, M.; Krogh-Jespersen, K.; Goldman, A. S. *J. Am. Chem. Soc.* **2000**, *122*,11017. (h) Goettker-Schnetmann, I.; White, P.; Brookhart, M. *J. Am. Chem. Soc.* **2004**, *126*, 1804. (h) Haenel, M. W.; Oevers, S.; Angermund, K.; Kaska, W. C.; Fan, H.; Hall, M. B. *Angew. Chem., Int. Ed. Eng.* **2001**, *40*, 3596. (i) Shen, C.; Garcia-Zayas, E. A.; Sen, A. *J. Am. Chem. Soc.* **2000**, *122*, 4029. (j) Lin, M.; Shen, C.; Garcia-Zayas, E. A.; Sen, A. *J. Am. Chem. Soc.* **2001**, *123*, 1000. (k) Baudry, D.; Ephritikine, M.; Felkin, H.; Zakrzewski, J. *J. Chem. Soc. Chem. Commun.*, **1982**, 1235; (l) Baudry, D.; Ephritikhine, M.; Felkin, H.; Zakrzewski, J. *Tet. Lett.* **1984**, *25*, 1283. (m) Burk, M. J.; Crabtree, R. H. *J. Am. Chem. Soc.* **1987**, *109*, 8025. (n) Sakakura, T.; Sodeyama, T.; Sasaki, K.; Wada, K.; Tanaka, M. *J. Am. Chem. Soc.* **1990**, *112*, 7221.

9 Kushch, K. A.; Lavrushko, V. V.; Misharin, Yu. S.; Moravsky, A. P.; Shilov, A. E. *New J. Chem.* **1983**, *7*, 729.

10 Xu, X.; Kua, J.; Periana, R. A.; Goddard, W. A., III. *Organometallics* **2003**, 22, 2057.

11 (a) Geftakis, S.; Ball, G. E. *J. Am. Chem. Soc.* **1998**, *120*, 9953. (b) Gross, C. L.; Girolami, G. S. *J. Am. Chem. Soc.* **1998**, *120*, 6605. (c) Gould, G. L.; Heinekey, M. *J. Am. Chem. Soc.* **1989**, *111*, 5502.

12 Northcutt, T. O.; Wick, D. D.; Vetter, A. J.; Jones, W. D. *J. Am. Chem. Soc.* **2001**, *123*, 7257.

13 Johansson, L.; Tilset, M. *J. Am. Chem. Soc.* **2001**, *123*, 739.

14 (a) Xu, W.; Rosini, G. P.; Krogh-Jespersen, K.; Goldman, A. S.; Gupta, M.; Jensen, C. M.; Kaska, W. C. *Chem. Commun.* **1997**, *23*, 2273. (b) Liu, F.; Pak, E. B.; Singh, B.; Jensen, C. M.; Goldman, A. S. *J. Am. Chem. Soc.* **1999**, *121*, 4086. (c) Kanzelberger, M.; Singh, B.; Czerw, M.; Krogh-Jespersen, K.; Goldman, A. S. *J. Am. Chem. Soc.* **2000**, *122*, 11017. (d) Goettker-Schnetmann, I.; White, P.; Brookhart, M. *J. Am. Chem. Soc.* **2004**, *126*, 1804. (e) Haenel, M. W.; Oevers, S.; Angermund, K.; Kaska, W.; C.; Fan, H.; Hall, M. B. *Angew. Chem., Int. Ed. Eng.* **2001**, *40*, 3596.

15 *Mechanisms of Inorganic Reactions, 2nd Ed.*, Basolo, F.; Pearson, R. G. Wiley, John and Sons, New York, USA, **1967**.

16 (a) Klei, S. R.; Golden, J. T.; Burger, P.; Bergman, R. G. *J. Mol. Cat. A: Chem.* **2002**, *189*, 79. (b) Tellers, D. M.; Yung, C. M.; Arndtsen, B. A.; Adamson, D. R.; Bergman, R. G. *J. Am. Chem. Soc.* **2002**, *124*, 1400. (c) Burger, P.; Bergman, R. G. *J. Am. Chem. Soc.* **1993**, *115*, 10462.

17 *Superacids*, Olah, G. A.; Prakash, G. K. S.; Sommer, J. Wiley, New Yor, USA, **1985**.

18 Ziatdinov, V. R.; Oxgaard, J.; Mironov, O. A.; Young, K. J. H.; Goddard, W. A., III; Periana, R. A. *J. Am. Chem. Soc.* **2006**, *128*, 7404.

19 Davies, D. L.; Donald, S. M. A.; Macgregor, S. A. *J. Am. Chem. Soc.* **2005**, *127*, 13754.

20 Ryabov, A. D.; Sakodinskaya, I. K.; Yatsimirsky, A. K. *J. Chem. Soc., Dalton Trans.* **1985**, 2629.

21 Kragten, D. D.; van Santen, R. A.; Neurock, M.; Lerou, J. J. *J. Phys. Chem. A* **1999**, *103*, 2756.

22 Dick, A. R.; Sanford, M. S. *Tetrahedron* **2006**, *62*, 2439 and citations therein.

23 (a) Arndtsen, B. A.; Bergman, R. G. *Science* **1995**, *270*, 1970. (b) Wong-Foy, A. G.; Bhalla, G.; Liu, X. L.; Periana, R. A. *J. Am. Chem. Soc.* **2003**, *125*, 14292. (c) Ben-Ari, E.; Gandelman, M.; Rozenberg, H.; Shimon, L. J. W.; Milstein, D. *J. Am. Chem. Soc.* **2003**, *125*, 4714.

24 Tenn, W. J.; Young, K. J. H.; Bhalla, G.; Oxgaard, J.; Goddard, W. A., III; Periana, R. A. *J. Am. Chem. Soc.* **2005**, *127*, 14173.

25 Klei, S. R.; Golden, J. T.; Tilley, T. D.; Bergman, R. G. *J. Am. Chem. Soc.* **2002**, *124*, 2092 and citations therein.

26 (a) Jones, W. D. *Acc. Chem. Res.* **2003**, *36*, 140. (b) Jones, W. D.; Feher, F. J. *J. Am. Chem. Soc.* **1986**, *108*, 4814. (c) Bhalla, G.; Liu, X. Y.; Oxgaard, J.; Goddard, W. A., III; Periana, R. A. *J. Am. Chem. Soc.* **2005**, *127*, 11372.

27 For further computational details, see supporting information.

28 Feng, Y.; Lail, M.; Barakat, K. A.; Cundari, T. R.; Gunnoe, T. B.; Peterson, J. L. *J. Am. Chem. Soc.* **2005**, *127*, 14174.

29 We are examining higher TON and reaction times to ensure that the system is under thermodynamic control.

30 Periana, R. A.; Mironov, O.; Taube, D.; Bhalla, G.; Jones, C. *Science* **2003**, *30*, 814.

31 The related reactions of metal alkoxo and amidos with H_2 and acidic hydrocarbons have been reported by (a) Conner, D.; Jayaprakash, K. N.; Cundari, T. R.; Gunnoe, T. B. *Organometallics*, **2004**, *23*, 2724. For a review of late transition metal alkoxo chemistry see: (b) Fulton, J. R.; Holland, A. W.; Fox, D. J.; Bergman, R G. *Acc. Chem. Res.* **2002**, *35*, 44.

32 Periana, R. A.; Liu, X. Y.; Bhalla, G. *Chem. Commun.* **2002**, 3000.

33 (a) Caulton, K. G. *New J. Chem.* **1994**, *18*, 25 and references therein. (b) Zhou, R.; Wang, C.; Hu, Y.; Flood, T. C. *Organometallics* **1997**, *16*, 434.

34 (a) Oxgaard, J.; Muller, R. P.; Goddard III, W. A.; Periana, R. A. *J. Am. Chem. Soc.* **2004**, *126*, 352. (b) Oxgaard, J.; Periana, R. A.; Goddard III, W. A. *J. Am. Chem. Soc.* **2004**, *126*, 11658.

35 (a) Yung, S. M.; Skaddan, M. B.; Bergman, R. G. *J. Am. Chem. Soc.* **2004**, *126*, 13033. (b) Haenel, M. W.; Oevers, S.; Angermund, K.; Kaska, W. C.; Fan, H.-J.; Hall, M. B. *Angew. Chem. Int. Ed.* **2001**,

40, 3596. (c) Bernskoetter, W. H.; Lobkovsky, E.; Chirik, P. J. *Chem. Comm.* **2004**, 764.

36 (a) Vaska, L.; Di Luzio, J. W. *J. Am. Chem. Soc.* **1962**, *84*, 4989. (b) Bryndza, H. E.; Tam, W. *Chem. Rev.* **1988**, *88*, 1163. (c) Bernard, K. A.; Rees, W. M.; Atwood, J. D. *Organometallics* **1986**, *5*, 390.

37 This is readily evident from the distinctive singlet resonance for $^{13}CH_3OH(D)$.

38 Kim, S.; Choi, D.; Lee, Y.; Chae, B.; Ko, J.; Kang, S. *Organometallics* **2004**, *23*, 559 and references therein.

39 (a) Matano, Y.; Northcutt, T. O.; Brugmann, J.; Bennett, S. L.; Mayer, J. M. *Organometallics* **2000**, *19*, 2781. (b) Brown, S.; Mayer, J. M. *J. Am. Chem. Soc.* **1996**, *118*, 12119.

40 Lersch, M.; Tilset, M. *Chem. Rev.* **2005**, *105*, 2471.

41 Smith, M. B. *Organic Synthesis*, McGraw-Hill: New York, **2004.**

42 (a) Kuhn, F. E., Scherbaum, A.; Herrmann, W. A. *J. Organomet. Chem.* **2004**, 4149. (b) Owens, G. S.; Arias, J., Abu-Omar, M. M. *Catalysis Today* **2000**, *55*, 317. (c) Espenson, J. H. *Chem. Comm.* **1999**, 479 and references therein.

43 Abu-Omar, M. M.; Hansen, P. J.; Espenson, J. H. *J. Am. Chem. Soc.* **1996**, *118*, 4966.

44 Lersch, M.; Tilset, M. *Chem. Rev.* **2005**, *105*, 2471.

45 A peroxo bond is a weak O-O high energy bond, $\Delta H = 33$ kcal/mol.

46 Solvent optimized B3LYP/LACVP** (with corrections for diffuse functions) enthalpies are in kcal mol^{-1}.

47 The rate of O-atom exchange between IO_4^- and MTO is slow compared to the rate of formation of methanol.

48 Edwards, P.; Wilkinson, G. *J. Chem. Soc. Dalton Trans.* **1984**, 2695.

49 Wang, W. D.; Espenson, J. H. *J. Am. Chem. Soc.* **1998**, *120*, 11335.

50 Conley, B. L.; Ganesh, S. K.; Gonzales, J. M.; Tenn, W. J. III; Young, K. J. H.; Oxgaard, J.; Goddard, W. A. III, Periana, R. A. *in press.*

8
Water Activation: Catalytic Hydrolysis

Lisa M. Berreau

8.1
Introduction

8.1.1
Water Activation

Metal-mediated reactions involving water are essential to life and catalytic industrial processes [1–3]. In biological systems, metalloenzymes containing various divalent metal ions catalyze the hydrolysis of amide, carboxylic ester and phosphate ester bonds using both mono- and multinuclear active-site structural motifs [4–6]. Mononuclear metal centers are also found within the active sites of enzymes that catalyze the hydration, or the addition of water, to CO_2 [Zn(II)] and nitriles [Co(III)/Fe(III)] [7–10]. In many of these processes, formation of a metal hydroxide moiety via deprotonation of a metal-coordinated water molecule is a key proposed step in the reaction pathway. Thus, a substantial amount of research over the past several years has been directed at delineating how the structural and electronic environments of biological metal ions influence the pK_a of a metal-bound water molecule. In this regard, studies directed at the preparation, characterization and elucidation of the reactivity of discrete metal aqua and hydroxo complexes have been paramount [11–13].

Notably, another area of interesting research in metal/water interactions involves studies of the acidity and water exchange properties of organometallic aqua ions [14]. Of particular current interest are technetium-containing organometallic aqua complexes for applications in radiopharmaceutical imaging [15].

8.1.2
Catalytic Hydrolysis

Significant effort has been put forth toward the development of synthetic metal complexes capable of hydrolytically cleaving biologically relevant substrates (amides, carboxylic and phosphate esters, DNA, RNA, etc.), both selectively and

Activation of Small Molecules. Edited by William B. Tolman

ISBN: 3-527-31312-5

with high catalytic efficiency [16–20]. In this regard, several recent studies have focused on evaluating how the incorporation of biomimetic secondary hydrogen-bonding (H-bonding) interactions in synthetic metal complexes influences phosphate ester coordination, activation and hydrolysis. The emergence of this line of research parallels efforts toward evaluating how the presence of H-bond donors influences a variety of small-molecule reactions, including dioxygen coordination and activation (Chapter 6) [21].

In this chapter, I summarize newly discovered fundamental chemistry of synthetic complexes that provides insight into how: (i) primary and secondary coordination sphere components influence the acidity and structural features of a metal-bound water molecule, and (ii) how secondary H-bonding interactions influence the metal coordination properties and hydrolytic reactivity of phosphate esters

8.2 Water Activation: Coordination Sphere Effects on M-OH$_2$ Acidity and Structure

8.2.1 Primary Coordination Environment

The most commonly encountered metal ion in hydrolytic metalloenzymes is Zn(II) [4]. Drawings of the zinc centers in selected mononuclear zinc hydrolytic enzymes are shown in Fig. 8.1 along with the pK_a value determined for the metal-bound water molecule [22–24]. These values are found over a range of around 3 pK_a units, indicating that the metal coordination environment plays an important role in influencing the acidity of the metal-bound water molecule. Theoretical calculations, reported by Bertini and coworkers in 1990, suggested that the pK_a of a zinc-bound water molecule should increase with increasing coordination number and the incorporation of anionic ligands [25]. This rationale can be used to explain the increase in pK_a in carboxypeptidase A versus carbonic anhydrase II, but does not provide a rationale for the magnitude of the change. In addition, it is difficult to explain the unusually low pK_a value found for the mutant Co(II)-substituted peptide deformylase. As outlined below, recent studies of model systems, while in some cases substantiating the aforementioned computational study, have revealed additional chemical factors that can significantly influence the pK_a value of a metal-bound water molecule [11]. These include the strength of zinc-ligand bonding and secondary H-bonding interactions.

Mononuclear biological zinc centers generally exhibit an overall coordination number of 4 and a tetrahedral or distorted tetrahedral geometry [4]. This structural motif enhances the Lewis acidity of the zinc center relative to an octahedral coordination environment. Thus, whereas $[Zn(H_2O)_6]^{2+}$ has a pK_a of 9.0 [26], the pK_a value for the zinc-coordinated water in carbonic anhydrase II is 6.8 (Fig. 8.1) [22]. During catalysis, the coordination number of the active site zinc center in a metallohydrolase can increase to 5 [13].

In synthetic $[(N_3)Zn\text{-}OH_2]^{2+}$-type complexes containing neutral nitrogen donor ligands (Fig. 8.2), a pK_a range of 6.2–9.2 is found for the metal-bound

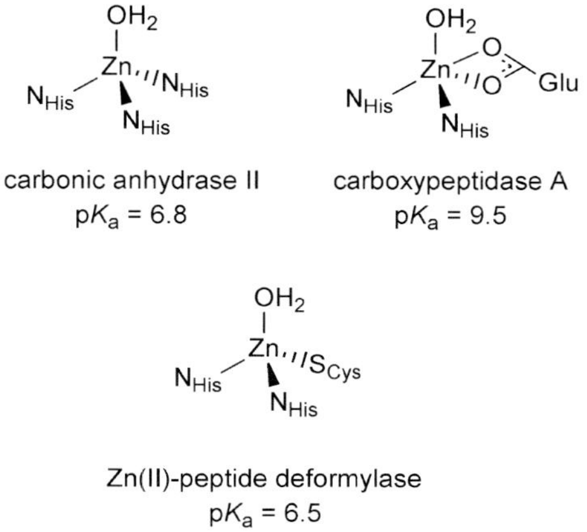

carbonic anhydrase II
pK_a = 6.8

carboxypeptidase A
pK_a = 9.5

Zn(II)-peptide deformylase
pK_a = 6.5

Fig. 8.1 Active-site coordination environments found in representative mononuclear zinc-containing enzymes. The pK_a value for peptide deformylase was measured using a Co(II)-substituted mutant enzyme (E133A) [23].

water [26–31], with most values falling in a narrower range of pK_a of 7.3–8.3. The unusually low pK_a value of 6.2 for $[(THB)Zn\text{-}OH_2]^{2+}$ was hypothesized to result from a hydrophobic effect involving the pseudopeptide ligand, although this remains speculative in the absence of definitive structural characterization of the complex [27]. The relatively high pK_a value of 8.9 for $[\{HN(C_2H_4NH_2)(C_3H_6NH_2)\}Zn\text{-}OH_2]^{2+}$ and $[\{HN(C_2H_4NH_2)_2\}Zn\text{-}OH_2]^{2+}$ has been ascribed to weaker coordination of the amine donor ligands in these complexes, as indicated by more negative values of ΔH_f for LZn-OH_2 versus that found, for example, in $[\{HN(C_3H_6NH_2)_2\}Zn\text{-}OH_2]^{2+}$ (pK_a=8.6) [29]. The acidity constant for $[(DMAM\text{-}PMHD)Zn\text{-}OH_2]^{2+}$ is a kinetic pK_a value [32], whereas all others shown in Fig. 8.2 are thermodynamic pK_a values determined by potentiometric titration.

For synthetic mononuclear zinc complexes having an overall coordination number of 5, and supported by four neutral nitrogen donors and a water molecule ($[(N_4)Zn\text{-}OH_2]^{2+}$; Fig. 8.3), a p$K_a$ range of 8.0–10.7 is found [13, 28, 33–36]. The increase in the pK_a value range relative to the four-coordinate structures is consistent with a reduced overall Lewis acidity for the zinc centers having a higher coordination number. Interestingly, comparison of the pK_a values for $[(N(CH_2CH_2NH_2)_3)Zn\text{-}OH_2]^{2+}$ (10.7) and $[(N(CH_2CH_2N(CH_3)_2)_3)Zn\text{-}OH_2]^{2+}$ (8.9) (Fig. 8.4) suggests that the more electron-rich methyl-substituted ligand produces a zinc center that is more Lewis acidic [34]. This seems counterintuitive, as the more electron-rich tertiary amine donor ligand should produce a less Lewis acidic zinc center. An alternative hypothesis is that the amino substituents of the tripodal $N(CH_2CH_2NH_2)_3$ ligand could stabilize the Zn-OH_2 moiety via H-bonding [34], thus resulting in a higher pK_a value. However, in light of studies that suggest that H-bonding interactions decrease the pK_a value of a metal-bound water molecule (*vide infra*), this rationale needs reconsideration. An alternative rationale is that differing interactions involving counterions and/or solvent may be responsible for producing the difference in pK_a values [11, 34].

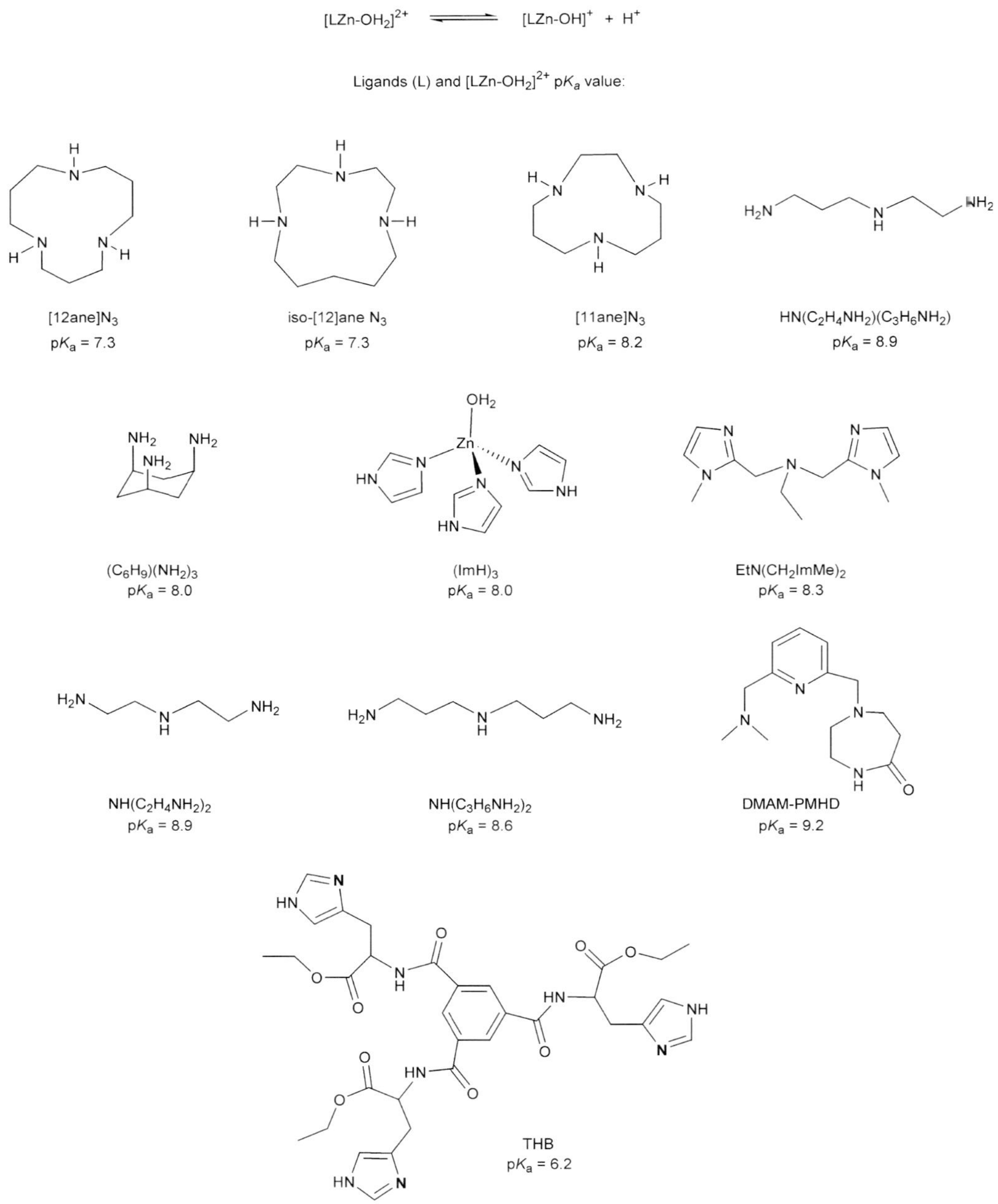

$$[LZn\text{-}OH_2]^{2+} \rightleftharpoons [LZn\text{-}OH]^{+} + H^{+}$$

Ligands (L) and $[LZn\text{-}OH_2]^{2+}$ pK_a value:

$[12ane]N_3$
pK_a = 7.3

iso-[12]ane N_3
pK_a = 7.3

$[11ane]N_3$
pK_a = 8.2

$HN(C_2H_4NH_2)(C_3H_6NH_2)$
pK_a = 8.9

$(C_6H_9)(NH_2)_3$
pK_a = 8.0

$(ImH)_3$
pK_a = 8.0

$EtN(CH_2ImMe)_2$
pK_a = 8.3

$NH(C_2H_4NH_2)_2$
pK_a = 8.9

$NH(C_3H_6NH_2)_2$
pK_a = 8.6

DMAM-PMHD
pK_a = 9.2

THB
pK_a = 6.2

Fig. 8.2 Tridentate N_3-donor chelate ligands and pK_a values for $[LZn\text{-}OH_2]^{2+}$ complexes [26–31]. The nitrogen donors for the zinc ion in the THB ligand are shown in bold.

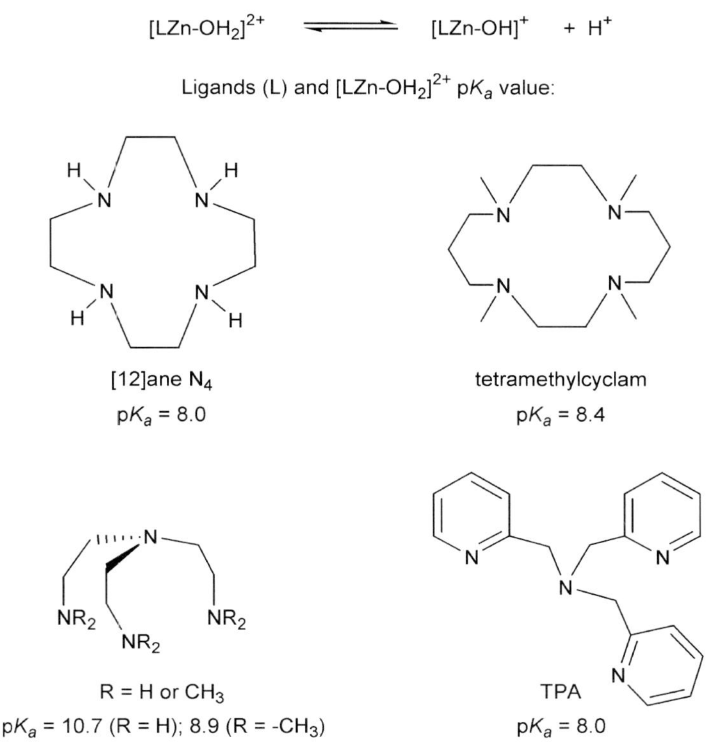

$$[LZn\text{-}OH_2]^{2+} \rightleftharpoons [LZn\text{-}OH]^{+} + H^{+}$$

Ligands (L) and $[LZn\text{-}OH_2]^{2+}$ pK_a value:

[12]ane N_4
$pK_a = 8.0$

tetramethylcyclam
$pK_a = 8.4$

R = H or CH_3
$pK_a = 10.7$ (R = H); 8.9 (R = $-CH_3$)

TPA
$pK_a = 8.0$

Fig. 8.3 Tetradentate N_4-donor chelate ligands and pK_a values for $[LZn\text{-}OH_2]^{2+}$ complexes [13, 28, 33–36].

However, as with many of the $LZn\text{-}OH_2$ complexes for which pK_a values have been reported, the solid state structures of $[(N(CH_2CH_2NH_2)_3)Zn\text{-}OH_2]X_2$ and $[(N(CH_2CH_2N(CH_3)_2)_3)Zn\text{-}OH_2]X_2$ have not been reported, and thus counterion and/or solvent interactions remain undefined.

Inclusion of one or more anionic ligands in the coordination sphere of a mononuclear zinc center is expected to raise the pK_a of metal-bound water [25]. This is consistent with the observed increase in the pK_a of the zinc-bound water

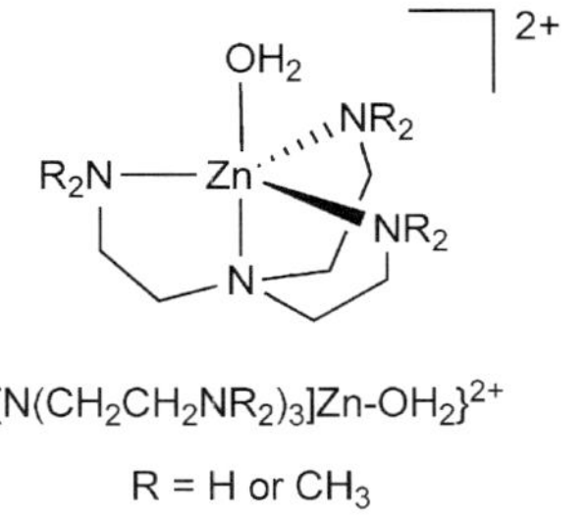

$\{[N(CH_2CH_2NR_2)_3]Zn\text{-}OH_2\}^{2+}$
R = H or CH_3

Fig. 8.4 Tripodal zinc aqua complexes. For R = H, the pK_a value for $[LZn\text{-}OH_2]^{2+}$ is 10.7; for R = CH_3, the pK_a for $[LZn\text{-}OH_2]^{2+}$ is 8.9 [34].

ligand of carboxypeptidase A (9.5) versus that of carbonic anhydrase II (6.8) (Fig. 8.1) [22, 24]. However, similar to studies involving N_3-ligated zinc complexes, studies of synthetic LZn-OH_2 complexes using a wide range of tripodal, tetradentate ligands (Fig. 8.5) suggest that multiple factors influence the pK_a of a bound water molecule [35]. For example, for ligands of identical charge, the magnitude of the zinc-binding constant, K_{LZn}, correlates with the acidity of the LZn-OH_2 water molecule. This indicates that the more tightly a zinc is held, the less Lewis acidic is the zinc center and implies that in a protein environment, a less tightly held zinc ion will be more Lewis acidic. With regard to type of donor moiety, for the series of ligands shown in Fig. 8.5, the pK_a of LZn-OH_2 increases as pyridyl groups are replaced with carboxylate donors. Interestingly, however, replacement of pyridyl groups with amino donors produces a similar effect, indicating that the charge on the carboxylate ligand is not the only factor influencing the pK_a value. Furthermore, replacement of carboxylate ligands with either 2-imidazolyl or 4-imidazolyl ligands produces little effect on the pK_a value of LZn-OH_2. These combined results further suggest that active site effects, including noncovalent interactions and the electrostatic environment surrounding the metal center, substantially influence LZn-OH_2 pK_a values [34, 37]. In this regard, computational studies indicate that the acidity of a zinc-bound water molecule increases with an increase in the dielectric constant of the surrounding environment [38].

$$[LZn\text{-}OH_2]^{2+/+/0} \rightleftharpoons [LZn\text{-}OH]^{+/0/-} + H^+$$

Ligands (L) and $[LZn\text{-}OH_2]^{2+/+/0}$ pK_a value:

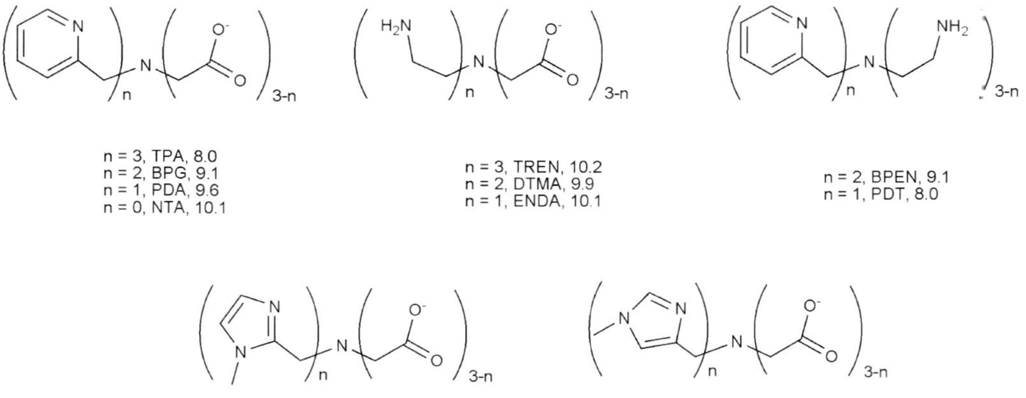

n = 3, TPA, 8.0
n = 2, BPG, 9.1
n = 1, PDA, 9.6
n = 0, NTA, 10.1

n = 3, TREN, 10.2
n = 2, DTMA, 9.9
n = 1, ENDA, 10.1

n = 2, BPEN, 9.1
n = 1, PDT, 8.0

n = 3, T21A, 8.7
n = 2, B21G, 9.0
n = 1, DA21m, 8.9

n = 3, T41A, 9.1
n = 2, B41G, 9.2
n = 1, DA41m, 8.5

Fig. 8.5 Hybrid tetradentate ligands and pK_a values for LZn-OH_2 complexes. Reprinted with permission from [35]. Copyright (2003) American Chemical Society.

$$[LZn\text{-}OH_2]^+ \rightleftharpoons [LZn\text{-}OH]^0 + H^+$$

Ligands (L) and $[LZn\text{-}OH_2]^+$ pK_a value:

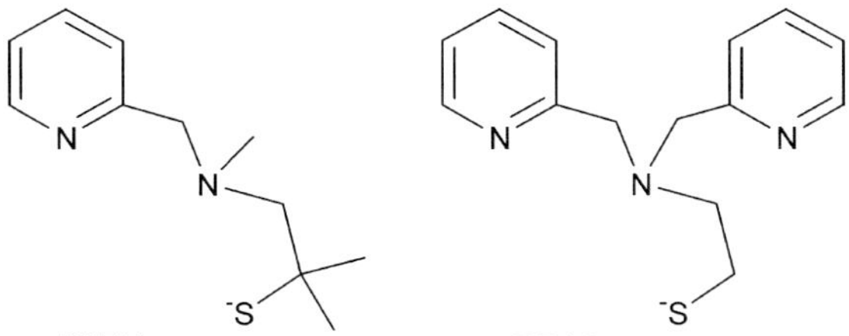

PATH
pK_a = 7.7

DPAS
pK_a = 8.6

Fig. 8.6 Nitrogen/sulfur(thiolate) ligands and pK_a values for $[LZn\text{-}OH_2]^+$ complexes [39, 41].

To date, only one zinc aqua complex supported by a N_2S-donor ligand (PATH, Fig. 8.6) and containing an aliphatic thiolate donor akin to that found in Zn(II)-containing peptide deformylase, has been examined in terms of the acidity of its zinc-bound water molecule [39]. The thermodynamic pK_a value determined for $[(PATH)Zn\text{-}OH_2]^+$ [7.7(1)] is slightly higher than that found for $\{([12ane]\text{-}N_3)Zn\text{-}OH_2\}^{2+}$ (7.3) [26], but is similar to that reported for $[(TPA)Zn\text{-}OH_2]^{2+}$ (8.0) [36] and $\{([12]aneN_4)Zn\text{-}OH_2\}^{2+}$ (7.9) [40]. This indicates that a N_2S(thiolate) ligand donor environment apparently produces a zinc center of similar Lewis acidity to a tetradentate neutral donor ligand. As may be expected, the pK_a of $[(PATH)Zn\text{-}OH_2]^+$ is well below that found for $[(DPAS)Zn\text{-}OH_2]^+$ (8.6) [41], a complex in which the supporting DPAS chelate ligand is tetradentate and contains a single anionic thiolate donor.

8.2.2
Secondary H-Bonding

The importance of secondary interactions in modulating the chemistry of a metal-bound water or hydroxide moiety has been discussed for metalloenzymes of varying function [42, 43]. For example, in zinc-containing carbonic anhydrase II, the zinc-bound hydroxide acts as a H-bond donor to a threonine residue (Thr199; Fig. 8.7) [37]. This interaction orients the lone pair of the hydroxide and reduces the entropic barrier for catalysis [44]. Notably, perturbation of this H-bonding interaction is reported to increase the Zn-OH_2 pK_a value by around 2 units [45]. The threonine residue also stabilizes the transition state via H-bonding and destabilizes the product (bicarbonate-bound) form of the enzyme. It is also interesting that an X-ray diffraction study of crystals of carbonic anhydrase II isolated at pH 7.8 revealed two water molecules that donate H-bonds to the zinc-bound hydroxide (Fig. 8.7) [46].

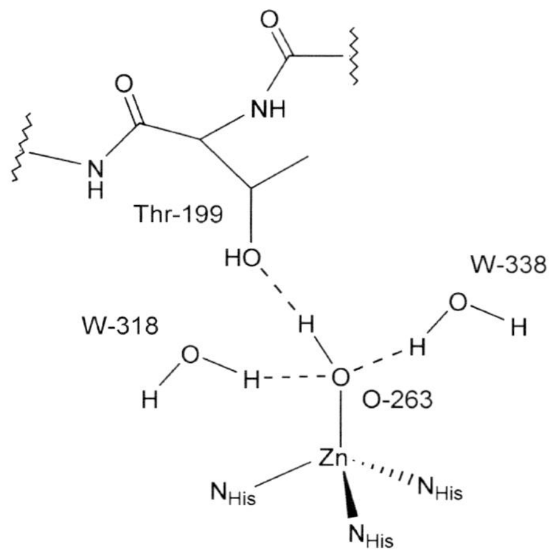

Fig. 8.7 Active site of carbonic anhydrase II including secondary interactions [46].

To gauge the influence of secondary H-bonding on the chemical properties of a metal-coordinated water molecule, several new tetradentate, tripodal ligands having one or more internal H-bond donors have been synthesized. Treatment of L1–L3 (Fig. 8.8) with $Zn(NO_3)_2$ in aqueous solution produces mononuclear zinc complexes with two pK_a values (pK_a^1 and pK_a^2), as determined by potentiometric titration, for zinc-bound groups [47]. These pK_a values are expected to be for a zinc-coordinated water molecule and the pendant ligand hydroxyl group, although each was not definitively assigned. Notably, the reduction of these values by around 0.6–0.7 units on going from L1 to L2/L3 led to the conclusion that intramolecular H-bonding interactions involving the ligand amino groups occur with both the metal-bound alcohol and water moieties. Such secondary interactions would withdraw electron density from the metal-bound alcohol or

$$[LZn\text{-}OH_2]^{2+/+} \rightleftharpoons [LZn\text{-}OH]^{+/0} + H^+$$

Ligands (L) and $[LZn\text{-}OH_2]^{2+/+}$ pK_a values:

L1
$pK^1_a = 8.6$
$pK^2_a = 11$

L2
$pK^1_a = 7.9$
$pK^2_a = 10$

L3
$pK^1_a = 8.0$
$pK^2_a = 10$

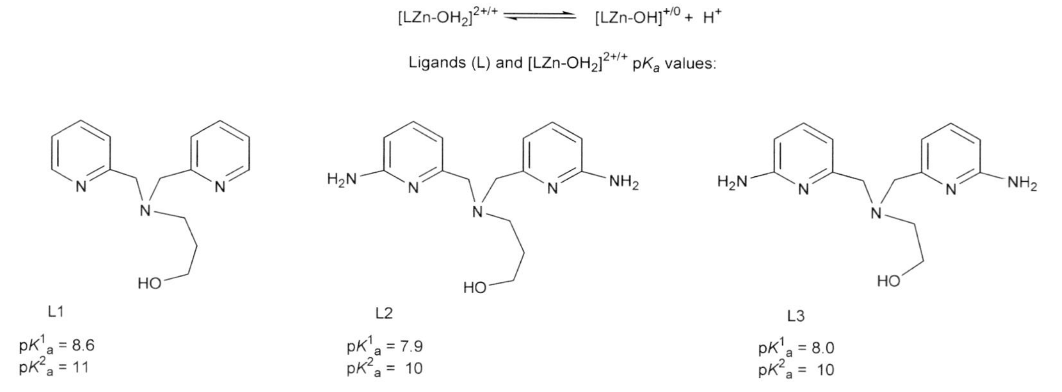

Fig. 8.8 N_3O-donor ligands having a varying number of H-bond donors [47]. Reproduced by permission of The Royal Society of Chemistry.

water ligand, thus resulting in a lowering of the pK_a value. However, as the zinc complexes of L1–L3 have not been further characterized by spectroscopic or X-ray crystallographic methods, the influence of H-bonding in these complexes remains to be fully defined.

A series of zinc complexes supported by tripodal N_4-donor ligands having variable numbers and types of internal H-bond donors (bpapa, bapapa, tapa, bpnpa, bnpapa, tnpa; Fig. 8.9) has been reported [48, 49]. The pK_a values for $[LZn\text{-}OH_2]^{2+}$ for this series of ligands are given in Fig. 8.9. For the primary amine-appended ligands, sequential addition of H-bond donors within the ligand structure results in a reduction of around 0.7–0.9 pK_a units for the zinc-bound water molecule per H-bond donor group. However, a different result was

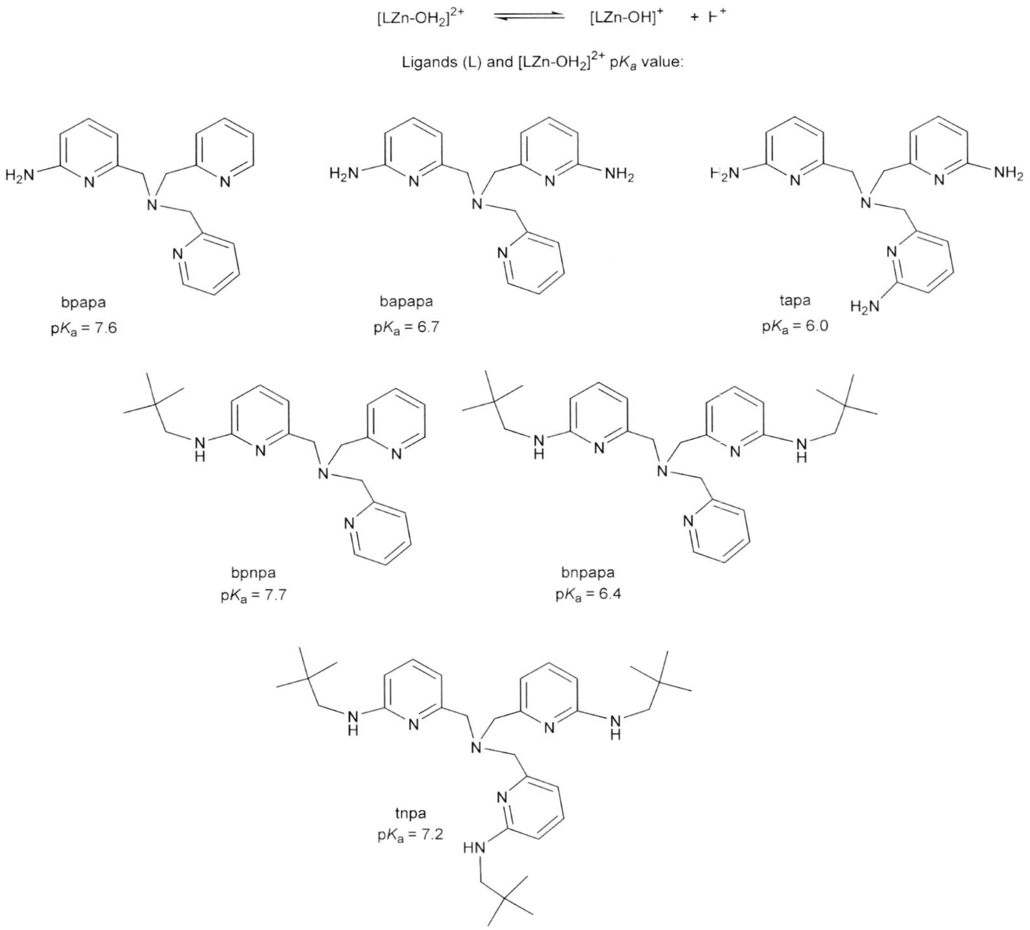

$$[LZn\text{-}OH_2]^{2+} \rightleftharpoons [LZn\text{-}OH]^{+} + H^{+}$$

Ligands (L) and $[LZn\text{-}OH_2]^{2+}$ pK_a value:

bpapa pK_a = 7.6

bapapa pK_a = 6.7

tapa pK_a = 6.0

bpnpa pK_a = 7.7

bnpapa pK_a = 6.4

tnpa pK_a = 7.2

Fig. 8.9 N_4-donor ligands having a varying number of H-bond donors; pK_a values for $[LZn\text{-}OH_2]^{2+}$ complexes [48, 49].

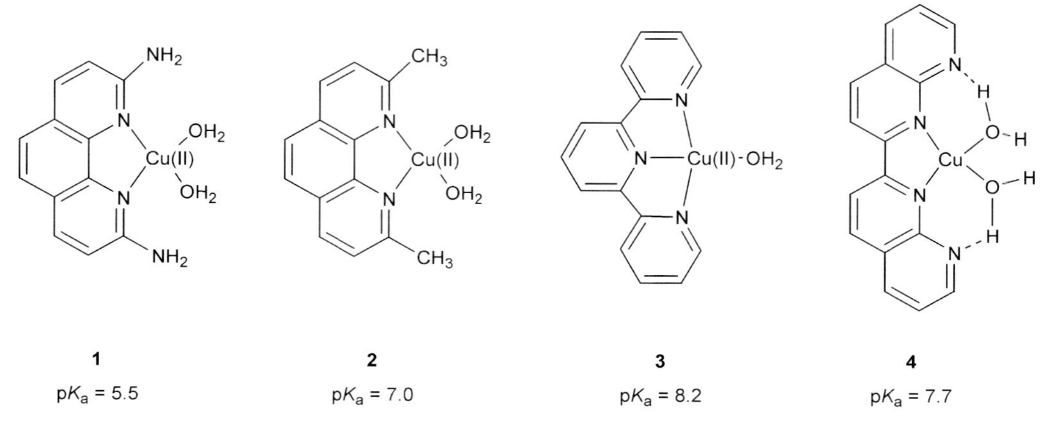

Fig. 8.10 pK_a values for Cu(II) aqua complexes having H-bond donor (**1**) and acceptor (**4**) appendages. Reprinted in part with permission from [50]. Copyright (1999) American Chemical Society.

obtained using neopentylamine-appended ligands (bpnpa, bnpapa, tnpa). Specifically, while the pK_a values for the bpnpa- and bnpapa-ligated complexes follow a similar trend, the pK_a value for $[(tnpa)Zn\text{-}OH_2]^{2+}$ is higher. This may due to increased steric congestion introduced by the third neopentyl amine substituent, and its effects on Zn-OH_2 and Zn-OH bonding (*vide infra*).

Studies of Cu(II) and Co(III) complexes having internal H-bond donors produced similar conclusions [50, 51]. For example, comparison of the pK_a values of the bound water molecules in [(2,9-diamino-*o*-phenanthroline)Cu$(OH_2)_2]^{2+}$ (**1**, pK_a=5.5), [(2,9-dimethyl-*o*-phenanthroline)Cu$(OH_2)_2]^{2+}$ (**2**, pK_a=7.0) and [(terpyridyl)Cu$(OH_2)]^{2+}$ (**3**, pK_a=8.2) (Fig. 8.10) revealed that the complex having internal H-bond donors (**1**) had the most acidic water molecule. This issue was further examined by analyzing the acidity of an analog complex having internal H-bond acceptor moieties (**4**, Fig. 8.10). For this Cu(II) aqua complex, a pK_a of

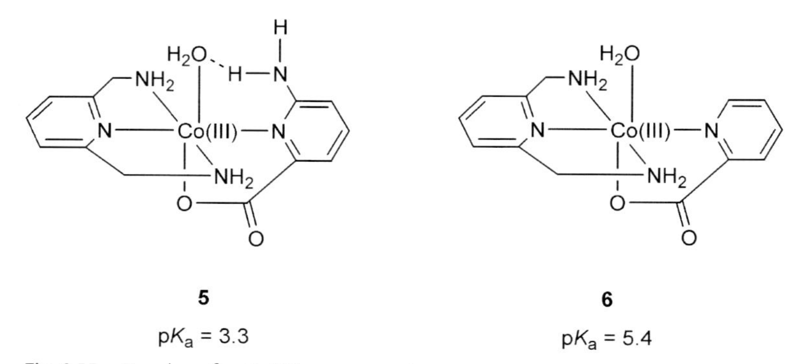

Fig. 8.11 pK_a values for Co(III) aqua complexes of 2-aminopicolinate (5) and picolinate ligands (6) [51].

7.7 was determined, a value that is 2.2 pK_a units above that found for the structurally similar H-bond donor system (**1**). For the Co(III) complexes outlined in Fig. 8.11, the ligand system incorporating an internal H-bond donor again produced the more acidic metal-bound water molecule.

8.2.3
Intramolecular H-Bonding and Mononuclear Zn-OH Stabilization

An important role for H-bonding in influencing the pK_a of a metal-bound water in synthetic complexes of ligands such as those shown in Figs. 8.8 and 8.9 may be stabilization of the Zn-OH moiety relative to the Zn-OH_2 species. In this regard, the X-ray crystal structures of three mononuclear zinc hydroxide complexes wherein the bound hydroxide accepts a H-bond from a supporting chelate ligand have been reported [49, 52, 53]. For two of these, [(bnpapa)Zn-OH]-ClO_4 and [(tnpa)Zn-OH]ClO_4 (Fig. 8.12), the pK_a value of the coordinated water molecule in the corresponding aqua complex has been determined (Fig. 8.9) [49, 53]. In [(bnpapa)Zn-OH]ClO_4, two H-bonding interactions involving the zinc-bound hydroxide are present. Both of these are moderate H-bonds [54], with heteroatom N···O distances of around 2.73 Å.

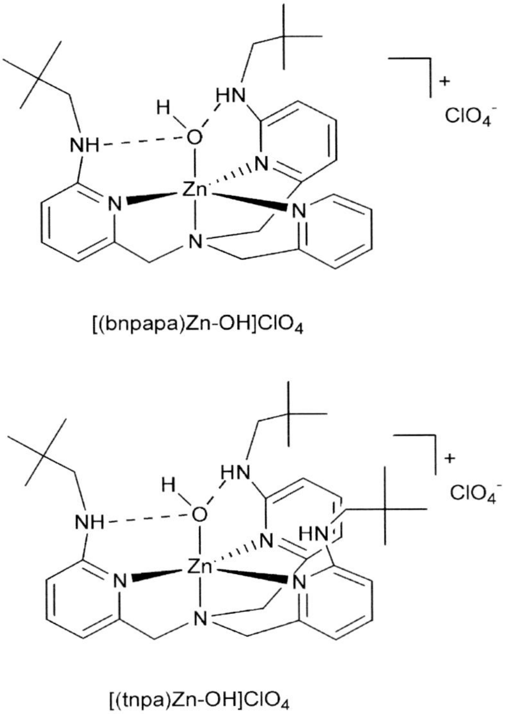

[(bnpapa)Zn-OH]ClO_4

[(tnpa)Zn-OH]ClO_4

Fig. 8.12 Structural drawings of [bnpapa)Zn-OH]ClO_4 and [(tnpa)Zn-OH]ClO_4 [49, 53].

Two X-ray crystal structures of [(tnpa)Zn-OH]ClO_4 have been reported [49, 53]. For both, the hydroxyl oxygen accepts two H-bonds from three possible secondary amine H-bond donor groups in the supporting ligand. The average heteroatom distance of these interactions ($N\cdots O_{avg}$ 2.82 Å and $N\cdots O_{avg}$ 2.88 Å) exceeds that found in [(bnpapa)Zn-OH]ClO_4. This is likely due to steric congestion in the tnpa-ligated complex. However, this difference has little affect on the Zn–O(H) bond distance ([(bnpapa)Zn-OH]ClO_4, 1.941(3) Å; [(tnpa)Zn-OH]ClO_4, 1.9315(8) and 1.957(2) Å) [53]. These Zn–O distances are around 0.1 Å longer than those found in structurally characterized tetrahedral zinc hydroxide complexes [55]. Multiple factors may be responsible for this elongation, including the higher coordination number of the zinc center. Furthermore, as there is not a reported X-ray structure of a mononuclear zinc hydroxide species of the parent TPA ligand, it is not possible at this point to evaluate to what extent the H-bonding interactions may elongate the Zn–O interaction. In this regard, it is worth noting that the Zn–O distance for a carbonyl oxygen of *N,N*-dimethylformamide coordinated to a mononuclear N_2S_2-ligated zinc center is elongated by around 0.025 Å when the amide oxygen participates in one intramolecular moderate H-bonding interaction [56]. Rivas and coworkers have noted an elongation of a similar magnitude for each H-bonding interaction involving the Zn-O(amide) unit present in complexes of amide-appended tripodal ligands akin to the bpnpa and bnpapa ligands (Fig. 8.9) [57]. Thus, as an initial estimate, the Zn–O bond in [(bnpapa)Zn-OH]ClO_4 (Fig. 8.12) may be elongated by as much as around 0.05 Å relative to that of a comparable system lacking the internal H-bond donors. The influence of this elongation on the acidity of the corresponding aqua complex remains to be elucidated.

8.2.4
Structural Effects Derived from M-OH_2 Acting as an Intramolecular H-Bond Donor to a Bound Phosphate Ester

A recent X-ray diffraction study has provided a glimpse into how intramolecular H-bonding involving a metal-bound water molecule acting as the H-bond donor influences the structural features of that bound water molecule. Specifically, in X-ray crystal structures of [Cu(dtbp)$_2$(phen)(OH_2)] (Fig. 8.13, dtbp = di-*tert*-butylphosphate anion, phen = 1,10-phenanthroline) collected at three different temperatures (93, 208 and 298 K) the involvement of both protons of the copper-bound water molecule in H-bonding interactions involving phosphate diester P=O groups was identified [58]. Notably, these secondary H-bonding interactions are substantially different, with O–H distances of 1.07 Å [O(9)–H(9a)] and 0.69 Å [O(9)–H(9b)], respectively. The elongation of the O(9)–H(9a) bond has been described in terms of intramolecular activation of the coordinated water molecule by the proximal P(1)=O(2) group. Such activation can be viewed as resulting in hydroxide and neutral dtbpH coordination to the Cu(II) center, as well as partial multiple bond character throughout the O(2)–P(1)–O(1) group. In support of this formulation, the P(1)–O(1) bond is notably shorter than P(2)–

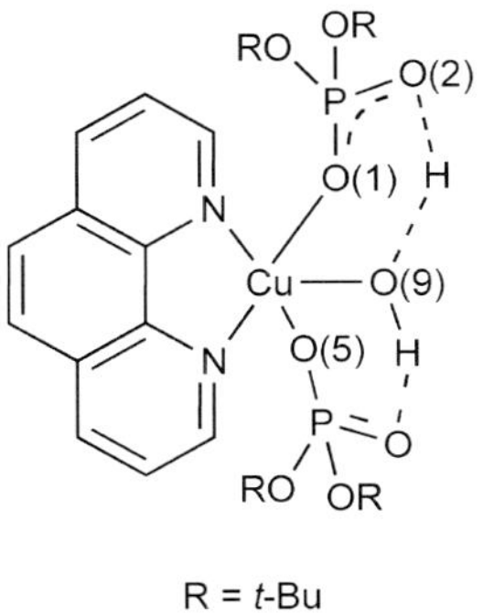

R = *t*-Bu

Fig. 8.13 Cu(II) complex having asymmetric intramolecular H-bonds [58]. Reproduced by permission of The Royal Society of Chemistry.

O(5) bond. Also consistent with the notion of partial neutral bonding character in the P(2)–O(1) dtbp ligand is a longer Cu–O bond [Cu(1)–O(1) 2.125(3) Å] than found for the other coordinated dtbp ligand [Cu(1)–O(5) 1.974(3) Å]. To date, the pK_a of the water molecule in [$Cu(dtbp)_2(phen)(OH_2)$] has not been reported.

8.2.5
Ligand Effects on the pK_a of a Metal-bound Water in Co(III) and Fe(III) Complexes

Several synthetic Co(III) and Fe(III) complexes have been generated as models for the active-site metal center in nitrile hydratases [8, 10]. However, few have been examined in terms of water coordination and acidity. Cobalt complexes supported by N_5-type donor ligands, with two carboxamido nitrogen donors, exhibit a pK_a near 7 (Fig. 8.14a and b) [59, 60]. Introduction of two thiolate sulfur donors into the Co(III) coordination sphere increases the pK_a of the bound water to 8.3 (Fig. 8.14c) [61]. Interestingly, oxidation of one of the sulfur donors to a S-bound sulfinate (Fig. 8.14d) reduces the pK_a by around 1 unit [62]. As the active-site metal center in nitrile hydratases contain oxidatively modified cysteine residues coordinated to the metal center [8], it has been suggested that the oxidized sulfur donors play a role in modulating the acidity of the metal-bound water molecule.

A model complex for Fe(III)-containing nitrile hydratase $Et_4N[Fe(PyPS)(H_2O)]$, Fig. 8.15) has been shown to exhibit a pK_a value around 2 units lower than that found for the Co(III) analog (Fig. 8.14c) [63]. Notably, this value is also well below that found for $[Fe(3,4\text{-}TDTA)(H_2O)]^-$ (pK_a=8.2), a compound in which TDTA is a tetra anionic ligand with a N_2O_4 donor set [64].

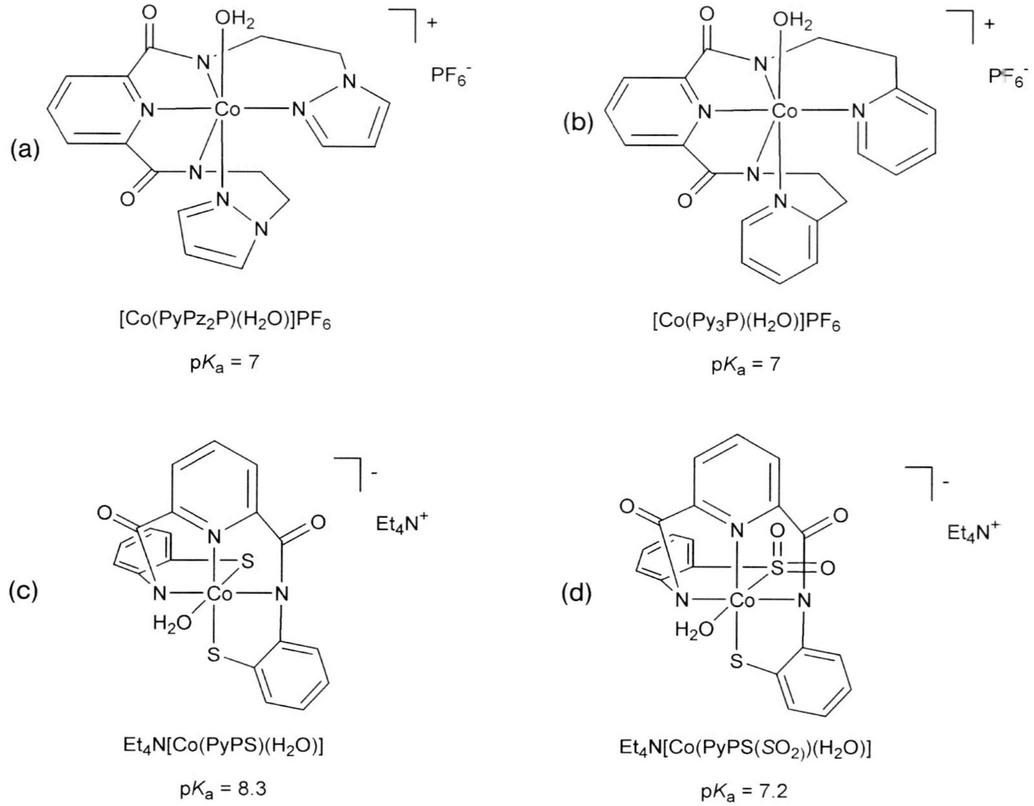

(a) $[Co(PyPz_2P)(H_2O)]PF_6$

$pK_a = 7$

(b) $[Co(Py_3P)(H_2O)]PF_6$

$pK_a = 7$

(c) $Et_4N[Co(PyPS)(H_2O)]$

$pK_a = 8.3$

(d) $Et_4N[Co(PyPS(SO_2))(H_2O)]$

$pK_a = 7.2$

Fig. 8.14 Co(III) aqua complexes and pK_a values [59–62].

8.2.6
Acidity and Water Exchange Properties of Organometallic Aqua Ions

Although the majority of studies reported over recent years concerning water activation by synthetic metal complexes have focused on systems of biological relevance, advances have also been made in understanding the water binding and activation properties of organometallic complexes. For example, several transition metal aqua complexes that feature only π-cyclic [e.g. η^6-arene, η^5-cyclopentadiene (Cp), pentamethyl-η^5-cyclopentadiene (Cp*)] or carbonyl ligands have been recently reported [65–74].

Several aqua complexes having a supporting pentamethylcyclopentadienyl ligand have been prepared and characterized in terms of acidity. For example, the Group 4 complex $[Cp^*_2Ti(H_2O)_2]^{2+}$ has a first pK_a value predicted to be around 0 [14]. The Co(III) derivative $[Cp^*Co(H_2O)_3]^{2+}$ has a first pK_a value of 5.9 [75], with $[Co(H_2O)_6]^{3+}$ having a reported first pK_a value that is around 3.5–5.0 units lower. This difference can be rationalized in part on the basis of the difference

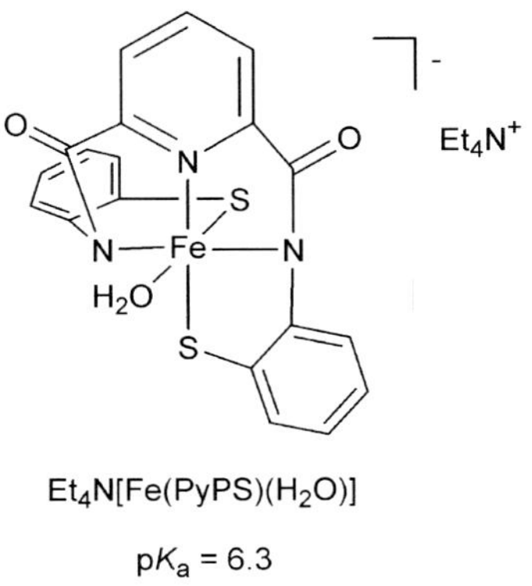

Et_4N^+

$Et_4N[Fe(PyPS)(H_2O)]$

$pK_a = 6.3$

Fig. 8.15 Fe(III) aqua complex and pK_a value [63].

in charge at the Co(III) center between the organometallic species and the hexaaqua ion. Interestingly, the Rh(III) analog $[Cp^*Rh(H_2O)_3]^{2+}$ has been shown to have pK_a values of 3.8, 5.7 and 8.3 [14]. Compared to reported first pK_a values for the hexaaqua ion $[Rh(H_2O)_6]^{3+}$ (pK_a=3.2 and 4.0 [76, 77]), the organometallic aqua complex is only slightly less acidic (maximum 0.6 pK_a units). Strong covalency in the M–Cp* interaction for the Rh(III) complex has been put forth as a rationale for why the difference in pK_a values between the organometallic and hexaaqua ions differs less for Rh(III) than for the Co(III) analog [14]. Notably, the pK_a of $[Cp^*Ir(bpy)(OH_2)]^{2+}$ is 6.6 [78, 79]. As this complex is similar to the $[Cp^*M(H_2O)_3]^{2+}$ (M=Co, Rh) derivatives, this suggests that moving down a group in a structurally similar set of Cp*-ligated complexes results in weaker acidity for the metal-bound aqua ligand.

For the carbonyl derivatives $[Re(CO)_3(OH_2)_3]^+$ and $[Re(CO)_3(OH)(OH_2)_2]$, pK_a values of 7.5(2) and 9.3(3) have been determined [67]. The acidity of these complexes is attributed to the π-electron-withdrawing effect of the CO ligands.

Although not directly relevant to the scope of this review, it is interesting to note that structural and water exchange studies of triaqua(benzene)M(II) derivatives $[(C_6H_6)M(H_2O)_3]^{2+}$ (M=Ru, Os) [65] and $[(\eta^5\text{-}Cp^*)M(H_2O)_3](OTf)_3$ (M=Rh, Ir) complexes [66, 73, 74] have revealed that the organometallic substituent imparts a strong labilizing effect on *trans* water molecules, resulting in an increased rate for water exchange as compared to that found for the hexaaqua ion of the same metal. However, it must be noted that this is not universally the case, as for the series of carbonyl aqua complexes $[(CO)_nRu(H_2O)_{6-n}]^{2+}$ ($1 \leq n \leq 3$), the water exchange rate is estimated to be 1–2 orders of magnitude slower than that found for $[Ru(H_2O)_6]^{2+}$ [80].

Organometallic aqua ions of technetium are of particular current interest for applications in radiopharmaceutical imaging [15]. Alberto's reagent, *fac*-$[^{99m}Tc(OH_2)_3(CO)_3]^+$, which is easily prepared from $[^{99m}TcO_4]^-$ in saline solution under 1 atm of CO [68], is a complex having labile water ligands that can be displaced by ligand groups to form technetium-labeled biomolecules. The half-life for water exchange for $[^{99m}Tc(OH_2)_3(CO)_3]^+$ is between 1 s and 1 min at 277 K in 2 M $HClO_4$ [70]. This exchange rate is similar to that reported for $[Ru(H_2O)_5\text{-}CO]^{2+}$ ($t_{1/2} \sim 3$ min) [81].

8.3
Secondary H-Bonding Effects on Substrate Coordination, Activation and Catalytic Hydrolysis Involving Phosphate Esters

Several recent review articles provide excellent summaries of the stoichiometric and catalytic reactivity of synthetic metal aqua and hydroxide complexes with carbon-centered electrophiles (esters, amides, peptides, CO_2, nitriles) [8, 10–13, 82–90] and phosphate derivatives (activated phosphate esters, DNA, RNA) [6, 11–13, 17–20, 82, 83, 85, 91–99]. In particular, these reviews provide insight into how various metal/ligand assemblies influence catalytic hydrolytic reactions.

In the second portion of this contribution, the scope of the work presented is limited to a burgeoning area of research directed at understanding how metal coordination and secondary H-bonding interactions may be cooperatively employed to affect stoichiometric and catalytic phosphate ester hydrolysis reactions. The presence of H-bond donors in organic receptors has been previously shown to enhance recognition of phosphate anions [100–103]. However, only recently have efforts been made to incorporate H-bond donor groups into metal complexes that are reactive with phosphate ester substrates. This is timely in that investigations are also underway to probe the influence of H-bonding on several other biologically relevant reactions involving a nonredox active metal such as zinc. These include reactions of zinc alkoxide species of relevance to liver alcohol dehydrogenase [52, 104] and reactions that probe the nucleophilicity and alkylation reactivity of Zn-SR (thiolate) complexes of relevance to zinc-containing thiolate-alkylating enzymes [105–107].

8.3.1
H-Bonding and Phosphate Ester Coordination to a Metal Center

Chin and coworkers have recently evaluated the anion binding properties of mononuclear Co(III) complexes as a function of the presence of an internal H-bond donor [51]. Using ^{31}P nuclear magnetic resonance (NMR), equilibrium binding constants for phosphate diester monoanion coordination to the Co(III) aqua complexes **5** and **6** shown in Fig. 8.16 were found to be 210 and 6.2 M^{-1}, respectively, in water at 80 °C. As monodentate coordination of phosphate diester monoanions to metal ions is typically weak in aqueous solution ($K < 10^1$ M^{-1}) [108], the incorporation of a H-bond donor clearly influences the anion coordination properties. In sum, the combination of metal coordination and H-bonding produces an equilibrium binding constant that is still well below that found for coordination of both phosphoryl oxygens of the diester to a dinuclear metal structure (around 400 M^{-1}) [109].

The X-ray structure of the phosphate anion adduct of **5a** was determined. The H-bonding interaction between the metal-coordinated 2-aminopicolinate ligand

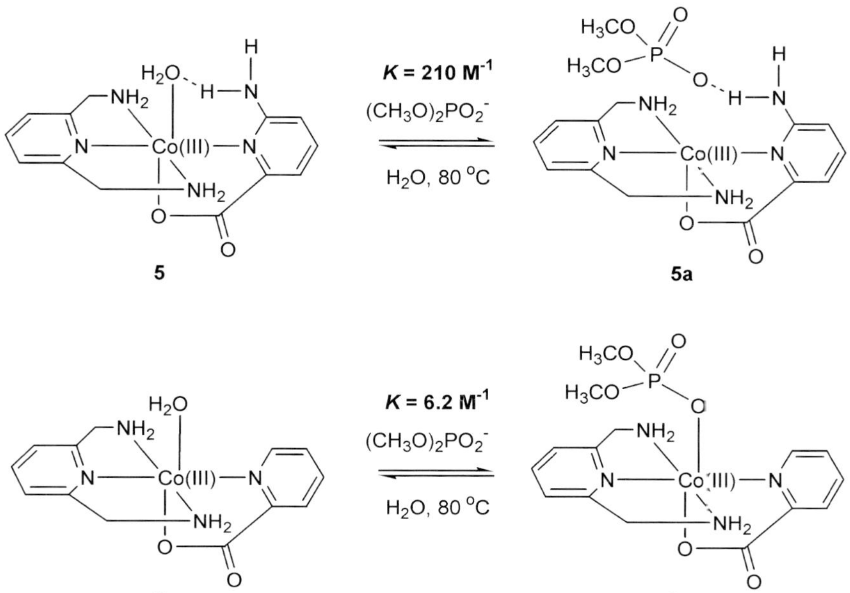

Fig. 8.16 Phosphate diester monoanion equilibrium binding to mononuclear Co(III) complexes. Reprinted in part with permission from [51]. Copyright (2002) American Chemical Society.

and the bound phosphate ester may be classified as moderate (N–H···O 2.77 Å), as is typically found in biological systems [54]. Preferential coordination of phosphate anion to **5**, versus water coordination, may be rationalized by the fact that the anion is a stronger H-bond acceptor than the neutral water molecule [110].

Mareque-Rivas and coworkers have demonstrated that the presence of internal H-bond donors will enhance the affinity of a phosphate ester to a zinc center [111]. Specifically, a monoaryl phosphate ester dianion [$PhOP(O)_3^{2-}$] in aqueous solution at pH 7 has been found to bind more strongly to a zinc cation having secondary intramolecular H-bonding capabilities ([bapapa)Zn-$OH_2]^{2+}$; log K= 3.6 ± 0.1) than to a structurally similar cation lacking such secondary interactions ([(TPA)Zn-$OH_2]^{2+}$, log K=4.4 ± 0.1) (Fig. 8.17) [111]. The X-ray crystal structure of [(bapapa)Zn(NO_3)]NO_3 shows that a zinc-bound nitrate can participate in two moderate H-bonding interactions [O(1)···N(amine)$_{avg}$ 2.93 Å] involving the supporting bapapa ligand. The bound phosphate ester dianion likely coordinates in a similar fashion. Notably, the magnitude of the improved affinity of [(bapapa)Zn]$^{2+}$ over [(TPA)Zn]$^{2+}$ for the phosphate ester dianion is similar to that found for dinuclear versus mononuclear zinc receptors [112]. Thus, secondary H-bonding can play a prominent role in facilitating anion coordination to a metal center.

A dinuclear zinc complex, [(bpapa)Zn(μ-DBP^-)$_2$Zn(bpapa)](PF_6)$_2$ (DBP^-=dibenzyl phosphate anion) with intramolecular H-bonding to the bridging phos-

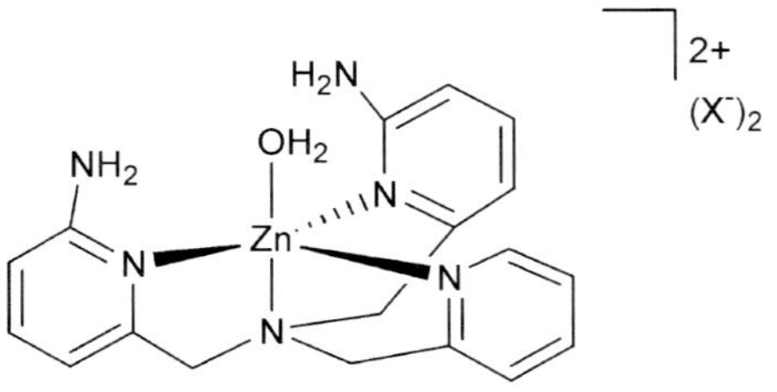

$[(bapapa)Zn]X_2$

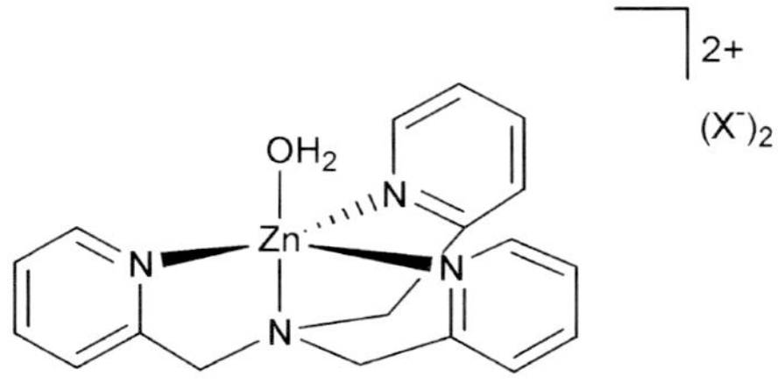

$[(TPA)Zn]X_2$

Fig. 8.17 Mononuclear zinc complexes used to evaluate the influence of H-bonding on phosphate ester coordination to a zinc cation [111].

phate diester has been characterized by X-ray crystallography (Fig. 8.18) [113]. To evaluate the affinity of DBP^- for the bpapa-ligated zinc center, titration experiments were performed in CD_3CN/D_2O (pD=7.4) and monitored by ^{31}P NMR spectroscopy. These studies provided further evidence, in terms of the magnitude of the upfield shift of the ^{31}P NMR signal of DBP^-, for a higher binding affinity for $[(bpapa)Zn]^{2+}$ than for $[(TPA)Zn]^{2+}$.

8.3.2
H-Bonding and Stochiometric and Catalytic Phosphate Ester Hydrolysis

Enhanced rates of phosphate ester cleavage have been recently noted for several systems having internal H-bond donors [47, 50]. For example, Ott and Krämer have demonstrated that inclusion of the tris(hydroxymethyl)aminomethane chelate ligand (Fig. 8.19 a) in weakly acidic solutions (pH 3.5) of Zr(IV) salts yielded a 1:1 complex that is 1.5 times more reactive than the free metal ion for the hydrolysis of bis(*p*-nitrophenyl)phosphate (BNPP, Fig. 8.19 b) and the DNA dinucleotide thymidyl-thymidine (Tpt, Fig. 8.19 c). The Tpt hydrolysis reaction mediated by the 1:1 Zr(IV) complex of the chelate ligand is 3×10^9 faster than the uncatalyzed reaction [114]. A structural hypothesis has been put forth suggesting tridentate coordination of the tris(hydroxymethyl)aminomethane ligand to the Zr(IV) ion, with the ligand amino group being protonated at pH 3.5. As no structural data is available to date on this system, it is unclear whether the ligand ammonium group can directly interact with the substrate.

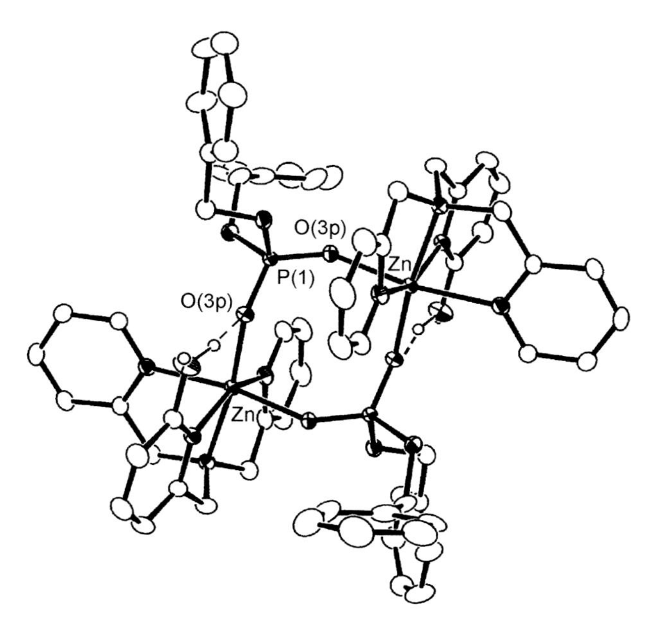

Fig. 8.18 Oak Ridge Thermal Ellipsoid Plot (ORTEP) representation of the cationic portion of $[(bpapa)Zn(\mu\text{-}DBP^{-})_2Zn(bpapa)](PF_6)_2$ [113]. Ellipsoids are drawn at the 35% probability level.

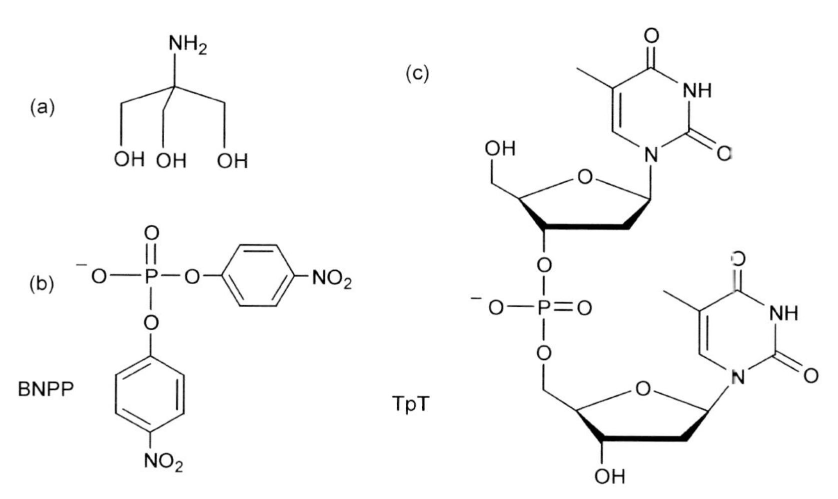

Fig. 8.19 (a) Tris(hydroxymethyl)aminomethane ligand, (b) BNPP and (c) thymidyl-thymidine (TpT) dinucleotide [114].

Krämer and coworkers have also noted significantly increased rates of phosphate ester hydrolysis using supporting chelate ligands wherein two ammonium groups are attached to a chelate 2,2-bipyridyl unit via alkyne linkers (L4, Fig. 8.20) [115, 116]. For comparison, studies of a tetraalkylammonium ligand (L5) were also performed. Treatment of $L4(NO_3)_2$ with $Cu(NO_3)_2 \cdot 3H_2O$ and disodium phenyl phosphate resulted in the isolation of the novel dinuclear phosphate ester complex, $[(L4)_2Cu_2(1,3\text{-}\mu\text{-}O_3POPh)_2(OH_2)_2](NO_3)_4$ (Fig. 8.21) which has a crystallographic center of symmetry. Each copper center exhibits a slightly distorted square pyramidal geometry ($\tau=0.14$) [117], with one of the bipyridyl nitrogens in the axial position. Intramolecular H-bonding interactions involving the coordinated phosphate monoester are present in this complex. As shown in Fig. 8.21 (bottom), one ammonium group forms a moderate H-bonding interaction [N(3)···O(4a), 2.67 Å] [54] with a noncoordinated phosphate ester oxygen. The same oxygen atom forms a second moderate H-bonding interaction with the copper-coordinated water molecule [O(4a)···O(1), 2.74 Å]. The oxygen atom of the coordinated water is 3.2 Å from the phosphorous atom. The relative positioning of the water and phosphate ester substrate suggests that if deprotonated, the water molecule could act a nucleophile to initiate phosphate ester hydrolysis.

Admixture of $L4(NO_3)_2$, $Cu(NO_3)_2 \cdot 3H_2O$, 2,4,6-trimethylpyridinium BNPP and 2,4,6-trimethylpyridinium nitrate in ethanol:water (19:1) results in phosphate ester hydrolysis to yield *p*-nitrophenyl phosphate. This reaction is 4×10^7 times faster than the uncatalyzed hydrolysis of BNPP under the same conditions. Of particular note is the fact that the reaction involving L4 is 1000 times faster than that of L5 and 3400 times faster than that found for a copper center ligated by the parent 2,2′-bipyridyl ligand. With a maximum rate being observed at pH ~6 for the L4-ligated system, a mechanism was proposed in which a copper-coordinated hydroxide anion acts as a nucleophile toward the bound phosphate diester (Fig. 8.22) which is oriented and activated via metal coordination and secondary H-bonding interactions. The reaction in this system is inhibited by coordination of the *p*-nitrophenolate product.

A zinc complex of the L4 ligand is only approximately 5 times more reactive for the hydrolysis of bis(*p*-nitrophenyl)phosphate than an analog complex ligated

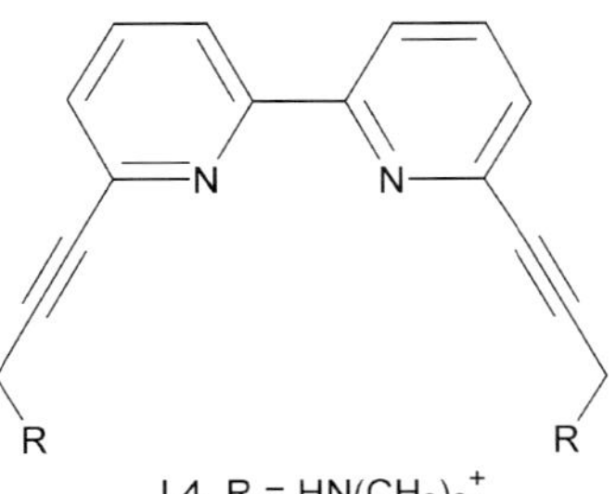

Fig. 8.20 Ligands used by Krämer and coworkers to examine secondary H-bonding effects on phosphate ester cleavage [115].

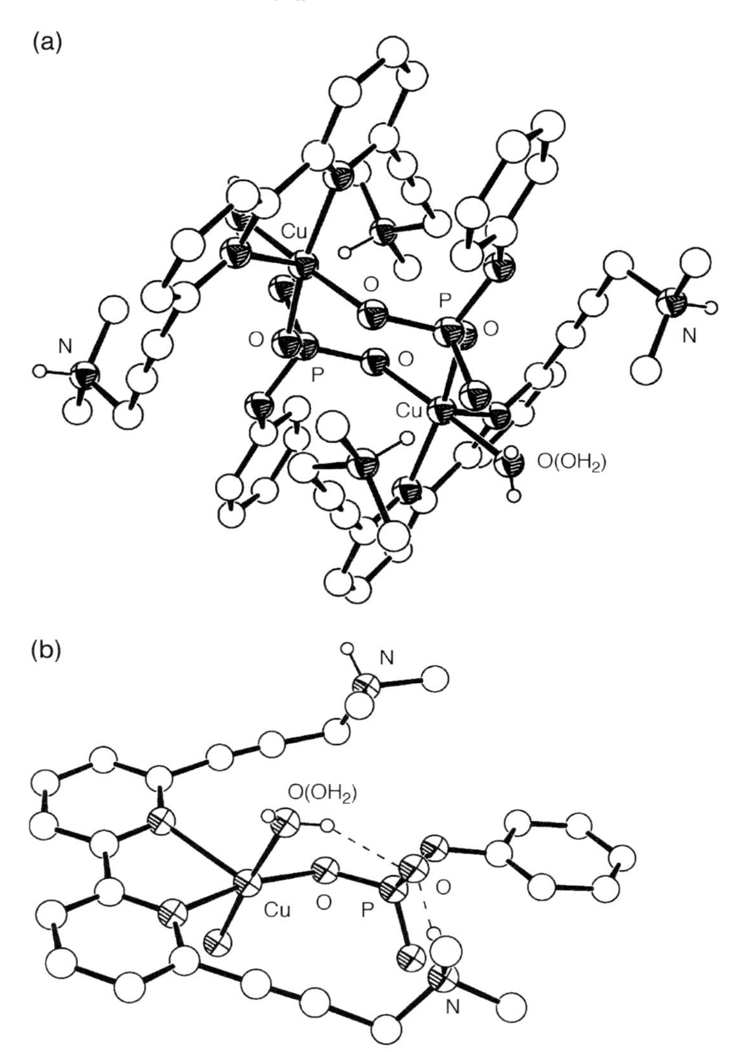

Fig. 8.21 (a) ORTEP representation of the cationic portion of $[(L4)_2Cu_2(1,3\text{-}\mu\text{-}O_3POPh)_2(OH_2)_2](NO_3)_4$ [115]. Ellipsoids are drawn at the 35% probability level. (b) Representation of one of the two copper centers of this complex showing secondary H-bonding interactions.

by the 2,2′-bipyridyl chelate ligand [118]. This is likely due to a different orientation of the bound ligands surrounding the zinc center. In the crystal structure of $[(L4)ZnCl_2]$, the zinc center is tetrahedral, with the chloride ligands being positioned above and below the plane of the bipyridyl and alkyne portions of L4. This difference in geometry, versus than found in distorted square pyramidal

Fig. 8.22 Proposed orientation between the phosphate ester substrate and copper hydroxide moiety in $[(L4)Cu(O_2P(C_6H_4NO_2))_2)(OH_2)(OH)](NO_3)_2$. Reprinted with permission from [115]. Copyright (1996) American Chemical Society.

copper complexes of L4, could be responsible for the reduced reactivity of the zinc derivative.

Chin and coworkers have compared the hydrolytic reactivity of Cu(II) derivatives of 2,9-diamino-*o*-phenanthroline, 2,9-dimethyl-*o*-phenanthroline, and terpyridine (**1**–**3**, Fig. 8.10) with 2′,3′-cAMP (Fig. 8.23) at 25 °C and pH 5. The results of this study demonstrate that complex **1** is approximately 2×10^4 more reactive than **2** and approximately 4×10^6 more reactive than **3**. The rate enhancement of **1** for hydrolysis of 2′,3′-cAMP has been rationalized in terms of H-bond stabilization of the transition state by the ligand amino groups (Fig. 8.23).

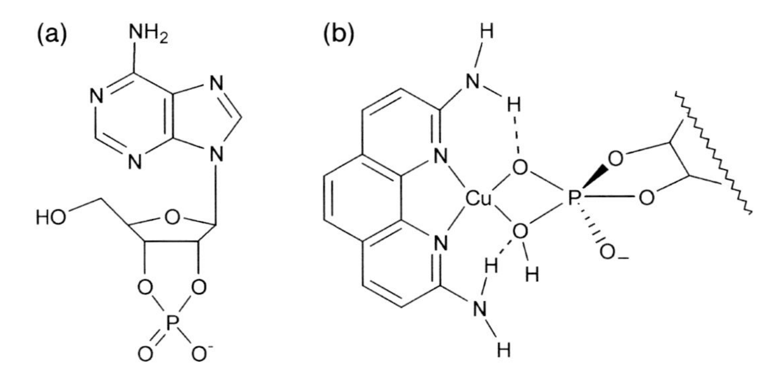

Fig. 8.23 (a) 2′,3′-cAMP. (b) Proposed H-bonding interactions involving ligand amino groups in the transition state of 2′,3′-cAMP hydrolysis catalyzed by **1** (Fig. 8.10). Reprinted in part with permission from [50]. Copyright (1999) American Chemical Society.

Ansyln and coworkers have explored the stoichiometric hydrolysis reactivity of zinc complexes of terpyridyl-type ligand having guanidinium or ammonium appendages (Fig. 8.24) using the RNA dimer adenylyl (3′ → 5′)phosphoadenine (ApA) [119]. The hydrolysis of this substrate yields adenosine 3′-monophosphate, adenosine 2′-monophosphate, adenosine 2′,3′-cAMP, and adenosine. The optimal rate of reaction of **Zn(II)-4** with ApA is found at pH ~7.5, which is near the pK_a value of a zinc-coordinated water molecule in this system (pK_a=7.3; determined independently by potentiometric titration). This suggests the involvement of a zinc hydroxide species as the active complex for ApA hydrolysis.

The pseudo-first-order rate constants for the reactions of the complexes shown in Fig. 8.24 with ApA at pH 7.4 (HEPES buffer, 37 °C) are 8.0×10^{-2} (**Zn(II)-4**), 9.0×10^{-3} (**Zn(II)-5**) and 2.4×10^{-5} h^{-1} (**Zn(II)-6**). Thus, **Zn(II)-4** is nine times more reactive than **Zn(II)-5** and greater than 3300 times more reactive than **Zn(II)-6**. In addition, **Zn(II)-4** is over 10^5 more reactive than **4** in the absence of the zinc center.

Proposed roles for the zinc center of **Zn(II)-4** in the hydrolysis of ApA have been put forth (Fig. 8.25). Coordination of the phosphate ester by the zinc center, coupled with interaction of a second P–O bond with a ligand guanidinium group, is proposed to orient and activate the phosphate ester for general base-promoted deprotonation of the 2′-OH group by a Zn-OH moiety. The zinc cen-

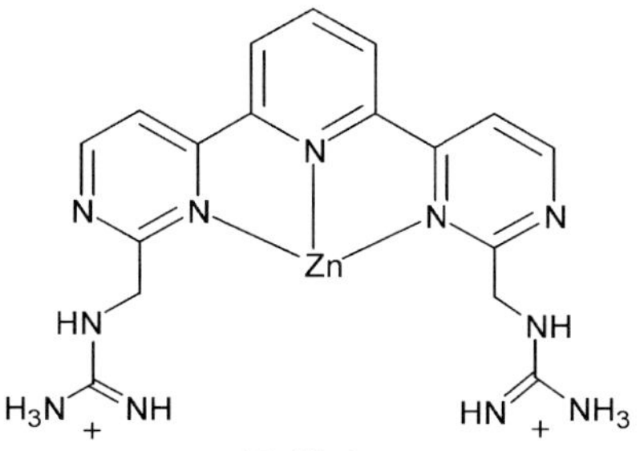

Zn(II)-4 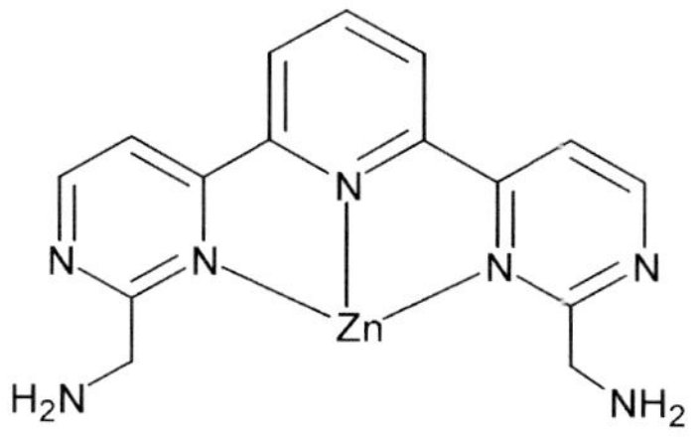 **Zn(II)-5** 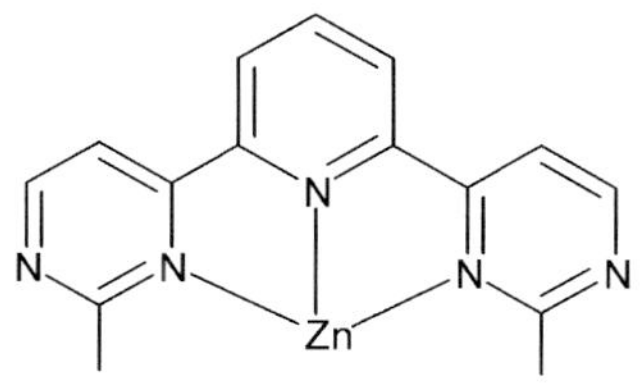 **Zn(II)-6**

Fig. 8.24 Zinc complexes employed by Anslyn and coworkers for hydrolysis of the RNA dimer adenylyl (3′ → 5′)phosphoadenine (ApA) [119]. Reproduced by permission of Wiley-VCH.

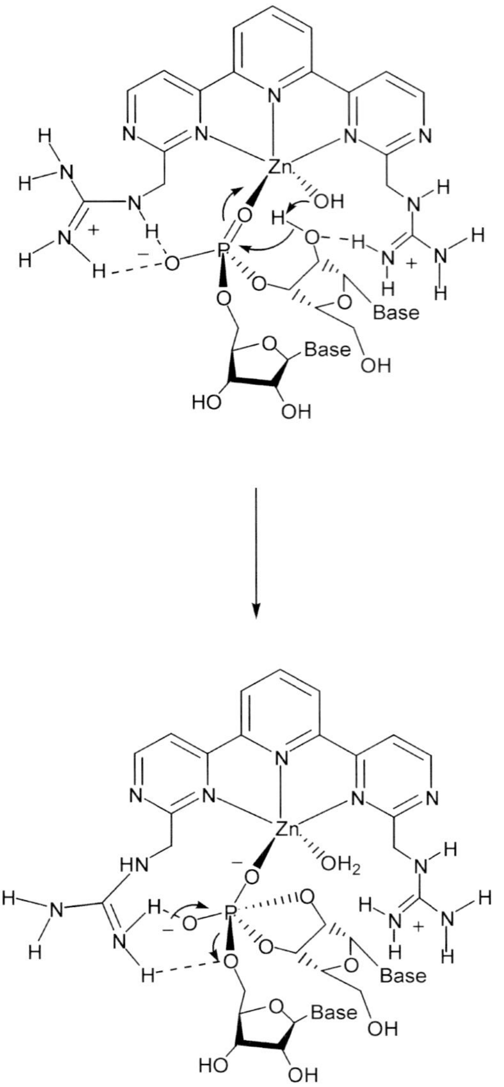

Fig. 8.25 Proposed roles of zinc center and guanidinium groups of Zn(II)-4 in the hydrolysis of ApA [119]. Reproduced by permission of Wiley-VCH.

ter and positively charged guanidinium groups also likely stabilize the transition state and may be involved in the protonation of the leaving group.

Incubation of Zn(II) complexes supported by the L1–L3 ligands (Fig. 8.8), which are comprised of three nitrogen donors and one alkoxide moiety, with bis-*p*-nitrophenyl phosphate yields variable products (Fig. 8.26) depending on the nature of the chelate ligand [47]. For L1 and L2, *p*-nitrophenol and *O*-phosphorylated chelate ligand were produced. These transesterification reaction products indicate the involvement of an alkoxide nucleophile in the reaction. For L3, only *p*-nitrophenol *p*-

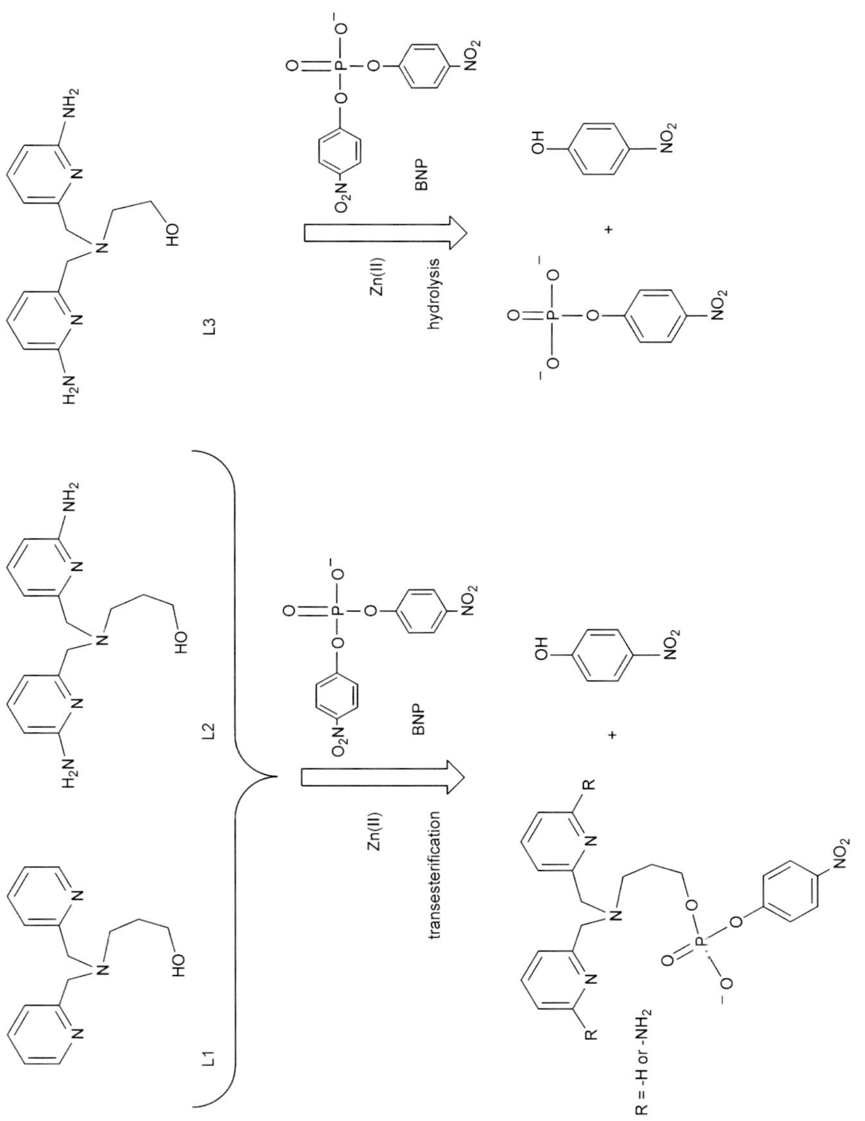

Fig. 8.26 Phosphate ester cleavage reactivity of L1–L3-ligated zinc complexes [47].

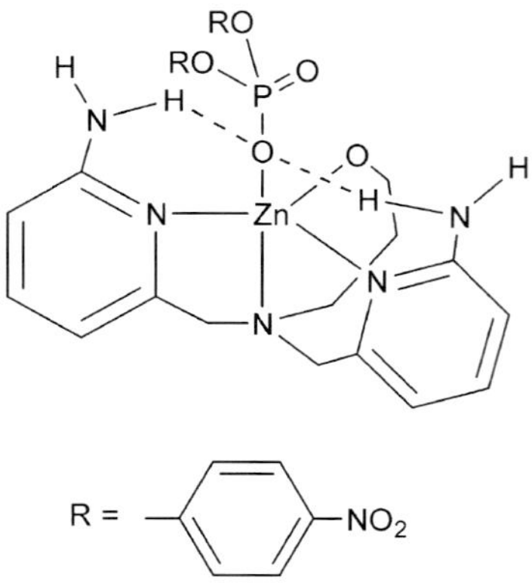

Fig. 8.27 Proposed H-bonding interactions in a BNPP-bound form of L2·Zn(II) [47]. Reproduced by permission of The Royal Society of Chemistry.

nitrophenyl phosphate monoester hydrolysis products were generated. For the reaction of the zinc complex of L2 with BNPP, a bell-shaped pH profile was determined, with maximum reactivity at pH 9–9.5. This is consistent with formation of a reactive monodeprotonated complex, which may be a zinc alkoxide species [47]. Notably, the zinc complex of L2 is much more reactive toward BNPP than those of L1 and L3, as evidenced by comparison of the second-order rate constants ($L1=4.2\times10^{-4}$, $L2=9.7\times10^{-2}$ and $L3=5.6\times10^{-3}\ M^{-1}\ s^{-1}$; determined by monitoring of *p*-nitrophenolate formation at 400 nm). The rate for L2 · Zn(II) represents an acceleration of approximately 6 orders of magnitude over the spontaneous hydrolysis of BNPP at pH 7.0 and 25 °C [108]. This rate enhancement is attributed to the involvement of an alkoxide nucleophile (L2 versus L3) and the presence of secondary H-bonding interactions (L1 versus L2). Comparison of the rate constants of the zinc complexes of L1 versus L2 reveals a 230-fold reactivity increase. Chin and coworkers rationalize this rate enhancement on the basis of formation of H-bonding interactions with the zinc-bound substrate (Fig. 8.27). Such interactions, similar to their effect on a zinc-coordinated water molecule (*vide supra*), withdraw electron density from the substrate thus providing additional Lewis acid activation. For the phosphate ester hydrolysis reaction catalyzed by L3 · Zn, the rate acceleration is around 10^5 over the uncatalyzed reaction. Thus, in both phosphate ester alcoholysis and hydrolysis reactions, rate accelerations are dramatically enhanced by the presence of a supporting chelate ligand having internal H-bond donors.

8.4 Summary and Future Directions

This contribution summarizes recent progress in the development of synthetic systems for evaluating ligand effects on water activation and catalytic hydrolysis reactions. These combined studies provide strong evidence that systematic examinations of model systems can provide important insight into the chemical details governing reactions of biological metal centers involved in hydrolysis or hydration reactions. Combined results to date in this area indicate that: (i) an increase in the coordination number of a supporting chelate ligand (L), or inclusion of an anionic donor within L, results in decreased acidity for $LZn\text{-}OH_2$ complexes, and (ii) inclusion of secondary H-bond donors which interact with a metal-bound water molecule increases the acidity of that water molecule by up to around 2 pK_a units. Intramolecular H-bonding interactions may also: (i) stabilize a metal hydroxide species relative to the metal aqua complex and (ii) differentiate O–H bonds within a metal-bound aqua ligand.

Despite these advances in our understanding of metal-mediated water activation, more work is needed. For example, despite significant efforts, few biomimetic ligand systems containing anionic sulfur donors akin to cysteine have been reported that enable the isolation and/or characterization of biologically relevant $Zn\text{-}OH/OH_2$ complexes [11]. Analysis of the impact of secondary H-bond acceptors, or hydrophobic groups, on the coordination properties and reactivity of a metal-bound water molecule also remains to be fully explored. In this regard, Carrano and coworkers recently reported the preparation of a new ester-appended ligand system ($Tp^{CO2Et,Me}$, Fig. 8.28 a) that enables the isolation of biologically relevant metal aqua complexes that are stabilized by intramolecular H-bond acceptors in the supporting chelate ligand [120]. A zinc complex of the formulation $[(Tp^{CO_2Et,Me})Zn(OAc)(H_2O)]$ (Fig. 8.28 b) was characterized by X-ray crystallography [121]. Intramolecular H-bonds are present between the coordinated water molecule and both a ligand ester group and the bound acetate anion. The interaction involving the ester moiety has a heteroatom O···O distance of 2.66 Å, indicating the presence of a moderate H-bond [54]. Notably, these secondary interactions may be responsible for two interesting structural features of $[(Tp^{CO_2Et,Me})Zn(OAc)(H_2O)]$. First, whereas for five-coordinate Tp-ligated zinc aqua complexes, the coordinated water molecule is typically found in an axial position of a pseudo trigonal bipyramidal structure [122–124], the water mole-

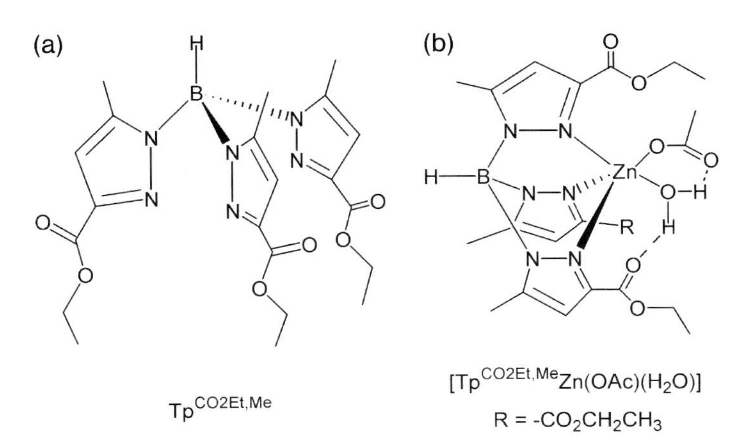

Fig. 8.28 (a) Ester-appended ligand. (b) Mononuclear zinc aqua complex having internal H-bond acceptors [120, 121].

cule is equatorial in [($Tp^{CO_2Et,Me}$)Zn(OAc)(H_2O)]. Second, the Zn-O(H_2O) bond distance [1.966(4) Å] is approximately 0.1 Å shorter than the corresponding bond in other five-coordinate Tp-ligated zinc aqua complexes [122–124]. This suggests that the presence of the H-bonding interactions stabilizes the Zn-OH_2 interaction, perhaps through partial intramolecular activation of the bound water (see Section 8.2.4). The pK_a of the zinc-bound water molecule in the $Tp^{CO_2Et,Me}$-ligated complex has not been reported to date.

As descriptions of enzymes often invoke several simultaneous interactions (for example, metal coordination, secondary H-bonding, etc.) between the substrate and active site as being essential for catalysis [85], recent studies have probed how the inclusion of secondary H-bond donor groups within a supporting chelate ligand framework influences phosphate ester coordination and hydrolysis. From studies reported to date, it can be concluded that the presence of intramolecular H-bond donors enhances metal coordination and the rate of hydrolysis of phosphate esters. However, while significant rate enhancement has been identified in several such systems, an important remaining challenge is the development of synthetic complexes that exhibit catalytic activity.

For organometallic systems, acidity values for a number of metal carbonyl aqua complexes have now been reported. In some cases, these complexes have enhanced rates for water exchange relative to the hexaaqua ion of the same metal. Ease of water exchange in organometallic aqua ions of technetium is essential toward the generation of ^{99m}Tc-labeled biomolecules for radiopharmaceutical imaging applications.

References

1 S.J. Lippard, J.M. Berg, *Principles of Bioinorganic Chemistry*, University Science Books, Mill Valley, CA, **1994**.
2 S.M. Thomas, R. DiCosimo, V. Nagarajan, *Trends Biotechnol* **2002**, *20*, 238–242.
3 A. Zaks, *Curr Opin Chem Biol* **2001**, *5*, 130–136.
4 W.N. Lipscomb, N. Strater, *Chem Rev* **1996**, *96*, 2375–2434.
5 D.E. Wilcox, *Chem Rev* **1996**, *96*, 2435–2458.
6 J. Weston, *Chem Rev* **2005**, *105*, 2151–2174.
7 B.C. Tripp, K. Smith, J.G. Ferry, *J Biol Chem* **2001**, *276*, 48615–48618.
8 J.A. Kovacs, *Chem Rev* **2004**, *104*, 825–848.
9 P.K. Mascharak, *Coord Chem Rev* **2002**, *225*, 201–214.
10 T.C. Harrop, P.K. Mascharak, *Acc Chem Res* **2004**, *37*, 253–260.
11 G. Parkin, *Chem Rev* **2004**, *104*, 699–768.
12 G. Parkin, *Met Ions Biol Syst* **2001**, *38*, 411–460.
13 H. Vahrenkamp, *Acc Chem Res* **1999**, *32*, 589–596.
14 U. Koelle, *Coord Chem Rev* **1994**, *135/136*, 623–650.
15 R. Alberto, *Top Curr Chem* **2005**, *252*, 1–44.
16 F. Mancin, P. Scrimin, P. Tecilla, U. Tonellato, *Chem Commun* **2005**, 2540–2548.
17 A. Sreedhara, J.A. Cowan, *J Biol Inorg Chem* **2001**, *6*, 337–347.
18 J.A. Cowan, *Curr Opin Chem Biol* **2001**, *5*, 634–642.
19 J.R. Morrow, *Met Ions Biol Syst* **1996**, *33*, 561–592.
20 C. Liu, M. Wang, T. Zhang, H. Sun, *Coord Chem Rev* **2004**, *248*, 147–168.
21 A.S. Borovik, *Acc Chem Res* **2005**, *38*, 54–61.

22 L. L. Kiefer, C. A. Fierke, *Biochemistry* **1994**, *33*, 15233–15240.
23 P. T. R. Rajagopalan, S. Grimme, D. Pei, *Biochemistry* **2000**, *39*, 779–790.
24 K. Zhang, D. S. Auld, *Biochemistry* **1993**, *32*, 13844–13851.
25 I. Bertini, C. Luchinat, M. Rosi, A. Sgamellotti, F. Tarantelli, *Inorg Chem* **1990**, *29*, 1460–1463.
26 I. Bertini, C. Luchinat, in *Bioinorganic Chemistry*, I. Bertini, H. B. Gray, S. J. Lippard, J. S. Valentine, (Eds.), University Science Books, Mill Valley, CA, **1994**.
27 M. Gelinsky, R. Vogler, H. Vahrenkamp, *Inorg Chem* **2002**, *41*, 2560–2564.
28 E. Kimura, T. Shiota, T. Koike, M. Shiro, M. Kodama, *J Am Chem Soc* **1990**, *112*, 5805–5811.
29 Y. Fujii, T. Itoh, K. Onodera, T. Tada, *Chem Lett* **1995**, 305–306.
30 H. Sigel, R. B. Martin, *Chem Soc Rev* **1994**, *23*, 83–91.
31 R. Jairam, P. G. Potvin, *J Org Chem* **1992**, *57*, 4136–4141.
32 J. T. Groves, R. R. Chambers, *J Am Chem Soc* **1984**, *106*, 630–638.
33 T. Itoh, Y. Fujii, T. Tada, Y. Yoshikawa, H. Hisada, *Bull Chem Soc Jpn* **1996**, *69*, 1265–1274.
34 J. W. Canary, J. Xu, J. M. Castagnetto, D. Rentzeperis, L. A. Marky, *J Am Chem Soc* **1995**, *117*, 11545–11547.
35 Y.-H. Chiu, J. W. Canary, *Inorg Chem* **2003**, *42*, 5107–5116.
36 G. Anderegg, E. Hubmann, N. G. Podder, F. Wenk, *Helv Chim Acta* **1977**, *60*, 123–140.
37 I. Bertini, C. Luchinat, S. Mangani, R. Pierattelli, *Comm Inorg Chem* **1995**, *17*, 1–15.
38 M. Sakurai, T. Furuki, Y. Inoue, *J Phys Chem* **1995**, *99*, 17789–17794.
39 R. C. diTargiani, S. Chang, M. H. Salter, R. D. Hancock, D. P. Goldberg, *Inorg Chem* **2003**, *42*, 5825–5836.
40 T. Koike, M. Takamura, E. Kimura, *J Am Chem Soc* **1994**, *116*, 8443–8449.
41 H. Kurosaki, T. Tawada, S. Kawasoe, Y. Ohashi, M. Goto, *Bioorg Med Chem Lett* **2000**, *10*, 1333–1337.
42 S. Karlin, Z.-Y. Zhu, *Proc Natl Acad Sci USA* **1997**, *94*, 14231–14236.
43 S. Karlin, Z.-Y. Zhu, K. D. Karlin, *Proc Natl Acad Sci USA* **1997**, *94*, 14225–14230.
44 D. W. Christianson, J. D. Cox, *Annu Rev Biochem* **1999**, *68*, 33–57.
45 J. F. Krebs, J. A. Ippolito, D. W. Christianson, C. A. Fierke, *J Biol Chem* **1993**, *268*, 27458–27466.
46 K. Hakansson, M. Carlsson, L. A. Svensson, A. Liljas, *J Mol Biol* **1992**, *227*, 1192–1204.
47 M. Livieri, F. Mancin, U. Tonellato, J. Chin, *Chem Commun* **2004**, 2862–2863.
48 J. C. Mareque-Rivas, R. Prabaharan, R. T. M. de Rosales, *Chem Commun* **2004**, 76–77.
49 J. C. Mareque-Rivas, R. Prabaharan, S. Parsons, *Dalton Trans* **2004**, 1648–1655.
50 M. Wall, B. Linkletter, D. Williams, A.-M. Lebuis, R. C. Hynes, J. Chin, *J Am Chem Soc* **1999**, *121*, 4710–4711.
51 J. Chin, S. Chung, D. H. Kim, *J Am Chem Soc* **2002**, *124*, 10948–10949.
52 D. K. Garner, S. B. Fitch, L. H. McAlexander, L. M. Bezold, A. M. Arif, L. M. Berreau, *J Am Chem Soc* **2002**, *124*, 9970–9971.
53 S. Yamaguchi, I. Tokairin, Y. Wakita, Y. Funahashi, K. Jitsukawa H. Masuda, *Chem Lett* **2003**, *32*, 406–407.
54 G. A. Jeffrey, *An Introduction to Hydrogen Bonding*, Oxford University Press, Oxford, **1997**.
55 C. Bergquist, T. Fillebeen, M. M. Morlok, G. Parkin, *J Am Chem Soc* **2003**, *125*, 6189–6199.
56 M. M. Makowska-Grzyska, P. C. Jeppson, R. A. Allred, A. M. Arif, L. M. Berreau, *Inorg Chem* **2002**, *41*, 4872–4887.
57 J. C. Mareque-Rivas, E. Salvagni, S. Parsons, *Dalton Trans* **2004**, 4185–4192.
58 R. Murugavel, M. Sathiyendiran, R. Pothiraja, R. J. Butcher, *Chem Commun* **2003**, 2546–2547.
59 F. A. Chavez, C. V. Nguyen, M. M. Olmstead, P. K. Mascharak, *Inorg Chem* **1996**, *35*, 6282–6291.
60 F. A. Chavez, M. M. Olmstead, P. K. Mascharak, *Inorg Chem* **1997**, *36*, 6323–6327.
61 J. C. Noveron, M. M. Olmstead, P. K. Mascharak, *J Am Chem Soc* **1999**, *121*, 3553–3554.

62 L.A. Tyler, J.C. Noveron, M.M. Olmstead, P.K. Mascharak, *Inorg Chem* **2003**, *42*, 5751–5761.
63 J.C. Noveron, M.M. Olmstead, P.K. Mascharak, *J Am Chem Soc* **2001**, *123*, 3247–3259.
64 J. Sanchez, S. Dominguez, A. Mederos, F. Brito, J.M. Arrieta, *Inorg Chem* **1997**, *36*, 4108–4114.
65 M. Stebler-Rothlisberger, W. Hummel, P.-A. Pittet, H.-B. Burgi, A. Ludi, A.E. Merbach, *Inorg Chem* **1988**, *27*, 1358–1363.
66 M.S. Eisen, A. Haskel, H. Chen, M.M. Olmstead, D.P. Smith, M.F. Maestre, R.H. Fish, *Organometallics* **1995**, *14*, 2806–2812.
67 A. Egli, K. Hegetschweiler, R. Alberto, U. Abram, R. Schibli, R. Hedinger, V. Gramlich, R. Kissner, P.A. Schubiger, *Organometallics* **1997**, *16*, 1833–1840.
68 R. Alberto, R. Schibli, A. Egli, A.P. Schubiger, U. Abram, T.A. Kaden, *J Am Chem Soc* **1998**, *120*, 7987–7988.
69 R. Alberto, R. Schibli, A.P. Schubiger, U. Abram, H.-J. Pietzsch, B. Johannsen, *J Am Chem Soc* **1999**, *121*, 6076–6077.
70 N. Aebischer, R. Schibli, R. Alberto, A.E. Merbach, *Angew Chem Int Ed* **2000**, *39*, 254–256.
71 U. Prinz, U. Koelle, S. Ulrich, A.E. Merbach, O. Maas, K. Hegetschweiler, *Inorg Chem* **2004**, *43*, 2387–2391.
72 B. Salignac, P.V. Grundler, S. Cayemittes, U. Frey, R. Scopelliti, A.E. Merbach, R. Hedinger, K. Hegetschweiler, R. Alberto, U. Prinz, G. Raabe, U. Kolle, S. Hall, *Inorg Chem* **2003**, *42*, 3516–3526.
73 S. Cayemittes, T. Poth, M.J. Fernandez, P.G. Lye, M. Becker, H. Elias, A.E. Merbach, *Inorg Chem* **1999**, *38*, 4309–4316.
74 S. Ogo, H. Chen, M.M. Olmstead, R.H. Fish, *Organometallics* **1996**, *15*, 2009–2013.
75 U. Koelle, B. Fuss, *Chem Ber* **1984**, *117*, 753–762.
76 P. Beutler, H. Gamsjager, *Chimia* **1975**, *29*, 525–532.
77 V. Plumb, G.M. Harris, *Inorg Chem* **1964**, *3*, 542–545.
78 S. Ogo, N. Makihara, Y. Kaneko, Y. Watanabe, *Organometallics* **2001**, *20*, 4903–4910.
79 T. Abura, S. Ogo, Y. Watanabe, S. Fukuzumi, *J Am Chem Soc* **2003**, *125*, 4149–4154.
80 U.C. Meier, R. Scopelliti, E. Solari, A.E. Merbach, *Inorg Chem* **2000**, *39*, 3816–3822.
81 N. Aebischer, E. Sidorenkova, M. Ravera, G. Laurenczy, D. Osella, J. Weber, A.E. Merbach, *Inorg Chem* **1997**, *36*, 6009–6020.
82 G. Parkin, *Chem Commun* **2000**, 1971–1985.
83 E.L. Hegg, J.N. Burstyn, *Coord Chem Rev* **1998**, *173*, 133–165.
84 D.A. Buckingham, C.R. Clark, *Met Ions Biol Syst* **2001**, *38*, 43–102.
85 R. Kramer, *Coord Chem Rev* **1999**, *182*, 243–261.
86 M. Komiyama, *Met Ions Biol Syst* **2001**, *38*, 25–41.
87 G. Allen, *Met Ions Biol Syst* **2001**, *38*, 197–212.
88 M. Kito, R. Urade, *Met Ions Biol Syst* **2001**, *38*, 187–196.
89 N.M. Milovic, N.M. Kostic, *Met Ions Biol Syst* **2001**, *38*, 145–186.
90 G.M. Polzin, J.N. Burstyn, *Met Ions Biol Syst* **2001**, *38*, 103–143.
91 J.K. Bashkin, *Curr Opin Chem Biol* **1999**, *3*, 752–758.
92 P. Molenveld, J.F.J. Engbersen, D.N. Reinhoudt, *Chem Soc Rev* **2000**, *29*, 75–86.
93 N. Strater, W.N. Lipscomb, T. Klabunde, B. Krebs, *Angew Chem Int Ed* **1996**, *35*, 2024–2055.
94 N.H. Williams, B. Takasaki, M. Wall, J. Chin, *Acc Chem Res* **1999**, *32*, 485–493.
95 N.H. Williams, *Biochem Biophys Acta* **2004**, *1697*, 279–287.
96 J. Suh, *Acc Chem Res* **2003**, *36*, 562–570.
97 E. Kimura, *Curr Opin Chem Biol* **2000**, *4*, 207–213.
98 J. Chin, *Curr Opin Chem Biol* **1997**, *1*, 514–521.
99 J.R. Morrow, O. Iranzo, *Curr Opin Chem Biol* **2004**, *8*, 192–200.
100 M.W. Hosseini, A.J. Blacker, J.M. Lehn, *J Am Chem Soc* **1990**, *112*, 3896–3904.

101 S. Watanabe, O. Onogawa, Y. Komatsu, K. Yoshida, *J Am Chem Soc* **1998**, *120*, 229–230.

102 K. A. Schug, W. Lindner, *Chem Rev* **2005**, *105*, 67–113.

103 M. D. Best, S. L. Tobey, E. V. Anslyn, *Coord Chem Rev* **2003**, *240*, 3–15.

104 D. K. Garner, R. A. Allred, K. J. Tubbs, A. M. Arif, L. M. Berreau, *Inorg Chem* **2002**, *41*, 3533–3541.

105 S.-J. Chiou, C. G. Riordan, A. L. Rheingold, *Proc Natl Acad Sci USA* **2003**, *100*, 3695–3700.

106 J. N. Smith, Z. Shirin, C. J. Carrano, *J Am Chem Soc* **2003**, *125*, 868–869.

107 J. N. Smith, J. T. Hoffman, Z. Shirin, C. J. Carrano, *Inorg Chem* **2005**, *44*, 2012–2017.

108 J. Chin, M. Banaszczyk, V. Jubian, X. Zou, *J Am Chem Soc* **1989**, *111*, 186–190.

109 D. Wahnon, A.-M. Lebuis, J. Chin, *Angew Chem Int Ed* **1995**, *34*, 2412–2414.

110 C. L. Perrin, J. B. Nielson, *Annu Rev Phys Chem* **1997**, *48*, 511–544.

111 J. C. Mareque-Rivas, R. T. M. de Rosales, S. Parsons, *Chem Commun* **2004**, 610–611.

112 E. Kimura, S. Aoki, T. Koike, M. Shiro, *J Am Chem Soc* **1997**, *119*, 3068–3076.

113 J. C. Mareque-Rivas, R. T. M. de Rosales, S. Parsons, *Dalton Trans* **2003**, 4385–4386.

114 R. Ott, R. Kramer, *Angew Chem Int Ed* **1998**, *37*, 1957–1960.

115 E. Kovari, R. Kramer, *J Am Chem Soc* **1996**, *118*, 12704–12709.

116 E. Kovari, J. Heitker, R. Kramer, *Chem Commun* **1995**, 1205–1206.

117 A. W. Addison, T. N. Rao, J. Reedijk, J. v. Rijn, G. C. Verschoor, *Dalton Trans* **1984**, 1349–1356.

118 E. Kovari, R. Kramer, *Chem Ber* **1994**, *127*, 2151–2157.

119 H. Ait-Haddou, J. Sumaoka, S. L. Wiskur, J. F. Folmer-Andersen, E. V. Anslyn, *Angew Chem Int Ed* **2002**, *41*, 4013–4016.

120 B. S. Hammes, M. W. Carrano, C. J. Carrano, *Dalton Trans* **2001**, *9*, 1448–1451.

121 B. S. Hammes, X. Luo, M. W. Carrano, C. J. Carrano, *Inorg Chim Acta* **2002**, *341*, 33–38.

122 K. Weis, H. Vahrenkamp, *Eur J Inorg Chem* **1998**, 271–274.

123 K. Weis, M. Rombach, M. Ruf, H. Vahrenkamp, *Eur J Inorg Chem* **1998**, 263–270.

124 K. Weis, M. Rombach, H. Vahrenkamp, *Inorg Chem* **1998**, *37*, 2470–2475.

9 Carbon Monoxide as a Chemical Feedstock: Carbonylation Catalysis

Piet W. N. M. van Leeuwen and Zoraida Freixa

9.1 Introduction

9.1.1 Heterogeneous Processes

Activation of carbon monoxide (CO) and industrial aspects thereof is certainly different from the activation of other small molecules such as methane, nitrogen and carbon dioxide. CO is rather reactive compared with these molecules. It is produced and used on a very large scale as a feedstock for the production of fuel and petrochemicals. Methane and nitrogen are also used in large-scale conversions in spite of their inertness, but the conditions are considerably milder for CO. CO will react readily with oxygen, chloride, bases and oxygen in the presence of other nucleophiles such as amines [1]. With the aid of catalysts the range is extended with important molecules such as hydrogen, alkenes, alkyl halides, alcohols, nitro compounds, etc. In contrast to methane and dinitrogen, CO reacts readily with coordination compounds and metals to form carbonyl complexes, and with bases, including metal alkyl compounds.

CO is made industrially as a mixture with hydrogen from coal and water [synthesis gas (syngas)], reforming of methane and water ("spaltgas"), partial oxidation of methane or a combination of the latter two [1]. Higher alkane gases are also used. Biomass, peat and braun-coal can be used as well. Subsequently syngas is converted to methanol, higher hydrocarbons and diesel fuel [2, 3]. Throughout the world hydrocarbon gases are converted to liquids this way [gas-to-liquid (GTL)] and in recent years all major oil companies have announced expansions of GTL capacity. The driving forces are a better use of the resources and the costs of transportation. Natural gas is 4 times more expensive to transport than oil. Converting remote natural gas into a liquid before transport is more cost-effective. Although coal and biomass may produce an impure syngas with COS as a major impurity, purification of gases is relatively simple and cheap. Thus, a pure starting material can be obtained and, for instance, the die-

Activation of Small Molecules. Edited by William B. Tolman

ISBN: 3-527-31312-5

sel fuel [2] made is environmentally much more attractive than the sulfur-containing fuel from petrochemical sources.

Coal is used as the carbon source to make syngas by Sasol in South Africa, but it has been announced that also here coal will be replaced by natural gas. On the other hand, in the US, where there are large amounts of coal accessible to surface mining, the possibility of coal-to-liquid processes (CTL) is a topic of study. The conversion of syngas to liquid products involves the Fischer-Tropsch process developed in Germany during the 1930s when it was anticipated that crude oil might not be available and local sources (coal) should be used for the production of petrol. Sasol uses high-temperature Fischer-Tropsch technology [3] in the SAS (Sasol Advanced Synthol) process to convert synthesis gas (derived from coal) into automotive and other fuels, as well as a wide range of light olefins. Synthesis feed gas is fed into reactors where hydrogen and CO react under pressure at a moderate temperature. A fluidized, iron-based catalyst is added. This yields a broad spectrum of hydrocarbons, mainly in the C_1–C_{20} range, containing large quantities of alkenes. α-Olefins pentene, hexene and octene are recovered, while the longer-chain olefins are introduced into the fuel pool. Separation technologies have enabled Sasol to become an international marketer of 1-pentene, 1-hexene and 1-octene. These technologies have also generated a phenols and cresols business. Oxygenates in the aqueous stream from the SAS process are separated and purified in the chemical work-up plant to produce alcohols, acetic acid and ketones, including acetone, methyl ethyl ketone (MEK) and methyl isobutyl ketone (MIBK).

Supported cobalt catalysts can also be used and while working with these catalysts Roelen found that alkenes undergo hydroformylation [4]. The active species for this reaction is $HCo(CO)_4$, a homogeneous catalyst, while the Fischer-Tropsch process requires heterogeneous catalysts. The catalysts for the more selective processes [GTL, Shell Middle Distillate Synthesis (SMDS)] to make predominantly diesel fuel have not been disclosed, but it would seem that they are cobalt based. In the diesel process less or no oxygenates are coproduced, the hydrocarbons are saturated and alkene byproducts are hydrogenated after the reaction. The lighter products are removed and used for other purposes. In all cases the Fischer-Tropsch reaction gives a broad distribution of molecular weights, typically a Schulz-Flory distribution, and yields to individual products never exceed a few percent. Nickel catalysts in general lead to methanation of CO. The lowest carbon numbers can be made with high selectivity, even though a Schulz-Flory distribution governs the selectivity. Rate constants for the lower members of a series may deviate from those of the higher homologs and selectivities may deviate from the Schulz-Flory distribution in either way.

An important product made from methane is methanol. It involves an indirect route, also via the reforming of methane with water and oxygen to form syngas. Direct partial oxidation of methane is not sufficiently selective to make methanol or formaldehyde. Methanol is made from syngas by passing it over a copper zinc oxide catalyst. The selectivity is very high and thus the impetus for development of other routes such as direct oxidation is not extremely high.

Methanol is converted to formaldehyde, acetic acid, acetic anhydride and various methyl ester solvents. Formaldehyde finds its major application in the synthesis of resins in condensation reactions with phenols.

9.1.2 Homogeneous Catalysts

The above two processes are the largest ones using CO, but as they use heterogeneous catalysts we will not expand further on these here. Homogeneous counterparts of these processes have been studied, especially after the first oil crisis in the 1970s. It was thought that homogeneous catalysts could perhaps be tuned in such a way that they would give selectively only one product, such as ethylene glycol, benzene or cyclohexene. The envisaged products should have features that might circumvent the Fischer-Tropsch "statistics", the geometric distribution of products. Thermodynamically, hydrogenation of CO to most products is allowed below a few hundred degrees centigrade and the reactions are also highly exothermic. Liquid-phase operation might present an advantage, it was thought, to transfer the heat from the reaction medium.

Interestingly, homogeneous catalysts with cobalt, ruthenium and rhodium that can convert CO into oxygenates exist, but the reaction conditions are much more demanding (nickel is an exception that yields methanol under mild conditions) than those of the heterogeneous processes giving alkanes! The reasons for this have been discussed by Maitlis [5]. The main difference resides in the ease by which an assembly of metal atoms can dissociate a CO molecule (it is not even the rate-determining step in Fischer-Tropsch synthesis). A homogeneous counterpart does not exist that can provide the many bonds needed to stabilize the dissociated carbon and oxygen atoms.

Homogeneous processes for methanol synthesis have been studied; in particular, a nickel-based catalyst, which has become known as the Brookhaven catalyst. The catalyst is composed of nickel carbonyl and potassium methoxide. The system can be operated at low temperature and low pressure, it gives high gas conversion per pass eliminating a recycle, and the space-time yield is high, as is the product selectivity (the space-time yield is an important industrial parameter expressing how much product can be made per reactor volume in a certain time) [6, 7]. At 100 °C, syngas conversion of 99% with turnover frequency of 66 h^{-1} has been reported. The reaction is truly catalytic in nickel as well as in base. A kinetic study of the $Ni(CO)_4$/KOMe catalyst system in triglyme-MeOH solvent mixtures has been reported. The kinetic expression includes terms of zero-order in H_2 and first-order in CO with syngas H_2:CO stoichiometry of around 2:1 and less than first-order in nickel, approximately third-order in base, and an exponential dependence on methanol concentration. A gas-phase infrared (IR) spectrum recorded at the end of each run showed an intense signature peak at 2060 cm^{-1} for $Ni(CO)_4$ [8]. The major problems related with these catalysts are probably the low stability, toxicity and volatility of nickel tetracarbonyl under the reaction conditions. The process has been in pilot-plant stage and worldwide several companies have worked on this system. As yet no commercial applications have been announced.

The main bulk chemical product made industrially from CO and hydrogen via a combination of homogeneous and heterogeneous catalysts is acetic acid from methanol and CO. Methanol is made with the aid of heterogeneous catalysts from CO and hydrogen. Further hydrogenation of acetic acid with hydrogen or homologation of methanol or formaldehyde with syngas to give ethanol have been studied, but these processes are not commercial [9–12].

In this chapter we will review the two main industrial processes using homogeneous catalysts, i.e. carbonylation of methanol to acetic acid and the hydroformylation of alkenes to aldehydes.

9.2 Rhodium-catalyzed Hydroformylation

9.2.1 Introduction

Until the 1970s the hydroformylation of alkenes was conducted with cobalt catalysts, the majority being unmodified cobalt carbonyl species. This catalyst required pressures up to 200 bar and catalyst recovery often required decomposition. The only industrial application of cobalt carbonyls modified by alkylphosphines as the ligand was the Shell system, invented by Slaugh [13]. The phosphine modified process uses higher temperatures, but lower pressures; it gives higher linearities in the alcohol products as in this process the aldehydes are hydrogenated. Many of these plants are still in operation and only where new plants were built for propene hydroformylation (with few exceptions for butene up to octene) is the more economic rhodium process is being used. For higher, internal alkenes cobalt is still the catalyst of choice.

The discovery of triphenylphosphine as the ligand in rhodium-catalyzed hydroformylation by Wilkinson and coworkers [14] has had a major impact on the industrial practice of hydroformylation and on academic research. The phosphine rhodium catalysts are much faster and more selective than the cobalt ones, and as a result they can be used at much lower temperatures and pressures. Since the 1970s commercial applications were brought on-stream by Celanese, Union Carbide Corporation (UCC; now Dow Chemical) and Mitsubishi Chemical Corporation for propene hydroformylation. Mitsubishi also operates a process for higher alkenes based on a phosphine-free catalyst system; this is the oldest rhodium-based process, in operation since 1970. In the mid-1980s a new process came on-stream, the Ruhrchemie/Rhône-Poulenc process, which is based on two-phase catalysis using trisulfonated triphenylphosphine as the water-soluble ligand. It can be used for propene and 1-butene, and its economics are slightly better than those of the triphenylphosphine process due to energy costs, selectivity and catalyst stability [15].

Ligand modification turned out to be extremely useful for rhodium-catalyzed hydroformylation and a great deal of the more than 6000 publications on hydroformylation known today are concerned with this, of which around 30% appeared after 1994. Phosphites, first used by Pruett and Smith [16], came into focus again after the peculiar behavior of bulky phosphites was reported by van Leeuwen [17]. In particular, diphosphites have been addressed by many authors after their favorable effect on rate and selectivity for many substrates had been discovered by Bryant and coworkers [18]. The effect of backbone and bite angle of bidentate phosphites followed this development, which was first discovered for diphosphines with the introduction of BISBI (**15**, *vide infra*) by Devon (Texas Eastman) [19], and examined by Casey and coworkers in relation with other ligands and bite angles [20]. The introduction of Xantphos-type ligands (**30**, *vide infra*) has enlarged the number of possibilities for variation of the bite angle enormously and has allowed the study of its effect on catalysis [21]. In the last decade the understanding of the mechanistic details has grown considerably as well. A monograph on rhodium-catalyzed hydroformylation appeared in 2000 [22] which discussed a large number of representative catalysts, industrial processes [17], laboratory procedures ranging from reactions of simple alkenes to multistep organic applications, new developments in separation, and issues concerning catalyst preparation and decomposition.

9.2.2
CO as the Ligand

For quite some time it has been known that rhodium carbonyls are very active catalysts for hydroformylation – more active than cobalt catalysts and, depending on the circumstances, also more active than the phosphine-modified rhodium catalysts. The first industrial application of rhodium-catalyzed hydroformylation was based on rhodium carbonyls. As is to be expected, CO dissociation from $HRh(CO)_4$ is faster than that from donor ligand-substituted complexes, while insertion reaction rates are not that much influenced. Thus, carbonyl catalysts are potentially fast. Oxidative addition of H_2 is slower for the less electron-rich rhodium carbonyls than the phosphorus donor-substituted complexes and this is now the rate-determining step of the overall process. In the last decade this has been demonstrated for a large number of substrates under mild conditions by Garland and coworkers [23]. The key factor diminishing the overall effectiveness of these catalysts (apart from selectivity) is the position of the equilibrium between $HRh(CO)_4$ and $Rh_4(CO)_{12}$ and/or $Rh_6(CO)_{16}$, which is strongly in favor of the latter clusters, thus diminishing the initial rate of reaction. Since the resting state under mild conditions is the rhodium-acyl species, potentially all rhodium participates in the catalytic cycle, and depending on the rates of equilibration between the carbonyl complexes and the competing rate of insertion of alkene, there may be no rhodium residing in the $Rh_4(CO)_{12}$ sink. Thus, even for a simple carbonyl system, in the absence of modifying phosphorus ligands the kinetics can be complicated.

A method for the potential commercial synthesis of isovaleraldehyde is hydroformylation of isobutene, which was considered by Solodar and coworkers at Monsanto [24]. However, the low production volume excluded the possibility of a continuous process and it was felt that catalyst handling losses would make a traditional rhodium/phosphine system too costly in a batch mode. This dilemma was solved by developing a batch process for the hydroformylation of isobutene, which employed a nonphosphine $Rh(CO)_2$acetylacetonate system at such low levels that rhodium could be considered a "throw-away" ingredient in the process.

The kinetics of the rhodium carbonyl catalyst has been studied extensively by Garland and coworkers, albeit at mild conditions instead of the harsh practical conditions. They studied as many as 20 alkenes [25], starting with $Rh_4(CO)_{12}$ as catalyst precursor in *n*-hexane solvent, using high-pressure *in situ* IR spectroscopy as the analytical tool. Five categories of alkenes were studied, i.e. cycloalkenes, symmetrical internal linear alkenes, terminal alkenes, methylenecycloalkanes and branched alkenes. The typical reaction conditions were 293 K, with hydrogen and CO pressures of 20 bar. In most experiments the precursor $Rh_4(CO)_{12}$ was converted in good yield, but at different rates to the corresponding observable mononuclear acyl Rh tetracarbonyl intermediate $RCORh(CO)_4$ of which 15 have been observed. To a first approximation, the primary differences in rates of hydroformylation are due to the conversion of $Rh_4(CO)_{12}$ and are not real turnover frequencies of substrate converted per molecule of active rhodium catalyst per unit of time. For cyclohexene, complete conversion of the precursor $Rh_4(CO)_{12}$ to the intermediate $C_6H_{11}CORh(CO)_4$ was never observed during the 8-h experiments. Instead, after approximately 30 min, a pseudo steady state was achieved between the species $Rh_4(CO)_{12}$ and $C_6H_{11}CORh(CO)_4$ (298 K, 20 bar H_2 pressure, 60 bar CO). There is no statistically significant contribution to aldehyde formation from a catalytic binuclear elimination (Rh-acyl + RhH) in the rhodium-catalyzed hydroformylation reaction under these conditions [26]. For styrene, hydrogenolysis of the two acyl species is the rate-determining step at temperatures of 298–313 K [27]. During hydroformylation of ethene at 293 K both propionylrhodium and ethylrhodium tetracarbonyl species were observed by IR spectroscopy [28].

9.2.3 Phosphites as Ligands

Soon after the discovery of Wilkinson that PPh_3 leads to active catalysts, phosphites were first used successfully [29]. A survey of phosphites has been reported recently showing that until the late 1980s only few of them showed remarkable results since selectivities for any product did not exceed 80% [30]. Two examples are worth mentioning, both from Shell; one concerns an electronic effect on the linear/branched ratio (*l*/*b*) with the use of ligand **1** [31], while the bulky *o*-^{t}Bu-phenyl phosphite **2** gave rise to extremely fast catalysts [32]. The latter were thought to lead to mono-ligand complexes under hydroformylation conditions. A recent spectroscopic study has also shown that a bis-ligand complex is involved [33]. Ligand **1** gives a *l/b* ratio of 24 and a low percentage of isomerization can be obtained; at low phosphite concentrations it is also a good isomerization catalyst and internal heptenes give linearities as high as 60%.

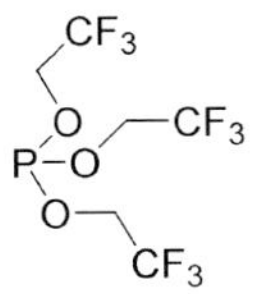

1

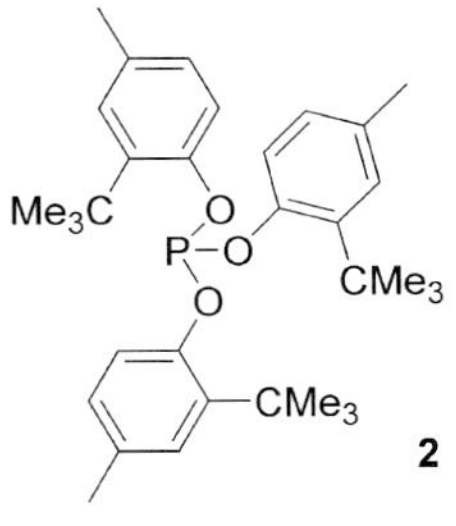

2

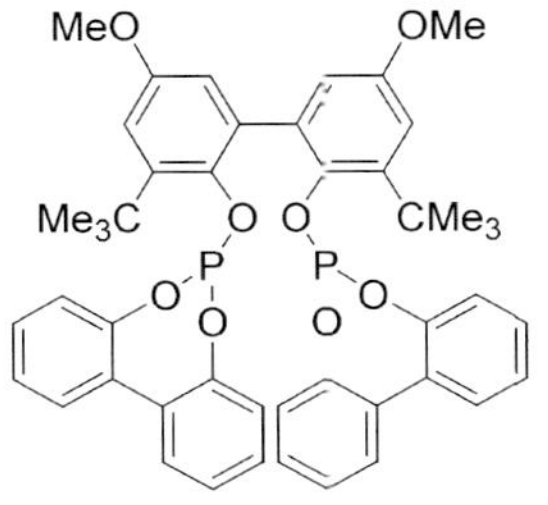

3

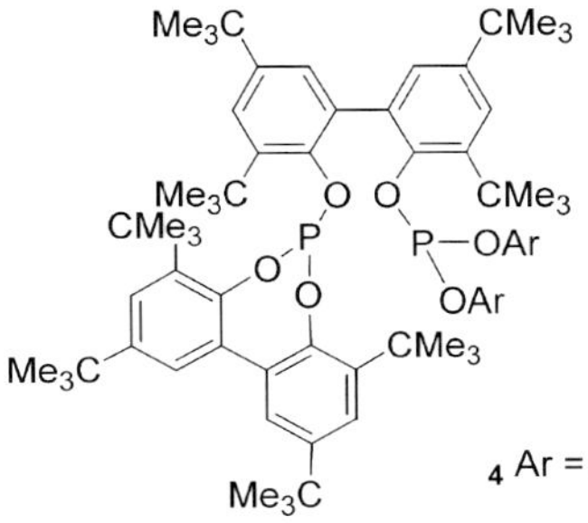

4 Ar = 4-C_6H_4Cl

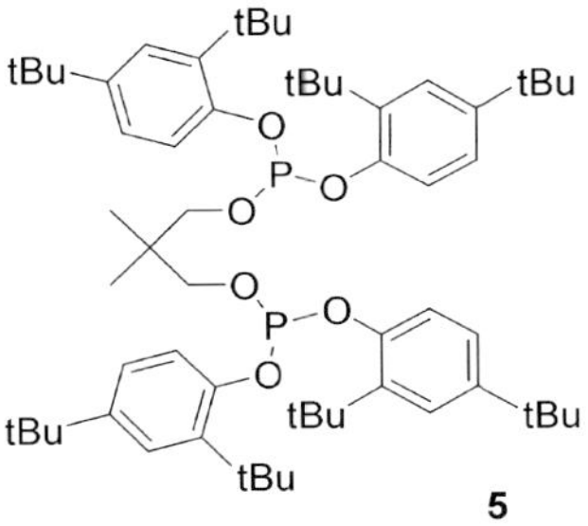

5

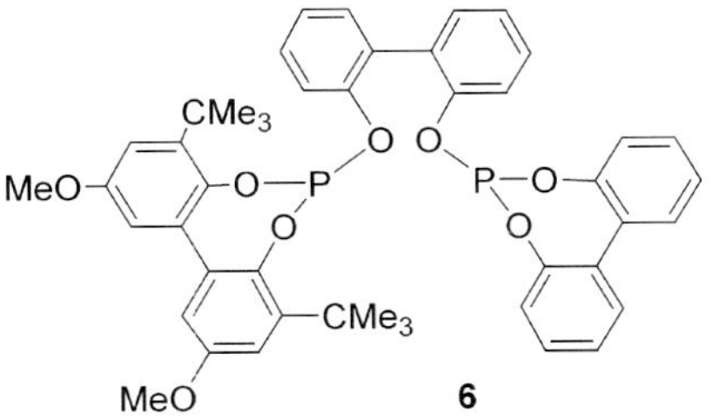

6

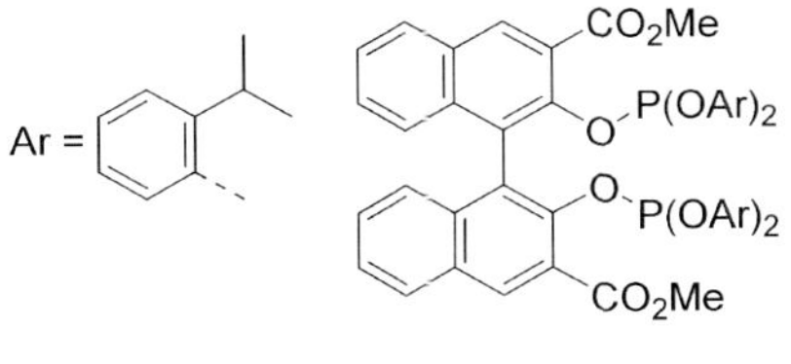

7

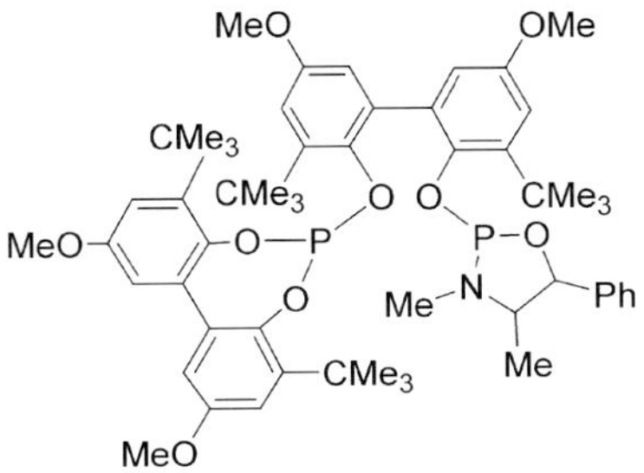

8

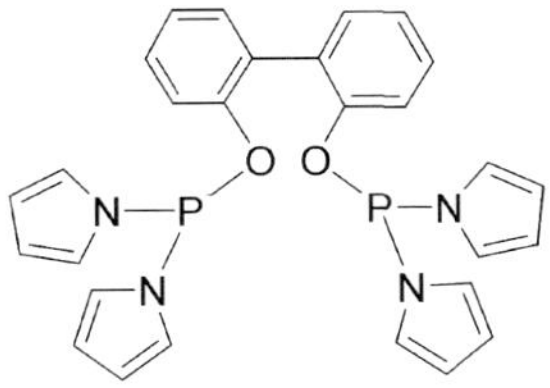

9

Phosphites are easier to make and may be more stable than phosphines towards oxidation (provided that water and alcohols are absent). Interestingly, the breakthroughs in the area of phosphites originate from industry. Pursuing the effect of bulky phosphite ligands reported by van Leeuwen [34], Bryant and coworkers at UCC turned to bidentates such as **3**, **4** and **6** [34], which led to an impressive number of ligands tested. Later, numerous chemical companies with an interest in hydroformylation observed related activities (i.e. ligands **5** and **7**) as the results with the bidentates were astounding. Several bidentates have been reported that give very high *l*/*b* ratios and high rates in the hydroformylation of terminal and internal alkenes, with or without functional groups. Typical examples taken from the patent literature have been collected in a book that appeared on rhodium-catalyzed hydroformylation [35]. A more readily accessible example can be found in the work of Cuny and Buchwald, who used one of the most successful UCC ligands **3** for the hydroformylation of a range of different α-alkenes [36], and in the work by van Rooy and coworkers who focused on 1-octene, styrene and cyclohexene as the substrates and several diphosphites [37].

The solution structures of such selective catalysts have been studied and most likely the active species $HRh(L\text{-}L)(CO)_2$ all contain the bidentate phosphite in the equatorial plane (structure $\mathbf{D_{e\text{-}e}}$ in Scheme 9.1) [38], especially bulky ligands lead to high *l*/*b* ratios [39]. The reverse is not true – not all bis-equatorial complexes are selective catalysts for making linear aldehyde! The bisphenol backbone is usually highly effective; it is the same structural motif as that of BISBI

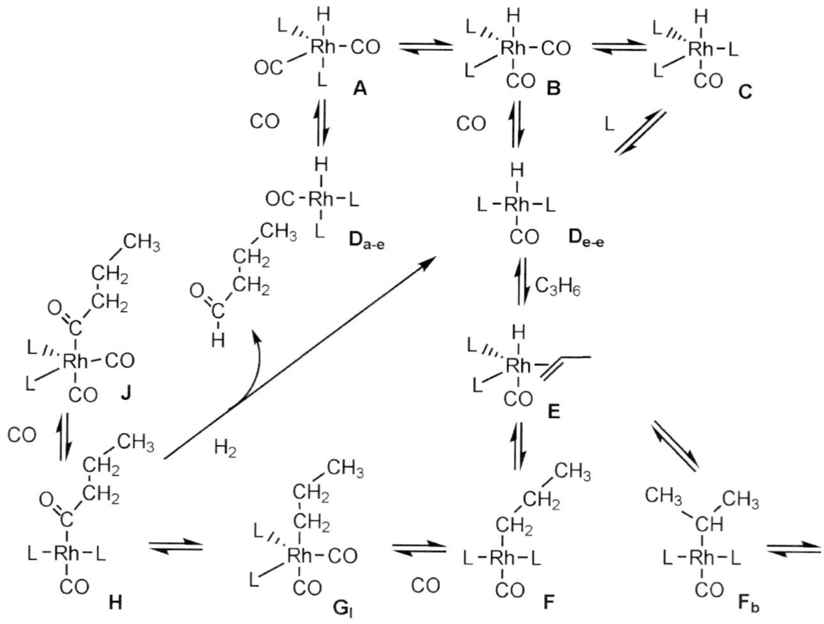

Scheme 9.1

(**15**, *vide infra*), the diphosphine analog of the bisphenol-based diphosphites. Other phosphorus ligands containing this basic structure often retain the high selectivity and the list of phosphonites, phosphoramidites **8** [40], etc., is still limited, leaving room for more variations. Pyrroles as substituents at phosphorus were found to result in strongly electron-withdrawing phosphorus ligands and they were used in hydroformylation [41]. When used as a bidentate ligand with a wide bite angle biphenyl backbone an extraordinary behavior was found (**9**) [42] – the catalyst formed is as fast as a bulky monophosphite catalyst and it produces 1-aldehyde only! Yet, the selectivity is not 100% as an additional around 10% of 2-alkene is formed which does not react under the circumstances tested. Thus, in addition to the bite angle effect, which has a steric origin, there is an electronic effect, as was mentioned above also for trifluoroethyl phosphite **1** [33], although this is still relatively rare for rhodium-catalyzed hydroformylation.

Calix[4]arenes have been used extensively as scaffolds for phosphines, phosphinites or aryl phosphites and many of them are suitable as ligands in hydroformylation [43]. In particular, the work by Paciello stands out as high *l*/*b* ratios were obtained for diphosphites based on a calix[4]arene with bulky substituents [44]. In these systems the four oxygen atoms of the calix[4]arene are used to bind two P-OAr groups, where OAr is a bulky phenol group, 2,6-di-^{t}Bu-4-methoxyphenol, e.g. **10**. Most likely the steric constraints allowed the formation of the linear rhodium-alkyl only, while rates were still relatively high as is typical of bulky phosphites. A disadvantage is that the highest selectivities for linear product are accompanied by formation of alkanes in yields up to 27%. Phosphites **11** utilizing three phenol groups of one calix[4]arene led to less exciting results [45]. Other structures **12** with $P(OAr)_2$ or PAr_2 groups connected to two adjacent or two alternate oxygen atoms of a calix[4]arene also gave modest results [46].

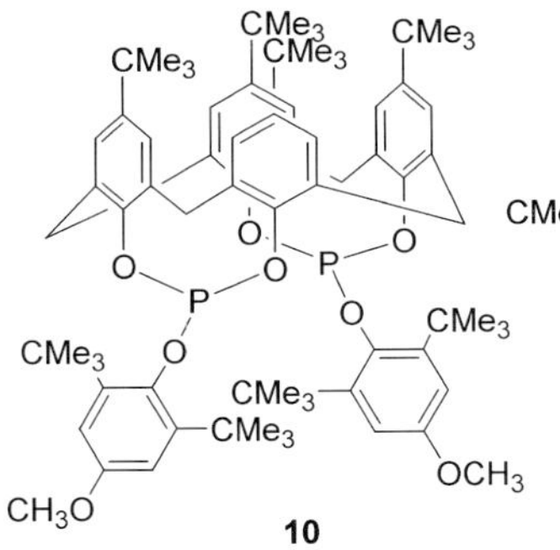

10 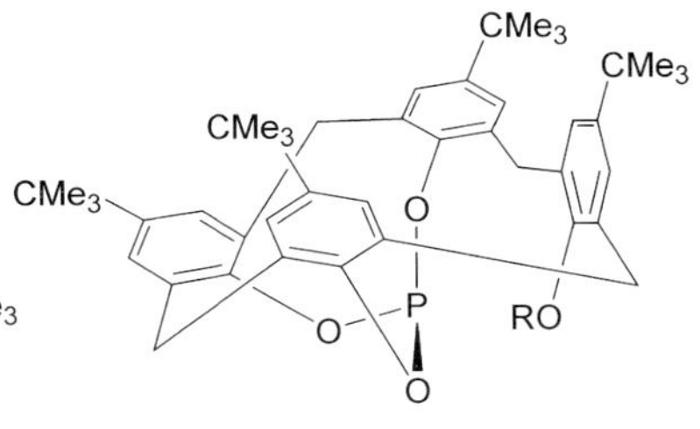 **11** 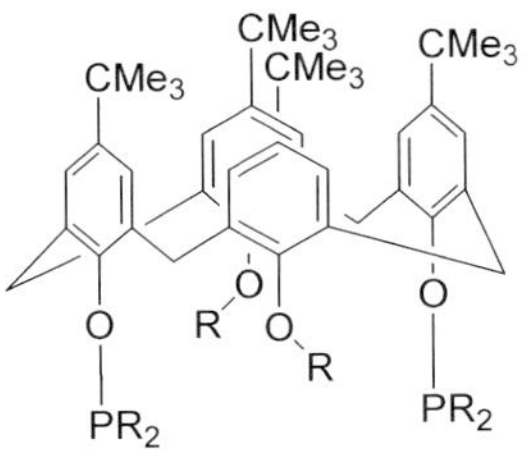 **12**

9.2.4
Arylphosphines as Ligands

9.2.4.1 Monophosphines

Triphenylphosphine and trisulfonated triphenylphosphine continue to be the most important monodentate phosphine ligands in rhodium-catalyzed hydroformylation, both in industrial processes and in applications in organic syntheses. The precursor introduced by Wilkinson was the complex $RhH(PPh_3)_3CO$, structure **13** (as **C** in Scheme 9.1), which actually was first reported by Bath and Vaska as early as 1963 [47], but its hydroformylation activity was only discovered 5 years later. At moderate pressures of CO and high phosphine concentrations (100 mM) the tris-phosphine species is the resting state of the catalyst. Since they are all in the equatorial plane of the trigonal bipyramid, dissociation of a phosphine always leads to a species having two phosphines *trans* to one another ($\mathbf{D_{e\text{-}e}}$ in Scheme 9.1) [32]. It is thought that this leads to the high *l*/*b* ratio in the UCC/Johnson Matthey process (which uses PPh_3) and the Ruhrchemie/Rhône-Poulenc system (using tppts, tris-sulfonated PPh_3 in the *meta* positions). At lower PPh_3 concentrations (10 mM) *l*/*b* ratios of around 2 are typically obtained and the resting state is $RhH(PPh_3)_2(CO)_2$, as a mixture of bis-equatorial and equatorial/apical species ($\mathbf{D_{e\text{-}e}}$ and $\mathbf{D_{a\text{-}e}}$ in Scheme 9.1) [48]. These observations inspired a number of groups to search for bidentate ligands that might enhance bis-equatorial coordination of diphosphines. *In situ* studies using ^{31}P nuclear magnetic resonance (NMR) spectroscopy confirmed the presence of $RhH(PPh_3)_2(CO)_2$ during a hydroformylation reaction of 1-hexene when $RhH(PPh_3)_3CO$ is used as the precursor [49]. Other intermediates were not observed. The inhibiting effect of CO pressure, known from kinetic studies, was also demonstrated in these studies [50]. In complexes of the type $HRh(CO)_x(L)_{4-x}$ the hydride usually occupies an apical position in the trigonal bipyramid.

When the steric bulk in the group R of the alkene R–CH=CH_2 is increased the preference for linear product increases, without much loss of velocity, provided that no direct substitution at the vinyl group is involved [51]. The effect of steric bulk R of the substrate was studied both with triphenylphosphine and with *o*-^{t}Bu-phenyl phosphite, **2**. The selectivity towards the linear aldehyde increases progressively with substitution, from 66% for 1-octene up to 100% for 3,3-dimethyl-1-butene.

Often the kinetics of rhodium-catalyzed hydroformylation do not follow the expected first-order dependence on the alkene concentration and the minus one order in CO pressure [24]. Severe or slight incubation has also been observed. It is known [52] that impurities such as 1,3-alkadienes, enones and terminal alkynes may be the cause of such behavior. In a mixture of 1-alkenes and butadiene, for instance, the latter is much more reactive, and will react preferentially with the rhodium hydride catalyst. The allylic rhodium species formed, however, reacts much more sluggishly with CO than alkyl rhodium complexes and thus the catalyst is tied up in this inactive sink. Detailed studies were not known un-

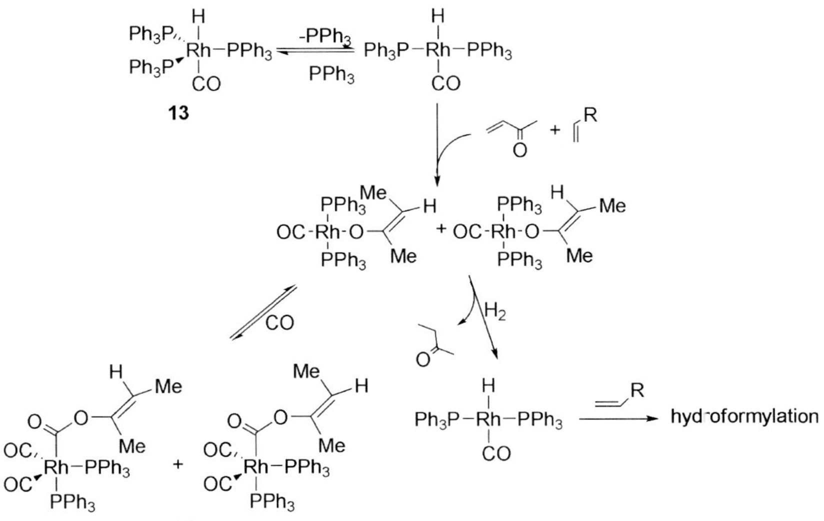

Scheme 9.2

til recently, although several workers in the field [53] took precautions to make sure that the alkene substrates were not contaminated with enones, dienes or alkynes. In an *in situ* IR study combined with high-pressure NMR studies Walczuk has shown how this temporary inhibition takes place and what intermediates are formed [54]. *In situ* IR clearly showed the disappearance of rhodium hydride catalyst when these "poisons" were added; the dormant states were slowly depopulated and after all impurities were converted the hydroformylation of 1-oct ene started, while simultaneously the hydride resting state of the catalyst was recovered. Enones gave rise to stable alkoxycarbonyl species **14** (Scheme 9.2) that were slowly converted to ketones after deinsertion of CO. The degree of inhibition depends both on the structure of the inhibitor and on the ligand; as yet, the latter has not been fully exploited.

9.2.4.2 Diphosphines

The UCC process uses $RhH(PPh_3)_3CO$, **13**, as the catalyst in the presence of an excess of PPh_3. Under such circumstances the precursor is also the resting state of the catalyst, which is relatively slow, i.e. turnover frequencies of a few hundred at 120 °C and 30 bar of syngas pressure, but it gives a high *l/b* product aldehyde ratio ($l/b = 12–16$). As mentioned above, dissociation of one of the phosphines always leads to a *trans*-like species and this was held responsible for the high regioselectivity. For example, the precursor $RhH(PPh_3)_2(CO)_2$ (structures $\mathbf{D_{e\text{-}e}}$ and $\mathbf{D_{a\text{-}e}}$ in Scheme 9.1) gives rise to several intermediate isomers and the selectivity is low ($l/b = 2–3$). Until 1987 all bidentate phosphines known gave *cis* complexes in square-planar complexes and apical-equatorial complexes in trigonal bipyramids. Perhaps it was for this reason that none of them seemed to give interesting results with respect to regioselectivity in rhodium-catalyzed hydroformylation of 1-alkenes. Brown and Kent [50] studied the intermediates occurring in the reaction of precursor **13** with 1-decene and styrene, and several intermediates were identified. Their studies suggested that bidentate diphosphines might be sufficiently flexible to allow wider bite angles during the catalytic cycle. At the time of this publication the important breakthrough with BISBI in bidentate ligands had already been obtained by Devon and coworkers at Eastman Kodak [55]. The introduction of BISBI, **15**, started a new era in rhodium-catalyzed hydroformylation. It showed unprecedented selectivity towards the linear aldehyde at satisfactory rates; $l/b=50$ and more recently values as high as 300 were reported [56]! In later studies Casey and Whiteker proved that key to this invention was the bis-equatorial mode of coordination of the diphosphine in trigonal bipyramid ($\mathbf{D_{e\text{-}e}}$ in Scheme 9.1) complexes (**15–18**). In the early 1990s, Casey and coworkers [57] and Yamamoto [58] designed several other ligands that might prefer wider bite angles, e.g. **19** (natural bite angle 126°), but the success was moderate and apparently the "calculated" bite angle is not the only criterion for a high selectivity. Note, however, that the successful bisphenol diphosphites have the same backbone structure and their bite angles are very similar. Molecular mechanics has proven to be a useful tool in the development of new bidentate diphosphines. The natural bite angle (β_n) and flexibility range of a bidentate ligand, introduced by Casey and Whiteker [22] are useful parameters that can be calculated using molecular modeling. In the actual calculation a "dummy"-type metal atom is used that has no defined geometry and a typical M–P bond length known from X-ray structures of similar complexes is enforced. The force constant for P–M–P bending is defined to be zero and consequently the structure of the complex is determined by the organic ligand only. The outcome of the calculations is dependent on the defined M–P distance, which is influenced by the metal of choice. In this way the natural bite angle can be calculated easily since the parameters for the metal are not needed in the actual calculations. Thus, it is the ligand-preferred angle of coordination,

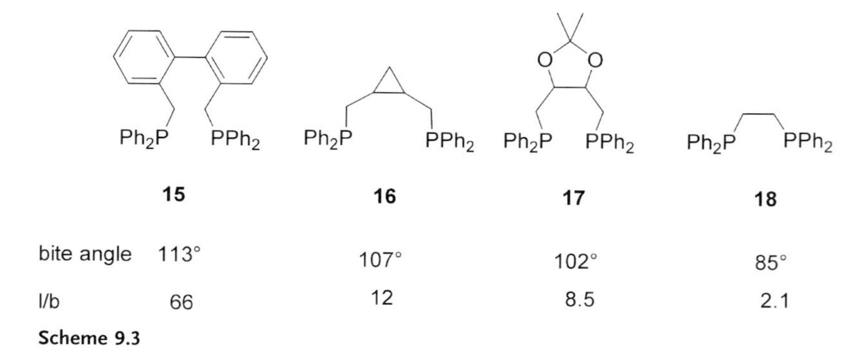

Scheme 9.3

without taking into account the angular preference of the metal. Other unknown parameters such dihedral and bending force constants in which the metal is involved may strongly influence the result, but usually these are ignored.

Casey and coworkers [22] found a trend between calculated bite angle and the linearity of the product of 1-alkene hydroformylation. In hindsight we can extend the list with a number of ligands that were tested long before these studies and it is seen that they fit quite nicely in the series [24, 59]. DIOP [60] (**17**) had been reported as early as 1973 by Consiglio to give $l/b=12$ and dppf [61] in 1981 by Unruh with $l/b=5$, clearly higher than PPh_3 systems. We have not included Xantphos, which will be discussed later.

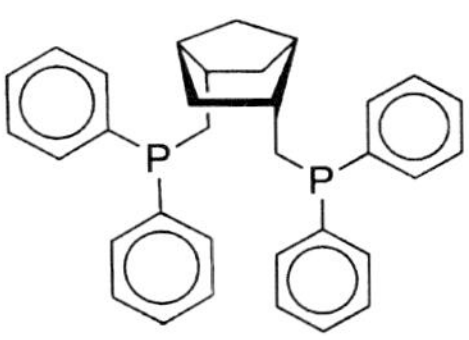

19 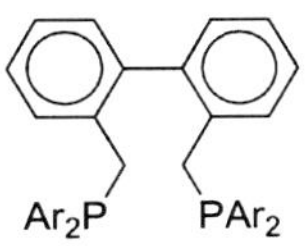 20  21 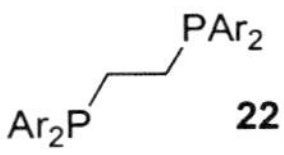 22 23

Ar = 3,5-bis(trifluoromethyl)phenyl

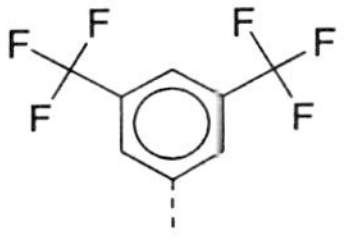

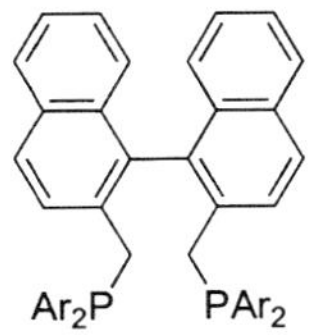

To investigate electronic effects of equatorial and apical phosphines, analogs of chelating diphosphines of BISBI, *trans*-dppm-cyp and dppe having electron-withdrawing substituents on the aryl rings were synthesized and used in hydroformylation [62] (see **20–22**). The introduction of electron-withdrawing substituents on the aryl rings of the diequatorial chelate **20** and **21** lead to an *increase* in linear aldehyde selectivity as well as rate. In contrast, introduction of electron-withdrawing substituents on the aryl rings of the apical-equatorial chelate **22** resulted in a *decrease* in linear aldehyde selectivity when compared with the phenyl-substituted dppe. Interestingly, electron-withdrawing groups in equatorial and apical positions had opposite effects on l/b regioselectivity.

Beller and coworkers used NAPHOS, a BISBI analog backbone, substituted with 3,5-bistrifluoromethylphenyl groups as the ligand in rhodium-catalyzed hydroformylation of internal alkenes and obtained high selectivities to the linear product [63].

Although water-soluble catalysts are not the topic of this chapter, a few highlights will be mentioned. The sulfonated analog of NAPHOS, named BINAS **24**, has been developed by Herrmann and coworkers [64]. It affords one of the most effective systems for propene hydroformylation in a two-phase system; the commercial process using tppts is the most effective process to date in terms of energy consumption, feedstock efficiency, and catalyst stability. In the Ruhrchemie/Rhône-Poulenc process a large excess of tppts is needed in order to obtain a high l/b ratio, but clearly for BISBI-type ligands this is not the case. In addition, the bidentate-based systems are much more reactive, because they require dissociation of a carbonyl ligand while in the monophosphine systems dissociation of a phosphine from a tris-phosphine complex is involved. The BINAS system based on **24** is about 10 times faster than the tppts system under comparable conditions.

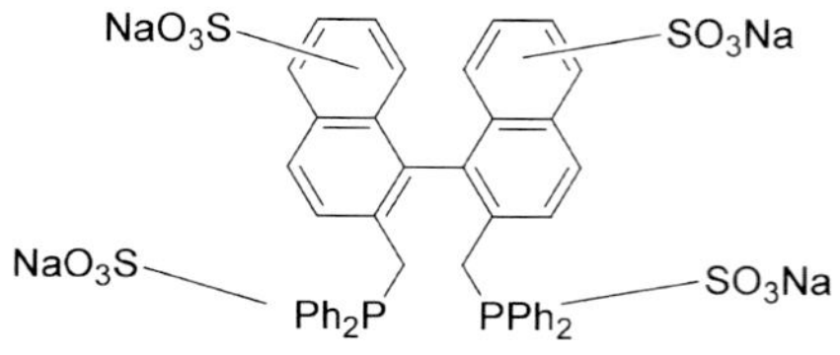

NaO_3S SO_3Na NaO_3S SO_3Na Ph_2P PPh_2

24

The success of the BISBI ligand has stimulated the design of new ligands with wide bite angles. Van Leeuwen and coworkers applied molecular mechanics methods for the design of a new range of xanthene-based ligands, the parent ligand being coined Xantphos [23]. By varying the bridge in the 10-position small variations in the bite angle were introduced. According to molecular mechanics (MM) calculations (Table 9.1), these ligands have natural bite angles ranging from 102 to 121° and a flexibility range of around 35°. It should be noted that the absolute values of the calculations will be dependent on the used force field, but the relative results and, therefore, the observed trends will be the same.

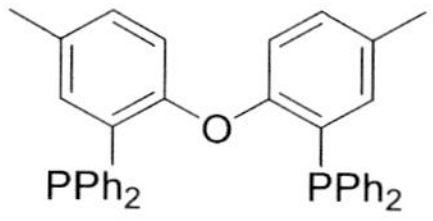

PPh_2 PPh_2 **25** · PPh_2 PPh_2 **26** · PPh_2 PPh_2 **27**

PPh_2 PPh_2 **28** · PPh_2 PPh_2 **29** · PPh_2 PPh_2 **30** Xantphos

PPh_2 PPh_2 **31** · PPh_2 PPh_2 **32** R = Bz, **33** R = H · PPh_2 PPh_2 **34**

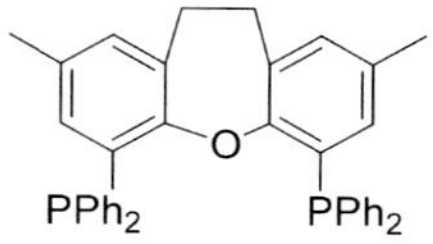

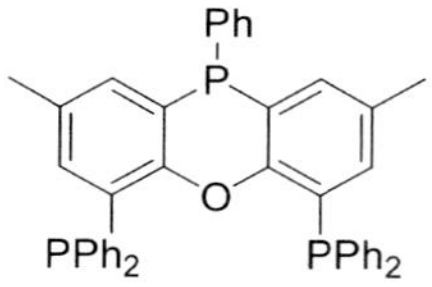

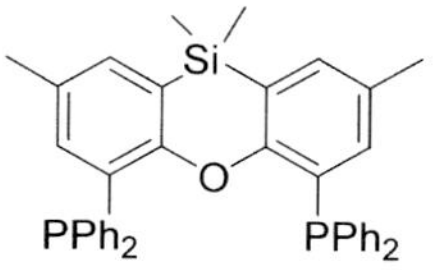

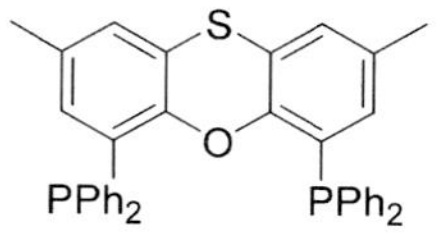

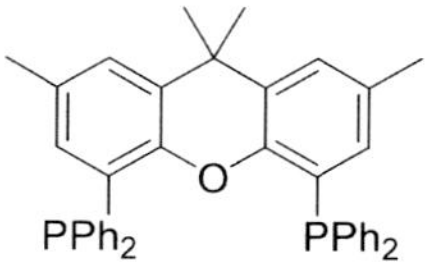

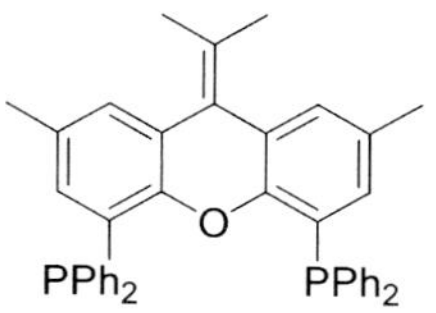

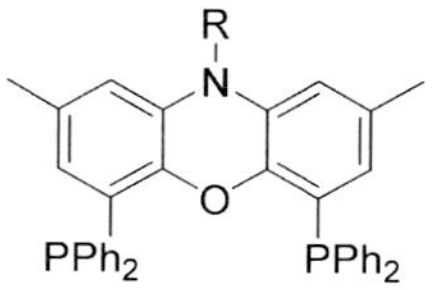

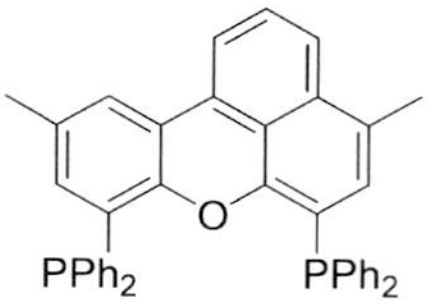

Table 9.1 Natural bite angles (β_n) and the flexibility range calculated for the Xantphos ligands

generic structure, X in position 10

X	Ligand	β_n (deg)	Flexibility range (deg)
H,H	25	102	86–120
C_2H_4	26	102	92–120
PhP	27	108	96–127
$Si(CH_3)_2$	28	109	96–130
S	29	110	96–133
$C(CH_3)_2$	30	111	97–133
$C{=}C(CH_3)_2$	31	113	98–139
NBz	32	114	99–139
NH	33	114	99–141
"benz"	34	121	102–146

The X-ray crystal structure of the free Xantphos ligand shows that only very little adjustment of the structure is necessary to form a bis-equatorial chelate; the orientation of the diphenyl-phosphine moieties is nearly ideal. The observed P···P distance in the free ligand is 4.080 Å, while MM studies indicate that a decrease of the P···P distance to 3.84 Å is necessary for chelation with a P–Rh–P angle of 112°, a decrease of only 0.24 Å. The P-atoms are brought together by means of a slight decrease of the angle between the two phenyl planes in the backbone of the ligand from around 166 to 158°. As a consequence these Xantphos-type ligands do not form bimetallic species, whereas the oxygen atom in the backbone prevents metallation of the ligand.

9.2.4.2.1 1-Alkenes

Ligands **25–34** have been applied in a variety of catalytic reactions but their initial purpose was to serve in rhodium-catalyzed hydroformylation. The selectivity of the Xantphos ligands in the rhodium-catalyzed hydroformylation was tested with the use of 1-octene (Table 9.2). DPEphos, having a calculated natural bite angle of 102°, induced an enhanced, although moderate selectivity (compared to most diphosphines), but no isomerization was detected. The ligands with an one-atom bridge between the aromatic rings of the backbone **27–34** have calculated natural bite angles near 110°, and showed a very high regioselectivity and a very low rate of isomerization to internal alkenes. The ultimate test for a catalyst to check its selectivity towards the linear aldehyde is the hydroformylation of styrene, since this is a substrate with a distinct preference for the formation of the branched aldehyde due to the stability of the 2-alkyl-rhodium species, in-

Table 9.2 Hydroformylation of 1-octene using Xantphos ligands [a)]

Ligand	β_n (deg)	*l*/*b* ratio [b)]	Linear aldehyde (%) [b)]	Isomer (%) [b)]	TOF [b, c)]
25 [d)]	102	6.7	87.0	0	250
26	102	8.5	88.2	1.4	37
27	108	14.6	89.7	4.2	74
28	109	34.6	94.3	3.0	81
29	110	50.0	93.2	4.9	110
30	111	52.2	94.5	3.6	187
31	113	49.8	94.3	3.8	162
32	114	50.6	94.3	3.9	154
33	114	69.4	94.9	3.7	160
34	121	50.2	96.5	1.6	343
15 [d)]	123	80.5	89.6	9.3	850

a) Conditions: $CO/H_2=1$, $P(CO/H_2)=20$ bar, ligand/Rh = 5, substrate/Rh = 637, [Rh] = 1.00 mM, number of experiments = 3. In none of the experiments was hydrogenation observed.
b) The *l*/*b*, percent linear aldehyde, percent isomerization to 2-octene and turnover frequency (TOF) were determined at 20% alkene conversion.
c) Turnover frequency = mol aldehyde mol Rh^{-1} h^{-1}.
d) $P(CO/H_2)=10$ bar.

duced by the formation of an η^3-benzyl complex. The rhodium-catalyzed hydroformylation of styrene using **30** (the most selective catalyst) resulted in relatively high selectivity for the linear aldehyde (a *l*/*b* ratio of up to 2.35 was obtained, whereas for monophosphines this may be as low as 0.01!).

Under these mild reaction conditions the selectivities toward the linear aldehyde observed for **28–34** and especially **34** are higher than that observed for BISBI, **15**. This is mainly due to the very low amount of isomerization of 1-octene. The *l*/*b* ratios of the ligands are very close to that of BISBI. Furthermore, no hydrogenation was observed. Even though the *l*/*b* ratio is 80.5 for BISBI, the selectivity towards the linear aldehyde amounts to only 89.6% due to the relatively high level of isomerization of 1-octene to 2-octene (under these conditions, as at lower temperatures BISBI shows no isomerization). Scheme 9.4 shows the mechanism for isomerization for 1-butene.

The catalytically active complexes could be synthesized by facile exchange of PPh_3 in $(PPh_3)_3Rh(H)(CO)$ with the diphosphines. Subsequent bubbling of CO through a solution of (diphosphine)Rh(H)(CO)(PPh_3) led to displacement of the remaining PPh_3. In situ high-pressure IR experiments (*vide infra*) have shown that ligands **25–34** form mixtures of bis-equatorial (e-e) and equatorial-apical (e-a) isomers, which rapidly equilibrate. Especially for ligands having small bite angles and electron-donating substituents, the proportion of the "unwanted" e-a isomer can be substantial, but the preference for linear aldehyde remains relatively high for such systems.

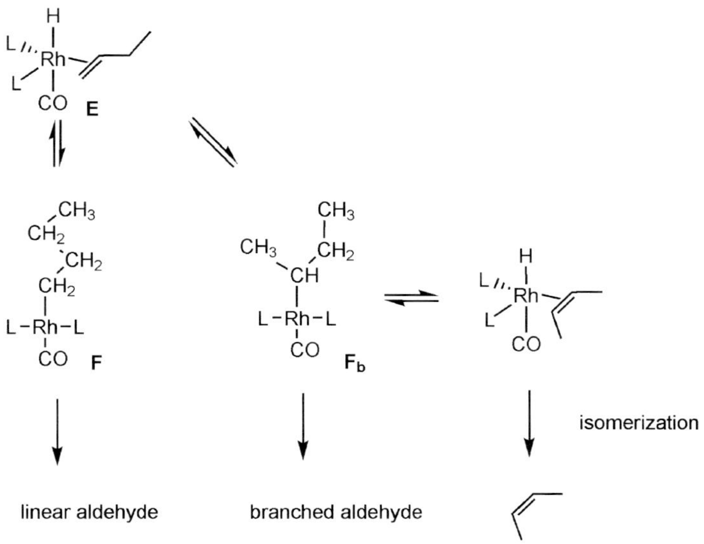

isomerization

linear aldehyde

branched aldehyde

Scheme 9.4

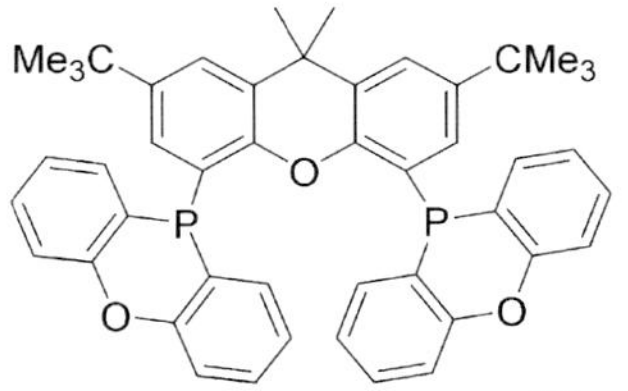

35

9.2.4.2.2 **2-Alkenes**

Conversion of internal alkenes to linear products with the use of rhodium catalysts remains a challenge. Industry still uses cobalt catalysts for these processes [65]. Rhodium catalysts that give high linearity for terminal alkenes and that show isomerization activity are candidate catalysts for the selective conversion of internal alkenes. Wide-bite-angle Xantphos ligands show potential. Xantphos ligands having still wider bite angles were obtained when rigid, cyclic substituents are used, as in **35**. In the hydroformylation of 1-octene the introduction of the phosphacyclic moieties leads to higher reaction rates [66]. More importantly, the dibenzophospholyl- and phenoxaphosphino-substituted Xantphos ligands exhibit unprecedented high activity and selectivity in the hydroformylation of *trans*-2- and 4-octene to linear nonanal. The high activities of the phosphacyclic Xantphos ligands are explained by the lower phosphine basicity and the wider natural bite angles of the phosphacyclic ligands. The extraordinarily high activity of the phenoxaphosphino-substituted Xantphos ligand can be attributed to the 4- to 6-fold higher rate of CO dissociation compared to the other Xantphos ligands [67]. The effect of the phosphacyclic moieties on the coordination chemistry in the (diphosphine)Rh$(CO)_2$H complexes was studied using NMR and IR spectroscopy. Both NMR and IR spectroscopy showed that the phosphacyclic Xantphos ligands exhibit an enhanced preference for diequatorial (e-e) chelation compared to the diphenylphosphino-substituted parent compound. CO dissociation rates from the (diphosphine) Rh$(CO)_2$H complexes were determined using ^{13}CO labeling in rapid-scan IR experiments. This was the first rhodium catalyst containing phosphines giving such high selectivity for linear aldehydes from internal alkenes.

9.2.4.2.3 **Mechanistic Studies**

The detailed mechanism governing the higher linearity for wider bite angles is still unclear. The rate-determining step in the hydroformylation of 1-octene with Xantphos-type ligands is in an early stage in the catalytic cycle. Dissociation of CO, alkene coordination and migratory insertion are relevant for the determination of the rate. So far, only the first step has been studied separately [68, 69]. An experimental setup was introduced enabling the measurement of the rate constants for CO dissociation from the (diphosphine)Rh$(CO)_2$H complexes. By changing the ^{12}CO ligands for isotopically pure ^{13}CO ligands, the carbonyl absorptions in the IR spectra of the (diphosphine)Rh$(CO)_2$H complexes shift 30–40 cm^{-1} to lower energy.

The rate constants k_1 for the dissociation of ^{13}CO from the (diphosphine)-Rh$(^{13}CO)_2$H complexes were determined by monitoring the exchange of ^{13}CO for ^{12}CO by exposing the ^{13}CO-labeled complexes to a large excess of ^{12}CO, so any dissociated ^{13}CO will be replaced quantitatively by ^{12}CO. Exchange of CO turned out to be about 100 times faster than hydroformylation. Thus, a fast pre-equilibrium with CO exists before complexation of alkene and insertion of alkene into the Rh–H bond.

The wide bite angle of **35** leads to a high propensity to isomerization, while the high selectivity for linear product is retained. As a result internal alkenes can now be hydroformylated to linear aldehydes. For 2-octene the l/b ratio is 9 and for 4-octene this value is 4.4 (conditions: 120 °C, 2 bar CO/H_2, ligand/Rh = 5, octene/Rh = 637, [Rh] = 1 mM). At higher pressures of CO less linear aldehyde is formed, because the rate of isomerization decreases. Further exploration of ligand **35** led to catalyst systems with good selectivity and rates for the hydroformylation of internal alkenes [70]. These catalysts received industrial interest as they were potential catalysts for the hydroformylation of butene mixtures as they come from the cracker ("Raffinate II"). For industrial applications the solubility of the ligand **35** in common solvents (the product or heavy ends formed thereof) was too low [71]. The solubility was increased easily by adding alkyl chains R^1 and R^2 to ligand **35** to give **36**, which did not deteriorate the catalytic performance.

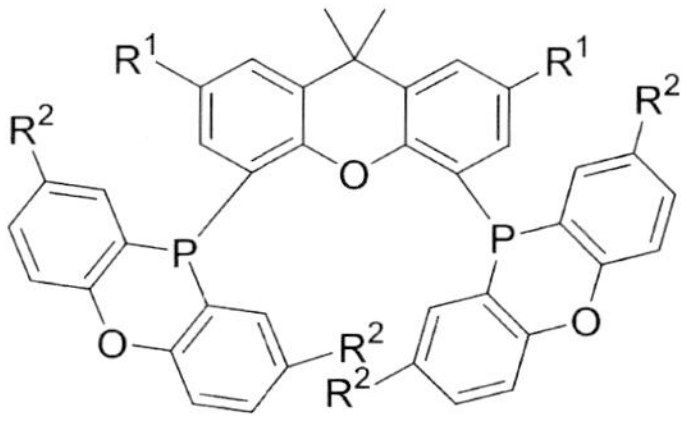

R^1 = H, Me, $CHMe_2$, CHMeEt, CMe_3, $CMe_3CH_2CH_2$, C_6H_{13}, $CMe_3CH_2CMe_2$, $C_{10}H_{21}$;
R^2 = H, Me, $CMe_3CH_2CH_2$, C_6H_{13}

36

For 1-octene *l*/*b* ratios above 100 were obtained and for internal octenes average rates of above 145 mol aldehyde mol Rh^{-1} h^{-1} (pressure = 3.6–10 bar, T = 120 °C, [Rh] = 1 mM, [alkene] = 640–928 mM). High regio-selectivities up to 91% toward the linear product were obtained (*l*/*b* = 10).

Compared to other wide-bite-angle diphosphine ligands, Xantphos-type ligands can be modified easily while retaining there favorable properties, especially in hydroformylation, and as a result many derivatives have been synthesized and used in fluorous-phase hydroformylation catalysis [72], aqueous-phase catalysis [73, 74], one-phase hydroformylation and catalyst extraction [75], catalysis in ionic liquids [76, 77], hydroformylation with immobilized catalysts [78, 79], and catalysis in supercritical CO_2 [80].

Ligands structurally related to the xanthene backbone such as generic **37** and triptycene derived **38** were shown to be just as effective as Xantphos in hydroformylation by Ahlers and coworkers [81]. Ligand **38** gave *l*/*b* ratios as high as 100 in the hydroformylation of 1-octene (100 °C, 5 bar).

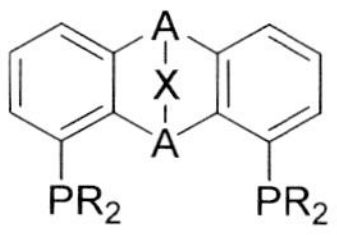

X = (triptycene)

37 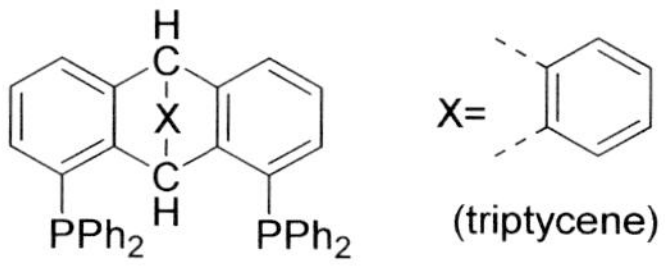 **38**

9.2.5 Alkylphosphines as Ligands

9.2.5.1 Monophosphines

Electron-rich phosphines were not expected to be good ligands in hydroformylation, because the complexes they form of the type $HRh(CO)_x(L)_{4-x}$ are very stable towards dissociation of either CO or L (L = alkylphosphine) due to a "push–pull" effect of the donor and acceptor ligand combination. Cole-Hamilton and coworkers [82] found activity for alkylphosphine complexes when $RhH(PEt_3)_3$ was used as the precursor, but at higher temperatures as expected. Under the reaction conditions $RhH(CO)(PEt_3)_3$ is the major species present. When the reactions are carried out in alcohols the product of the reaction is the alcohol rather than the aldehyde. Deuteration studies show that the alcohol is the primary product and the reaction sequence is not the common one for hydroformylation. A new mechanism for this direct hydrocarbonylation was proposed in which the key acyl intermediate becomes protonated by the alcohol solvent because of the high electron density it bears as a result of the presence of the electron-donating trialkylphosphines. High-pressure NMR studies showed that a rhodium carbene-like intermediate $[Rh\{C(O\cdots HOEt)Et\}(CO)_2(PEt_3)_2]$ is present during catalytic hydrocarbonylation of ethene in ethanol. Ethene gave rates as high as 54 000 turnovers per hour.

9.2.5.2 Dirhodium Tetraphosphine

The majority of rhodium-based hydroformylation catalysts have the general structure $HRh(CO)_x(L)_{4-x}$ for the key intermediate and no ionic intermediates are found. If under acidic conditions Rh^+ is formed, this usually leads to hydrogenation of the alkene and the aldehyde as a side-reaction [81]. Stanley and coworkers [83] discovered a new type of catalyst, which also contains electron-rich phosphines, but which is a tetraphosphine that gives bimetallic catalysts. This system involves cationic rhodium species, which due to the strong donor phosphines are sufficiently electron-rich for CO binding and hydroformylation activity [84]. Several potential intermediates have been identified spectroscopically in the complex catalyst mixture, and several may have close metal–metal distances and bridging CO's. For the catalyst precursor $Rh_2(nbd)_2$(rac-**39**)$(BF_4)_2$ the X-ray structure determination shows that the two rhodium atoms have a square-planar geometry forming five-membered rings with the ligand and a metal–metal distance of 5.5 Å. The catalytic activity in 1-hexene hydroformylation of the bimetallic species was compared with several mononuclear complexes of ligands such as **41**, which is one half of ligands **39** and **40** and lacks the propensity for the formation of dimeric species. The results are surprising, as $Rh_2(nbd)_2$(rac-**39**)$(BF_4)_2$ forms a catalyst superior to all other systems; a TOF of 600 mol mol^{-1} h^{-1} (at 6 bar, 90 °C) and a l/b ratio of 27 was found in the aldehyde product (selectivity to aldehyde 85%). Surprisingly, meso-**40** was 10 times less active and gave a l/b of 14. All other ligands (such as **41**) gave TOFs of 1–6 only and "normal" l/b ratios of 2–3! The large difference in activity between the bimetallic $Rh_2(nbd)_2$(rac-**39**)$(BF_4)_2$ and monometallic Rh(nbd)(**41**)$^+$ strongly suggests that the former reacts via a bimetallic mechanism. Most likely the intermediates are dicationic species, of which **42** is key in the bimetallic mechanism.

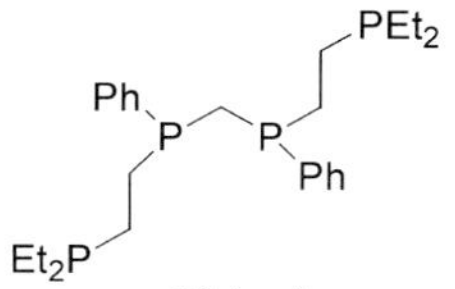

39 (rac) 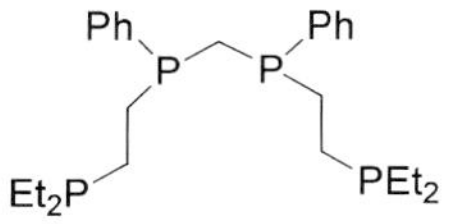 **40** (meso)  **41** 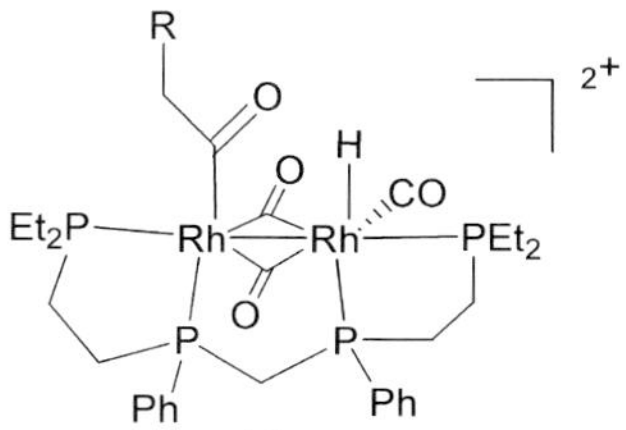 **42**

9.3
Methanol Carbonylation

9.3.1
Introduction

Acetic acid is a bulk chemical commodity industrially used for a wide range of applications, not only as a solvent, but also in the manufacture of cellulose acetate (for cigarette filters and photographic films) and vinyl acetate (paints, adhesives and textiles), as a bleach activator or in the manufacture of food, pharmaceuticals, pesticides, etc. [85]. The annual use of 7 million tons equals that of hydroformylation products. The production is lower as about 20% results from recycling.

In 1966, the rhodium/iodide process for the catalytic production of acetic acid was initiated at the Monsanto laboratories [86]. The first production plant based on this technology started operating in 1970 in Texas City with an initial capacity of 135 000 tons per year. In 1986 BP Chemicals (British Petroleum) acquired and further developed the technology while extending it all over the world. Nowadays, the global production of acetic acid reaches 8 million tons per year and it is led by two companies, Celanese (using "Monsanto" technology) and BP Chemicals, with processes based on both rhodium and iridium [87].

The high price of rhodium and its complex recovery process have continuously driven the development of new systems that overcome these disadvantages. Increasing the activity and the stability of the catalyst or immobilizing it to facilitate rhodium recuperation are still current challenges.

In the late 1990s BP introduced a new, more economic process using iridium as the catalyst – the CATIVA process. Since then many plants have been retrofitted for the use of this catalyst and new plants built are based on the new technology.

9.3.2
Mechanism and Side-reactions of the Monsanto Process

Mechanistically, the Monsanto process actually involves two interrelated catalytic cycles – one organometallic based on rhodium ionic species and an organic one where iodide should be considered as the true catalyst (Scheme 9.5) [88]. The two main components (rhodium and iodide) can be added in many forms. Under the reaction conditions RhI_3 is reduced by H_2O and CO to monovalent rhodium active species **43** and methanol is converted to the iodo form, MeI.

The organometallic cycle comprises oxidative addition of MeI to $[RhI_2(CO)_2]^-$ **43** (generated *in situ*), which is considered to be the rate-limiting step of the Monsanto process. Ligand migration to generate the acetyl complex **45**, CO coordination and reductive elimination of acetyl iodide regenerates the rhodium-active species **43**. Acetyl iodide enters the "organic" cycle where it is hydrolyzed to give acetic acid and HI which transforms MeOH into the more electrophilic MeI that enters the organometallic cycle.

It is important to note that the two "organic" reactions (1) and (2) involving iodide are quantitative. As a result, all of the iodide in the system occurs as a methyl iodide and, due to reaction (1), the rate of the catalytic cycle is independent of the methanol concentration [89].

$$CH_3OH + HI \rightarrow H_2O + CH_3I \tag{1}$$

$$CH_3COI + H_2O \rightarrow CH_3COOH + HI \tag{2}$$

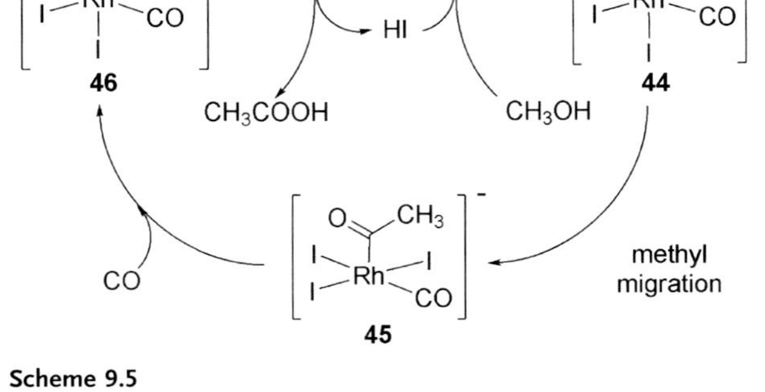

Scheme 9.5

One of the main drawbacks of the process is the loss of the expensive metal due to the formation of inactive Rh(III) species which, in areas of low CO pressure at the end of the reaction, precipitate as RhI_3. The inactive species $[RhI_4(CO)_2]^-$ can be produced by reaction of compounds **43** and **44** with HI according to:

$$[Rh(CH_3)I_3(CO)_2]^- + HI \rightarrow CH_4 + [RhI_4(CO)_2]^- \quad (3)$$

$$[RhI_2(CO)_2]^- + 2HI \rightarrow H_2 + [RhI_4(CO)_2]^- \quad (4)$$

In the Monsanto process this problem is solved by keeping the water content relatively high, because it inhibits Rh precipitation by regenerating the Rh(I)-active species from the labile $[RhI_4(CO)_2]^-$:

$$[RhI_4(CO)_2]^- + H_2O + CO \rightarrow [RhI_2(CO)_2]^- + 2HI + CO_2 \quad (5)$$

H_2 can also accomplish this goal (6) and it can be generated in the reactor at high water concentrations through the water–gas shift reaction (7). In the absence of water, added H_2 could also perform this role, as it does in the related Eastman acetic anhydride process [90, 91].

$$[RhI_4(CO)_2]^- + H_2 \rightarrow [RhI_2(CO)_2]^- + 2HI \quad (6)$$

$$CO + H_2O \rightarrow H_2 + CO_2 \quad (7)$$

In fact, the presence of water is controversial; it is needed to maintain the rhodium in the active Rh(I) form, but there are also several problems associated with keeping it at high concentrations:

(i) The reaction of H_2O with CO through the water–gas shift reaction causes the loss of one of the feedstocks. Even if the selectivity of the overall process in MeOH is in the high 90% range, the selectivity in CO can be as low as 90%.

(ii) For the synthesis of acetic anhydride, water needs to be removed from the product by distillation at the end of the reaction, increasing the production costs considerably.

In the late 1980s, a major technological advance was developed by Celanese in order to circumvent the water problem [92]. It consists of the use of salts (LiI or LiOAc) as promoters to maintain the reaction at low water levels. When using low water concentrations, the reduction of Rh(III) back to Rh(I) is slower, but so is the formation of $[RhI_4(CO)_2]^-$ due to the lower HI content available for reactions (3) and (4). The promotional effect of Li salts is attributed to the coordination of either acetate or iodide to **43** forming a highly nucleophilic intermediate dianion, $[Rh(CO)_2I_2X]^{2-}$ (X=I, OAc) and also to the kinetic and thermodynamic parameters of this new organic cycle, in which water has been replaced by LiOAc [93, 94].

9.3.3
Oxidative Addition of MeI to Rhodium – The Rate-limiting Step

The oxidative addition of MeI to **43** is the rate-determining step of the rhodium/iodide-catalyzed process. It is generally accepted that it follows a two-step S_N2 mechanism; nucleophilic attack by the metal on the methyl carbon to displace iodide, presumably with inversion of configuration at the carbon atom, and subsequent iodide coordination to the five-coordinate rhodium complex to give the methyl complex **44** [95, 96]. The product of this reaction has been fully characterized spectroscopically [88a, 97–99]. Observations that support this mechanism are the second-order rate law observed (first-order in both MeI and **43**) and the large negative activation entropies that indicate highly organized transition states.

A general approach to facilitate the oxidative addition of MeI to the active rhodium species is to increase the nucleophilicity of the metal center by means of donor ligands. In order to obtain insight into the activity of related ligand-modified systems, independent kinetic studies pertaining only to this elementary step can be performed by using IR spectroscopy. In Table 9.3 we have presented the values of second-order rate constants and activation parameters for MeI-oxidative addition to rhodium complexes when using several ligands. They have been calculated by measuring pseudo-first-order constants (k_{obs}) (using at least 10-fold excess of MeI in CH_2Cl_2) at different temperatures and MeI concentrations.

In all cases, a first-order dependence on both MeI and rhodium complex is observed. Depending on the relative rates of oxidative addition/methyl migration, most often the corresponding acetyl compound is observed directly, but occasionally (*vide infra*) the alkyl intermediate, the product of the oxidative addition, can be detected by IR spectroscopy or even isolated as the sole product of the reaction (Scheme 9.6).

Table 9.3 ν(C–O)-IR of the complexes [RhI(CO)L] and their reactivity towards MeI

References	Added ligand (L)	ν(C–O) (cm^{-1}) [RhI(CO)L]	$10^3\ k_1$ ($M^{-1}\ s^{-1}$)	$\Delta H^{\neq}$ (kJ mol^{-1})	$\Delta S^{\neq}$ (J $mol^{-1}\ K^{-1}$)
99, 101	–[a)]	2055, 1985	0.0293	50 ± 1	–165 ± 4
105	2 PEt_3	1961	1.37	56 ± 13	–112 ± 44
103	**47**	1981	no reaction		
100	**48a**	1970	no reaction		
104	2 **49a**[b)]	1943	0.0441	36 ± 1	–189 ± 3
111, 112	dppms **51**	1987	1.19	47 ± 1	–144 ± 2
111, 110	dppmo **52**	1983	1.14	34 ± 4	–188 ± 13
111, 112	dppe **18**	2011	1.41	40 ± 1	–167 ± 2

a) Neat MeI.
b) [MeI] = 1.6 M, CH_2Cl_2, data concerning only the forward reaction.

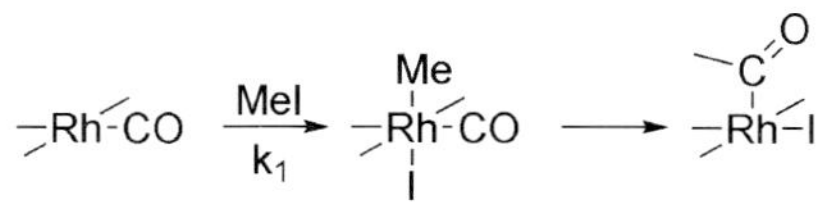

$$-Rh-CO \xrightarrow[k_1]{MeI} Me-Rh(I)-CO \longrightarrow (MeC(O))-Rh-I$$

Scheme 9.6

The frequency of the ν(C–O) of the catalytically active complex [RhI(CO)L] is usually considered as a measure of the electron density on the metal center, and consequently its nucleophilicity and reactivity toward MeI. It is generally accepted that the greater the electron density on the metal, the lower the ν(C–O), due to electron back-donation into the π^* orbital of the coordinated CO. Nevertheless, some examples provide evidence that a direct correlation between the frequency of the IR stretch vibration of the coordinated CO and the reactivity of the rhodium complex versus MeI cannot be established, and that several other factors need to be taken into account. First, the relative disposition of the ligands around the rhodium center plays a role. If we compare [RhI(CO)dppe] with [RhI(CO)(PEt_3)$_2$], which exhibit similar reactivity towards MeI, the corresponding ν(C–O) values differ by around 50 cm^{-1}. This can be attributed to the fact that a carbonyl *trans* to iodide experiences considerably more back-donation than one *trans* to a phosphine. Secondly, not only electronic, but also steric effects play an important role in this elementary step. Ligands **47** and **48** (SPANphos) represent extreme cases [100, 101]. The corresponding metal complexes [RhI(CO)**47**] and *trans*-[RhI(CO)**48**] show ν(CO) values similar to other active systems, but they do not react with MeI via oxidative addition. This difference in reactivity is attributed to the fact that the backbone of the ligand eclipses one of the axial coordination sites of the rhodium center, inhibiting oxidative addition.

In the case of ligands **49** [102], despite the higher nucleophilicity of the corresponding Rh-biscarbene complex predicted on the basis of the low value of their corresponding ν(C–O), they display a very low reactivity with MeI. In this case, the characteristic ligand disposition is such that the NCN plane is perpendicular to the Rh(I) coordination plane, so the R substituents partially block the rhodium axial sites and inhibit nucleophilic attack on MeI. When using ligand **49a**, the corresponding Rh(III) acetyl compound is obtained at the end of the reaction, but it is unstable and decomposes to [RhI(CO)**49**$_2$] by elimination of MeI. The thermodynamic parameters for the forward reaction are similar to those of more active systems. When these carbene ligands are applied in methanol carbonylation, they show no activity enhancement compared to Monsanto process and they dissociate from the rhodium center to generate [$RhI_2(CO)_2$]$^-$.

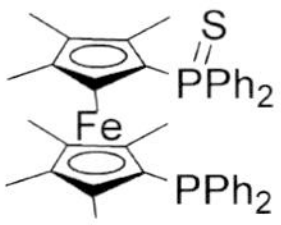

47

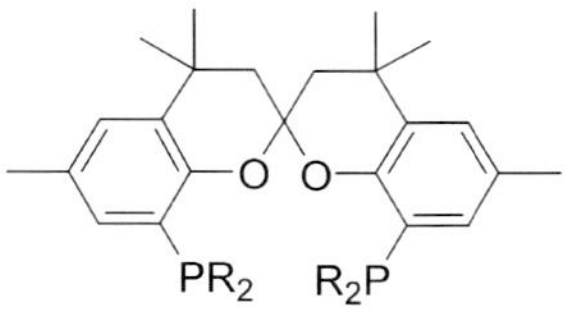

48a PR_2= PPh_2
48b PR_2= PCy_2
48c PR_2= 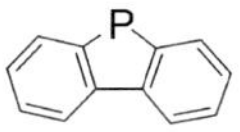
48d PR_2=

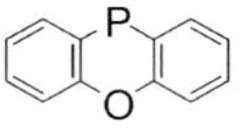

49a: R=Me
49b: R=Et
49c: R=CH_2Ph
49d: R=4-pentenyl
49e: Mesityl

In summary, in order to accelerate the rate of MeI-oxidative addition (and consequently the rate of the overall process) electron-donating ligands need to be used. The IR stretch vibration of the CO coordinated to the metal can be indicative of the electron density on the rhodium center, but steric effects play a key role when relating it to reactivity versus MeI.

9.3.4
Ligand Design

In the last 20 years much effort has been directed to develop ligands able to increase the activity of the rhodium/iodide-based Monsanto system. In some cases, even if they do not contribute to a significant reaction improvement, they are crucial for a better understanding of the reaction mechanism and constitute basic pillars for further developments. Some representative examples are mentioned in this section.

In 1997, Ranking and coworkers reported a PEt_3-rhodium-modified system [103], and when employing this basic phosphine, the rate of oxidative addition of MeI to $[RhI(CO)(PEt_3)_2]$ is around 47 times higher than that reported for **43** at 25 °C. The corresponding alkyl compound $[RhI_2(CO)(CH_3)(PEt_3)_2]$ was isolated and crystallographically characterized, exhibiting a *trans* configuration of the phosphine ligands about the metal center. Although in comparison to the Monsanto system there is a large increase in the rate of oxidative addition, it represents only an increase in activity of 1.8 under catalytic conditions (150 °C). At this temperature, loss of activity is observed after around 10 min, due to degradation of the catalytically active system to $[Rh(CO)_2I_2]^-$. Decomposition proceeds via reaction of $[RhI_2(CO)(CH_3)(PEt_3)_2]$ with HI to generate $[RhI_3(CO)(PEt_3)_2]$, which reductively eliminates $[PEt_3I]^+$ leading to $OPEt_3$.

Chelating symmetrical and unsymmetrical ligands have been introduced in order to overcome the instability of the monophosphines under the harsh conditions of the process. The use of symmetrical diphosphines has been covered by the patent literature [104, 105] and when using Xantphos derivatives (**25–34**) as ligands for methanol carbonylation they are stable systems that show a slightly higher activity than the one obtained with the Monsanto catalyst. In this case a terdentate P–O–P coordination of the ligand is claimed.

Carraz and coworkers [106] reported the use of asymmetrically substituted 1,2-ethanediyl diphosphines **50** that are very stable under the reaction conditions, although they do not represent an improvement on the catalytic activity when compared with the Monsanto process. At the end of the reaction a mixture of diphosphine Rh(III) carbonyl complexes can be isolated and characterized, and a second run can be performed without loss of activity.

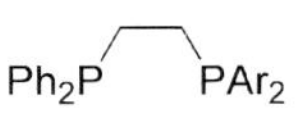

Ph_2P PAr_2

50a Ar= C_6H_4OMe-4
50b Ar= C_6H_4F-3
50c Ar= $C_6H_4CF_3$-3
50d Ar= $C_6H_3F_2$-3,5
50e Ar= $C_6H_2F_3$-3,4,5
50f Ar= $C_6H_3(CF_3)_2$-3 5

An alternative strategy to increase the activity and stability of the systems is the use of hemilabile phosphine ligands (PX; X=P, O, S). They are thought to stabilize the complex via the chelate effect and to increase the nucleophilicity of the rhodium by coordination of a hetero-donor atom.

A comparative study of PS, PO and PP ligands able to form five-membered chelate rings (**51** [107], **52** [108] and **18**) showed that the corresponding [RhI(CO)(PX)] complexes react between 35 and 50 times faster with MeI than **43** – an activity similar to the one observed with PEt_3 [109, 110]. This enhancement in activity is attributed to the good donor properties of the ligands. Unlike dppe and PEt_3 systems, for which the corresponding alkyl intermediate can be isolated, use of dppms **51** results in a fast subsequent migratory insertion to give the corresponding acetyl compound.

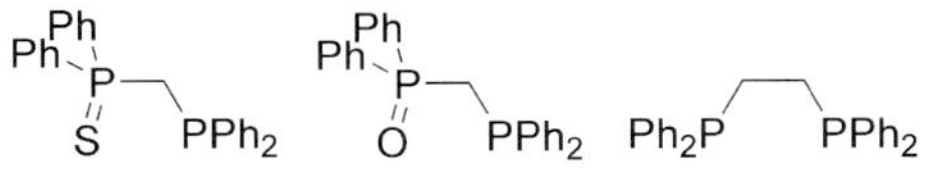

Ph Ph P S PPh_2 Ph Ph P O PPh_2 Ph_2P PPh_2

dppms **51** dppmo **52** dppe **18**

Another example of this family of ligands using oxygen as a heteroatom is methyl-2-diphenylphosphinobenzoate **53** [111]. Depending on the stoichiometry of the reaction with $[Rh(CO)_2Cl]_2$, the corresponding rhodium species [RhCl(CO)**53**], **54**, and [RhCl(CO)$\mathbf{53}_2$], **55**, are formed. In complex **54**, the ligand acts as a chelate through the P- and O-atoms, whereas in **55** two phosphines coordinate in a *trans* fashion and secondary Rh–O interactions are suggested. Due to the increased nucleophilicity of the rhodium center, both systems show an activity that is around 1.5 times higher than Monsanto's system under the same catalytic conditions. This increase in nucleophilicity is in agreement with the low values of ν(C–O) observed (1979 and 1949 cm^{-1}, respectively). Nevertheless complex **55** shows slightly lower reaction rates, which is attributed to the increase in steric hindrance about the metal center on account of the presence of two ligands.

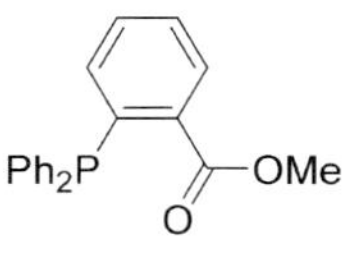

Hemilabile ether-phosphine ligands Ph_2PCH_2OMe (**56**) and $PPh_2(CH_2)_2OEt$ (**57**) react with half an equivalent of $[Rh(CO)_2Cl]_2$ to form the corresponding complexes $[RhCl(CO)_2$**56**] and $[RhCl(CO)_2$**57**] with the ligands coordinating only through the phosphorus atom [κ(P), Scheme 9.7] [112]. These rhodium complexes undergo oxidative addition with MeI to yield the six-coordinate acetyl species in which the ligand acts as a chelate with the oxygen bound to the metal *trans* to the acetyl moiety. Under catalytic conditions, ligands **56** and **57** show an activity around 1.2 and 3.3 times higher, respectively, than the Monsanto system. The difference in activity is attributed to the higher basicity and stability of the five-membered chelate ring of the acetyl complex formed when using **57**, in comparison to the four-membered chelate ring obtained with **56**. In both cases a "closing–opening" mechanism is proposed, and a catalytic cycle involving species in which the ligands act in both monodentate κ(P) and bidentate κ(P)–κ(O) ways has been proposed.

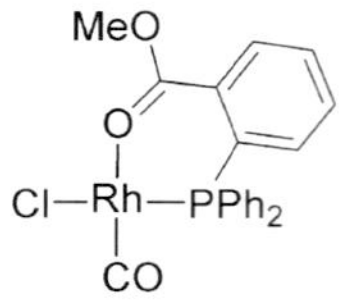

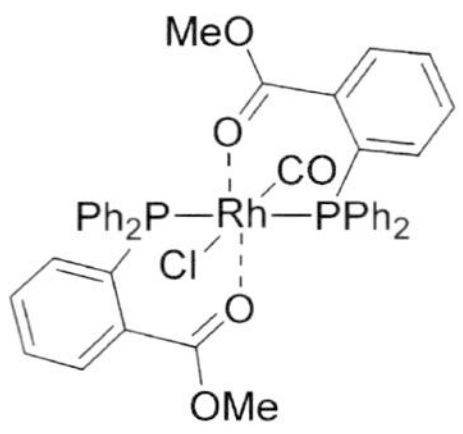

When using dppeo [110], $Ph_2P(CH_2)_2P(O)Ph_2$ (**58**), a ligand quite active for methanol carbonylation under mild conditions (80 °C, 50 psig CO, TOF = 400 mol mol^{-1} h^{-1}), the same type of "closing–opening" mechanism is thought to be operative. This is in contrast with dppmo, which forms a five-membered chelate ring, and the ligand acts as a bidentate chelate during the entire catalytic cycle.

Monodentate nitrogen donor ligands (i.e. *ortho*, *meta* and *para* amino benzoic acids and pyridine aldehydes) have also been tested in the carbonylation of methanol [113, 114]. The corresponding rhodium complexes with the ligand coordinating by the nitrogen atom $[Rh(CO)_2ClL]$ have been characterized and additional coordination of the oxygen atom to generate a bidentate ligand is not observed in any of the cases. The reaction with MeI yields directly the acetyl derivatives $[Rh(CO)(COCH_3)ClL]$, providing evidence that fast migration occurs after oxidative addition has taken place. The activities observed under catalytic conditions are around double than those obtained when employing $[Rh(CO)_2I_2]^-$ and,

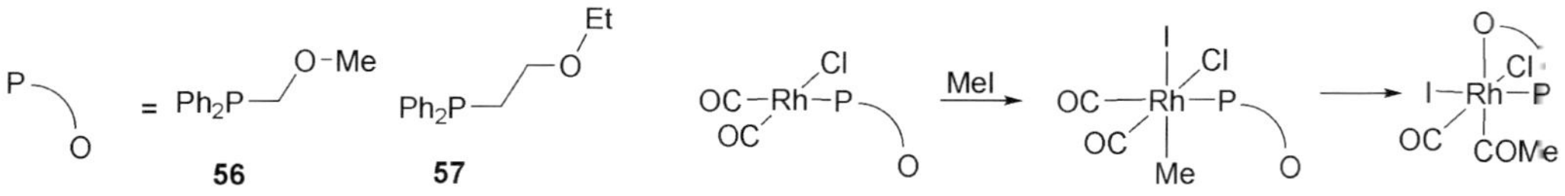

Scheme 9.7

in the case of aminobenzoic acids, the catalyst can be recycled without significant loss in activity, proving its thermal stability.

9.3.5
Trans-diphosphines in Methanol Carbonylation – Dinuclear Systems?

Phosphorus-based systems constitute an important family of ligands in homogeneous catalysis. Even though the use of basic phosphines, which increases the nucleophilicity of the rhodium center, was initially considered a very promising strategy [105], their instability under the harsh conditions of the process constituted a major drawback for their application. Attempts to circumvent the decomposition were undertaken by using bidentate diphosphines, which stabilize the active rhodium species by the chelate effect (*vide supra*).

In an effort to develop new systems able to combine both stability and activity, Süss-Fink and coworkers reported a family of *trans*-spanning, but flexible diphosphines [115–117]. These ligands form stable complexes due to the chelate effect which resemble the ones obtained with PEt_3 (*trans*-$[RhI(CO)(PEt_3)_2]$) with the two phosphorus atoms in a *trans* disposition. The authors proposed that these ligands, as benzoic acid diphenylphosphine derivatives, increase the electron density on the rhodium center due to a weak Rh–O interaction; alternatively, one might suggest that the highest occupied molecular orbital (HOMO) energy of the d_z^2 orbital is raised by this interaction.

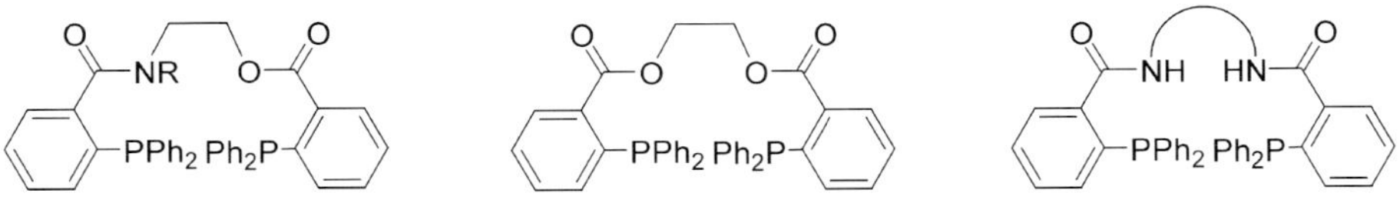

59a R=H
59b R=Ph

60

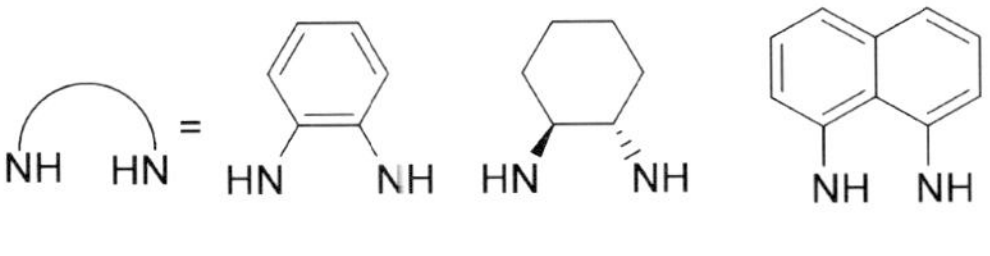

61a **61b** **61c**

When the corresponding rhodium complexes are tested in methanol carbonylation the activity obtained is around 2.5 times higher than that obtained with the Monsanto system under the same conditions. The stability of the systems is evident by the ability to perform several runs without any noticeable loss in activity. From the residue of the reaction when ligand **59b** is employed, apart from the corresponding [RhI(CO)**59b**] complex, the dinuclear isomeric compounds **62a** and **62b** were isolated and characterized by X-ray diffraction (XRD).

The proposed catalytic cycle is analogous to that for the Monsanto process, but it involves neutral species with a *trans*-chelating diphosphine. Presumably, the isolated dinuclear compounds **62a** and **62b** act as resting states for the ace-

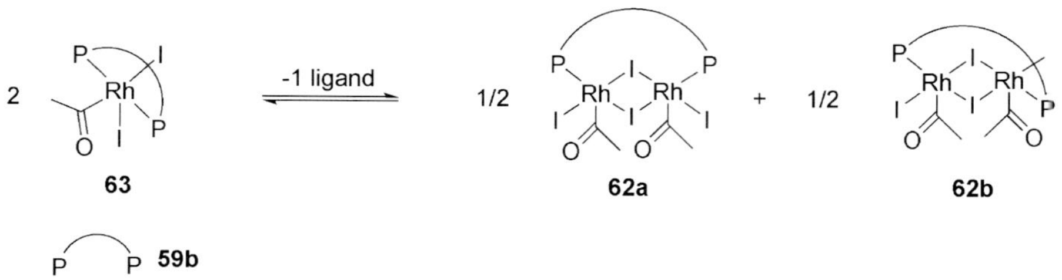

Scheme 9.8

tyl complex [$RhI_2(COMe)$**59b**] **63**, which arises from it by loss of diphosphine (Scheme 9.8).

Another family of diphosphines displaying a *trans* coordination mode, SPANphos **48 (a–d)**, has also been tested in this reaction [102]. Surprisingly, even if the corresponding *trans*-[RhI(CO)**48**] render systems as active as the Süss-Fink analogs, when tested in an elementary step, they show no reaction in the oxidative addition of MeI (*vide supra*). The observed catalytic activity has been attributed to dinuclear species formed under catalytic conditions. The dinuclear compounds [$Rh_2(\mu\text{-}Cl)_2(CO)_2$**48a**], **64**, react with MeI at 25 °C with a k_1 value of around 0.025 s^{-1} M^{-1} and currently they represent the fastest phosphine-based systems reported for methanol carbonylation. According to spectroscopic and gas chromatography/mass spectroscopy analysis, the products of the MeI-oxidative addition are dinuclear monoacetyl derivatives **65** (Scheme 9.9).

The reactivity of dinuclear rhodium compounds toward MeI has been a subject of study since the late 1970s [118, 119]. Unfortunately, no kinetic data is available, but the corresponding mono- and bis-acetyl derivatives have been detected. The fact that the oxidative addition/migration on the second metal center becomes more difficult after the first one has taken place is a common feature, especially when the "open-book" conformation of the complexes is enforced by the (bridging) ligands [120].

In fact, the possible involvement of dinuclear species in the Monsanto process was proposed in the earliest studies, where the dimerization equilibrium of the acetyl derivative $[RhI_3(COMe)(CO)]^-$ (**45**) was considered [99]. The corresponding dinuclear species $[Rh_2I_6(CO)_2(COMe)_2]^{2-}$ was isolated in the form of its tr-

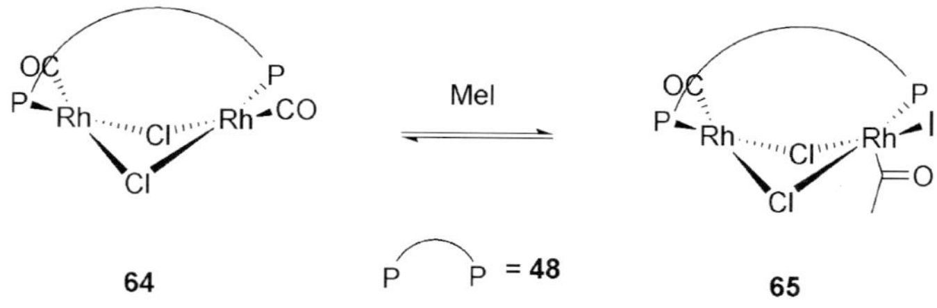

Scheme 9.9

methylphenylammonium salt and characterized by XRD. The possibility that neutral dinuclear methyl and acetyl Rh(III) species play an important role in the Monsanto process has very recently been reconsidered [121]. These neutral species should lead to faster CO migratory insertions and reductive eliminations than their anionic analogs, and surely in the future new ligands able to generate more active dinuclear systems for methanol carbonylation will be developed.

9.3.6 Iridium Catalysts

As has happened with many catalytic processes, many years of research on the carbonylation of methanol has led to the discovery of a new catalytic system, now based on iridium. Since the early patents of Monsanto it was known that iridium also forms active catalysts, but until the mid-1990s rhodium was the champion metal. Other metals are also active under similar conditions using the same iodide-based chemistry, such as nickel and palladium [122]. Slight, yet important, modifications by BP have made iridium the best catalyst now available in a system known as the CATIVA process. It is similar to the rhodium-based Monsanto process, which after 25 years of successful operation is gradually being replaced by the CATIVA process [123]. In 2003 four plants were in operation using this new catalyst.

In the CATIVA process the active catalyst is $[Ir(CO)_2I_2]^-$. Due to similar chemistry to the Monsanto process the same chemical plant may be used, which makes a retrofitting commercially highly attractive. Initial studies by Monsanto had shown the iridium complex to be a less-active catalyst than the rhodium complex. However, subsequent research showed that the iridium catalyst could be promoted using ruthenium and/or other salts, and this combination leads to a more-active and more-selective catalyst than the rhodium compound.

In summary, the economic reasons for switching from rhodium to iridium are:

- The iridium catalyst works at low water concentrations, which means that one can reduce the number of drying columns.
- Less propionic acid byproduct is produced, which also reduces the work-up train.
- A higher concentration of catalyst can be used, which increases the space-time yield.

As a result investments are reduced by 30%. Also, iridium is several times cheaper than rhodium (on average by a factor of 10 per kilogram over the last 5 years, which should be corrected for the atomic weight!); the prices of both show an enormous fluctuation [124]. The catalyst system exhibits high stability allowing a wide range of process conditions and compositions to be accessed without catalyst precipitation.

Oxidative addition to iridium [125] is much faster than that to the corresponding rhodium complexes. Also the equilibrium is on the side of the trivalent state.

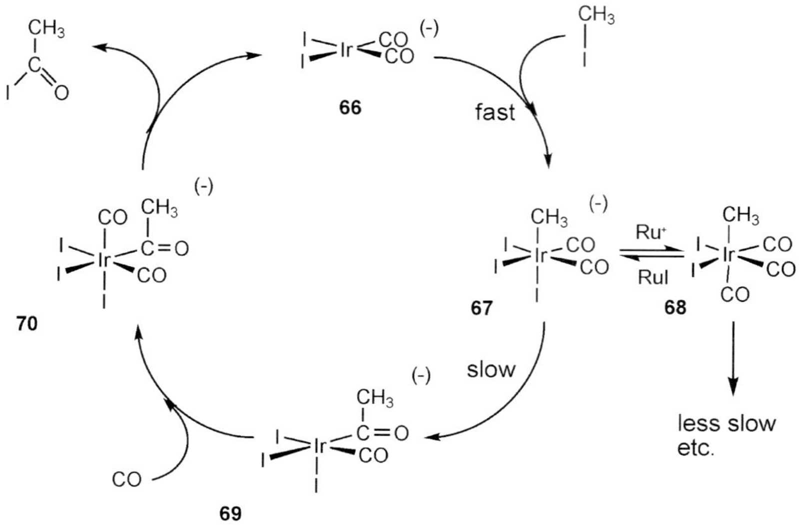

Scheme 9.10

Thus, as expected the reaction in Scheme 9.10 (**66** → **67**) is much faster for iridium. In and of itself this does not mean that the iridium catalyst is therefore faster than the rhodium catalyst. As we have learned before, the reductive elimination may be slower for iridium. Apparently, this is not the case. Migration is now the slowest step (**67** → **69**) [126]. In contrast to the rhodium process, the most abundant iridium species, the catalyst resting state, in the BP process is not **66**, but the product of the oxidative addition of MeI to this complex, **67**.

Two distinct classes of promoters have been identified for the reaction: simple iodide complexes of zinc, cadmium, mercury, indium and gallium, and carbonyl complexes of tungsten, rhenium, ruthenium and osmium. The promoters exhibit a unique synergy with iodide salts, such as lithium iodide, under low water conditions. Both main group and transition metal salts can influence the equilibria of the iodide species involved. A rate maximum exists under low water conditions and optimization of the process parameters gives acetic acid with a selectivity in excess of 99% based upon methanol. IR spectroscopic studies have shown that the salts abstract iodide from the ionic methyl iridium species and that in the resulting neutral species the migration is 800 times faster [127].

The levels of liquid byproducts formed are a significant improvement over those achieved with the conventional high water rhodium-based catalyst system and the quality of the product obtained under low water concentrations is exceptional [125].

Transition metal complexes added might also play a role in aiding the reduction of inactive Ir(III) species back to active Ir(I), but no evidence for this has been reported.

9.4 Concluding Remarks

The two reactions discussed above show that CO is a highly reactive molecule used on a large scale with homogeneous catalysts. A few smaller applications, volume-wise, such as the palladium-catalyzed carbonylation of benzylic halides, *en route* to Ibuprofen for instance, have not been discussed. In carbonylation reactions the C–O bond remains intact. In alcohol homologation chemistry it can be envisaged that the C–O bond is cleaved, but as yet no commercial reactions are known utilizing this. In heterogeneous catalysis, however, rupture of the C–O bond is very common and actually it is key in the Fischer-Tropsch process. In the metallic catalysts the bond is cleaved and a carbene-like species is formed while oxygen ends up as water. In spite of the high reactivity of CO there remain several reactions that are highly desirable targets in industry that cannot be carried out yet with sufficiently high selectivity and catalyst stability. We mentioned homologation of methanol to ethanol in the introduction as an example and we may add the direct production of ethylene glycol from syngas, which is feasible with rhodium catalysts [11], but has not been commercialized.

With more than 1 million tons per year provided, an important product made from CO by oxidation with chlorine is phosgene [128] (*cf.* hydroformylation products amount to 7 million tons per year). The highly toxic phosgene is produced in continuous reactors, "on-line", and directly used in the next step in order to minimize the risks. Its major use is for the conversion of aromatic amines to isocyanates, the starting materials for polyurethanes. Coproduced HCl is recycled via oxygen oxidation in the Deacon process. Aromatic amines are made via hydrogenation of nitro compounds. A more direct route to isocyanates is the conversion of nitro compounds with CO [129], requiring drastic conditions, but a milder route is available that involves nitro reduction with CO in methanol affording carbamates as intermediates [130, 131]. Recent advances have made this route rather attractive [132]. Likewise in polyester, polyamide, and polyketone manufacturing more direct and cleaner alternatives may become available, utilizing low-priced CO as a reactive feedstock.

References

1 H.A. Wittcoff, B.G. Reuben, J.S. Plotkin, *Industrial Organic Chemicals*, 2nd edn, Wiley, New York, **2004**.
2 For example, Shell has a GTL plant at Bintulu, Malaysia (SMDS). The plant began in 1993 and can produce up to 4000 barrels per day of GTL fuel for worldwide sales. In Qatar, Sasol-Chevron-Qatar Petroleum is building a huge GTL plant (Oryx GTL) to serve the world market with 34000 barrels per day.
3 A. Steynberg, M. Dry (Eds.), *Studies in Surface Science and Catalysis 152: Fischer-Tropsch Technology*, Elsevier, Amsterdam, **2004**.
4 O. Roelen, *Chem Zentr* **1953**, 927.
5 P.M. Maitlis, *J Mol Catal A* **2003**, *204/205*, 55–62.

6 D. Mahajan, R. S. Sapienza, W. Slegeir, T. E. O'Hare, *US Patent 4935395*, **1990**.
7 J. E. Wegrzyn, D. Mahajan, M. Gurevich, *Catal Today* **1999**, *50*, 97–108.
8 D. Mahajan, V. Krisdhasima, R. D. Sproull, *Can J Chem* **2001**, *79*, 848–853.
9 G. Braca (Ed.), *Catalysis by Metal Complexes 16: Oxygenates by Homologation or CO Hydrogenation with Metal Complexes*, Kluwer, Dordrecht, **1994**.
10 M. J. Chen, H. M. Feder, J. W. Rathke, S. A. Roth, G. D. Stucky, *Ann NY Acad Sci* **1983**, *415*, 152.
11 G. S. Koerner, W. E. Slinkard, *Ind Eng Chem Prod Res Dev* **1978**, *17*, 231.
12 J. F. Knifton, J. J. Lin, *Appl Organomet Chem* **1989**, *3*, 557.
13 L. H. Slaugh, R. D. Mullineaux, *US Patents 3239569* and *3239570*, 1966 (to Shell), *Chem Abstr* **1964**, *64*, 15745 and 19420; L. H. Slaugh, R. D. Mullineaux, *J Organomet Chem* **1968**, *13*, 469–477.
14 (a) F. H. Jardine, J. A. Osborn, G. Wilkinson, J. F. Young, *Chem Ind (London)* **1965**, 560; (b) D. Evans, J. A. Osborn, G. Wilkinson, *J Chem Soc A* **1968**, 3133.
15 P. Arnoldy, in *Catalysis by Metal Complexes 22: Rhodium Catalyzed Hydroformylation*, P. W. N. M. van Leeuwen, C. Claver (Eds.), Kluwer, Dordrecht, **2000**, pp. 203–231.
16 R. L. Pruett, J. A. Smith, *J Org Chem* **1969**, *34*, 327.
17 (a) P. W. N. M. van Leeuwen, C. F. Roobeek, *J Organomet Chem* **1983**, *258*, 343; (b) P. W. N. M. van Leeuwen, C. F. Roobeek, *British Patent 2068377*, 1980 (to Shell), *Chem Abstr* **1984**, *101*, 191142.
18 E. Billig, A. G. Abatjoglou, D. R. Bryant, *US Patent 4668651* and *Eur. Patent Appl. 213639*, 1987 (to Union Carbide), *Chem Abstr* **1987**, *107*, 7392.
19 (a) T. J. Devon, G. W. Phillips, T. A. Puckette, J. L. Stavinoha, J. J. Vanderbilt, *US Patent 4694109*, 1987 (to Texas Eastman), *Chem Abstr* **1988**, *108*, 7890; (b) T. J. Devon, G. W. Phillips, T. A. Puckette, J. L. Stavinoha, J. J. Vanderbilt, *US Patent 5,332,846*, 1994 (to Texas Eastman), *Chem Abstr* **1994**, *121*, 280, 879.
20 C. P. Casey, G. T. Whiteker, M. G. Melville, L. M. Petrovich, J. A. Gavney, Jr., D. R. Powell, *J Am Chem Soc* **1992**, *114*, 5535–5543.
21 M. Kranenburg, Y. E. M. van der Burgt, P. C. J. Kamer, P. W. N. M. van Leeuwen, *Organometallics* **1995**, *14*, 3081–3089.
22 P. W. N. M. van Leeuwen, C. Claver (Eds.), *Catalysis by Metal Complexes* 22: *Rhodium Catalyzed Hydroformylation*, Kluwer, Dordrecht, **2000**.
23 (a) W. Chew, E. Widjaja, M. Garland, *Organometallics* **2002**, *21*, 1982; (b) E. Widjaja, C. Li, M. Garland, *Organometallics* **2002**, *21*, 1991; (c) C. Li, E. Widjaja, M. Garland, *J Am Chem Soc* **2003**, *125*, 5540.
24 A. J. Solodar, E. Sall, L. E. Stout, Jr., *Chem Ind (Dekker)* **1996**, *68*, 119–132.
25 G. Liu, R. Volken, M. Garland, *Organometallics* **1999**, *18*, 3429–3436.
26 J. Feng, M. Garland, *Organometallics* **1999**, *18*, 1542–1546.
27 J. Feng, M. Garland, *Organometallics* **1999**, *18*, 417–427.
28 G. Liu, M. Garland, *J Organomet Chem* **2000**, *613*, 124–127.
29 R. L. Pruett, J. A. Smith, *South African Patent 6804937*, 1968 (to Union Carbide Corporation), *Chem Abstr* **1969**, *71*, 90819.
30 P. W. N. M. van Leeuwen, C. P. Casey, G. T. Whiteker, in *Catalysis by Metal Complexes 22: Rhodium Catalyzed Hydroformylation*, P. W. N. M. van Leeuwen, C. Claver (Eds.), Kluwer, Dordrecht, **2000**, 63–105.
31 P. W. N. M. van Leeuwen, C. F. Roobeek, *US Patent 4330678* (to Shell), **1982**.
32 A. van Rooy, E. N. Orij, P. C. J. Kamer, P. W. N. M. van Leeuwen, *Organometallics* **1995**, *14*, 34–43.
33 R. Crous, M. Datt, D. Foster, L. Bennie, C. Steenkamp, J. Huyser, L. Kirsten, G. Steyl, A. Roodt, *Dalton Trans* **2005**, 1108–1116.
34 (a) E. Billig, A. G. Abatjoglou, D. R. Bryant, *US Patent 4668651* and *Eur. Patent Appl. 213639*, 1987 (to Union Carbide), *Chem Abstr* **1987**, *107*, 7392; (b) E. Billig, A. G. Abatjoglou, D. R. Bryant, R. E. Murray, J. M. Maher, *US Patent 4599206*, 1986 (to Union Carbide), *Chem Abstr* **1988**, *109*, 233177.

35 P. C. J. Kamer, P. W. N. M. van Leeuwen, J. N. H. Reek, in *Catalysis by Metal Complexes 22: Rhodium Catalyzed Hydroformylation*, P. W. N. M. van Leeuwen, C. Claver (Eds.), Kluwer, Dordrecht, **2000**, 35–62
36 G. D. Cuny, S. L. Buchwald, *J Am Chem Soc* **1993**, *115*, 2066–2068; J. R. Johnson, G. D. Cuny, S. L. Buchwald, *Angew Chem Int Ed* **1995**, *34*, 1760–1761.
37 A. van Rooy, P. C. J. Kamer, P. W. N. M. van Leeuwen, K. Goubitz, J. Fraanje, N. Veldman, A. L. Spek, *Organometallics* **1996**, *15*, 835–847.
38 (a) P. W. N. M. van Leeuwen, G. J. H. Buisman, A. van Rooy, P. C. J. Kamer, *Rec Trav Chim Pays-Bas* **1994**, *113*, 61–62; (b) B. Moasser, W. L. Gladfelter, D. C. Roe, *Organometallics* **1995**, *14*, 3832–3838; (c) A. van Rooy, P. C. J. Kamer, P. W. N. M. van Leeuwen, N. Veldman, A. L. Spek, *J Organomet Chem* **1995**, *494*, C15–C18.
39 J. R. Briggs, G. T. Whiteker, *Chem Commun* **2001**, 2174–2175.
40 (a) A. van Rooy, D. Burgers, P. C. J. Kamer, P. W. N. M. van Leeuwen, *Rec Trav Chim Pays-Bas* **1996**, *115*, 492–498; (b) E. Wissing, A. J. J. M. Teunissen, C. B. Hansen, P. W. N. M. van Leeuwen, A. van Rooy, D. Burgers, *PCT Int. Appl. WO 9616923* and *EP 94-203434* (to DSM and E. I. Du Pont de Nemours), **1996**.
41 (a) K. G. Moloy, J. L. Petersen, *J Am Chem Soc* **1995**, *117*, 7696; (b) A. M. Trzeciak, T. Glowiak, R. Grzybek, J. J. Ziolkowski, *Dalton Trans Inorg Chem* **1997**, 1831–1837.
42 S. C. van der Slot, J. Duran, J. Luten, P. C. J. Kamer, P. W. N. M. van Leeuwen, *Organometallics* **2002**, *21*, 3873–3883.
43 C. Kunze, D. Selent, I. Neda, M. Freytag, P. G. Jones, R. Schmutzler, W. Baumann, A. Börner, *Z Anorg Allgem Chem* **2002**, *628*, 779–787.
44 R. Paciello, L. Siggel, M. Roper, *Angew Chem Int Ed* **1999**, *38*, 1920–1923; R. Paciello, L. Siggel, H.-J. Kneuper, N. Walker, M. Roper, *J Mol Catal A* **1999**, *143, 85–97.*
45 (a) F. J. Parlevliet, C. Kiener, J. Fraanje, K. Goubitz, M. Lutz, A. L. Spek, P. C. J. Kamer, P. W. N. M. van Leeuwen, *Dalton* **2000**, 1113–1122; (b) C. J. Cobley, D. D. Ellis, A. G. Orpen, P. G. Pringle, *Dalton* **2000**, 1109–1112; (c) C. Kunze, D. Selent, I. Neda, R. Schmutzler, A. Spannenberg, A. Börner, *Heteroatom Chem* **2001**, *12*, 577–585; (d) S. Steyer, C. Jeunesse, D. Matt, R. Welter, M. Wesolek, *Dalton Trans* **2002**, 4264–4274.
46 (a) C. Kunze, D. Selent, I. Neda, M. Freytag, P. G. Jones, R. Schmutzler, W. Baumann, A. Borner, *Z Anorg Allg Chem* **2002**, *628*, 779–787; (b) S. Steyer, C. Jeunesse, J. Harrowfield, D. Matt, *Dalton Trans* **2005**, 1301–1309; (c) I. A. Bagatin, D. Matt, H. Thoennessen, P. G. Jones, *Inorg Chem* **1999**, *38*, 1585–1591.
47 S. S. Bath, L. Vaska, *J Am Chem Soc* **1963**, *85*, 3500–3501.
48 J. M. Brown, A. G. Kent, *Perkin Trans II*, **1987**, 1597.
49 C. Bianchini, H. M. Lee, A. Meli, F. Vizza, *Organometallics* **2000**, *19*, 849–853.
50 E. K. van den Beuken, W. G. J. de Lange, P. W. N. M. van Leeuwen, N. Veldman, A. L. Spek, B. L. Feringa, *Dalton Trans Inorg Chem* **1996**, 3561–3569.
51 A. Van Rooy, J. N. H. de Bruijn, K. F. Roobeek, P. C. J. Kamer, P. W. N. M. van Leeuwen, *J Organomet Chem* **1996**, *507*, 69–73.
52 (a) E. Billig, A. G. Abatjoglou, D. R. Bryant, R. E. Murray, J. M. Maher, *US Patent 4717775*, 1988 (to Union Carbide Corporation), *Chem Abstr* **1989**, *109*, 233177; (b) K. F. Muilwijk, P. C. J. Kamer, P. W. N. M. van Leeuwen, *J Am Oil Chem Soc* **1997**, *74*, 223–228.
53 C. Fyhr, M. Garland, *Organometallics* **1993**, *12*, 1753.
54 E. B. Walczuk, P. C. J. Kamer, P. W. N. M. van Leeuwen, *Angew Chem Int Ed* **2003**, *42*, 4665–4669.
55 T. J. Devon, G. W. Phillips, T. A. Puckette, J. L. Stavinoha, J. J. Vanderbilt, *US Patent 4694109*, 1987 (to Texas Eastman); *Chem Abstr* **1988**, *108*, 7890.
56 T. J. Devon, G. W. Phillips, T. A. Puckette, J. L. Stavinoha, J. J. Vanderbilt, *US Patent 5,332,846*, 1994 (to Texas Eastman), *Chem Abstr* **1994**, *121*, 280879.
57 C. P. Casey, G. T. Whiteker, *J Org Chem* **1990**, *55*, 1394.
58 K. Yamamoto, S. Momose, M. Funahashi, M. Miyazawa, *Synlett* **1990**, 711.

59 O.R. Hughes, J.D. Unruh, *J Mol Catal* **1981**, *12*, 71–83.
60 G. Consiglio, C. Botteghi, C. Salomon, P. Pino, *Angew Chem* **1973**, *85*, 665.
61 J.D. Unruh, J.R. Christenson, *J Mol Cat* **1982**, *14*, 19–34.
62 C.P. Casey, E.L. Paulsen, E.W. Beuttenmueller, B.R. Proft, L.M. Petrovich, B.A. Matter, D.R. Powell, *J Am Chem Soc* **1997**, *119*, 11817–11825.
63 H. Klein, R. Jackstell, K.-D. Wiese, C. Borgmann, M. Beller, *Angew Chem Int Ed* **2001**, *40*, 3408–3411.
64 W.A. Herrmann, C.W. Kohlpaintner, R.B. Manetsberger, H. Bahrmann, H. Kottmann, *J Mol Cat A Chem* **1995**, *97*, 65–72.
65 P.W.N.M. van Leeuwen, in *Homogeneous Catalysis: Understanding the Art*, Kluwer, Dordrecht, **2004**, pp. 125–138.
66 L.A. van der Veen, P.C.J. Kamer, P.W.N.M. van Leeuwen, *Angew Chem Int Ed* **1999**, *38*, 336–338.
67 R.P.J. Bronger, P.C.J. Kamer, P.W.N.M. van Leeuwen, *Organometallics* **2003**, *22*, 5358–5369.
68 L.A. van der Veen, P.C.J. Kamer, P.W.N.M. van Leeuwen, *Organometallics* **2000**, *19*, 872–883.
69 L.A. van der Veen, M.D.K. Boele, F. Bregman, P.C.J. Kamer, P.W.N.M. van Leeuwen, K. Goubitz, J. Fraanje, H. Schenk, C. Bo, *J Am Chem Soc* **1998**, *120*, 11616–11626.
70 R.P.J. Bronger, J.P. Bermon, J. Herwig, P.C.J. Kamer, P.W.N.M. van Leeuwen, *Adv Synth Catal* **2004**, *346*, 789–799.
71 H. Bohnen, J. Herwig, *PCT Int. Appl. WO 2002-EP 1379, 20020209*, 2002 (to Celanese Chemicals Europe); *Chem Abstr* **2002**, *137*, 203024.
72 D.J. Adams, D.J. Cole-Hamilton, D.A.J. Harding, E.G. Hope, P. Pogorzelec, A.M. Stuart, *Tetrahedron* **2004**, *60*, 4079–4085.
73 L. Leclercq, F. Hapiot, S. Tilloy, K. Ramkisoensing, J.N.H. Reek, P.W.N.M. van Leeuwen, E. Monflier, *Organometallics* **2005**, *24*, 2070–2075.
74 M. Schreuder Goedheijt, B.E. Hanson, J.N.H. Reek, P.C.J. Kamer, P.W.N.M. van Leeuwen, *J Am Chem Soc* **2000**, *122*, 1650–1657.
75 A. Buhling, P.C.J. Kamer, P.W.N.M. van Leeuwen, J.W. Elgersma, K. Goubitz, J. Fraanje, *Organometallics* **1997**, *16*, 3027–3037.
76 P. Wasserscheid, H. Waffenschmidt, P. Machnitzki, K.W. Kottsieper, O. Stelzer, *Chem Commun* **2001**, 451–452.
77 S.M. Silva, R.P.J. Bronger, Z. Freixa, J. Dupont, P.W.N.M. van Leeuwen, *New J Chem* **2003**, *27*, 1294–1296.
78 A. Riisager, R. Fehrmann, S. Flicker, R. van Hal, M. Haumann, P. Wasserscheid, *Angew Chem Int Ed* **2005**, *44*, 815–819.
79 A.J. Sandee, J.N.H. Reek, P.C.J. Kamer, P.W.N.M. van Leeuwen, *J Am Chem Soc* **2001**, *123*, 8468–8476.
80 R.P.J. Bronger, J.P. Bermon, J.N.H. Reek, P.C.J. Kamer, P.W.N.M. van Leeuwen, D.N. Carter, P. Licence, M. Poliakoff, *J Mol Catal A* **2004**, *224*, 145–152.
81 W. Ahlers, M. Roeper, P. Hofmann, D.C.M. Warth, R. Paciello, *PCT Int. Appl. WO 2001-EP 1422, 20010209* (to BASF), 2001; *Chem Abstr* **2001**, *135*, 168206.
82 J.K. MacDougall, M.C. Simpson, M.J. Green, D.J. Cole-Hamilton, *Dalton Trans Inorg Chem* **1996**, 1161–1172.
83 M.E. Broussard, B. Juma, S.G. Train, W.-J. Peng, S.A. Laneman, G.G. Stanley, *Science* **1993**, *260*, 1784–1788.
84 W.-J. Peng, S.G. Train, D.K. Howell, F.R. Fronczek, G.G. Stanley, *Chem Commun* **1996**, 2607–2608.
85 A. Haynes, *Educ Chem* **2001**, *38*, 99–10[illegible].
86 (a) D. Forster, *J Am Chem Soc* **1976**, *98*, 846–848; (b) D. Forster, *Adv Organomet Chem* **1979**, *17*, 255.
87 Q. Smejkal, D. Linke, M. Baerns, *Chem Eng Process* **2005**, *44*, 421–428.
88 P.W.N.M. van Leeuwen, in *Homogeneous Catalysis: Understanding the Art*, Kluwer Dordrecht, **2004**, pp. 109–124.
89 C. Claver, P.W.N.M. van Leeuwen, *Comprehensive Coordination Chemistry II*, Elsevier, Amsterdam, **2003**, vol. 9.
90 P.M. Maitlis, A. Haynes, B.R. James, M. Catellani, G.P. Chiusoli, *Dalton Trans* **2004**, 3409–3419.
91 J.R. Zoeller, V.H. Agreda, S.L. Cook, N.L. Lafferty, S.W. Polichnowski, D.M. Pond, *Catal Today* **1992**, *13*, 73–91.

92 (a) B. L. Smith, G. P. Torrence, M. A. Murphy, A. Aguiló, *J Mol Catal* **1987**, *39*, 115–136; (b) B. L. Smith, G. P. Torrence, A. Aguiló, J. S. Alder, *US Patent 5144068* (to Hoechst Celanese Corporation), **1991**.

93 M. A. Murphy, B. L. Smith, G. P. Torrence, A. Aguiló, *J Organomet Chem.* **1986**, *303*, 257–272.

94 T. Kinnunen, K. Laasonen, *J Mol Struct (Teochem)* **2001**, *542*, 273–288.

95 T. R. Griffin, D. B. Cook, A. Haynes, J. M. Pearson, D. Monti, G. Morris, *J Am Chem Soc* **1996**, *118*, 3029–3030.

96 V. Chauby, J.-C. Daran, C. S.-B. Berre, F. Malbosc, P. Kalck, O. D. Gonzalez, C. E. Haslam, A. Haynes, *Inorg Chem* **2002**, *41*, 3280–3290.

97 W. Adamson, J. J. Daly, D. Forster, *J Organomet Chem* **1974**, *1*, C17–C19.

98 A. Haynes, B. E. Mann, D. J. Gulliver, G. E. Morris, P. M. Maitlis, *J Am Chem Soc* **1991**, *113*, 8567–8569.

99 A. Haynes, B. E. Mann, G. E. Morris, P. M. Maitlis, *J Am Chem Soc* **1993**, *115*, 4093–4100.

100 Z. Freixa, P. J. C. Kamer, M. Lutz, A. L. Spek, P. W. N. M. van Leeuwen, *Ang Chem Int Ed* **2005**, *44*, 4385–4388.

101 R. Broussier, M. Laly, P. Perron, B. Gautheron, I. E. Nifant'ev, J. A. K. Howard, L. G. Kuz'mina, P. Kalck, *J Organomet Chem.* **1999**, *587*, 104–112.

102 H. C. Martin, N. H. James, J. Aitken, J. A. Gaunt, H. Adams, A. Haynes, *Organometallics* **2003**, *22*, 4451–4458.

103 (a) J. Rankin, A. D. Poole, A. C. Benyei, D. J. Cole-Hamilton, *J Chem Commun* **1997**, 1835–1836; (b) J. Rankin, A. C. Benyei, A. D. Poole, D. J. Cole-Hamilton, *Dalton Trans* **1999**, 3771–3782.

104 C. M. Bartish, *US Patent 4,102,920* (to Air Products & Chemicals, Inc.), **1977**.

105 (a) S. Gaemers, J. G. Sunley, *PCT Int. Appl. WO 2004/101487* (to BP Chemicals Limited), **2004**; (b) S. Gaemers, J. G. Sunley, *PCT Int. Appl. WO 2004/101488* (to BP Chemicals Limited), **2004**.

106 (a) C.-A. Carraz, E. J. Ditzel, A. G. Orpen, D. D. Ellis, P. G. Pringle, G. J. Sunley, *Chem Commun* **2000**, 1277–1278; (b) M. J. Baker, C.-A. Carraz, E. J. Ditzel, P. G. Pringle, G. J. Sunley, *UK Patent Appl. 2336154*, **1999**.

107 M. J. Baker, M. F. Giles A. G. Orpen, M. J. Taylor, R. J. Watt, *Chem Commun* **1995**, 197–198.

108 R. W. Wegman, A. G. Abatjoglou, A. M. Harrison, *Chem Commun* **1987**, 1891–1892.

109 L. Gonsalvi, H. Adams G. J. Sunley, E. Ditzel, A. Haynes, *J Am Chem Soc* **2002**, *124*, 13597–13612.

110 L. Gonsalvi, H. Adams G. J. Sunley, E. Ditzel, A. Haynes, *J Am Chem Soc* **1999**, *121*, 11233–11234.

111 D. K. Dutta, J. D. Woollins, A. M. Z. Slawin, D. Konwar, P. Das, M. Sharma, P. Bhattacharyya, S. M. Autcott, *Dalton Trans* **2003**, 2674–2679.

112 P. Das, M. Sharma, N. Kumari, D. Konwar, D. K. Dutta, *Appl Organometal Chem* **2002**, *16*, 302–306.

113 M. Sharma, N. Kumari, P. Das, P. Chutia, D. K. Dutta, *J Mol Catal A* **2002**, *188*, 25–35.

114 N. Kumari, M. Sharma, P. Chutia, D. K. Dutta, *J Mol Catal A* **2004**, *222*, 53–58.

115 S. Burger, B. Therrien, G. Süss-Fink, *Helv Chim Acta* **2005**, *88*, 478–486.

116 C. M. Thomas, G. Süss-Fink, *Coord Chem Rev* **2003**, *243*, 125–142.

117 C. M. Thomas, R. Mafia, B. Therrien, E. Rusanov, H. Stœckli-Evans, G. Süss-Fink, *Chem Eur J* **2002**, *8*, 3343–3352.

118 A. Mayanza, J-J. Bonnet, J. Galy, P. Kalck, R. Poilblanc, *J Chem Res (S)* **1980**, 146.

119 M. J. Doyle, A. Mayanza, J.-J. Bonnet, P. Kalck, R. Poilblanc, *J Organomet Chem* **1978**, *146*, 293–310.

120 M. V. Jiménez, E. Sola, M. A. Egea, A. Huet, A. C. Francisco, F. J. Lahoz, L. A. Oro, *Inorg Chem* **2000**, *39*, 4868–4878.

121 A. Haynes, P. M. Maitlis, I. A. Stanbridge, S. Haak, J. M. Pearson, H. Adams, N. A. Bailey, *Inorg Chim Acta* **2004**, *357*, 3027–3037.

122 P. W. N. M van Leeuwen, C. F. Roobeek, *Eur. Patent Appl. EP 133331* (to Shell Internationale Research Maatschappij), **1985**.

123 J. H. Jones, *Platinum Met Rev* **2000**, *44*, 94.

124 See www.engelhard.com/eibprices.
125 (a) P. R. Ellis, J. M. Pearson, A. Haynes, H. Adams, N. A. Bailey, P. M. Maitlis, *Organometallics* **1994**, *13*, 3215; (b) T. R. Griffin, D. B. Cook, A. Haynes, J. M. Pearson, D. Monti, G. E. Morris, *J Am Chem Soc* **1996**, *118*, 3029.
126 (a) G. J. Sunley, D. J. Watson, *Catal Today* **2000**, *58*, 293; (b) T. Ghaffar, J. P. H. Charmant, G. J. Sunley, G. E. Morris, A. Haynes, P. M. Maitlis, *Inorg Chem Commun* **2000**, *3*, 11.
127 A. P. Wright, Abstracts of Papers, *222nd ACS National Meeting*, Chicago, IL, **2001** CATL-044. *Chem Abstr* AN **2001**, 637430.
128 T. A. Ryan, C. Ryan, E. A. Seddon, K. R. Seddon, *Phosgene and Related Carbonyl Halides*, Elsevier, Amsterdam, **1996**.
129 S. Cenini, F. Ragaini (Eds.), *Catalysis by Metal Complexes 20: Catalytic Reductive Carbonylation of Organic Nitro Compounds*, Kluwer, Dordrecht, **1996**.
130 E. Drent, P. W. N. M. van Leeuwen, *Eur. Patent Appl. GB 81-36371* (to Shell Internationale Research Maatschappij), **1983**.
131 (a) P. Wehman, G. C. Dol, E. R. Moorman, P. C. J. Kamer, P. W. N. M. van Leeuwen, L. Fraanje, K. Goubitz, *Organometallics* **1994**, *13*, 4856–4869; (b) P. Wehman, V. E. Kaasjager, F. Hartl, P. C. J. Kamer, P. W. N. M. van Leeuwen, L. Fraanje, K. Goubitz, *Organometallics* **1995**, *14*, 3751–3761.
132 (a) F. Ragaini, C. Cognolato, M. Gasperini, S. Cenini, *Angew Chem Int Ed* **2003**, *42*, 2886–2889; (b) F. Ragaini, C. Attilio, S. Cenini, M. Gasperini, *Adv Synth Catal* **2004**, *346*, 63–71.

Subject Index

Activation of Small Molecules. Edited by William B. Tolman

ISBN: 3-527-31312-5

d

e